Lancia Beta Owners Workshop Manual

P M Methuen

Models covered:

UK: Lancia Beta Saloon, Coupe, HPE and Spider; Series A and B models with 1297 cc, 1301 cc, 1438 cc, 1585 cc, 1592 cc, 1756 cc and 1995 cc engines

USA: Lancia Beta Sedan, Coupe and HPE models with 107 cu in (1.8 liter) and 122 cu in (2 liter) engines. Covers most features of Zagato models.

Does not cover Monte Carlo/Scorpion models

ISBN 0 85696 533 2

© Haynes Publishing Group 1981

All rights reserved. No part of this book may be reproduced or transmitted in any form or by any means, electronic or mechanical, including photocopying, recording or by any information storage or retrieval system, without permission in writing from the copyright holder

Printed in England

**HAYNES PUBLISHING GROUP
SPARKFORD YEOVIL SOMERSET ENGLAND**
distributed in the USA by
**HAYNES PUBLICATIONS INC
861 LAWRENCE DRIVE
NEWBURY PARK
CALIFORNIA 91320
USA**

Acknowledgements

Thanks are due to Lancia (England) Limited for the supply of technical information and certain illustrations. Castrol Limited supplied the lubrication details and the Champion Sparking Plug Company supplied the illustrations showing the various spark plug conditions.

The bodywork repair photographs used in this manual were provided by Holt Lloyd Limited who supply 'Turtle Wax', 'Dupli-colour Holts' and other Holts range products.

About this manual

Its aim

The aim of this manual is to help you get the best value from your car. It can do so in several ways. It can help you decide what work must be done (even should you choose to get it done by a garage), provide information on routine maintenance and servicing, and give a logical course of action and diagnosis when random faults occur. However, it is hoped that you will make full use of the manual by tackling the work yourself. On simpler jobs it may even be quicker than booking the car into a garage, and having to go there twice, to leave and collect it. Perhaps most important, a lot of money can be saved by avoiding the costs the garage must charge to cover its labour and overheads.

The manual has drawings and descriptions to show the function of the various components so that their layout can be understood. Then the tasks are described and photographed in a step-by-step sequence so that even a novice can do the work.

Its arrangement

The manual is divided into twelve Chapters, each covering a logical sub-division of the vehicle. The Chapters are each divided into consecutively numbered Sections and the Sections into paragraphs (or sub-sections), with decimal numbers following on from the Section they are in, eg 5.1, 5.2, 5.3 etc.

It is freely illustrated in those parts where there is a detailed sequence of operations to be carried out. There are two forms of illustration: figures and photographs. The figures are numbered in sequence with decimal numbers, according to their position in the Chapter: eg, Fig. 6.4 is the 4th drawing/illustration in Chapter 6. Photographs are numbered (either individually or in related groups) the same as the Section or sub-section of the text where the operation they show is described.

There is an alphabetical index at the back of the manual as well as a contents list at the front.

References to the 'left' or 'right' of the vehicle are in the sense of a person in the driver's seat facing forwards.

Unless otherwise stated nuts and bolts are tightened by turning clockwise and removed by turning anti-clockwise.

Vehicle manufacturers continually make changes to specifications and recommendations, and these when notified are incorporated into our manuals at the earliest opportunity.

Whilst every care is taken to ensure that the information in this manual is correct, no liability can be accepted by the authors or publishers for loss, damage or injury caused by any errors in, or omissions from, the information given.

Introduction to the Lancia Beta Series

The Lancia Beta series now consists of four different models, but when the original Lancia Beta Saloon was introduced into the UK in July 1973 it was a totally new design and shape to come from the Lancia factory. Although a hatchback shape it is not in fact that type of body as the luggage compartment has no direct access to the rear seating area.

In its original guise the Saloon model was available in 1400, 1600 or 1800 engine form; the engine being a twin overhead camshaft transversely mounted 4-cylinder unit. In April 1974 the 1800 ES (special equipment) model was added to the range, and then in September 1974 the second model in the Beta series was introduced. This was the Coupe, a totally different 2 + 2 sports car.

The Coupe on introduction was available with a 1600 or 1800 engine. Later engine options were to come but not until the other two now familiar models had been introduced.

In July 1975 the Lancia Beta Spider was introduced, but in 1600 form only. Unlike most Italian Spider versions of cars such as Alfa Romeo it does not have a fully folding soft top. Rather it has a wide rollover bar with a 'targa' top between the windscreen and bar and a small folding rear section.

In October 1975 the HPE (High Performance Estate) model was announced. Although this model shares much of the looks of the Coupe and Spider at the front it is a totally different concept, providing a small sporting estate car, but based on the floor pan of the Saloon model to give more room inside. Initially it was available in 1600 form only. At the same time as the HPE was announced the 1400 saloon model was replaced by the 1300.

The next major change was the arrival of the 'B' series Lancia Betas. This happened in February 1978 when the 2000 engine was introduced as an addition to the Spider and HPE range and in May 1976 the same thing happened for the Saloon and Coupe when the 2000 replaced the 1800 engines. At this time the 1600 engine was also modified to give a capacity of 1585 cc rather than the 1592 cc of the earlier models.

The 1300 Coupe was introduced in June 1977 and was the last new model to be introduced. Other changes have taken place since then. In late 1978 electronic ignition was announced as a standard fitting and by February of the following year all models were fitted with it. At the same time the 1300 engine was enlarged from 1297 cc to 1301 cc.

In 1979 the interior design and the facia layout were drastically altered in the Coupe, Spider and HPE models, giving a much clearer and more pleasing interior.

There is an automatic gearbox optionally available on 1600 and 2000 versions of the Saloon, Coupe and HPE, but not the Spider. This was introduced in September 1979.

Power steering is not fitted to right-hand drive cars due to the configuration of the engine bay. However, it is a standard item on all left-hand drive 2000 models.

Vehicles exported to the USA have been less confusing as only a limited engine range has been used; the 1800 and then later on the 2000. All are fitted with the statutory anti-pollution equipment. In the USA the Spider is known as the Zagato.

Contents

	Page
Acknowledgements	2
About this manual	2
Introduction	2
General dimensions and weights	6
Use of English	7
Buying spare parts and vehicle identification numbers	8
Tools and working facilities	10
Jacking and towing	12
Recommended lubricants and fluids	13
Safety first!	14
Routine maintenance	15
Chapter 1 Engine	19
Chapter 2 Cooling system	65
Chapter 3 Fuel, emission control and exhaust systems	72
Chapter 4 Ignition system	94
Chapter 5 Clutch	102
Chapter 6 Manual gearbox and automatic transmission	110
Chapter 7 Differential unit	130
Chapter 8 Driveshafts and CV joints	138
Chapter 9 Braking system	144
Chapter 10 Electrical system	157
Chapter 11 Suspension and steering	224
Chapter 12 Bodywork and subframe	250
Conversion factors	280
Index	281

1978 Lancia Beta 2000ES Saloon – UK model

1978 Lancia Beta 1600 HPE – UK model

1979 Lancia Beta Coupe – US model

1979 Lancia Beta HPE – US model

General dimensions and weights

Overall length
UK models:
    Saloon .... 168.9 in (4290 mm)
    Coupe .... 157.3 in (3995 mm)
    HPE .... 168.7 in (4285 mm)
    Spider .... 159.1 in (4040 mm)
USA models:
    Saloon .... 178.0 in (4520 mm)
    Coupe .... 166.5 in (4230 mm)
    HPE .... 178.5 in (4535 mm)
    Zagato .... 165.0 in (4191 mm)

Overall width
UK models:
    Saloon .... 66.5 in (1690 mm)
    Coupe and HPE .... 65.0 in (1650 mm)
    Spider .... 64.8 in (1646 mm)
USA models:
    Saloon .... 66.5 in (1690 mm)
    Coupe, HPE and Zagato .... 65.0 in (1650 mm)

Overall height
UK models:
    Saloon .... 55.1 in (1400 mm)
    Coupe and HPE .... 50.6 in (1285mm)
    Spider .... 49.7 in (1263 mm)
USA models:
    Saloon .... 55.1 in (1400 mm)
    Coupe .... 50.6 in (1285 mm)
    HPE .... 51.5 in (1310 mm)

Wheelbase
Saloon and HPE .... 100.0 in (2540 mm)
Coupe, Spider and Zagato .... 92.5 in (2350 mm)

Turning circle
UK models:
    Saloon .... 429.9 in (10920 mm)
    Coupe and HPE .... 417.3 in (10600 mm)
    Spider .... 401.6 in (10200 mm)
USA models:
    Saloon .... 429.9 in (10920 mm)
    Coupe .... 401.5 in (10198 mm)

Kerb weight
UK models:
    Saloon
        1400 engine .... 2370 lb (1075 kg)
        1600 engine .... 2392 lb (1085 kg)
        1800 engine .... 2414 lb (1095 kg)
    Coupe .... 2183 lb (990 kg)
    HPE .... 2337 lb (1060 kg)
    Spider .... 2315 lb (1050 kg)
USA models (without optional extras):
    Saloon .... 2467 lb (1119 kg)
    Coupe .... 2398 lb (1088 kg)

Use of English

As this book has been written in England, it uses the appropriate English component names, phrases, and spelling. Some of these differ from those used in America. Normally, these cause no difficulty, but to make sure, a glossary is printed below. In ordering spare parts remember the parts list will probably use these words:

English	American	English	American
Aerial	Antenna	Layshaft (of gearbox)	Countershaft
Accelerator	Gas pedal	Leading shoe (of brake)	Primary shoe
Alternator	Generator (AC)	Locks	Latches
Anti-roll bar	Stabiliser or sway bar	Motorway	Freeway, turnpike etc
Battery	Energizer	Number plate	License plate
Bodywork	Sheet metal	Paraffin	Kerosene
Bonnet (engine cover)	Hood	Petrol	Gasoline
Boot lid	Trunk lid	Petrol tank	Gas tank
Boot (luggage compartment)	Trunk	'Pinking'	'Pinging'
Bottom gear	1st gear	Propeller shaft	Driveshaft
Bulkhead	Firewall	Quarter light	Quarter window
Cam follower or tappet	Valve lifter or tappet	Retread	Recap
Carburettor	Carburetor	Reverse	Back-up
Catch	Latch	Rocker cover	Valve cover
Choke/venturi	Barrel	Roof rack	Car-top carrier
Circlip	Snap-ring	Saloon	Sedan
Clearance	Lash	Seized	Frozen
Crownwheel	Ring gear (of differential)	Side indicator lights	Side marker lights
Disc (brake)	Rotor/disk	Side light	Parking light
Drop arm	Pitman arm	Silencer	Muffler
Drop head coupe	Convertible	Spanner	Wrench
Dynamo	Generator (DC)	Sill panel (beneath doors)	Rocker panel
Earth (electrical)	Ground	Split cotter (for valve spring cap)	Lock (for valve spring retainer)
Engineer's blue	Prussian blue	Split pin	Cotter pin
Estate car	Station wagon	Steering arm	Spindle arm
Exhaust manifold	Header	Sump	Oil pan
Fast back (Coupe)	Hard top	Tab washer	Tang; lock
Fault finding/diagnosis	Trouble shooting	Tailgate	Liftgate
Float chamber	Float bowl	Tappet	Valve lifter
Free-play	Lash	Thrust bearing	Throw-out bearing
Freewheel	Coast	Top gear	High
Gudgeon pin	Piston pin or wrist pin	Trackrod (of steering)	Tie-rod (or connecting rod)
Gearchange	Shift	Trailing shoe (of brake)	Secondary shoe
Gearbox	Transmission	Transmission	Whole drive line
Halfshaft	Axleshaft	Tyre	Tire
Handbrake	Parking brake	Van	Panel wagon/van
Hood	Soft top	Vice	Vise
Hot spot	Heat riser	Wheel nut	Lug nut
Indicator	Turn signal	Windscreen	Windshield
Interior light	Dome lamp	Wing/mudguard	Fender

Miscellaneous points

An 'oil seal' is fitted to components lubricated by grease!

A 'damper' is a 'shock absorber', it damps out bouncing, and absorbs shocks of bump impact. Both names are correct, and both are used haphazardly.

Note that British drum brakes are different from the Bendix type that is common in America, so different descriptive names result. The shoe end furthest from the hydraulic wheel cylinder is on a pivot; interconnection between the shoes as on Bendix brakes is most uncommon. Therefore the phrase 'Primary' or 'Secondary' shoe does not apply. A shoe is said to be 'Leading' or 'Trailing'. A 'Leading' shoe is one on which a point on the drum, as it rotates forward, reaches the shoe at the end worked by the hydraulic cylinder before the anchor end. The opposite is a 'Trailing' shoe, and this one has no self servo from the wrapping effect of the rotating drum.

Buying spare parts and vehicle identification numbers

Buying spare parts

Spare parts are available from many sources, for example ; Lancia garages, other garages and accessory shops, and motor factors. Our advice regarding spare part sources is as follows:

Officially appointed Lancia garages – This is the best source of parts which are peculiar to your car and are otherwise not generally available (eg. complete cylinder heads, internal gearbox components, badges, interior trim etc). It is also the only place at which you should buy parts if your car is still under warranty – non-Lancia components may invalidate the warranty. To be sure of obtaining the correct parts it will always be necessary to give the storeman your car's vehicle identification number, and if possible, to take the 'old' part along for positive identification. Remember that many parts are available on a factory exchange scheme – any parts returned should always be clean! It obviously makes good sense to go straight to the specialists on your car for this type of part for they are best equipped to supply them.

Other garages and accessory shops – These are often very good places to buy materials and components needed for the maintenance of your car (eg. oil filters, spark plugs, bulbs, drivebelts, oils and greases, touch-up paint, filler paste etc). They also sell general accessories, usually have convenient opening hours, charge lower prices and can often be found not far from home.

Motor factors – Good factors will stock all of the more important components which wear out relatively quickly (eg. clutch components, pistons, valves, exhaust systems, brake cylinders/pipes/hoses/seals/shoes and pads etc). Motor factors will often provide new or reconditioned components on a part exchange basis – this can save a considerable amount of money.

Vehicle identification numbers

The car identification number is stamped on the engine compartment front rail.

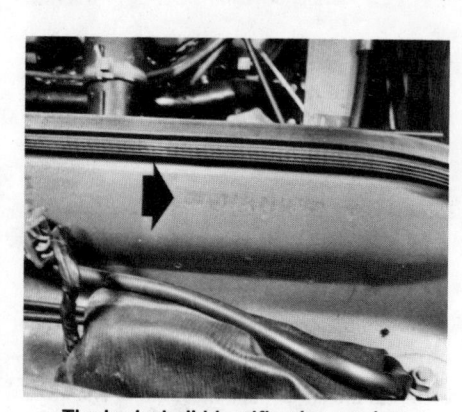

The bodyshell identification marker

The car data plate – right-hand inner wing

The engine number is engraved on the side of the crankcase above the oil filter housing

Buying spare parts and vehicle identification numbers

The car data plate is fixed to the left-hand side of the engine compartment; it contains information such as engine number, paint code, weight limits and other data required by the licensing authorities in the destination country. A further data plate may be fitted to the boot wall.

Major assemblies such as engine and transmission have their identification numbers stamped on the outside of their casings.

Car identification number (1) and data plates (2 to 4). Position and number of data plates may vary according to territory

The cylinder block and crankcase type

Tools and working facilities

Introduction

A selection of good tools is a fundamental requirement for anyone contemplating the maintenance and repair of a motor vehicle. For the owner who does not possess any, their purchase will prove a considerable expense, offsetting some of the savings made by doing-it-yourself. However, provided that the tools purchased are of good quality, they will last for many years and prove an extremely worthwhile investment.

To help the average owner to decide which tools are needed to carry out the various tasks detailed in this manual, we have compiled three lists of tools under the following headings: *Maintenance and minor repair*, *Repair and overhaul*, and *Special*. The newcomer to practical mechanics should start off with the *Maintenance and minor repair* tool kit and confine himself to the simpler jobs around the vehicle. Then, as his confidence and experience grow, he can undertake more difficult tasks, buying extra tools as, and when, they are needed. In this way, a *Maintenance and minor repair* tool kit can be built-up into a *Repair and overhaul* tool kit over a considerable period of time without any major cash outlays. The experienced do-it-yourselfer will have a tool kit good enough for most repair and overhaul procedures and will add tools from the *Special* category when he feels the expense is justified by the amount of use these tools will be put to.

It is obviously not possible to cover the subject of tools fully here. For those who wish to learn more about tools and their use there is a book entitled *How to Choose and Use Car Tools* available from the publishers of this manual.

Maintenance and minor repair tool kit

The tools given in this list should be considered as a minimum requirement if routine maintenance, servicing and minor repair operations are to be undertaken. We recommend the purchase of combination spanners (ring one end, open-ended the other); although more expensive than open-ended ones, they do give the advantages of both types of spanner.

Combination spanners - 10, 11, 12, 13, 14 & 17 mm
Adjustable spanner - 9 inch
Spark plug spanner (with rubber insert)
Spark plug gap adjustment tool
Set of feeler gauges
Brake bleed nipple spanner
Screwdriver - 4 in long x $\frac{1}{4}$ in dia (flat blade)
Screwdriver - 4 in long x $\frac{1}{4}$ in dia (cross blade)
Combination pliers - 6 inch
Hacksaw (junior)
Tyre pump
Tyre pressure gauge
Grease gun
Oil can
Fine emery cloth (1 sheet)
Wire brush (small)
Funnel (medium size)

Repair and overhaul tool kit

These tools are virtually essential for anyone undertaking any major repairs to a motor vehicle, and are additional to those given in the *Maintenance and minor repair* list. Included in this list is a comprehensive set of sockets. Although these are expensive they will be found invaluable as they are so versatile - particularly if various drives are included in the set. We recommend the $\frac{1}{2}$ in square-drive type, as this can be used with most proprietary torque spanners. If you cannot afford a socket set, even bought piecemeal, then inexpensive tubular box wrenches are a useful alternative.

The tools in this list will occasionally need to be supplemented by tools from the *Special* list.

Sockets (or box spanners) to cover range in previous list
Reversible ratchet drive (for use with sockets)
Extension piece, 10 inch (for use with sockets)
Universal joint (for use with sockets)
Torque wrench (for use with sockets)
Mole wrench - 8 inch
Ball pein hammer
Soft-faced hammer, plastic or rubber
Screwdriver - 6 in long x $\frac{5}{16}$ in dia (flat blade)
Screwdriver - 2 in long x $\frac{5}{16}$ in square (flat blade)
Screwdriver - 1$\frac{1}{2}$ in long x $\frac{1}{4}$ in dia (cross blade)
Screwdriver - 3 in long x $\frac{1}{8}$ in dia (electricians)
Pliers - electricians side cutters
Pliers - needle nosed
Pliers - circlip (internal and external)
Cold chisel - $\frac{1}{2}$ inch
Scriber
Scraper
Centre punch
Pin punch
Hacksaw
Valve grinding tool
Steel rule/straight-edge
Allen keys
Selection of files
Wire brush (large)
Axle-stands
Jack (strong scissor or hydraulic type)

Special tools

The tools in this list are those which are not used regularly, are expensive to buy, or which need to be used in accordance with their manufacturers' instructions. Unless relatively difficult mechanical jobs are undertaken frequently, it will not be economic to buy many of these tools. Where this is the case, you could consider clubbing together with friends (or joining a motorists' club) to make a joint purchase, or borrowing the tools against a deposit from a local garage or tool hire specialist.

The following list contains only those tools and instruments freely available to the public, and not those special tools produced by the vehicle manufacturer specifically for its dealer network. You will find occasional references to these manufacturers' special tools in the text of this manual. Generally, an alternative method of doing the job without the vehicle manufacturers' special tool is given. However, sometimes, there is no alternative to using them. Where this is the case and the relevant tool cannot be bought or borrowed you will have to entrust the work to a franchised garage.

Valve spring compressor
Piston ring compressor
Balljoint separator
Universal hub/bearing puller
Impact screwdriver
Micrometer and/or vernier gauge
Dial gauge

Stroboscopic timing light
Dwell angle meter/tachometer
Universal electrical multi-meter
Cylinder compression gauge
Lifting tackle (photo)
Trolley jack
Light with extension lead

Buying tools

For practically all tools, a tool dealer is the best source since he will have a very comprehensive range compared with the average garage or accessory shop. Having said that, accessory shops often offer excellent quality tools at discount prices, so it pays to shop around.

Remember, you don't have to buy the most expensive items on the shelf, but it is always advisable to steer clear of the very cheap tools. There are plenty of good tools around at reasonable prices, so ask the proprietor or manager of the shop for advice before making a purchase.

Care and maintenance of tools

Having purchased a reasonable tool kit, it is necessary to keep the tools in a clean serviceable condition. After use, always wipe off any dirt, grease and metal particles using a clean, dry cloth, before putting the tools away. Never leave them lying around after they have been used. A simple tool rack on the garage or workshop wall, for items such as screwdrivers and pliers is a good idea. Store all normal spanners and sockets in a metal box. Any measuring instruments, gauges, meters, etc, must be carefully stored where they cannot be damaged or become rusty.

Take a little care when tools are used. Hammer heads inevitably become marked and screwdrivers lose the keen edge on their blades from time to time. A little timely attention with emery cloth or a file will soon restore items like this to a good serviceable finish.

Working facilities

Not to be forgotten when discussing tools, is the workshop itself. If anything more than routine maintenance is to be carried out, some form of suitable working area becomes essential.

It is appreciated that many an owner mechanic is forced by circumstances to remove an engine or similar item, without the benefit of a garage or workshop. Having done this, any repairs should always be done under the cover of a roof.

Wherever possible, any dismantling should be done on a clean flat workbench or table at a suitable working height.

Any workbench needs a vice: one with a jaw opening of 4 in (100 mm) is suitable for most jobs. As mentioned previously, some clean dry storage space is also required for tools, as well as the lubricants, cleaning fluids, touch-up paints and so on which become necessary.

Another item which may be required, and which has a much more general usage, is an electric drill with a chuck capacity of at least $\frac{5}{16}$ in (8 mm). This, together with a good range of twist drills, is virtually essential for fitting accessories such as wing mirrors and reversing lights.

Last, but not least, always keep a supply of old newspapers and clean, lint-free rags available, and try to keep any working area as clean as possible.

Spanner jaw gap comparison table

Jaw gap (in)	Spanner size
0.250	$\frac{1}{4}$ in AF
0.276	7 mm
0.313	$\frac{5}{16}$ in AF
0.315	8 mm
0.344	$\frac{11}{32}$ in AF; $\frac{1}{8}$ in Whitworth
0.354	9 mm
0.375	$\frac{3}{8}$ in AF
0.394	10 mm
0.433	11 mm
0.438	$\frac{7}{16}$ in AF
0.445	$\frac{3}{16}$ in Whitworth; $\frac{1}{4}$ in BSF
0.472	12 mm
0.500	$\frac{1}{2}$ in AF
0.512	13 mm
0.525	$\frac{1}{4}$ in Whitworth; $\frac{5}{16}$ in BSF
0.551	14 mm
0.563	$\frac{9}{16}$ in AF
0.591	15 mm
0.600	$\frac{5}{16}$ in Whitworth; $\frac{3}{8}$ in BSF
0.625	$\frac{5}{8}$ in AF
0.630	16 mm
0.669	17 mm
0.686	$\frac{11}{16}$ in AF
0.709	18 mm
0.710	$\frac{3}{8}$ in Whitworth, $\frac{7}{16}$ in BSF
0.748	19 mm
0.750	$\frac{3}{4}$ in AF
0.813	$\frac{13}{16}$ in AF
0.820	$\frac{7}{16}$ in Whitworth; $\frac{1}{2}$ in BSF
0.866	22 mm
0.875	$\frac{7}{8}$ in AF
0.920	$\frac{1}{2}$ in Whitworth; $\frac{9}{16}$ in BSF
0.938	$\frac{15}{16}$ in AF
0.945	24 mm
1.000	1 in AF
1.010	$\frac{9}{16}$ in Whitworth; $\frac{5}{8}$ in BSF
1.024	26 mm
1.063	$1\frac{1}{16}$ in AF; 27 mm
1.100	$\frac{5}{8}$ in Whitworth; $\frac{11}{16}$ in BSF
1.125	$1\frac{1}{8}$ in AF
1.181	30 mm
1.200	$\frac{11}{16}$ in Whitworth; $\frac{3}{4}$ in BSF
1.250	$1\frac{1}{4}$ in AF
1.260	32 mm
1.300	$\frac{3}{4}$ in Whitworth; $\frac{7}{8}$ in BSF
1.313	$1\frac{5}{16}$ in AF
1.390	$\frac{13}{16}$ in Whitworth; $\frac{15}{16}$ in BSF
1.417	36 mm
1.438	$1\frac{7}{16}$ in AF
1.480	$\frac{7}{8}$ in Whitworth; 1 in BSF
1.500	$1\frac{1}{2}$ in AF
1.575	40 mm; $\frac{15}{16}$ in Whitworth
1.614	41 mm
1.625	$1\frac{5}{8}$ in AF
1.670	1 in Whitworth; $1\frac{1}{8}$ in BSF
1.688	$1\frac{11}{16}$ in AF
1.811	46 mm
1.813	$1\frac{13}{16}$ in AF
1.860	$1\frac{1}{8}$ in Whitworth; $1\frac{1}{4}$ in BSF
1.875	$1\frac{7}{8}$ in AF
1.969	50 mm
2.000	2 in AF
2.050	$1\frac{1}{4}$ in Whitworth; $1\frac{3}{8}$ in BSF
2.165	55 mm
2.362	60 mm

A Haltrac hoist and gantry in use during a typical engine removal sequence

Jacking and towing

The jack which is supplied with the car should be used whenever it is necessary to raise the car to change a roadwheel (photo).

Jacking points are provided, one on each side of the car. Make sure that the jack is fully inserted before trying to lift the car up. If the jacking point is full of road dirt then clean it out first using the toolkit provided, which also includes a wheel stud spanner. Always apply the handbrake before jacking up the car.

For maintenance or repair work a proper hydraulic jack, trolley jack or scissor jack must be used. Never attempt maintenance work under the car using the roadwheel jack. Use axle stands as well.

When the car is being raised using a hydraulic jack, trolley jack or scissor jack, place the jack under the front subframe members or the jacking plate at the front of the body. Jacking up at the rear is harder but it is safe to use the rear towing bracket. Always use a wooden packing piece between the body and jack when lifting the car. This prevents any damage being done to the bodywork.

Towing brackets are provided at the front and rear of the car. At the front the eye is integral with the front subframe by the right-hand front mounting bolt and the bracket is attached to the lower rear body panel at the rear (photo).

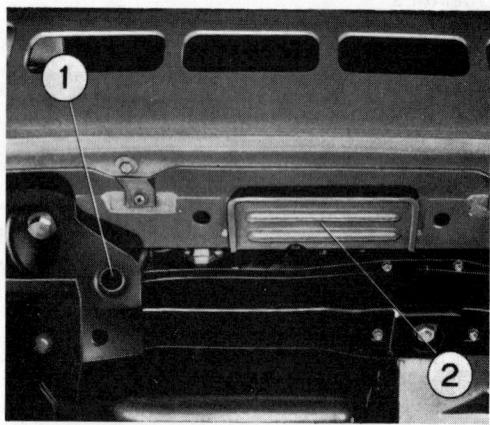

Front jacking and towing brackets

1 Towing bracket   2 Jacking plate

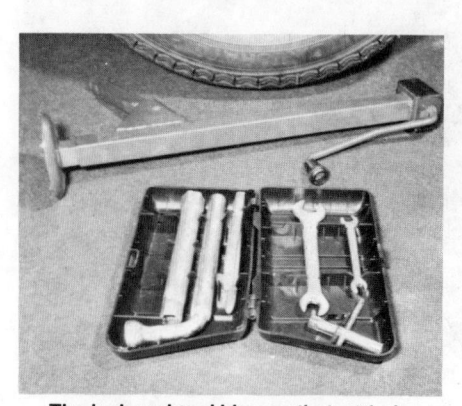

The jack and tool kit supplied with the car

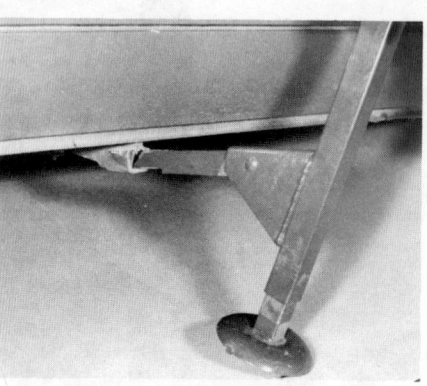

Inserting the jack into the jacking point in the side of the car

The rear towing bracket

Recommended lubricants and fluids

Component or system	Lubricant type or specification	Castrol product
Engine (1)	SAE 20W/50 engine oil	Castrol GTX
Manual gearbox/differential unit (2)	SAE 85W/50 oil	Castrol GP 50
Automatic transmission/differential unit (3)	Automatic transmission fluid. Dexron type	Castrol TQ Dexron ® II
Power steering system (4)	Automatic transmission fluid. Dexron type	Castrol TQ Dexron ® II
Driveshaft constant velocity joints (5)	Molykote BR2	No equivalent
Brake fluid reservoir (6)	SAE J1703	Castrol Girling Universal Clutch and Brake Fluid
Wheel bearings (7)	Multi-purpose grease	Castrol LM grease

Safety first!

Professional motor mechanics are trained in safe working procedures. However enthusiastic you may be about getting on with the job in hand, do take the time to ensure that your safety is not put at risk. A moment's lack of attention can result in an accident, as can failure to observe certain elementary precautions.

There will always be new ways of having accidents, and the following points do not pretend to be a comprehensive list of all dangers; they are intended rather to make you aware of the risks and to encourage a safety-conscious approach to all work you carry out on your vehicle.

Essential DOs and DON'Ts

DON'T rely on a single jack when working underneath the vehicle. Always use reliable additional means of support, such as axle stands, securely placed under a part of the vehicle that you know will not give way.

DON'T attempt to loosen or tighten high-torque nuts (e.g. wheel hub nuts) while the vehicle is on a jack; it may be pulled off.

DON'T start the engine without first ascertaining that the transmission is in neutral (or 'Park' where applicable) and the parking brake applied.

DON'T suddenly remove the filler cap from a hot cooling system — cover it with a cloth and release the pressure gradually first, or you may get scalded by escaping coolant.

DON'T attempt to drain oil until you are sure it has cooled sufficiently to avoid scalding you.

DON'T grasp any part of the engine, exhaust or catalytic converter without first ascertaining that it is sufficiently cool to avoid burning you.

DON'T syphon toxic liquids such as fuel, brake fluid or antifreeze by mouth, or allow them to remain on your skin.

DON'T inhale brake lining dust — it is injurious to health.

DON'T allow any spilt oil or grease to remain on the floor — wipe it up straight away, before someone slips on it.

DON'T use ill-fitting spanners or other tools which may slip and cause injury.

DON'T attempt to lift a heavy component which may be beyond your capability — get assistance.

DON'T rush to finish a job, or take unverified short cuts.

DON'T allow children or animals in or around an unattended vehicle.

DO wear eye protection when using power tools such as drill, sander, bench grinder etc, and when working under the vehicle.

DO use a barrier cream on your hands prior to undertaking dirty jobs — it will protect your skin from infection as well as making the dirt easier to remove afterwards; but make sure your hands aren't left slippery.

DO keep loose clothing (cuffs, tie etc) and long hair well out of the way of moving mechanical parts.

DO remove rings, wristwatch etc, before working on the vehicle — especially the electrical system.

DO ensure that any lifting tackle used has a safe working load rating adequate for the job.

DO keep your work area tidy — it is only too easy to fall over articles left lying around.

DO get someone to check periodically that all is well, when working alone on the vehicle.

DO carry out work in a logical sequence and check that everything is correctly assembled and tightened afterwards.

DO remember that your vehicle's safety affects that of yourself and others. If in doubt on any point, get specialist advice.

IF, in spite of following these precautions, you are unfortunate enough to injure yourself, seek medical attention as soon as possible.

Fire

Remember at all times that petrol (gasoline) is highly flammable. Never smoke, or have any kind of naked flame around, when working on the vehicle. But the risk does not end there — a spark caused by an electrical short-circuit, by two metal surfaces contacting each other, or even by static electricity built up in your body under certain conditions, can ignite petrol vapour, which in a confined space is highly explosive.

Always disconnect the battery earth (ground) terminal before working on any part of the fuel system, and never risk spilling fuel on to a hot engine or exhaust.

It is recommended that a fire extinguisher of a type suitable for fuel and electrical fires is kept handy in the garage or workplace at all times. Never try to extinguish a fuel or electrical fire with water.

Fumes

Certain fumes are highly toxic and can quickly cause unconsciousness and even death if inhaled to any extent. Petrol (gasoline) vapour comes into this category, as do the vapours from certain solvents such as trichloroethylene. Any draining or pouring of such volatile fluids should be done in a well ventilated area.

When using cleaning fluids and solvents, read the instructions carefully. Never use materials from unmarked containers — they may give off poisonous vapours.

Never run the engine of a motor vehicle in an enclosed space such as a garage. Exhaust fumes contain carbon monoxide which is extremely poisonous; if you need to run the engine, always do so in the open air or at least have the rear of the vehicle outside the workplace.

If you are fortunate enough to have the use of an inspection pit, never drain or pour petrol, and never run the engine, while the vehicle is standing over it; the fumes, being heavier than air, will concentrate in the pit with possibly lethal results.

The battery

Never cause a spark, or allow a naked light, near the vehicle's battery. It will normally be giving off a certain amount of hydrogen gas, which is highly explosive.

Always disconnect the battery earth (ground) terminal before working on the fuel or electrical systems.

If possible, loosen the filler plugs or cover when charging the battery from an external source. Do not charge at an excessive rate or the battery may burst.

Take care when topping up and when carrying the battery. The acid electrolyte, even when diluted, is very corrosive and should not be allowed to contact the eyes or skin.

If you ever need to prepare electrolyte yourself, always add the acid slowly to the water, and never the other way round. Protect against splashes by wearing rubber gloves and goggles.

Mains electricity

When using an electric power tool, inspection light etc which works from the mains, always ensure that the appliance is correctly connected to its plug and that, where necessary, it is properly earthed (grounded). Do not use such appliances in damp conditions and, again, beware of creating a spark or applying excessive heat in the vicinity of fuel or fuel vapour.

Ignition HT voltage

A severe electric shock can result from touching certain parts of the ignition system, such as the HT leads, when the engine is running or being cranked, particularly if components are damp or the insulation is defective. Where an electronic ignition system is fitted, the HT voltage is much higher and could prove fatal.

Routine maintenance

Regular maintenance is essential to ensure the safety of the car, and to retain the economy and performance of the car.

The maintenance tasks and instructions listed are those recommended by the manufacturer and the majority are visual checks; they are supplemented by additional tasks which, from practical experienece need to be carried out. The additional tasks are primarily of a preventative nature – they will assist in eliminating the unexpected failure of a component

Weekly, before a long journey or every 250 miles (400 km)

Depress and release the engine oil level gauge transmitter push button on the dashboard with the engine stopped. Check the level reading on the gauge. For a more accurate reading remove the dipstick and check the engine oil level which should be up to the MAX mark. Top up the oil in the sump if necessary. On no account allow the oil to fall below the MIN mark on the dipstick.
Check the battery electrolyte level and top up as necessary with distilled water. Make sure the top of the battery is kept clean and dry. If the battery requires regular topping up refer to Chapter 10
When the engine is cold inspect the level of the coolant in the translucent plastic system expansion tank. The level must be up to the mark on the tank. Top up the tank with water and antifreeze solution if necessary. Should the tank require topping up on consecutive checks, a week or less apart, refer to Chapter 2: Should the reservoir be found empty, remove the reservoir cap and fill the reservoir to the aforementioned level. Refit the cap and inspect the system hoses for leaks – see Chapter 2
Check the tyre pressures with an accurate gauge, and alter as necessary. Inspect the tyre walls and tread for damage such as cuts and blisters. Regular inspection of tyres is essential to safety since damage can quickly develop to a catastrophic degree if not attended to. Remember that the law requires at least 1 mm deep tread across three quarters of the width of the tread, around the whole circumference of the tyre.
Refill the windscreen washer container with water. Add an antifreeze solution satchet in cold weather. Do not use ordinary engine antifreeze as it corrodes paintwork. Check that the washer jets are aligned effectively
Remove the wheel trims and check all wheel bolts for tightness – do not overtighten; the correct torque on the wheel bolts is given in Chapter 11
Ensure that all lights and electrical systems are working properly. Chapter 10 includes fault diagnosis information

Every month or 1500 miles (2400 km)

Carry out those checks listed in the weekly Section
Lubricate the electropneumatic horn compressor by pouring a few drops of oil in the oiler after removing the cap on the top of the compressor
Check the level of hydraulic fluid in the brake fluid reservoir(s) and top up as necessary

Topping up the engine oil

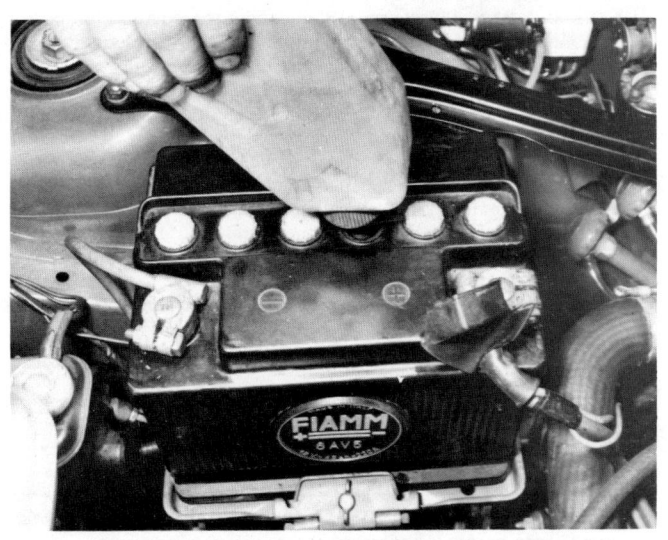

Topping up the battery

Topping up the cooling system

Checking the tyre pressures

Checking the windscreen washer reservoir (bag type)

Lubricating the horn compressor

Checking the brake fluid reservoir level

Remove the gearbox differential unit dipstick and check the oil level

Every 3 months or 3000 miles (4800 km)

Carry out those tasks listed in the Sections for weekly and monthly maintenance

Wipe the area around the gearbox oil level dipstick/filler on the transmission housing and remove the dipstick to check the oil level. It should be between the MIN and MAX marks on the dipstick. Top up with oil if necessary and refit the dipstick (photos)

As well as checking the hydraulic fluid level in the brake fluid reservoir, the hoses, system pipes and joints should be brushed clean and inspected for signs of corrosion, cracks or leaks. The metal pipes may be sprayed with a de-watering wax to maintain them between inspections

The three nuts which secure the cleaner top cover to the cleaner canister should be removed and the air cleaner cartridge lifted out. Shake the cartridge repeatedly to free the dust and then blow through with an air line. If dusty conditions prevail the cleaner cartridge may need to be renewed at 3000 mile (4800 km) intervals

The spark plugs should be removed and carbon deposits on the electrodes and porcelain insulation cleaned off. This is best done at your local garage which will probably have a spark plug sand blasting machine which also checks the plug electrically. Finally before refitting the plugs into the engine, the electrode gap should be checked. See Chapter 4.

Check the tension of the alternator and water pump drivebelt and adjust the tension if necessary as described in Chapter 10

Check the free travel of the clutch cable at the butterfly adjusting nut as described in Chapter 5

Check the wear of all the brake pads, as described in Chapter 9

Every 6 months or 6000 miles

Carry out those tasks listed for maintenance at weekly, monthly and three monthly intervals

Run the engine until thoroughly warmed up, and then remove the protective panel underneath the front of the engine. Place a container of 1 Imp gal (1.2 US gal, 5 litres) under the engine sump drain plug located in the middle of the rear edge of the sump. Remove the drain plug and allow the old oil to drain out. Whilst the oil is draining, the oil filter cartridge located on the front of the engine can be unscrewed and discarded. Smear the rubber seal on the new cartridge with a little engine oil and then refit to the filter head. Once the seal of the new filter contacts the filter head, tighten a further $\frac{1}{2}$ turn. Clean the oil filler cap and refit the sump drain plug

Refill the engine with the correct quantity of oil and clean off any oil which may have been spilt over the engine or its components. Restart the engine and run for a few minutes, then inspect the cartridge oil filter joint for leaks, and check the level of oil in the sump. Finlly refit the protective panel underneath the front of the engine

The interval between oil changes should be reduced in very hot or dusty conditions or during cool weather with much slow and stop/start driving

Check the valve clearance as described in Chapter 1

Renew the air filter

Check and adjust the contact breaker points gap and apply a little grease to the cam, as described in Chapter 4

Check and adjust the handbrake. The handbrake system should be brushed clean and lubricated with engine oil. The handbrake should apply the rear brakes after 4 notches have been passed on the handle ratchet. If the handbrake handle travel is excessive refer to Chapter 9 for the adjustment procedure

Check the rubber boots on the driveshaft CV joints for cracking or tears. If any signs of leakage exists, it must be dealt with promptly.

Check the fluid level in the automatic transmission after running the engine for 20 to 30 seconds

Check the battery electrolyte levels. The terminals and clamp connections should be separated, cleaned and refitted after being given a liberal coat of vaseline

Carefully examine the cooling and heater systems for signs of leaks. Make sure that all hose clips are tight and that none of the hoses have cracked or perished. Do not attempt to repair a leaking

Routine maintenance

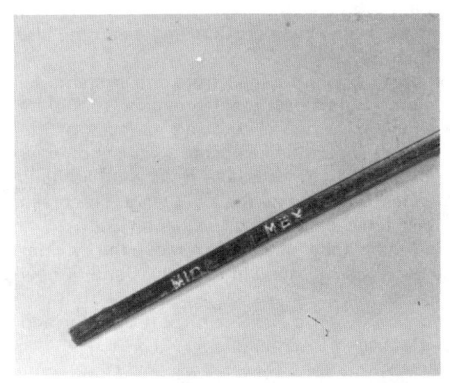

The level should be between the MIN and MAX marks

Topping up the gearbox differential unit with oil

The sump drain plug

hose, always fit a new one. Generally inspect the exterior of the engine for signs of leaks or stains. The method of repair will depend on its location. This check is particularly important before filling the cooling system with antifreeze as it has a greater searching action than water and will find any weak spots

Boot and bonnet hinges and mechanisms all require drops of engine oil at 6 monthly intervals to minimize wear. Door hinges, locks, tiltable seat articulations and limiting arms need lubricating with engine oil also, but the striker blocks, latches, and seat rails need smearing lightly with grease. Avoid a too liberal lubrication of items around the door and inside the car because it may get transferred to passengers' clothing

Inspect the seat belts for damage to the webbing and check that the anchorages are secure

Wash the bodywork and chromium fittings and clean out the interior of the car. Wax polish the bodywork including all the chromium and bright trim. Force wax polish into any joints in the body work to prevent rust formation. Clean out all the door and body drain holes

Every 12 months or 12 000 miles (19 000 km)

In addition to the weekly, monthly, 3 monthly and 6 monthly checks the following inspection tasks should be carried out

Refer to Chapter for full instructions on cleaning and tuning the carburettor

Clean the crankcase emission control system as described in Chapter 3

The condition of the elastic mountings of the exhaust pipe system should be checked, and each joint should be secure. Minor holes and corrosion may be repaired with one of the many proprietary compounds, but in the majority of cases it is cheaper in the long run to renew the relevant section of pipe

Check the general condition and tension of the timing belt as described in Chapter 1

Jack up the front of the car until the wheels are off the ground and place chassis stands underneath the front subframe. Grasp the top and bottom of the wheel and move to check for wear in the wheel bearing and sloppiness in any of the front suspension joints. Finally grasp the wheel by its forward and rearward edges and move to check for wear and slackness in the steering mechanism. Wear in the wheel bearing will also be noticed with this last check. Examine each suspension and steering joint for corrosion and deterioration of the dust boots or seals. Steering joints should be checked for looseness individually. Also check the rear suspension shock absorbers for leaks and check all mountings for tightness. Undo the hose clamps and renew the in-line fuel filter, which lies between the fuel pump and carburettor

On models fitted with electronic ignition distributor remove the reductor and inspect and clean it. Check the distributor cap for cracks or carbon deposits. Also check the ignition advance setting

Check that the electric fan for the cooling system operates correctly, as described in Chapter 2

Check the power steering fluid level and top up as necessary

Always pay particular attention to the manner in which the front tyres wear. Uneven wear is a definite indication of wheel misalignment and once noticed, the car should be taken to a Lancia agent who will have the specialised equipment to make an accurate job of wheel alignment

Check that the heating system works correctly

Carry out an annual check on the air conditioning system to ensure that it is operating correctly

Thoroughly inspect and ensure the integrity of all joints of mechanical units to the body shell. Rusted suspension anchorages are a common cause of failure of the MoT vehicle test in Britain. Remember the steering system and damper anchorages, failure of these would also be catastrophic. It is worth while having the underside of the car steam cleaned so that the condition of the undersealing and chassis can be determined more thoroughly. Always entrust chassis repairs and renovation to a reputable garage. Check the condition and security of the subframe mounting bolts. For early models, mainly A series Saloons, the subframe mounting points in the bodyshell have been subject to rusting. If this happens then professional advice shuld be taken as it may be that the car is unroadworthy. More detailed information on this subject is given in Chapter 12.

Every 18 months or 18 000 miles (29 000 km)

In addition to those checks listed for weekly, monthly, three monthly and six monthly interval services the following tasks should be performed

Drain the oil from the gearbox and differential unit. Refill with the specified oil and check the level is correct using the dipstick

Drain the oil from the automatic transmission/differential unit. drain the fluid when it is hot, and let it drain thoroughly before

Using a thin piece of wire to clear out the door drain holes (don't forget the body ones)

refitting the plug and refilling the unit with new fluid. Pour the fluid through the dipstick tube next to the expansion tank for the cooling system

The starter motor should be removed from the car and cleaned. Refer to Chapter 10 for cleaning and inspection of the motor commutator and the brushes. Renew the brushes if necessary. The drive pinion and slide, and shaft bushes should be lubricated with light oil as directed in Chapter 10

Every 2 years or 24 000 miles (39 000 km)

Carry out all those maintenance tasks listed for weekly, monthly, 3 monthly, 6 monthly and yearly servicing
Renew the timing belt. Chapter 1 details the procedure for belt renewal with the engine in the car

Remove the alternator and inspect the brushes and clean the slip rings as directed in Chapter 10

Completely drain the brake hydraulic fluid from the system. All seals and flexible hoses throughout the braking circuits should be examined and preferably renewed. The working surfaces of the master cylinder and calipers should be inspected for signs of wear or scoring and new parts fitted as necessary. Refill the hydraulic system with recommended brake fluid. See Chapter 9. The brake servo unit should be inspected as directed in Chapter 9

Test the cylinder compression and if necessary remove the cylinder head and decarbonise. If necessary grind in the valves and fit new valve springs. See Chapter 1

Drain and flush the cooling system. Refill with 50/50 water and antifreeze mixture, as described in Chapter 2.

Chapter 1 Engine

Contents

Ancillary components – refitting (Stage One)	30
Ancillary components – refitting (Stage Two)	36
Ancillary driveshaft – refitting	26
Ancillary driveshaft and oil pump/distributor drivegear – examination	19
Big-end bearings, pistons, connecting rods and oil pump – removal and refitting in situ	41
Big-end and main bearing shells – examination and renovation	22
Cylinder bores – examination and remedial action	15
Cylinder head – reassembly	31
Cylinder head – refitting to cylinder block	33
Cylinder head, camshaft and housing – removal and refitting (engine in situ)	39
Cylinder head, valve gear and camshafts – overhaul	11
Crankshaft – examination and renovation	21
Crankshaft – refitting	24
Engine – initial start-up after overhaul	38
Engine and gearbox – refitting	37
Engine and gearbox – removal	5
Engine dismantling procedure	10
Engine dismantling – general	7
Engine reassembly – general	23
Engine shock absorbers – renewal	43
Fault diagnosis – engine	46
Flywheel – examination, and renewal of the starter ring gear	20
Flywheel and clutch – refitting	29
General description	1
Gudgeon pin – renewal	17
Lubrication system – general description	12
Major operations possible with engine in situ	2
Major operations possible with engine removed	3
Main engine mounting – renewal	42
Methods and equipment for engine removal	4
Notes for cars fitted with air condition systems	44
Notes for USA models	45
Oil filter and housing – general	14
Oil pump – overhaul	13
Oil pump – refitting	27
Oil sump – refitting	28
Oil sump – removal and refitting in situ	40
Pistons, connecting rods and big-end bearings – refitting	25
Pistons and piston rings – examination and renewal	16
Removing the ancillary engine components	8
Small-end bush – renewal (except 1400 models)	18
Subframe, engine and gearbox – removal and refitting	6
Timing belt – refitting, alignment and tensioning	35
Timing belt and cover (engine out of car) – removal	9
Timing wheels and belt – examination	34
Valve/tappet clearance setting (engine in or out of car)	32

Specifications

Engine (general)

Type	4-cylinder, transverse mounted, water cooled, twin overhead camshafts
Firing order	1 – 3 – 4 – 2 (No 1 cylinder at timing belt end)
Engine timed on	No 4 cylinder (flywheel end)

Engine data

	Bore	Stroke	Engine cubic capacity
1300 (up to 1979)	76 mm (2.992 in)	71.5 mm (2.838 in)	1297.9 cc
1300 (1979 onwards)	76.1 mm (2.996 in)	71.4 mm (2.811 in)	1301 cc
1400	76 mm (2.992 in)	71.5 mm (2.83 in)	1437.6 cc
1600 A.000	80 mm (3.14 in)	71.5 mm (2.83 in)	1437.6 cc
1600 AC.000	80 mm (3.14 in)	79.2 mm (3.11 in)	1592.4 cc
1600 Series B	80 mm (3.14 in)	79.2 mm (3.11 in)	1592.4 cc
1800 A1.000	84 mm (3.30 in)	71.5 mm (2.83 in)	1585 cc
1800 AC1.000	84 mm (3.30 in)	79.2 mm (3.11 in)	1755.6 cc
1800 USA	84 mm (3.30 in)	79.2 mm (3.11 in)	1755.6 cc
2000	84 mm (3.30 in)	79.2 mm (3.11 in)	1755.6 cc
	84 mm (3.30 in)	90 mm (3.54 in)	1995 cc

Engine data (continued)	Compression ratio	Maximum horsepower (DIN)	Maximum torque (DIN)
1300 (up to 1979)	8.9 : 1	82 BHP at 6200 rpm	81.1 lbf ft (11 kgf m) at 3300 rpm
1300 (1979 onwards)	8.9 : 1	85 BHP at 5800 rpm	79.5 lbf ft (10.7 kgf m) at 3300 rpm
1400	8.9 : 1	90 BHP at 6000 rpm	87.0 lbf ft (11.8 kgf m) at 3800 rpm
1600 A.000	8.9 : 1	100 BHP at 6000 rpm	96.6 lbf ft (13.1 kgf m) at 3000 rpm
1600 AC.000	9.8 : 1	108 BHP at 6000 rpm	101.7 lbf ft (13.8 kgf m) at 4500 rpm
1600 Series B	9.4 : 1	100 BHP at 5800 rpm	100.3 lbf ft (13.6 kgf m) at 3000 rpm
1800 A1.000	8.9 : 1	110 BHP at 6000 rpm	108.4 lbf ft (14.7 kgf m) at 4500 rpm
1800 AC1.000	9.8 : 1	120 BHP at 6200 rpm	112.8 lbf ft (15.3 kgf m) at 2800 rpm
1800 USA	8.0 : 1	Information not available at time of writing	
2000	8.9 : 1	119 BHP at 5500 rpm	130.5 lbf ft (17.7 kgf m) at 2800 rpm

Piston types and sizes

Engine	Piston type	Class A	Class B	Class C	Class D	Class E	Pistons to be measured "x" mm (in) from piston crown Measured at 90° to gudgeon pin axis
1300 engines	MONDIAL	75.93 to 75.94 mm (2.9893 to 2.9897 in)	75.94 to 75.95 mm (2.9897 to 2.9901 in)	75.95 to 75.96 mm (2.9901 to 2.9905 in)	75.96 to 75.97 mm (2.9905 to 2.9909 in)	75.97 to 75.98 mm (2.9909 to 2.9913 in)	52.95 mm (2.0846 in)
1400 engines and 1600 AB engines	MONDIAL	79.925 to 79.935 mm (3.1466 to 3.1470 in)	79.935 to 79.945 mm (3.1470 to 3.1474 in)	79.945 to 79.955 mm (3.1474 to 3.1478 in)	79.955 to 79.965 mm (3.1478 to 3.1482 in)	79.965 to 79.975 mm (3.1482 to 3.1486 in)	52.25 mm (2.0570 in)
1400 engines and 1600 A engines	FIAT	79.920 to 79.930 mm (3.1464 to 3.1468 in)	79.930 to 79.940 mm (3.1468 to 3.1472 in)	79.940 to 79.950 mm (3.1472 to 3.1476 in)	79.950 to 79.960 mm (3.1476 to 3.1480 in)	79.960 to 79.970 mm (3.1480 to 3.1484 in)	52.25 mm (2.0570 in)
1600 A engines	MONDIAL	79.940 to 79.950 mm (3.1472 to 3.1476 in)	79.950 to 79.960 mm (3.1476 to 3.1480 in)	79.960 to 79.970 mm (3.1480 to 3.1484 in)	79.970 to 79.980 mm (3.1484 to 3.1488 in)	79.980 to 79.990 mm (3.1488 to 3.1492 in)	50.20 mm (1.9763 in)
1600 B series, 1800 AC and 2000 engines	MONDIAL	83.940 to 83.950 mm (3.3047 to 3.3051 in)	83.950 to 83.960 mm (3.3051 to 3.3055 in)	83.960 to 83.970 mm (3.3055 to 3.3059 in)	83.970 to 83.980 mm (3.3059 to 3.3063 in)	83.980 to 83.990 mm (3.3062 to 3.3066 in)	52.95 mm (2.0846 in)
1800 A1 and 1800 USA	MONDIAL	83.950 to 83.960 mm (3.3051 to 3.3055 in)	83.960 to 83.970 mm (3.3055 to 3.3060 in)	83.970 to 83.980 mm (3.3060 to 3.3062 in)	83.980 to 83.990 mm (3.3062 to 3.3066 in)	83.990 to 84.000 mm (3.3066 to 3.3070 in)	53.00 (2.0866 in)– UK 47.25 (1.8602 in)– USA
2000	MONDIAL	83.940 to 83.950 mm (3.3047 to 3.3051 in)	83.950 to 83.960 mm (3.3051 to 3.3055 in)	83.960 to 83.970 mm (3.3055 to 3.3059 in)	83.970 to 83.980 mm (3.3059 to 3.3063 in)	83.980 to 83.990 mm (3.3062 to 3.3066 in)	52.95 (2.0846 in)

Standard pistons supplied as spares ...... Class A, C and E only
Oversizes available ...... 0.2, 0.4 and 0.6 mm (0.0078, 0.0157 and 0.0236 in)

Cylinder bore sizes

Engine	Class A	Class B	Class C	Class D	Class E
1300 engine	76.000 to 76.010 mm (2.992 to 2.9925 in)	76.010 to 76.020 mm (2.9925 to 2.9929 in)	76.020 to 76.030 mm (2.9929 to 2.9933 in)	76.030 to 87.040 mm (2.9933 to 2.9936 in)	76.040 to 76.050 mm (2.9936 to 2.9940 in)
1400 and 1600 Series 'A'	80.00 to 80.010 mm (3.1496 to 3.1499 in)	80.010 to 80.020 mm (3.1499 to 3.1503 in)	80.020 to 80.030 mm (3.1503 to 3.1507 in)	80.030 to 80.040 mm (3.1507 to 3.1511 in)	80.040 to 80.050 mm (3.1511 to 3.1515 in)
1600 – Series 'B' 1800, 1800 USA and 2000	84.00 to 84.010 mm (3.3070 to 3.3074 in)	84.010 to 84.020 mm (3.3074 to 3.3078 in)	84.020 to 84.030 mm (3.3078 to 3.3082 in)	84.030 to 84.040 mm (3.3082 to 3.3086 in)	84.040 to 84.050 mm (3.3086 to 3.3090 in)

Chapter 1 Engine

	1800 and 1600 series A	1300, 1400 and 1600 series B	2000
Cylinder bore gauging			
A (top) – distance below top of bore	10 mm (0.393 in)	9.5 mm (0.374 in)	13.5 mm (0.531 in)
B (centre) – distance below top of bore	55 mm (2.16 in)	50.5 mm (2 in)	63.5 mm (2.49 in)
C (bottom) – distance below top of bore	100 mm (4 in)	91.5 mm (3.60 in)	113.5 mm (4.46 in)

Note: *Cylinder bores must be gauged in the above three position at right angles to the engine centreline*

Piston clearance in bore

	Piston type	Clearance in bore
1300 engines	MONDIAL	0.060 to 0.080 mm (0.0023 to 0.0031 in)
1400 engines	MONDIAL	0.065 to 0.085 mm (0.0025 to 0.0033 in)
	FIAT	0.070 to 0.090 mm (0.0027 to 0.0035 in)
1600 engines – Series A	MONDIAL	0.050 to 0.070 mm (0.0019 to 0.0027 in)
1600 engines – Series B	MONDIAL	0.050 to 0.070 mm (0.0019 to 0.0027 in)
1800 engines – Series A1	MONDIAL	0.040 to 0.060 mm (0.0015 to 0.0023 in)
1800 engines – Series AC	MONDIAL	0.050 to 0.070 mm (0.0019 to 0.0027 in)
1800 engines – USA version	MONDIAL	0.040 to 0.060 mm (0.0015 to 0.0023 in)
2000 engines	MONDIAL	0.050 to 0.070 mm (0.0019 to 0.0027 in)

Piston ring data

	Top ring	2nd ring*	Oil ring
Thickness of ring	1.478 to 1.490 mm (0.0581 to 0.0586 in)	1.980 to 2.00 mm (0.0779 to 0.0787 in)	3.925 to 3.937 mm (0.1545 to 0.1549 in)
Width of ring groove in piston	1.535 to 1.555 mm (0.060 to 0.061 in)	2.030 to 2.050 mm (0.0799 to 0.0807 in)	3.967 to 3.987 mm (0.1565 to 0.1569 in)
Clearance of ring in groove	0.045 to 0.077 mm (0.0017 to 0.0030 in)	0.030 to 0.070 mm (0.0011 to 0.0027 in)	0.030 to 0.062 mm (0.0011 to 0.0024 in)

* Not 1300 engine

1300 engine (2nd ring):
- Thickness of ring .... 1.978 to 1.990 mm (0.0778 to 0.0783 in)
- Clearance of ring in groove .... 0.040 to 0.072 mm (0.0015 to 0.0025 in)

Piston ring gaps in bore

	Top ring	2nd ring	Oil ring
1300	0.30 to 0.50 mm (0.0118 to 0.0196 in)	0.30 to 0.50 mm (0.0118 to 0.0196 in)	0.20 to 0.35 mm (0.0078 to 0.0137 in)
1400, 1600 series A and 1800 engines	0.30 to 0.45 mm (0.0118 to 0.0177 in)	0.20 to 0.35 mm (0.0078 to 0.0137 in)	0.20 to 0.35 mm (0.0078 to 0.0137 in)
1600 series B and 2000 engines	0.30 to 0.45 mm (0.0118 to 0.0177 in)	0.30 to 0.45 mm (0.0118 to 0.0177 in)	0.25 to 0.40 mm (0.0098 to 0.0157 in)

Piston ring oversizes .... As per pistons

Gudgeon pins

	Class 1	Class 2
Gudgeon pin size:		
1300, 1600, 1800 and 2000 engines	21.991 to 21.994 mm (0.8657 to 0.8659 in)	21.994 to 21.997 mm (0.8659 to 0.8660 in)
1400 engines	21.970 to 21.974 mm (0.8649 to 0.8651 in)	21.974 to 21.978 mm (0.8651 to 0.8652 in)
Gudgeon pin hole in piston:		
1300, 1600, 1800 and 2000 engines	21.996 to 21.999 mm (0.8659 to 0.8661 in)	21.999 to 22.002 mm (0.8661 to 0.8662 in)
1400 engine	21.984 to 21.988 mm (0.8655 to 0.8656 in)	21.988 to 21.992 mm (0.8656 to 0.8658 in)

Clearance fit between gudgeon pin and piston:
   1300, 1600, 1800 and 2000 engines .... 0.002 to 0.008 mm (0.000078 to 0.00031 in)
   1400 engines .... 0.010 to 0.018 mm (0.00039 to 0.00070 in)

Connecting rods
Small-end type:
   1400 engines .... Gudgeon pin is an interference fit — there is no small-end bush
   1300, 1600, 1800 and 2000 engines .... Small-end has a pressed in bush
Clearance fit between gudgeon pin and small-end .... 0.010 to 0.016 mm (0.00039 to 0.00062 in)
Big-end bore without bearing shells:
   1400 engines .... 48.630 to 48.646 mm (1.914 to 1.931 in)
   1300 and 1600 Series B engines .... 51.330 to 51.346 mm (2.0208 to 2.0214 in)
   1600 Series A, 1800 and 2000 engines .... 53.897 to 53.913 mm (2.1219 to 2.1225 in)
Big-end endfloat on crankpin:
   1300 and 1600 series B engines .... 0.12 to 0.45 mm (0.0047 to 0.0177 in)
   1400 engines .... 0.12 to 0.45 mm (0.0047 to 0.0177 in)

	Crankpin diameter	Class	Bearing shell thickness to be fitted 1300 engines — up to No 128065 1600 B engines — up to No 128025	Bearing shell thickness to be fitted 1300 engines — from No 128066 1600 B series engine from No 128026	Bearing shell colour code
Big-end bearing shell sizes and classifications:					
1300 engines and 1600 series B engines	48.224 to 48.244 mm	A – Standard	1.516 to 1.520 mm (0.0596 to 0.0598 in)	1.524 to 1.528 mm (0.0599 to 0.060)	Red
	(1.8985 to 1.8993 in)	B – Standard	1.520 to 1.524 mm (0.0598 to 0.0599 in)	1.529 to 1.533 mm (0.061 to 0.0603 in)	Blue*
	48.097 to 48.117 mm	A – Standard	1.579 to 1.583 mm (0.0621 to 0.0623 in)	1.587 to 1.591 mm (0.0624 to 0.0626 in)	Red
	(1.8935 to 1.5943 in)	B – Standard	1.583 to 1.587 mm (0.0623 to 0.0624 in)	1.591 to 1.596 mm (0.0626 to 0.0628 in)	Blue*
Undersizes available:					
1			1.643 to 1.651 mm (0.0646 to 0.0649 in)	1.651 to 1.659 mm (0.0649 to 0.0653 in)	
2			1.770 to 1778 mm (0.0696 to 0.0699 in)	1.778 to 1.786 mm (0.0699 to 0.0703 in)	
3			1.897 to 1.905 mm (0.0746 to 0.0749 in)	1.905 to 1.913 mm (0.0749 to 0.0753 in)	
4			2.024 to 2.032 mm (0.0796 to 0.0799 in)	2.032 to 2.040 mm (0.0799 to 0.0810 in)	

	Crankpin diameter	Class	Bearing shell thickness to be fitted	Bearing shell thickness to be fitted	Bearing shell colour code
1600 series A engines and 1800, 1800 USA and 2000 engines			2000 engines — up to engine No 114193 1800 USA version — up to engine No 112067 1800 California — up to engine No 112062	2000 engines — from engine No 114194 1800 USA version — from engine No 112068 1800 California — from engine No 112063	
	50.782 to 50.802 mm (1.999 to 2.000 in)	A – Standard	1.521 to 1.525 mm (0.0598 to 0.0600 in)	1.528 to 1.532 mm (0.0601 to 0.0603 in)	Red
		B – Standard	1.525 to 1.529 mm (0.0600 to 0.0601 in)	1.533 to 1.537 mm (0.06035 to 0.0605 in)	Blue*
	50.655 to 50.675 mm (1.994 to 1.995 in)	A – standard	1.584 to 1.588 mm (0.0623 to 0.0625 in)	1.591 to 1.595 mm (0.0626 to 0.0627 in)	Red
		B – Standard	1.588 to 1.592 mm (0.0625 to 0.0626 in)	1.596 to 1.600 mm (0.0627 to 0.06299 in)	Blue*

Chapter 1 Engine

Undersizes available:
1 ....
2 ....
3 ....
4 ....

1.648 to 1.656 mm   1.655 to 1.663 mm
(0.06488 to 0.0651 in)(0.0651 to 0.0654 in)
1.775 to 1.783 mm   1.782 to 1.790 mm
(0.0698 to 0.0701 in) (0.0701 to 0.0704 in)
1.902 to 1.910 mm   1.909 to 1.917 mm
(0.0748 to 0.0751 in) (0.0751 to 0.0754 in)
2.029 to 2.037 mm   2.036 to 2.044 mm
(0.0798 to 0.0801 in) (0.08015 to 0.0804 in)

* **Note:** *Not available as spares*

	Crankpin diameter	Class	Shell thickness to be fitted
1400 engines	45.508 to 45.528 mm (1.7916 to 1.7934 in)	Standard	1.531 to 1.538 mm (0.0602 to 0.0605 in)
Undersizes available:			
1	45.381 to 45.401 mm (1.7866 to 1.7874 in)		1.594 to 1.601 mm (0.0627 to 0.0630 in)
2	45.254 to 45.274 mm (1.7816 to 1.7824 in)		1.658 to 1.665 mm (0.0652 to 0.0655 in)
3	45.000 to 45.020 mm (1.7716 to 1.7724 in)		1.785 to 1.792 mm (0.0702 to 0.0705 in)
4	44.746 to 44.766 mm (1.7616 to 1.7624 in)		1.912 to 1.919 mm (0.0752 to 0.0755 in)
5	44.492 to 44.512 mm (1.7516 to 1.7524 in)		2.039 to 2.046 mm (0.0802 to 0.0805 in)

Crankshaft main bearings

No of main bearings 5
Crankshaft endfloat .... Governed by half thrust washers on either side of rear main journal
Thrust washer thicknesses:
  Standard .... 2.310 to 2.360 mm (0.0909 to 0.0929 in)
  Oversize .... 2.437 to 2.487 mm (0.0959 to 0.0979 in)
  Endfloat permissible .... 0.055 to 0.305 mm (0.00216 to 0.0120 in)

Crankshaft main bearing journal sizes/classification/and bearing shell sizes

	Classification	Journal diameter	Bearing shell size
1400 engines	Standard	50.775 to 50.795 mm (1.9990 to 1.9997 in)	1.825 to 1.831 mm (0.0718 to 0.0720 in)
	Undersize – 1	50.648 to 50.668 mm (1.9940 to 1.9947 in)	1.888 to 1.894 mm (0.0743 to 0.0745 in)
	Undersize – 2	50.521 to 50.541 mm (1.9890 to 1.9897 in)	1.952 to 1.958 mm (0.0768 to 0.0770 in)
	Undersize – 3	50.267 to 50.287 mm (1.9790 to 1.9797 in)	2.079 to 2.083 mm (0.0818 to 0.0820 in)
	Undersize – 4	50.013 to 50.033 mm (1.9690 to 1.9697 in)	2.206 to 2.212 mm (0.0868 to 0.0870 in)
	Undersize – 5	49.759 to 49.779 mm (1.9590 to 1.9597 in)	2.333 to 2.339 mm (0.0918 to 0.0920 in)
1600 Series A, 1800, 1800 USA and 2000 engines .... (1800 USA up to engine No 97477) (2000 BB engines up to engine No 105818)	Standard	52.985 to 53.005 mm (2.0860 to 2.0868 in)	1.825 to 1.831 mm (0.0718 to 0.0720 in)
	Undersize – 1	52.858 to 52.878 mm (2.0810 to 2.0818 in)	1.888 to 1.894 mm (0.743 to 0.0745 in)
	Undersize – 2	52.731 to 52.751 mm (2.0760 to 2.0768 in)	1.852 to 1.958 mm (0.0768 to 0.0770 in)
	Undersize – 3	52.477 to 52.487 mm (2.0660 to 2.0664 in)	2.079 to 2.085 mm (0.0818 to 0.0820 in)
	Undersize – 4	52.223 to 52.243 mm (2.0560 to 2.0568 in)	2.206 to 2.12 mm (0.0868 to 0.0870 in)
	Undersize – 5	51.969 to 51.989 mm (2.0460 to 2.0468 in)	2.333 to 2.339 mm (0.0918 to 0.920 in)

	Classification	Journal diameter	Bearing shells for models listed below*	Bearing shells for models listed below**
1300 engines, 1600 Series B, 1800 USA and 2000 engines	Standard Class 1	5.985 to 53.005 mm (2.0860 to 2.0868 in)	1.829 to 1.835 mm (0.0720 to 0.0722 in)	1.834 to 1.840 mm (0.0722 to 0.0724 in)
	Class 2		1.834 to 1.840 mm (0.0722 to 0.0724 in)	1.839 to 1.845 mm (0.07240 to 0.0726 in)
	Class 1	52.858 to 52.878 mm (2.0810 to 2.0818 in)	1.892 to 1.898 mm (0.0744 to 0.0747 in)	1.897 to 1.903 mm (0.0746 to 0.0749 in)

Chapter 1 Engine

Class 2		1.897 to 1.903 mm (0.0747 to 0.0749 in)	1.902 to 1.908 mm (0.07488 to 0.0751 in)
Undersize – 1	52.731 to 52.751 mm (2.0760 to 2.0768 in)	1.952 to 1.958 mm (0.0768 to 0.0770 in)	1.961 to 1.967 mm (0.0772 to 0.0774 in)
Undersize – 2	52.477 to 52.497 mm (2.0660 to 2.0668 in)	2.079 to 2.085 mm (0.0818 to 0.0820 in)	2.088 to 2.094 mm (0.0822 to 0.08244 in)
Undersize – 3	52.223 to 52.243 mm (2.0560 to 2.0568 in)	2.206 to 2.212 mm (0.0868 to 0.08702 in)	2.211 to 2.217 mm (0.08704 to 0.08728 in)
Undersize – 4	51.969 to 51.989 mm (2.0460 to 2.0468 in)	2.333 to 2.339 mm (0.0918 to 0.9208 in)	2.338 to 2.344 mm (0.09204 to 0.09228 in)

* These shells can be fitted as follows:
    1300 engines – up to engine No 128005
    1300 BB engines – up to engine No 128025
    1600 B series engines – up to engine No 128025
    2000 BB engines – from engine No 105819 to No 114193
    USA version 1800 – from engine No 97478 to engine No 112067
    California version 1800 – up to engine No 122062

** These shells can be fitted as follows:
    1300 engines – from engine No 128006
    1600 B series engines – from engine No 128026
    2000 BB engines – from engine No 114194
    USA version 1800 – from engine No 112068
    California version 1800 – from engine No 112063

Flywheel

	1300, 1400 and 1600	1800 and 2000
Thickness when new	22.9 to 23.1 mm (0.901 to 0.909 in)	24.6 to 24.8 mm (0.968 to 0.976 in)
Minimum thickness after refacing	22.5 mm (0.885 in)	24.2 mm (0.9527 in)

Lubrication system

Type .... Wet sump – pressure fed
Pressure at 85°C .... 4.5 to 6.0 kgf/cm² (64 to 85 lbf/in²) at 6000 rpm
Oil pump .... Gear type
Oil sump capacity .... 1.09 Imp gals (5 litres/1.32 US gals) (including filter)
Oil filter type .... Full flow disposable cartridge
Oil pump gear clearances:
  Between teeth .... 0.15 mm (0.0059 in)
  Between gears and cover:
    1300, 1400, 1600, 1800 engines ..... 0.031 to 0.116 mm (0.00122 to 0.00456 in)
    2000 engines .... 0.055 to 0.147 mm (0.0021 to 0.0057 in)
  Between gears and body of housing ... 0.063 to 0.118 mm (0.00248 to 0.00464 in)

Oil relief valve spring data:

	Pump part No	Spring part No	Spring free length
1600 and 1800 early models and 1400 models	82308927	4153891 or 82329232	40.2 mm (1.582 in) 36.1 mm (1.421 in)
1600 and 1800 later models. 1300 and 1800 USA models	82317492	82325838	26.3 mm (1.035 in)
2000	82334025	82325476	44.4 mm (1.748 in)

Camshafts

Journal diameter:
  Front journal .... 29.944 to 29.960 mm (1.1788 to 1.1795 in)
  Centre journal .... 45.744 to 45.771 mm (1.8013 to 1.8020 in)
  Rear journal .... 46.155 to 46.171 mm (1.8171 to 1.8177 in)
Journal clearance:
  Front journal .... 0.049 to 0.090 mm (0.00192 to 0.00354 in)
  Centre journal .... 0.029 to 0.070 mm (0.00114 to 0.00275 in)
Rear journal clearance .... 0.029 to 0.070 mm (0.00114 to 0.00275 in)
Camshaft endfloat .... 0.100 to 0.195 mm (0.00393 to 0.00767 in)
Camshaft drivebelt tension/deflection ...... 15° to 18° from vertical under a load of 31 to 38 kg (68 to 84 lbs)

Valves

	Inlet	Exhaust
Valve clearances (cold):		
1300, 1400, 1600, 1800 and 2000 models	0.39 to 0.45 mm (0.0153 to 0.0177 in)	0.45 to 0.51 mm (0.0177 to 0.0200 in)
1800 USA models	0.41 to 0.49 mm (0.0161 to 0.0192 in)	0.46 to 0.54 mm (0.0181 to 0.0212 in)
Valve timing:		
1300 and 1600 B series:		
Opens	17° BTDC	48° BBDC
Closes	37° ABDC	6° ATDC
1400 models:		
Opens	12° BTDC	46° BBDC
Closes	42° ABDC	8° ATDC

Chapter 1 Engine

1600 A series, 1800 and 2000 models:
   Opens .... 13° BTDC   49° BBDC
   Closes .... 45° ABDC   9° ATDC
1800 USA models:
   Opens .... 5° BTDC   53° BBDC
   Closes .... 53° ABDC   5° ATDC
Clearance adjustment .... Variable sized shims
Shim thicknesses available:
   1400, 1600, A series and 1800 models .... 3.25 to 4.70 mm (0.1279 to 0.1850 in) in increments of 0.05 mm (0.00196 in)
   1300, 1600B and 2000 models .... 3.25 to 4.70 mm (0.1279 to 0.1850 in) in increments of 0.10 mm (0.00393 in) from 3.30 mm (0.1299 in) upwards
Machining tolerance .... ± 0.012 mm (0.00047 in)
Tappets:
   Type .... Bucket with recess in top for shims
   Bucket diameter .... 36.975 to 36.995 mm (1.455 to 1.456 in)
   Clearance of tappet in bore in housing .... 0.005 to 0.050 mm (0.000196 to 0.00196 in)
Valve seat angle .... 45° 25' to 45° 35'

Valve head diameter: Inlet   Exhaust
   1300 models .... 37.30 to 37.70 mm   33.35 to 33.95 mm
                                                                                                      (1.468 to 1.484 in)   (1.312 to 1.336 in)
   1400 models .... 41.2 to 41.6 mm   35.85 to 36.45 mm
                                                                                                          (1.622 to 1.637 in)   (1.411 to 1.435 in)
   1600, 1800 and 2000 models:
      Early engines .... 42.20 to 42.60 mm   35.85 to 36.45 mm
                                                                                                  (1.661 to 1.677 in)   (1.411 to 1.435 in)
      Late engines .... 41.60 to 42.0 mm   35.85 to 36.45 mm
                                                                                                  (1.637 to 1.653 in)   (1.411 to 1.433 in)
   1800 USA models (up to 1977) .... 41.2 to 41.6 mm   35.85 to 36.45 mm
                                                                                                  (1.622 to 1.637 in)   (1.411 to 1.435 in)
   1800 USA models (1978 on) .... 41.60 to 42.0 mm   35.85 to 36.45 mm
                                                                                                  (1.637 to 1.653 in)   '1.411 to 1.435 in)
Valve stem diameter .... 7.974 to 7.992 mm (0.3139 to 0.3146 in)
Valve guide bore .... 8.022 to 8.040 mm (0.3158 to 0.3165 in)
Valve stem clearance in bore .... 0.030 to 0.066 mm (0.00118 to 0.00259 in)
Valve guide outer diameter:
   Type A .... 15.040 to 15.058 mm (0.5921 to 0.5928 in)
   Type B .... 15.240 to 15.258 mm (0.5999 to 0.6007 in)
Oversize guides available (outer diameters):
   Type A .... +0.02 mm (0.0078 in), +0.05 (0.001968 in), +0.10 mm (0.00393 in)
   Type B .... +0.05 mm (0.00196 in)

Note: For type A guides oversizes only are available as spares. For type B a standard size spare is available

Valve seat diameter: Inlet   Exhaust
   1300 .... 36.5 mm (1.437 in)   32.5 mm (1.279 in)
   1400 .... 40.4 mm (1.590 in)   35 mm (1.377 in)
   A1600/1800/2000 .... 41.4 mm (1.629 in)   35 mm (1.377 in)
   B1600/2000 .... 40.8 mm (1.606 in)   35 mm (1.377 in)
   1800 USA (up to 1977) .... 40.4 mm (1.590 in)   35 mm (1.377 in)
   1800 USA (1978 onwards) .... 40.8 mm (1.606 in)   35 mm (1.377 in)
Valve springs:
   Free length
      Inner .... 41.80 to 42.80 mm (1.645 to 1.685 in)
      Outer .... 53.9 mm (2.122 in)

Ancillary driveshaft

Number of journals 2
Front journal diameter .... 48.013 to 48.038 mm (1.890 to 1.891 in)
Front journal clearance .... 0.046 to 0.091 mm (0.0018 to 0.0035 in)
Rear journal diameter .... 38.929 to 38.954 mm (1.532 to 1.533 in)
Rear journal clearance .... 0.070 to 0.220 mm (0.00275 to 0.00866 in)

Torque wrench settings
                                                                                                                                                        lbf ft   Nm
Inlet and exhaust manifolds to engine 18   24
Oil filter mounting block to engine:
   Series A 31   43
   Series B 36   49
Front main bearing cap bolts 59   80
Main bearing cap bolts:
   Except 1400 models 83   113
   1400 models 59   80
Big-end cap nuts:
   except 2000 and 1800 USA models 37   51
   2000 and 1800 USA models 54   74
Camshaft carrier to cylinder head bolts 16   21

Chapter 1 Engine

	lbf ft	Nm
Cylinder head to block bolts (dry)	61	83
Flywheel:		
Except 2000 and 1800 USA models	61	83
2000 and 1800 USA models	108	142
Crankshaft pulley:		
Except 1600 A series and 1800 models	144	196
1600 A series and 1800 models	180	245
Camshaft drivegear	86	118
Camshaft drivebelt stretcher to engine	32.5	44
Main engine mounting to right-hand subframe member	9	12
Engine to engine mounting	9	12
Engine right-hand shock absorber to body	18	25
Engine right-hand shock absorber to bracket	44	60
Left-hand shock absorber to body	18	25
Left-hand shock absorber to gearbox	44	60
Main engine stabiliser bar to body	13	18
Main engine stabiliser bar to engine bracket	13	18
Subframe to body:		
A series	66	90
B series	93	127
Alternator adjusting bracket to engine:		
Series A engines	38	52
1300 and 1600 Series B engines	31	43
2000 and 1800 USA engines	16	22

1 General description

All engines fitted to the Lancia Beta series models are basically the same design. This is not surprising when it is realised that the engine is of the twin overhead camshaft design. Engine sizes in the Lancia range from 1300 to 2000. Early models included the 1400 engine which was later replaced by the 1300 unit. The other units in the range include the 1600 and 1800 versions. In the UK market the 1800 has been superseded by the 2000.

The engine sizes themselves have also undergone some changes. In 1976 the 1600 unit was changed from 1592 to 1585 cc and in 1979 the 1300 unit was enlarged from 1297 cc to 1301 cc. This means that great care is required when checking the Specifications Section of this Chapter for information regarding specific parts of the engine, as there are many variations.

The engine is transversely mounted at the front with the gearbox in-line with the engine. The differential unit which provides the drive from the gearbox to the front wheels via driveshafts is integral with the gearbox.

The carburation system comprises a single twin choke or Weber downdraught carburettor. The inlet manifold feeds four inlet ports on the cylinder head and it has fittings to provide for the vacuum operated servo assisted brakes. The inlet ports lead to valves angled at 45° and a hemispherical combustion chamber in the cylinder head. The exhaust valves lead to four exhaust ports and a cast iron exhaust manifold. The two sets of valves are operated by separate camshafts situated on the cylinder head and the camshafts are driven by a toothed belt running on a pulley mounted on the front end of the crankshaft. Belt tension is maintained by an idler wheel. The same toothed belt also drives an auxiliary shaft mounted on the right-hand side of the cylinder block. This shaft drives the fuel pump, oil pump and distributor. On Californian version models however the distributor is situated on the rear left-hand corner of the engine and is driven by the exhaust camshaft.

The statically and dynamically balanced steel crankshaft rotates in five renewable main bearings which are individually fed with oil at pressure from the oil pump. Drillings in the crankshaft take oil from the main journals to the crankpin/big-end bearings. Drillings in the cylinder block cum crankcase take oil from the crankshaft main bearings to the camshaft bearings.

The auxiliary driveshaft bearings are also lubricated with oil direct from the pump. The oil pressure relief valve is incorporated in the oil pump cover in the sump.

The whole engine and gearbox assembly is mounted on three rubber mountings with two shock absorbers (one at each end of the engine and gearbox) and a central horizontal stabiliser bar.

Because the layout of the engine and transmission assemblies is a transverse one there is very little spare room in the engine compartment. All the components are packed in very tightly, which makes any repair task more difficult than one might expect, due to the fact that a certain amount of ancillary components always have to be removed to reach the problem area. This is even more true with later models and especially those with air conditioning units and those produced for the North American market with complicated exhaust emission control systems. Separate sections cover these models.

2 Major operations – engine in situ

1 The following major operations can be carried out on the engine with it in place in the car:

   (a) Removal and refitting of cylinder head assembly
   (b) Removal and refitting of sump (requires lifting engine with hoist)
   (c) Removal and refitting of timing belt
   (d) Removal and refitting of big end bearing shells
   (e) Removal and refitting of camshafts
   (f) Removal and refitting of water pump
   (g) Removal and refitting of oil pump
   (h) Removal and refitting of pistons and connecting rods

2 It must be noted that to perform some of these operations does require a great deal of preparatory stripping.

3 Major operations – engine removed

1 The following operations can only be carried out with the engine out of the car and on a bench or the floor:

   (a) Removal and refitting the main bearings
   (b) Removal and refitting the crankshaft
   (c) Removal and refitting the auxiliary driveshaft
   (d) Removal and refitting the flywheel

4 Methods and equipment for engine removal

1 Because of the engine/gearbox layout and the severe lack of space in the engine bay, the only way in which the engine can be removed is with the gearbox/differential unit attached. There are two ways of doing this. Either the engine and gearbox are separated from the ancillary components and are lifted out of the engine bay or the whole assembly can be removed attached to the subframe. In other words the subframe is detached from the bodyshell and the body is lifted away from the engine/gearbox and transmission/suspension assemblies. Unless the subframe needs to be removed the former method is the easier and more logical one to use.

2 Essential equipment includes chassis stands to retain the car at a height sufficient for access to the exhaust and steering systems in particular.

Chapter 1 Engine

3 The next piece of essential equipment is a hoist, the engine and gearbox together weigh approximately 200 lb (90 kg) therefore ensure that all the hoist equipment is sufficient for that load. A hydraulic jack or comparable lift scissor jack will be required to support the engine or gearbox during removal operations, especially when undoing the mounting bolts.

5 Engine and gearbox – removal

Before attempting this operation read through the whole of this Section and ensure that you have adequate facilities and tools for the job in hand. It is vital to have an assistant (or two if possible), when it comes to the actual removal phase of the operation, and one at the beginning for removing the bonnet.

Lay out all the parts as they are removed and put them in order, with the nuts and bolts etc in suitable containers.

1 Disconnect the battery leads and remove the battery. Place it to one side in a safe place. Then remove the battery tray.
2 Scribe the bonnet hinge positions. Remove the bonnet stay retainer and the two lower stay clamps. At this point the assistant must be supporting the bonnet.
3 With the stay removed the bonnet can now be removed. Depending on the model, either undo the three bolts on each side or remove the circlip on each of the hinge pins and remove the pins. The bonnet can now be lifted away.
4 Place the bonnet in a safe place where it will not get knocked. It is best to stand it up vertically against the wall, but place a rag or a similar soft pad under each corner so that the paintwork is protected. Also protect the top corners.
5 Undo the two nuts at the rear and the nut and bolt at the front of each body bracer and remove them. The rear nuts are also the front suspension strut upper mounting units.
6 Undo the nut and bolt at the front mounting bracket and the bolt at the rear of the main engine stabiliser bar and remove it.
7 Remove the air cleaner as described in Chapter 3.
8 Disconnect the plastic link arm from the carburettor linkage to the carburettor operating lever, as described in Chapter 3. Undo the three bolts which secure the cable adjuster and linkage bracket to the camshaft housing and cover.
9 Disconnect the manual choke, if one is fitted, from the carburettor control arm.
10 Remove the radiator as described in Chapter 2.
11 Disconnect the brake servo unit hose from the inlet manifold and tuck it away.
12 Slacken the clamps on the feed and return fuel pipes where the flexible pipes meet the fixed pipes next to the right-hand inner wing. Pull the pipes off.
13 Disconnect the HT lead from the coil to the distributor cap at the distributor and tuck it out of the way. There is no need to disconnect any of the other wires attached to the coil.
14 Disconnect the oil temperature gauge, low level warning light and transmitter wires from the oil filter mounting block. Note which wire fits where.

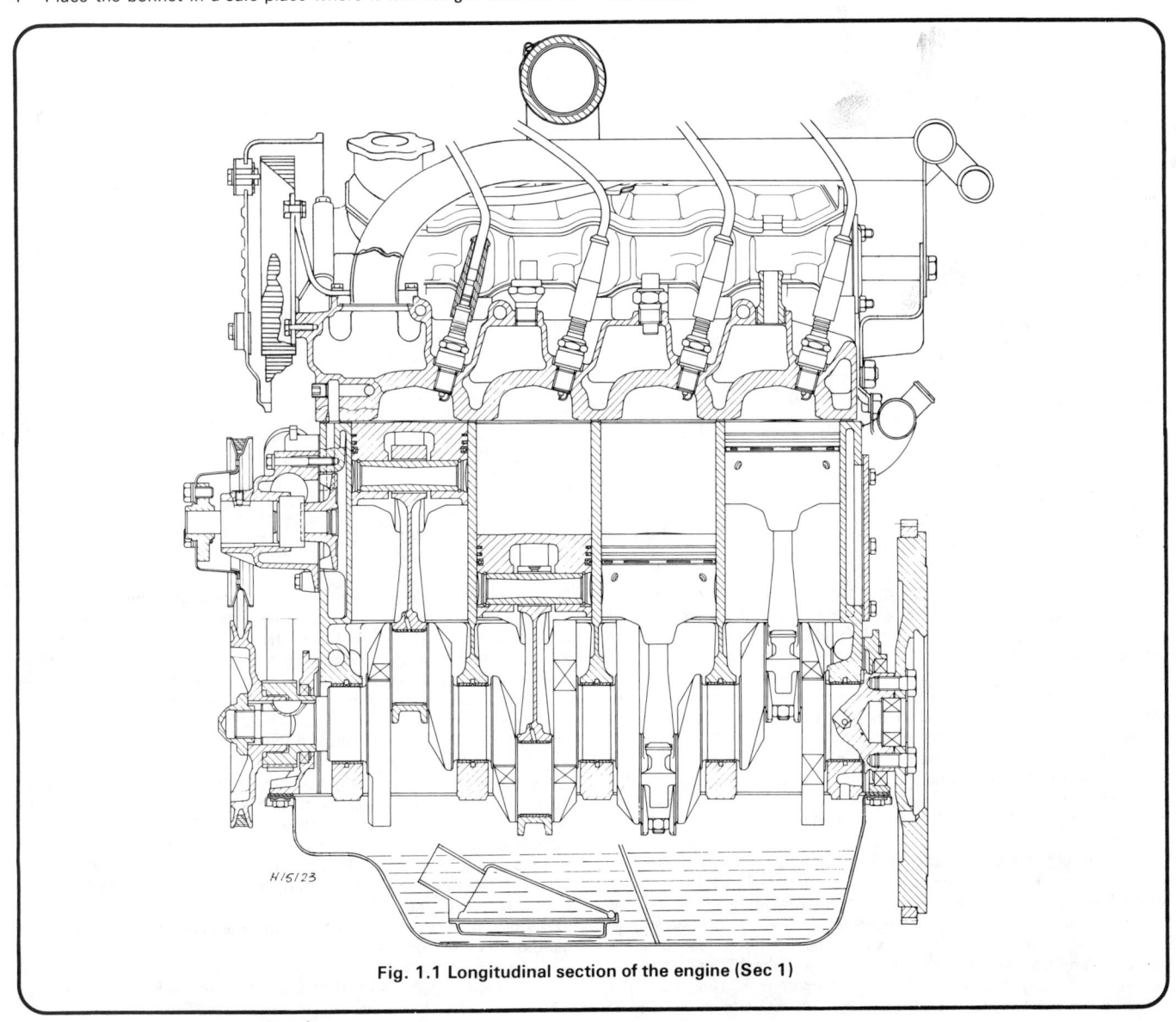

Fig. 1.1 Longitudinal section of the engine (Sec 1)

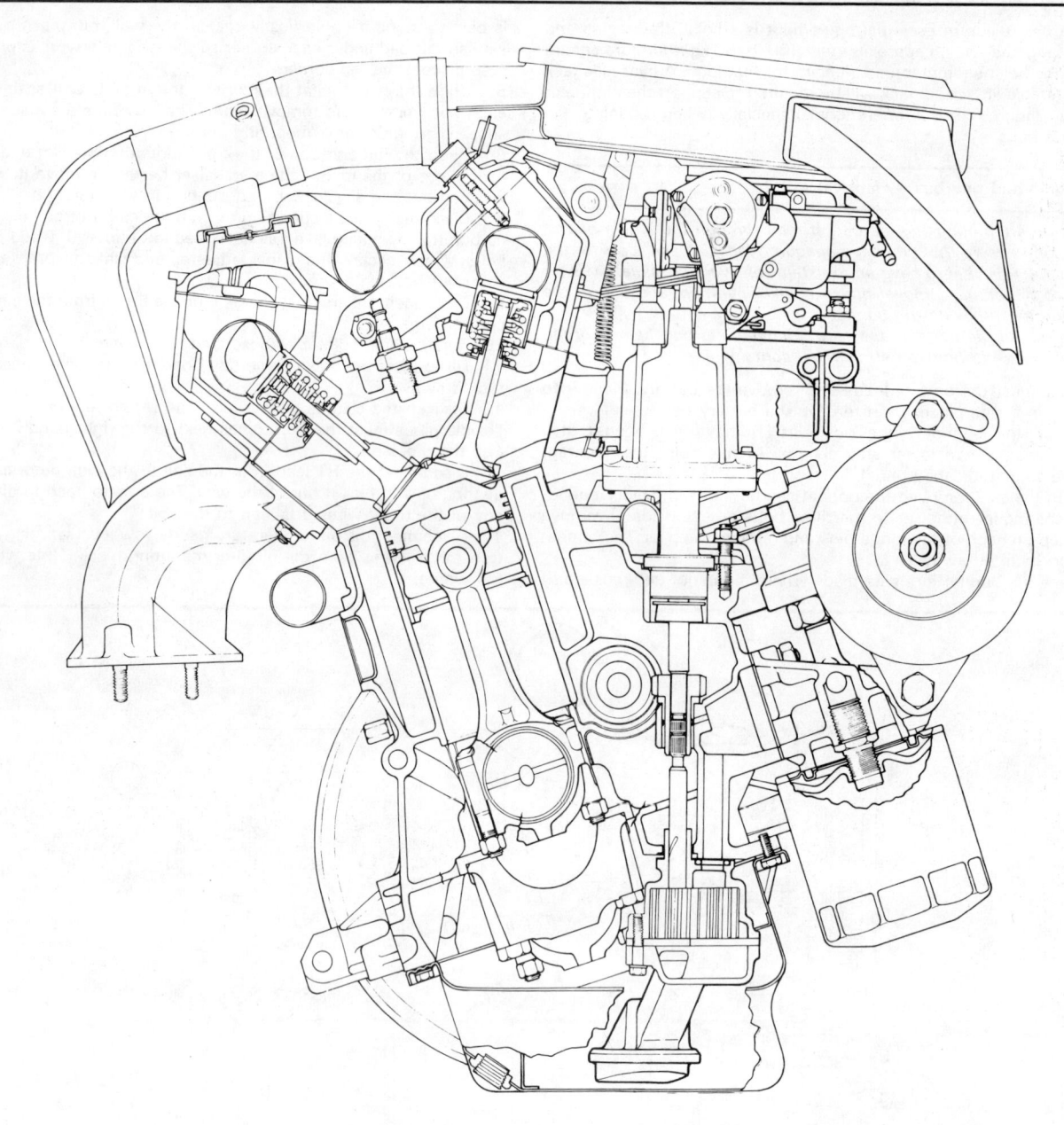

Fig. 1.2 End-on section of the engine (Sec 1)

15 Disconnect the wiring to the alternator. This will differ depending on alternator type and may be one large multi plug connector clipped in to the rear of the unit or several terminals connected to bolts or Lucas connectors. Note which wires fit where for the particular car you are working on, and draw a diagram if necessary.

16 Disconnect the wiring to the starter motor. There should be one single wire to the solenoid in addition to the main heavy duty feed cable.

17 Disconnect the plastic pipe from the oil level cum dipstick pipe and tuck it out of the way. The plastic pipe is connected to the dashboard gauge oil level checking device (photo).

18 Disconnect the wiring from the carburettor solenoid at the crossmember.

19 Disconnect the wiring from the temperature gauge transmitter and the engine overheating switch in the cylinder head. The black and white wire runs to the right-hand unit in the cylinder head and the red and blue wire to the transmitter between No 3 and 4 cylinders.

20 Undo the two screws and remove the distributor cap. Disconnect the HT leads from the spark plugs. Disconnect the LT lead from the distributor (orange/black to grey) at the connector. Then tuck the whole assembly out of the way inside the right-hand inner wing area.

21 Undo the heater feed hose clamp where it joins the rear water rail and pull off the hose, as covered in Chapter 2. The return hose is difficult to reach. Disconnect it as the engine is lifted out later on.

22 Disconnect and remove the top and bottom radiator hoses where they join the upper and rear water rails respectively. The top hose has a bracket attached to it which carries the small hose from the radiator to the expansion tank. This needs to be disconnected at the expansion tank first, then it can be removed with the top hose.

23 Remove the expansion tank as described in Chapter 2.

24 Undo the clutch operating cable butterfly adjusting nut and release it from the clutch operating lever. Release the cable from the bracket and tuck it out of the way. Finally undo the return spring from operating lever, remove the operating lever circlip and pull off the lever and place it to one side. This is covered in greater detail in Chapter 5.

25 Undo the 2 bolts and nut and remove the clutch cable bracket from the gearbox housing.

26 Disconnect the wires from the reverse light switch on top of the gearbox. (see Chapter 10).

27 Disconnect the earth wire from the gearbox housing.

28 Disconnect and remove the gearbox to body shock absorber (photo).

1.10 View of the tightly packed engine bay on a Series B HPE

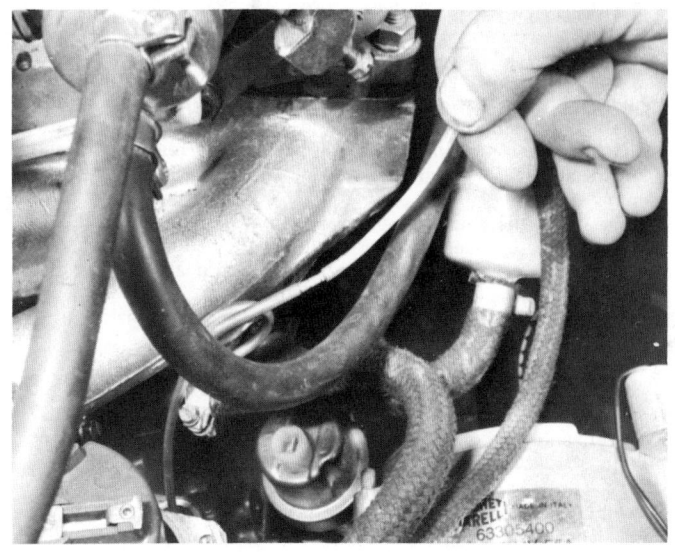

5.17 The oil level dipstick is hollow and connected by the plastic tube to the oil level gauge on the dashboard

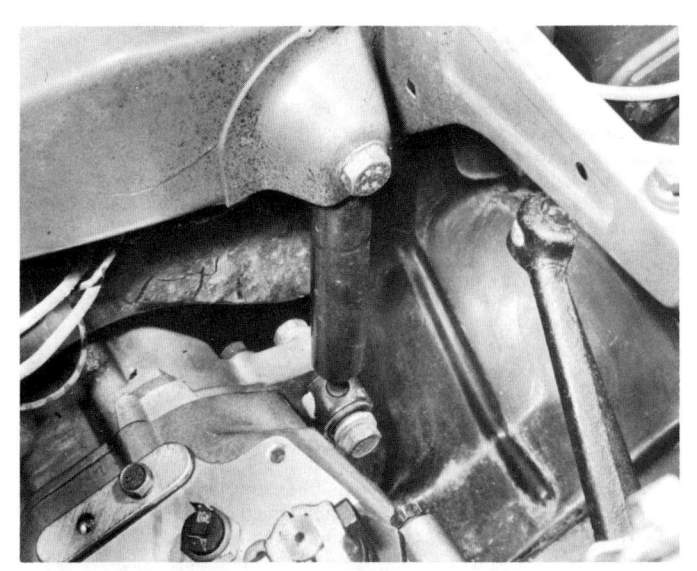

5.28 The gearbox to body shock absorber is easy to reach now

5.31 The speedometer cable is attached to the gearbox by a knurled ring (arrowed)

5.43 Removing the main engine mounting nuts and bolts

5.49 Lifting the engine out carefully – it is a very tight fit

Chapter 1 Engine

29 The next stage is to disconnect the gear linkage front control shaft at the front end as described in Chapter 6. Take the assembly off as a complete unit with the front stay rod, disconnecting it from the bellhousing lower mounting stud. Disconnect the rear end of the frontshaft later on when you are working under the car; then lift the assembly out.
30 Disconnect the engine earthing wire from the engine at the left-hand end of the cylinder head, where it leads to the rear bulkhead.
31 Disconnect the speedometer cable from the gearbox/differential unit by undoing the knurled locking ring (photo).
32 Jack up the front of the car and support it securely on axle stands, having loosened the front roadwheel bolts.
33 Remove the front roadwheels.
34 Pull the steering tie-rod rubber boots inwards to gain access to the driveshaft flange to inner CV joint nuts and bolts. Undo the inner CV joint to drive flange bolts and nuts as described in Chapter 8; however the bolts should be left in position in the CV joints so that they don't become lost.
35 With the CV joints separated from the drive flanges move the outer driveshafts as far to the rear as they will go.
36 Refit the front roadwheels, but don't tighten the bolts fully.
37 Remove the sump plug and drain the engine oil. Place a suitable container underneath the sump to catch the oil.
38 Drain the gearbox/differential unit as described in Chapter 6.
39 Disconnect the exhaust pipe from the exhaust manifold and remove the front exhaust pipe stay, as described in Chapter 3.
40 Now detach the rear end of the front gear linkage control shaft from the central idler arm and the small connector rod from the front control shaft.
41 Undo and remove the 2 gearbox rear mounting bolts as described in Chapter 6.
42 Then undo and remove the front gearbox mounting unit and bolt, also described in Chapter 6.
43 Undo the main engine mounting bolts and remove the bolts (photo).
44 Disconnect the right-hand engine mounting shock absorber at the bottom end where it is bolted onto the driveshaft centre section bearing housing.
45 Remove the four small bolts which hold the flywheel cover plate in position. It is far easier to remove the cover plate before the engine is removed from the car. Once the assembly has been lifted out and placed on the bench it is much harder to carry out this fairly simple task before the engine and gearbox assemblies can be separated.
46 The engine and gearbox assemblies should now be free to be lifted out of the car. Check carefully to ensure that everything has been separated and that there are no leads still attached between the engine/gearbox assembly and the body. The only hose that should still be attached is the heater unit water return hose.
47 The engine lifting eye attached to the right-hand end of the exhaust manifold is not only very difficult to reach in order to attach a lifting chain hook to it, but it is very small as well. We found from practical experience that it is easier to make up a sling which can be attached round the right-hand branch of the exhaust manifold which can then be attached to a lifting chain hook end.
48 We found also that the lifting eye on the inlet manifold side of the engine was very small and had to modify it by adding a shackle to fit a normal sized lifting hook.
49 With the lifting gear attached to the engine in a secure fashion and with a jack place under the gearbox start to raise the unit. The right-hand end of the engine (camshaft drivebelt cover end) must lift up first and also at a very steep angle, otherwise the complete assembly will not come out (photo). This is where it is important to have at least two people, as one must raise the lifting gear as the other supports the rear end of the gearbox to ensure that it does not snag the body. As the engine comes up there is sufficient clearance for the heater return hose to be disconnected from the rear water rail, and check again that nothing else is still attached.
50 Even when the right-hand end of the engine is clear of the suspension strut housing, the gearbox end cover can still become caught under the left-hand inner wing. If difficulty is experienced then it is better to lower the unit and re-arrange the lifting chain so that it is nearer the right-hand side of the engine. It is easier in many ways if the person supporting the gearbox end during the initial lift has a rope slung round it and controls the gearbox from above (photo).
51 It is easier still if you can acquire a third helper to guide the camshaft drivebelt cover end of the unit as this can also snag the body

5.50 It is very important to push down on the gearbox end to ensure that it doesn't snag the body as the 'front' of the engine comes clear

as it comes out (especially the engine mounting bracket).
52 The engine has to be lifted a considerable height to clear the front body panel. However, once the sump is clear the gearbox end can be lifted up so that the assembly can be pulled away from the car.
53 Pull the engine/gearbox assembly away from the car and lower it onto the floor or workbench, ready to separate the two units. Having refitted the front roadwheels to the car it can be rolled carefully out of the way if working space is limited.
54 To separate the gearbox and bellhousing assembly from the engine begin by removing the starter motor as described in Chapter 10.
55 Next, make sure that both the engine and gearbox are well supported before removing the retaining bolts.
56 Undo the four bolts which secure the bellhousing for the crankcase. There are two on top and one on each side. The bolt on the rear side has a unit on the front end of it.
57 Withdraw the bellhousing and gearbox from the engine. Take care that the primary shaft splines and clutch driven plate are not damaged.
58 The driveshaft centre section can now be removed complete with the bearing housing by undoing the two nuts and bolt which secure it to the crankcase (photo). This only applies to cars with a 'three-piece' drive shaft layout.

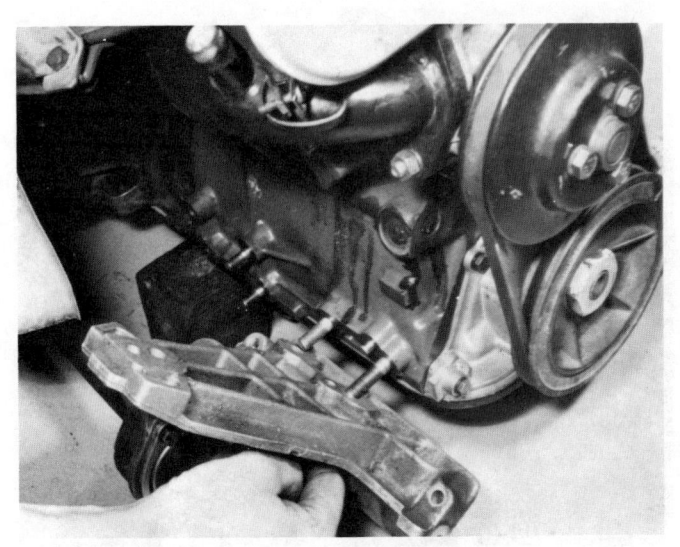

5.58 Removing the driveshaft centre section complete with bearing housing

6 Subframe, engine and gearbox – removal and refitting

1 The engine and gearbox assembly can also be removed by dropping the complete subframe out of the bodyshell with the engine, gearbox, differential, driveshafts and suspension arms attached to it.

2 Much of the procedure for doing this job is identical to the work which is necessary to lift the engine and gearbox assemblies out, as described in Section 5.

3 To make the operation simpler the tasks which are also covered in the previous Section are not given in detail here, but are given the Section 5 reference instead. Only those tasks which are specific to this operation are actually detailed in the following paragraphs.

4 Begin by ensuring that you have a wheeled platform or trolley which is capable of taking the combined weight of the whole engine and gearbox assembly, when the bodyshell is lifted from the subframe using a hoist.

5 Start by jacking the car up and removing the front roadwheels. Then support the front of the car on stands or blocks. Then drain the engine sump oil and gearbox oil.

6 Disconnect the front exhaust section as described in Chapter 3.

7 Detach the rear gear linkage control shaft from the central idler arm and the connector rod from the control shaft, as described in Chapter 6.

8 Drain the engine cooling system as described in Chapter 2, and disconnect the heater return pipe from the rear water rail.

9 Having removed the body bracers as described in Section 5, also remove the other two upper suspension strut mounting units. It is necessary to make up two plates with lifting eyes that can then be attached to the top of each suspension turret. The Lancia method is shown in Fig 1.3.

10 Remove the engine splash guard from underneath.

11 Detach the headlamp self levelling system sensor link arm from the front left-hand wishbone, as described in Chapter 10.

12 Separate the steering tie-rod end ball-joints from the steering arms on the hub carriers using a balljoint separator, as described in Chapter 11.

13 Remove both front disc brake caliper assemblies and tie them up out of the way in the front wheel arches, as described in Chapter 9. This should only leave the hubs and discs.

14 Now place the trolley bogey under the subframe. It must be wide enough to carry the wishbones as well as the subframe and have sufficient depth to support both the front and rear crossmembers (Fig 1.4).

15 Jack the car up and remove the stands, then carefully lower it into position on the trolley and bogey. Ensure that it sits squarely on the trolley, so that when the body is lifted up the assembly cannot slide off the trolley. The Lancia trolley shown in Fig 1.4 has lugs on it which locate in certain areas of the crankcase to help prevent any forward or sideways movement.

16 Undo and remove the front suspension strut to hub carrier mounting bolts, as covered in Chapter 11.

17 Refer to Section 5 and follow the procedure from paragraph 1 to paragraph 31 with the following comments. Omit paragraph 5, 25 and 29. The return feed hose has already been disconnected (paragraph 21). Only the lower mounting bolt which secures the shock absorber to the gearbox need to be removed (paragraph 28).

18 Remove the four bolts which secure the subframe to the body (photo). This is covered in detail in Chapter 12.

19 Attach the lifting gear to the brackets attached to the front suspension turrets and carefully raise the front of the body. Check as this is done that all hoses, cables and leads are tucked out of the way and have been disconnected.

20 Lift the car body up sufficiently high so that the trolley with the subframe and engine and gearbox on it can be rolled out (Fig 1.5).

21 Lower the body and rest it on stands or blocks.

22 The next task is to separate the assemblies from the subframe as necessary. If a major overhaul is being undertaken then remove the gear linkage point control shaft and stay rod and disconnect the inner CV joints from the differential unit drive flanges. Undo the engine mounting bolts and use a crane to lift the engine and gearbox assemblies onto a bench where they can be separated. All this is covered in Section 5.

23 Refitting is the reverse procedure to removal. Tighten all nuts and bolts to the torque wrench settings given in the Specifications.

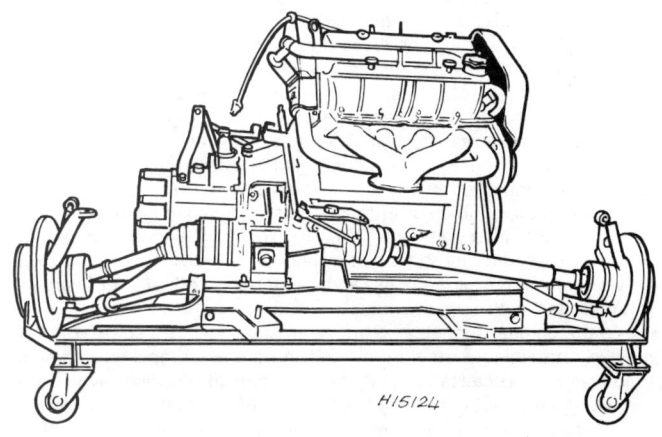

Fig. 1.4 Subframe and engine/gearbox trolley as used by Lancia (Sec 6)

Fig. 1.3 Typical lifting tackle and brackets required to lift the front of the body to drop out the subframe and engine/gearbox assemblies (Sec 6)

Fig. 1.5 Lift the body up far enough so that the subframe and engine/gearbox can be rolled out on the trolley (Sec 6)

88087030 – Trolley

6.18 The subframe is secured to the bodyshell by four bolts like this (N/S front)

7 Engine dismantling – general

1 Ideally the engine is mounted on a proper stand for overhaul but it is anticipated that most owners will have a strong bench on which to place it. If a sufficiently large strong bench is not available then the work can be done at ground level. It is essential, however, that some form of substantial wooden surface is available. Timber should be at least ¾ inch thick, otherwise the weight of the engine will cause projections to punch holes straight through it.
2 It will save a great deal of time later if the engine is thoroughly cleaned down on the exterior before any dismantling begins. This can be done by using paraffin and a stiff brush or more easily, probably, by the use of water soluble solvent which can be brushed on and then the dirt swilled off with a water jet. This will dispose of all the heavy grease and grit once and for all so that later cleaning of individual components will be a relatively clean process and the paraffin bath will not become contaminated with abrasive metal.
3 As the engine is stripped down clean each part as it comes off. Try to avoid immersing parts with oilways in paraffin as pockets of liquid could remain and cause oil dilution in the critical first few revolutions after reassembly. Clean oilways with wire, or preferably an air jet.
4 Where possible, avoid damaging gaskets on removal, especially if new ones have not been obtained. They can be used as patterns if new ones have to be specially cut.
5 It is helpful to obtain a few blocks of wood to support the engine whilst it is in the process of dismantling. Start dismantling at the top of the engine and then turn the block over and deal with the sump and crankshaft etc, afterwards.
6 Nuts and bolts should be refitted in their locations where possible to avoid confusion later. As an alternative keep each group of nuts and bolts (all the timing gear cover bolts for example) together in a jar or tin.
7 Many items which are removed must be refitted in the same position, if they are not being renewed. These include valves, tappets, pistons, bearings and connecting rods. Some of these are marked on assembly to avoid any possibility of mixing them up during overhaul. Others are not, and it is a great help if adequate preparation is made in advance to classify these parts. Suitably labelled tins or jars and for small items, egg trays, tobacco tins and so on, can be used. The time spent in this operation will be amply repaid later.
8 With the engine out of the car it is much easier to view and discuss it as though one was looking at the 'front' when standing in front of the crankshaft pulley in line with the cylinder block and this is the way the engine is referred to in the overhaul procedure.

8 Removing the ancillary engine components

1 Before basic engine dismantling begins it is necessary to strip it of ancillary components as follows:

Alternator
Distributor
Water rails – top and rear (including the thermostat)
Oil filter cartridge
Exhaust manifold
Inlet manifold and carburettor (not 1400 model)
Crankcase breather (not 1400 models)
Fuel pump (not 1400 models)

2 Presuming the engine is to be out of the car and on a bench and that the items mentioned are still on the engine follow the procedure described in the following paragraphs.
3 Slacken off the alternator retaining bolts and nuts, and remove the drivebelt. Remove the alternator together with its retaining nuts and bolts.
4 To remove the distributor first set the engine to TDC No 4 cylinder on its firing stroke. The timing mark on the crankshaft pulley must be in line with the large mark on the camshaft drivebelt cover and the holes in the camshaft drive wheels must be in line with the lugs on the camshaft carriers (photos).
5 Undo the nut which secures the distributor clamp plate having scribed a mark showing the relationship of the distributor to the crankcase and noted the rotor arm position. In the USA version with the camshaft mounted distributor the same basic procedure applies.
6 The rear water rail is mounted on to the water pump inlet flange by two nuts and to the main engine stabiliser mounting bracket by two bolts and nuts. Undo these and disconnect the hose to the thermostat housing and the hose to the carburettor choke thermostat housing, and the rear water rail can be lifted away.
7 The top water rail runs from the front end of the cylinder head to the rear of the engine where it is secured to the cylinder head by a bolt common to the rear water rail. Undo the four bolts which secure the front end of the water rail to the pump outlet in the cylinder head and the single bolt through the camshaft drivebelt cover bracket and the water rail, by-pass hose and thermostat can all be removed (photo).
8 The next item to be removed is the oil filter and housing. The filter and housing is secured to the cylinder block by four bolts. Once the bolts have been unscrewed the whole assembly can be removed. It comes away complete with the various transmitters (photo).
9 Remove the oil level pipe cum dipstick from the crankcase. It simply pulls out. The rubber seating plug is normally supplied as part of the overhaul gasket set, which will be needed if a complete engine overhaul is being undertaken.
10 Remove the nuts which secure the exhaust manifold in place and remove it and the gasket. There is no need to detach the heat deflector plate or heater box.
11 Remove the inlet manifold and carburettor. It is not possible to perform this operation on the 1400 engine as the inlet manifold will jam on the fuel pump and oil breather housing before it is released from the studs in the cylinder head. Therefore the inlet manifold and the fuel pump and oil breather housing can only be removed after the cylinder head has been removed as described in Section 16. For the other models the same procedure can be adopted or proceed as follows.
12 Undo the four bolts and two nuts which secure the inlet manifold to the cylinder head. Check that the fuel line from the pump to the carburettor is free and remove the manifold and the gasket, complete with the carburettor and fuel lines. Note that the alternator adjusting bracket is also attached to the front end of the manifold.
13 Take care of the inlet manifold gasket as the replacement one must be the same. These gaskets are shaped to specific engine blocks and if the wrong gasket is used when the engine is rebuilt then it may create a water leak.
14 Undo the central mounting bolt and unscrew the small right angled pipe clamps and lift the oil breather housing away, with the small pipe attached. This pipe should be renewed during an engine rebuild (photo).
15 The fuel pump is removed after undoing the two bolts which secured it to the side of the crankcase; keep the spacer and bolts with the fuel pump. This is also covered in Chapter 3.
16 The engine is now stripped of ancillary components and ready for the major dismantling tasks to begin.

9 Timing belt and cover (engine out of car) – removal

1 Remove the three cover securing bolts and remove the timing belt

Chapter 1 Engine

8.4a Line up the timing mark on the crankshaft pulley with the large mark on the timing belt cover ...

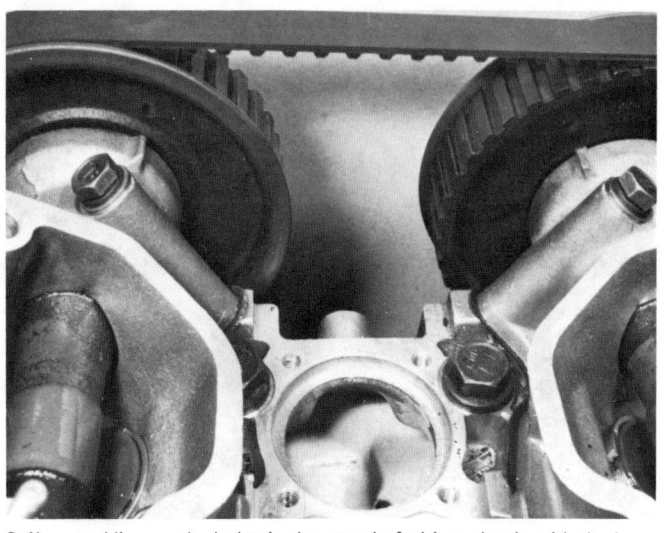

8.4b ... and line up the holes in the camshaft drive wheels with the lugs on the carriers to set the engine at TDC, No 4 cylinder firing (the camshaft covers are removed for clarity)

8.7 Removing the top water rail from the cylinder head (note the O-ring seal)

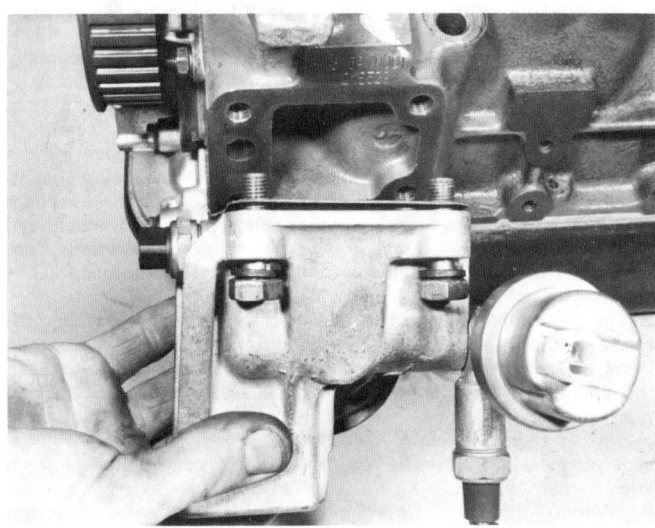

8.8 Undo the four bolts and remove the oil filter and housing

8.14 Renew the small pipe between the oil breather housing and crankcase

cover. Set the engine to TDC No 4 cylinder on the firing stroke.

2   If a major overhaul is being undertaken then slacken the ancillary drive wheel retaining bolt, before slackening the timing belt.

3   Lock the flywheel using a bracket or a bolt screwed into one of the clutch mounting holes, with the head against the crankcase so that the crankshaft cannot turn and undo the crankshaft pulley retaining nut. Ensure whilst doing this that the crankshaft does not rotate. If it moves then it must be reset, before the belt tensioner is loosened.

4   Check before the belt tensioner is loosened that the ancillary driveshaft wheel is correctly aligned. The hole in the wheel must be lined up with the belt tensioner spring securing bolt (photo).

5   Remove the crankshaft front pulley.

6   Slacken the belt tensioner idler wheel lock nut and lever the tensioner towards the water pump, then lock the nut. The timing belt is now slackened.

7   The timing belt can now be removed from the drive wheels.

10 Engine dismantling procedure

1   With the ancillary components removed as described in Section 8 and the timing belt removed as described in Section 9, the next stage is to remove the cylinder head.

2   Start by removing the camshaft drive wheel pointer bracket from the front of the engine (photo).

9.4 Line up the ancillary shaft drive wheel with the hole in line with the spring securing bolt

9.5 Removing the crankshaft pulley

3 Remove the camshaft housing covers, if this has not already been done. Two large knurled nuts hold them in place.
4 Undo and remove the ten cylinder head retaining bolts in the reverse order to that given in Fig. 1.17 and lift the cylinder head away (photo). The cylinder head assembly overhaul is covered later in Section 11. There are two schools of thought when it comes to removing cylinder heads. Once school prefers to remove all the ancillary parts, including the inlet and exhaust manifolds, before removing the cylinder head, and the other school prefers to leave them attached to help in the breaking of the seal between the cylinder head and block as it gives much greater purchase and leverage. Both manifolds can then be removed from the cylinder head once it has been separated from the block.
5 In the case of the Lancia the inlet manifold, especially on the 1400 model, cannot be removed until the head has been separated from the block, because of the lack of clearance between the inlet manifold and the fuel pump and oil breather housing. Because the camshaft housing is set at an angle the studs on which the inlet manifold is mounted are also at an angle. Therefore when the manifold is released it will not slide off the studs before it hits the other components and jams up against them.
6 Remove the cylinder head gasket and leave the dowels in the block.
7 Remove the clutch assembly from the flywheel as described in Chapter 5.
8 Undo the flywheel retaining bolts. It is necessary to make up a bracket, or fit a bolt through one of the holes in the flywheel, so that it locks against the crankcase in order to release them.
9 Then lift the flywheel off the crankshaft rear end (photo).
10 Turn the engine over so that it lies on its side or upper face and remove the sump retaining bolts.
11 Next remove the sump itself. It is usually well stuck to the gasket with a sealant compound and has to be levered off carefully. Take care as there is probably a bit of oil and sludge in the sump.
12 Clean all traces of the old gasket off the crankcase mating surface and the sump itself. Use a blunt scraper so as not to damage the metal surfaces. The old gasket does tend to be stuck on pretty well.
13 Undo the two larger bolts and remove the oil pump and pickup (photo). This can be placed to one side for further stripping and overhaul if necessary, which is covered in Section 13.
14 Undo the two bolts which retain the oil return pipe, from the oil separator in the crankcase breather housing, to the crankcase and withdraw the pipe (photo).
15 Undo the retaining nut, for the ancillary shaft drive wheel, which has previously been loosened, and remove the drive wheel. Note that there is a locating stud in the front end of the ancillary shaft and a corresponding hole in the rear of the wheel so that it can only be refitted in one way (photo).
16 Undo the two bolts and one nut which retain the ancillary shaft front seal housing in position and remove the housing and gasket behind it.

17 Slide the crankshaft timing bent drive cog off the crankshaft. Note that there is a key fitted into the keyway in the crankshaft. Tap it out and place it to one side (photo).
18 Undo the five bolts and remove the crankshaft front seal housing complete with seal, which can be renewed later on. Also remove the gasket if it has not come off attached to the housing. Clean off both mating surfaces.
19 Undo the bolts that retain the crankshaft rear oil seal housing and remove it complete with oil seal. Clean off the old gasket from both mating surfaces.
20 Undo the mounting bolts and lift off the water pump from the front face of the cylinder block, if required. Unless it needs overhauling there is no other reason to remove it. The water pump is dealt with in detail in Chapter 2.
21 The timing belt tensioner idler can now also be removed if desired as can the tension spring and mounting plate for the idler. It is held in place by the control unit and washer and spacer. The central hole in the idler is much bigger than the mounting bolt (photo).
22 The next stage is to remove the big-end caps and remove the pistons and connecting rods. Turn the engine on its side for this part of the dismantling procedure.
23 Start by slackening off all the big-end bearing cap nuts. Then take each cylinder in turn and remove the nuts, withdraw the bearing cap and shell and then push the connecting rod and piston down the

10.2 Remove the single bolt and lift the pointer arm away

10.4 The cylinder head bolts are the large black ones between the camshaft boxes

10.9 Removing the flywheel

10.13 Undoing the oil pump retaining bolts

10.14 Withdrawing the oil return pipe from the crankcase breather valve

10.15 Removing the ancillary shaft drive wheel (note the locating stud)

10.17 Slide the drivebelt cog off the crankshaft and remove the Woodruff key

10.21 Removing the timing belt tensioner idler

10.23 Removing No 1 cylinder big-end bearing cap and shell

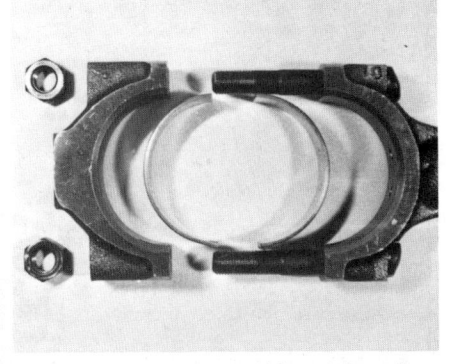

10.25 Each big-end assembly consists of the connecting rod, bearing shells, bearing cap, bolts and nuts

10.28 The main bearing caps are marked with notches for identification

10.30 Use a small screwdriver to remove the half thrust washer on either side of the rear main bearing position

10.31 Lifting out the crankshaft

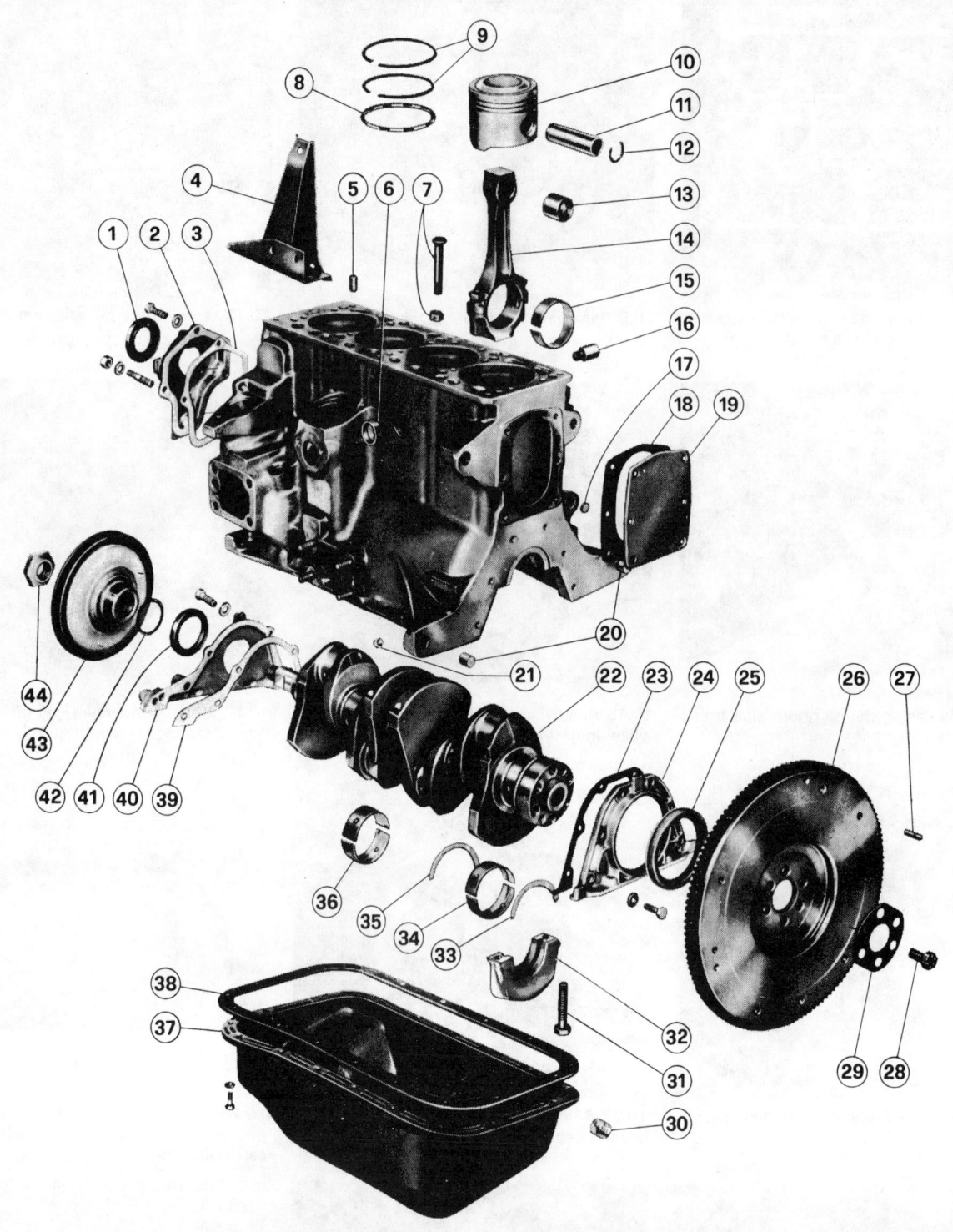

Fig. 1.6 Exploded view of crankcase and components (Sec 10)

1 Ancillary driveshaft oil seal
2 Cover
3 Gasket
4 Engine main mounting bracket (early models)
5 Cylinder head dowel
6 Core plug
7 Big-end cap bolt and nut
8 Oil scraper ring
9 Compression rings
10 Piston
11 Gudgeon pin
12 Circlip
13 Bush
14 Connecting rod
15 Big-end shell bearing
16 Drain plug for cylinder water jacket
17 Plug
18 Gasket
19 Water jacket cover
20 Dowels
21 Plug
22 Crankshaft
23 Rear oil seal cover gasket
24 Rear oil seal housing
25 Rear oil seal
26 Flywheel
27 Clutch dowel
28 Flywheel retaining bolt
29 Locking plate
30 Sump drain plug
31 Main bearing cap bolt
32 Main bearing cap
33 Half thrust washer
34 Main bearing shell
35 Half thrust washer
36 Main bearing shell
37 Sump
38 Sump gasket
39 Front oil seal gasket
40 Front oil seal housing
41 Front oil seal
42 O-ring
43 Crankshaft pulley
44 Pulley retaining nut

Chapter 1 Engine

10.32 Remove the inner bearing shells now that the crankshaft has been removed

10.33a Undo the bolts and remove the locking yoke which holds the ancillary driveshaft in position ...

10.33b ... and withdraw the shaft

cylinder bore and retrieve it as it emerges at the top of the cylinder block (photo).

24 Take great care as you start to push the assembly down the bore that the big-end bolts do not touch the crankpins or the cylinder bores otherwise they will score and damage them.

25 When the piston and connecting rod have each been withdrawn in turn refit the bearing cap and nuts, so that the bearing caps do not become mixed up (photo). The bearing caps are numbered 1 to 4 from front to rear. Lay the assemblies out in the correct order as it makes refitting a much easier process. When refitting the bearing cap to the connecting rod note that the tabs in the bearing inner and outer shells must marry up. In this engine the tabs in the shells are on the oil pump mounting side of the engine.

26 With all the piston and connecting rod assemblies removed, put together preferably in a sectioned box, and laid out in order, turn the engine so that it rests on the top of the cylinder block.

27 Now slacken all the main bearing cap bolts. Note that the front cap has one smaller bolt and is a slightly different shaped cap. This is to allow for the fitting of the oil pump. It has no other identifying mark.

28 Remove the bolts and caps and lay them out in order. The bearing caps are marked with notches from the second cap to the fifth cap (photo).

29 The centre cap (two notches) has no oil channel in the bearing shell and is also wider than the others.

30 With the caps removed use a small screwdriver to push out the two half thrust washers from the inner faces of the rear main bearing position. Note that both have oil grooves in them which face away from the bearing seating (photo).

31 Now hold both ends of the crankshaft and lift it out cleanly (photo).

32 With the crankshaft removed the inner main bearing shells can now be removed and placed with their appropriate caps for inspection (photo).

33 The cylinder block and crankcase are now fully stripped apart from the ancillary equipment driveshaft which can now be removed. Undo the two bolts which retain the locking yoke in position in the groove in the front end of the shaft and remove it (photo). The driveshaft can now be withdrawn straight out of the front of the block as the rear journal is smaller than the front (photo).

34 The front and rear bearing shells for the ancillary driveshaft can then be renewed if necessary.

35 The engine is now fully dismantled for overhauling purposes.

36 With the engine stripped and all parts thoroughly cleaned, every component should be examined for wear. The items listed in the sections following should receive particular attention and where necessary, renewed or renovated.

37 So many measurements on engine components require accuracies down to tenths of a thousandth of an inch. It is advisable therefore to either check your micrometer against a standard gauge occasionally to ensure that the instrument zero is set correctly, or use the micrometer as a comparative instrument. This last method however necessitates that a comprehensive set of slip and bore gauges is available.

11 Cylinder head, valve gear and camshafts – overhaul

1 If the cylinder head has been removed from the engine complete, first remove the inlet manifold complete with the carburettor and then the exhaust manifold, as described in Chapter 3. Two knurled retaining nuts secure the camshaft covers. Tap the covers off carefully if they are stuck to the gaskets. Note that the cover with the oil filter cap fits on the rear housing.

2 Remove the gaskets from the housings (photo).

3 Turn the whole assembly upside down and drain the oil out the camshaft housings.

4 Undo the ten camshaft housing retaining bolts or nuts and lift off the camshaft housing complete with the camshaft and drive cog. There is no need to remove the camshaft drive cog unless the camshaft itself if to be removed from the housing.

5 As the housing is lifted off push the tappets down the housing bore onto their respective valves. Then place the camshaft in its housing to one side.

6 Make up a box with four large sections in it for all the valve parts and then remove the tappets and shims and place them in order in the box sections (photo).

7 Remove the gasket. Note that when a new gasket is fitted the tappet clearances will have to be checked and new shims fitted as necessary, as described in Section 31.

8 Carry out the same procedure for the other camshaft housing.

9 Remove the spark plugs, and thermal transmitter if necessary.

10 With the camshaft housings lifted away the valves can now be removed. Both the inlet and exhaust valves have double springs and are all removed in the same way.

11 Use a standard valve removing tool to compress the spring, and then lift out the two split collets (photo).

12 Release the tool and lift off the upper collar, then remove the outer and inner valve springs, cup and washer (photos).

13 Now withdraw the valve having turned the cylinder head over, and then remove the valve guide oil seal. Repeat this procedure for the remaining valves (photos).

14 Place all the components in the appropriate section of the box which already has the tappets and shims in it.

15 Only remove the valve guides if after inspection they have been found to have worn excessively. The correct bore size is given in the Specifications.

16 The valve guides are an interference fit in the cylinder head block. If it is wished to remove the valve guides they can be removed from the cylinder head in the following manner.

17 Place the cylinder head upside down on a bench on two stout blocks of wood ensuring that the studs in the cylinder head do not contact the bench.

18 With a suitable diameter drive, (maximum diameter 14 mm, minimum 12 mm) carefully drive the guides from the cylinder head.

19 The inlet and exhaust valve guides are not interchangeable. It will be seen that the exhaust valve guides are different from the inlet valve guides as they are threaded throughout their length for lubrication. New guides must be the same class as the ones which have been removed, as detailed in the Specifications.

20 When fitting new guides the drift used must have a spigot to locate the new guide in the guide bore. Carefully drive new guides in from the top until the circlips locate on the face of the cylinder head. It will help if the cylinder head is first heated in boiling water.

21 With the cylinder head stripped, carefully remove with a wire brush and blunt scraper all traces of carbon deposits from the

11.2 Remove the camshaft housing cover gaskets

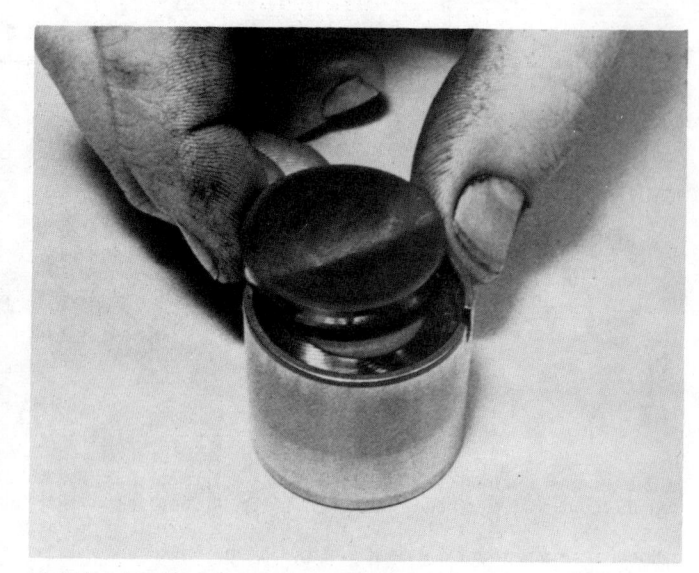

11.6 The shim fits in the top of the tappet

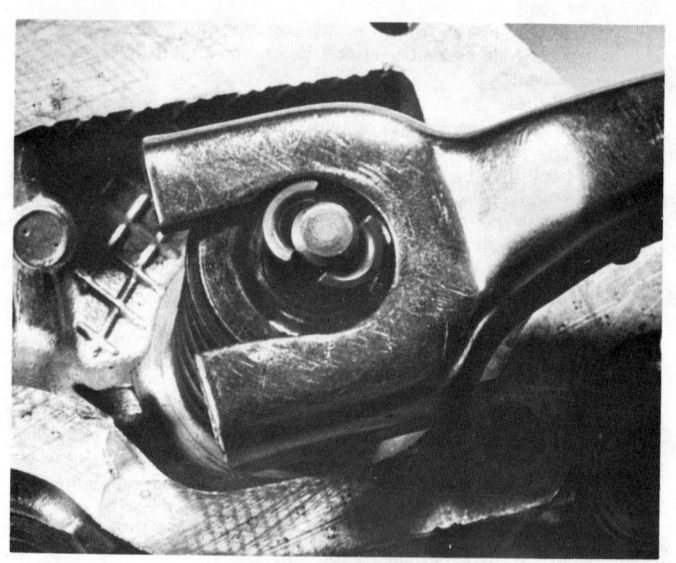

11.11 Clamp the valve spring and remove the collets

11.12a Release the spring clamp and remove the collar ...

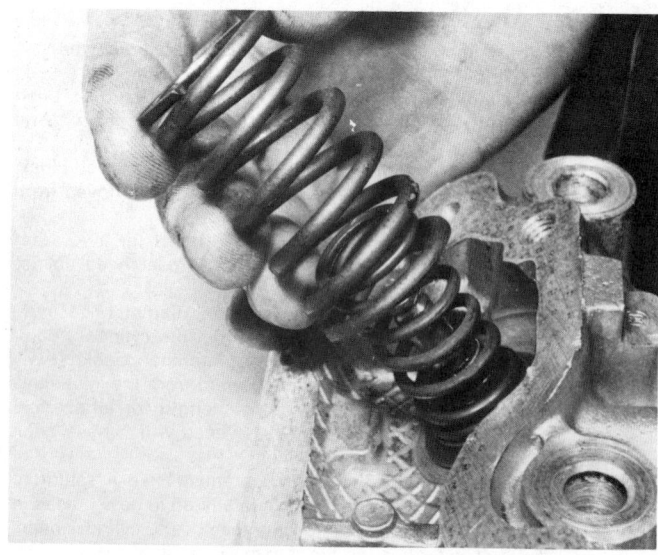

11.12b ... then the outer and inner valve springs ...

11.12c ... followed by the lower cup ...

11.12d ... and washer

11.13a Turn the cylinder head over and remove the valve ...

11.13b ... then remove the oil seal from the valve guide

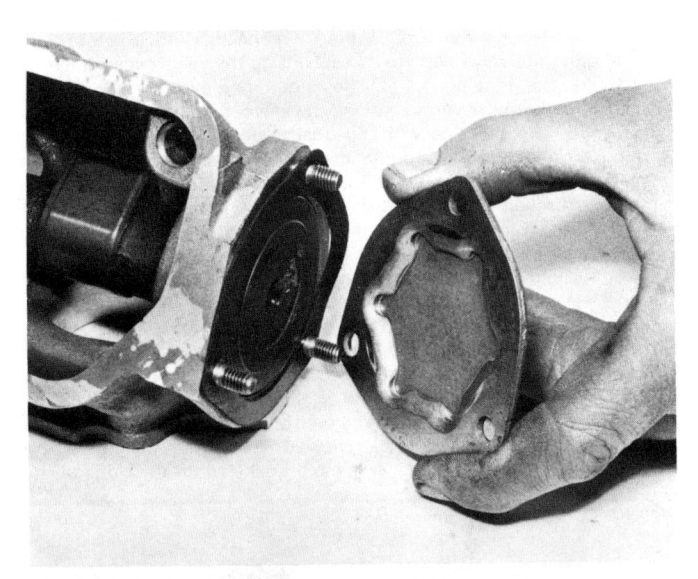

11.31a Removing the camshaft rear cover plate ...

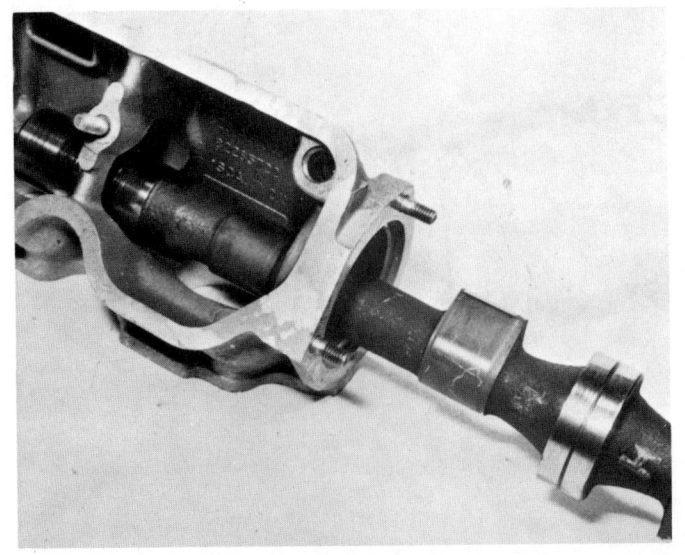

11.31b ... so that the camshaft can be withdrawn rearwards

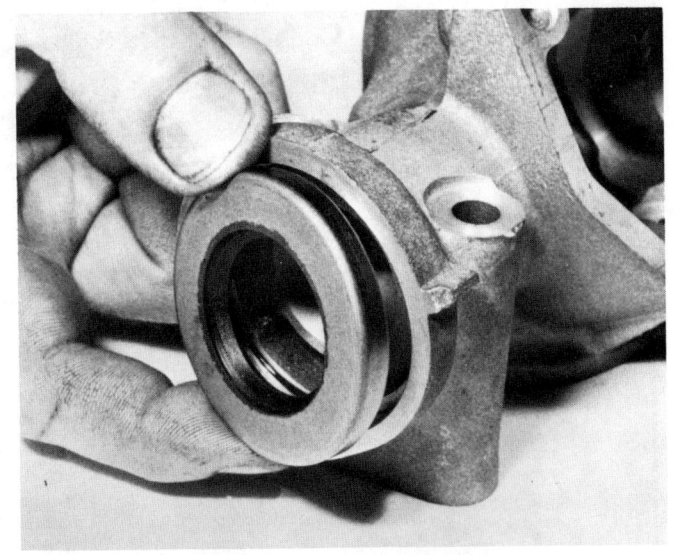

11.33 Offering up a new camshaft front oil seal to the housing

combustion spaces and the ports. The valve stems and valve guides should also be freed from any carbon deposits. Wash the combustion spaces and posts down with petrol and scrape the cylinder head surface of any foreign matter with the side of a steel rule or a similar article. Take care not to scratch the surface.

22 Examine the heads of the valves for pitting and burning especially the heads of the exhaust valves. The valve seatings should be examined at the same time. If the pitting on the valves and seats is very slight the marks can be removed by grinding the seats and valves together with coarse, and then fine, valve grinding paste. Where bad pitting has occurred to the valve seats it will be necessary to recut them to fit new valves. If the valve seats are so worn that they cannot be recut, then it will be necessary to fit new valve seat inserts. These latter two jobs should be entrusted to the local Lancia agent or automobile engineering works. In practice it is very seldom that the seats are so badly worn that they require renewal. Normally, it is the valve that is too badly worn to be refitted, and the owner can easily purchase a new set of valves and match them to the seats by valve grinding.

23 Valve grinding is carried out as follows: Place the cylinder head upside down on a bench, with a block of wood at each end to give clearance for the valve stems. Alternatively, place the head at 45° to a wall with the combustion chambers facing away from the wall.

24 Smear a trace of coarse carborundum paste of the seat face and apply a suction grinder tool to the valve head. With a semi-rotary action, grind the valve head to its seat, lifting the valve occasionally to redistribute the grinding paste. When a dull matt even surface finish is produced on both the valve seat and the valve, then wipe off the paste and repeat the process with fine carborundum paste, lifting and turning the valve to redistribute the paste as before. A light spring placed under the valve head will greatly ease this operation. When a smooth unbroken ring of light grey matt finish is produced, on both valve and the valve seat faces, the grinding operation is complete.

25 Scrape away all carbon from the valve head and the valve stem. Carefully clean away every trace of grinding compound, taking great care to leave none in the ports or in the valve guides. Clean the valves and valve seats with a paraffin soaked rag, then with a clean rag, and finally, if an airline is available, blow the valve, valve guides and valve ports clean.

26 Examine the bearing surface of the tappets which lie on the camshaft. Any indentation in this surface or any cracks indicate serious wear, and the tappets should be renewed. Thoroughly clean them out removing all traces of sludge. It is most unlikely that the sides of the tappets will be worn, but, if they are a very loose fit in their bores and can be readily rocked, they should be discarded and new tappets fitted. It is very unusual to find worn tappets and any wear present is likely to occur only at very high mileages.

27 The last stage of the cylinder head assembly overhaul is to remove and check the camshafts.

28 Since the camshafts are withdrawn from the rear end of the housings, it will be necessary to begin by removing the camshaft wheel at the forward end.

29 The wheel is retained on the shaft by a single bolt which is locked by the tab on the washer. Tap back the tab, grasp the wheel gently in a soft jawed vice and undo the bolt.

30 Remove the bolt and washer which keeps the wheel/shaft alignment dowel in place. Ease the wheel off the shaft and take care not to lose the alignment dowel.

31 Next at the rear end of the housing, the camshaft retaining plate is held on by three nuts. Once removed the camshaft can be carefully withdrawn from the housing (photos).

32 Be careful not to damage the camshaft bearing bores in the housing when removing the shaft. The housing is of aluminium alloy whereas the shaft is steel.

33 With the camshaft removed the camshaft front oil seal can be driven out from inside the housing. Fit a new seal with the lips facing into the housing and tap it evenly and fully home using a flat piece of wood as protection for the seal and to keep it level (photo).

34 Carefully examine camshaft bearings for wear. If the bearings are obviously worn or pitted or the metal underlay just showing through, then they must be renewed. This is an operation for your local Lancia agent or automobile engineering works, as it demands the use of specialised equipment. The bearings are removed using a special drift after which the new bearings are pressed in, care being taken that the oil holes in the bearings line up with those in the housing. With another special tool the bearings are then reamed in position.

35 The camshaft itself should show no signs of wear, but, if very slight scoring marks on the cams are noticed, the score marks can be removed by very gentle rubbing down with a very fine emery cloth or an oil stone. The greatest care should be taken to keep the cam profiles smooth.

36 Check that the clearances in the journals are not greater than those given in the Specifications.

37 Lubricate the bearings and journals before refitting the camshaft to the housing and also smear the inside lips of the camshaft front oil seal with oil.

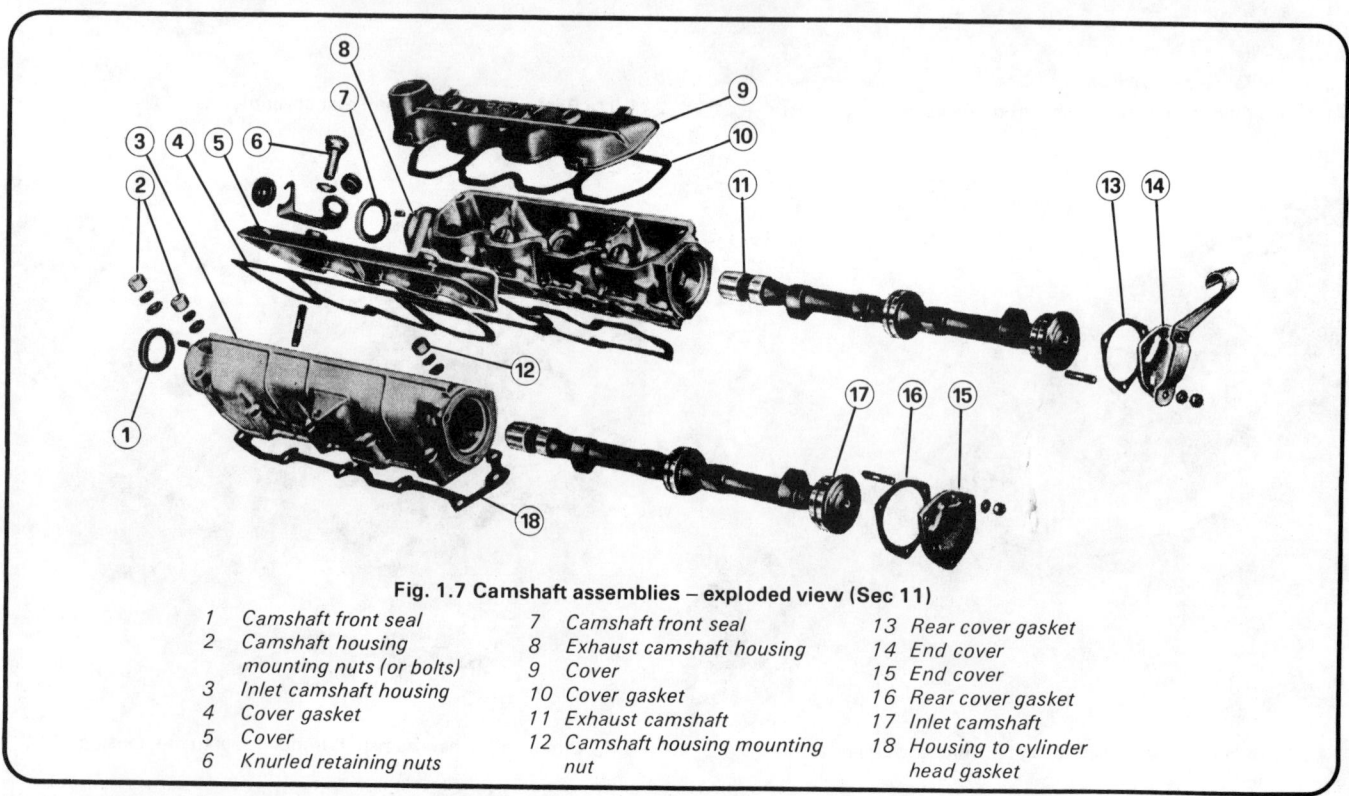

Fig. 1.7 Camshaft assemblies – exploded view (Sec 11)

1 Camshaft front seal
2 Camshaft housing mounting nuts (or bolts)
3 Inlet camshaft housing
4 Cover gasket
5 Cover
6 Knurled retaining nuts
7 Camshaft front seal
8 Exhaust camshaft housing
9 Cover
10 Cover gasket
11 Exhaust camshaft
12 Camshaft housing mounting nut
13 Rear cover gasket
14 End cover
15 End cover
16 Rear cover gasket
17 Inlet camshaft
18 Housing to cylinder head gasket

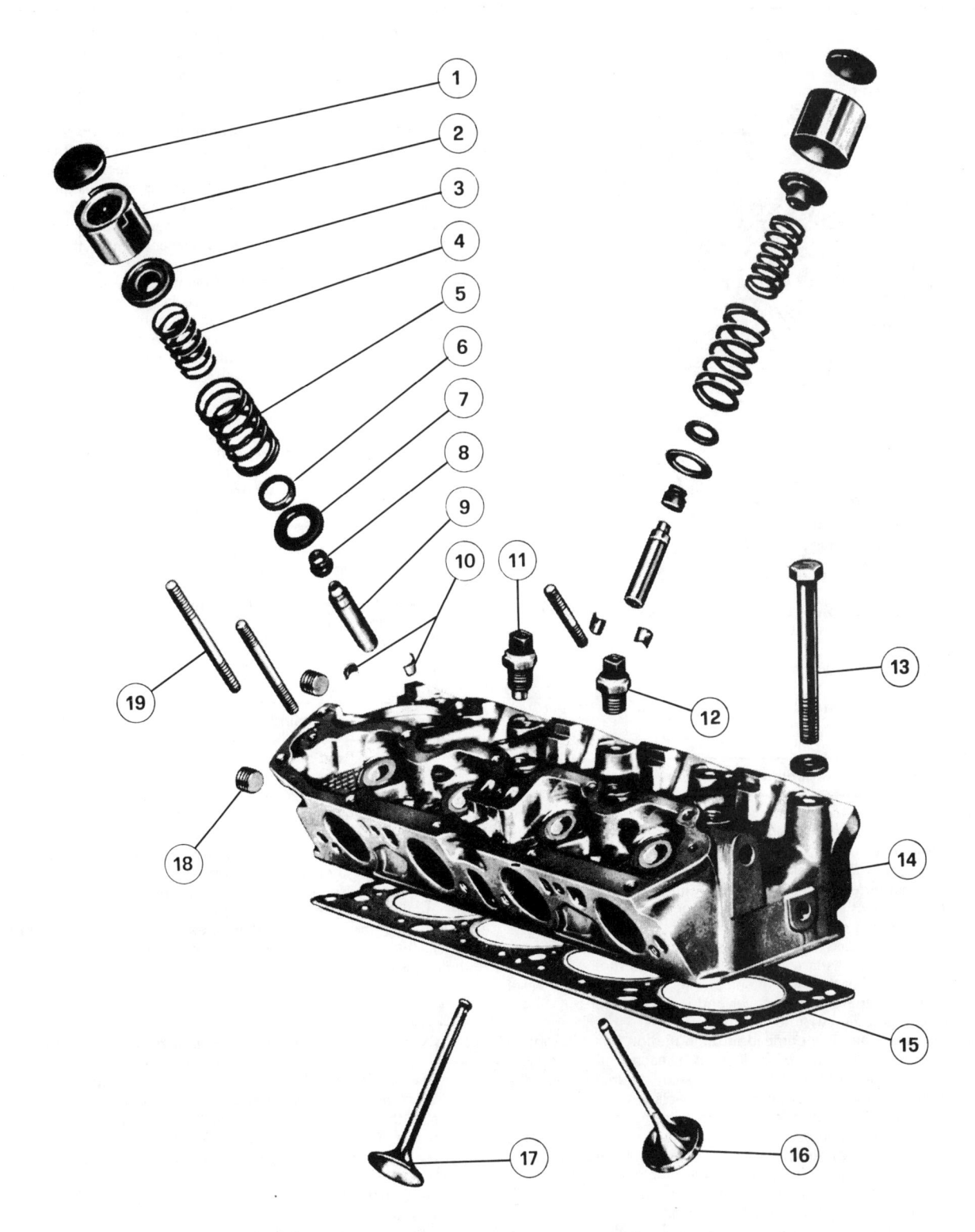

Fig. 1.8 Cylinder head and valve gear – exploded view (Sec 1)

1. Shim
2. Tappet bucket
3. Collar
4. Inner spring
5. Outer spring
6. Cup
7. Washer
8. Valve guide oil seal
9. Valve guide
10. Collets
11. Water temperature gauge transmitter
12. Engine overheating light switch
13. Cylinder head bolt
14. Cylinder head
15. Cylinder head gasket
16. Inlet valve
17. Exhaust valve
18. Plug
19. Camshaft housing mounting stud (on later models bolts were used instead)

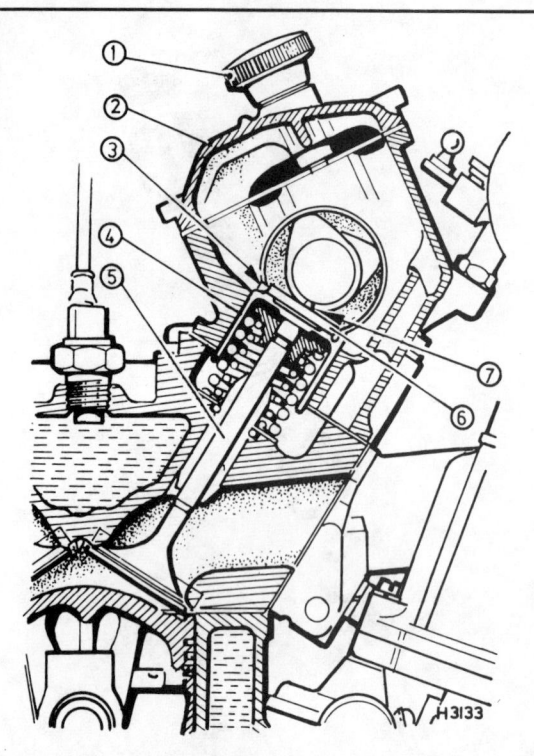

Fig. 1.9 Cross section of the inlet side of the cylinder head (Sec 11)

1 Camshaft cover retaining bolt
2 Camshaft cover
3 Notch in tappet for extracting capping plate
4 Tappet
5 Inlet valve
6 Tappet disc plate
7 Cam/tappet gap

12 Lubrication system – general description

The lubrication is of a standard pattern with one interesting novelty. This is a system which enables the driver to check the oil level in the sump without opening the bonnet and pulling out the dipstick. A thin plastic pipe is connected from the top end of the hollow dipstick pipe to the instrument panel right-hand push button device. By pressing the button when the engine is not running the level of the oil in the sump is shown on the gauge.

The oil pump, located in the sump, is driven by a gear at the front end of the ancillary driveshaft. The pump sucks in oil through a gauze mesh filter in the end of the pick-up and pumps it via an oil relief valve which is located in the oil pump cover and it provides for the excess oil to be discharged directly into the sump.

On all engines the oil is pumped through the full flow filter into a major gallery of oil drillings which direct it to each of the five main crankshaft bearings, the auxiliary drive shaft bearings and the camshaft bearings adjacent to the timing wheels.

Drilled galleries in the cylinder block tap oil from the main bearings on the crankshaft and feed it to the camshaft bearings increasing the supply direct from the pump to the forward camshaft bearings.

Drillings in the crankshaft tap oil from the main journls to supply it to the big-end bearings, and holes in the connecting rod allow oil to squirt from the big-ends onto the cylinder walls.

Holes drilled into the camshaft bearing housings allow oil to spray out from the bearings on to the cams and tappets.

Finally the oil pressure transducer and low pressure indicator switch are screwed into the oil filter fitting and monitor the pressure of oil after it has flowed through the filter.

In later models there is also an oil temperature transducer mounted onto the oil filter mounting block.

13 Oil pump – overhaul

1 With the oil pump removed from the engine proceed as follows.
2 The oil pump assembly is held together by three short studs and the two long bolts which secure it to the inside of the crankcase. Having already removed the two long bolts all that remains to dismantle the pump is the three studs.
3 The pump comprises the main body, two gears – housed in the main body, a cover plate incorporating the pressure relief valve and finally the suction horn with a metal filter screen across the aperture.
4 Once the retaining bolts have been removed all of these oil pump parts may be separated for cleaning and inspection (photos).
5 The relief valve is of caged ball spring design. Examination necessitates inspection of the ball and orifice for signs of wear and checking that the relief valve spring operates correctly and has not become weak. Check its free length according to the Specifications (photo).
6 Thoroughly clean all the component parts in petrol and then check the gear endfloat and tooth clearance in the following manner.
7 With the two gears in position in the pump, place the straight edge of a steel rule across the joint face of the housing, measure the gap between the bottom of the straight edge and the top of the gears with a feeler gauge. The clearance should not exceed that stated in the Specifications (photo).
8 Check the backlash between the two gears and this should again not exceed that stated in the Specifications (photo).
9 Check the clearance between the gears and pump body and this again should not exceed that stated in the Specifications (photo).
10 Inspect the gears for scoring or wear, especially if the clearances are greater than they should be (photo).
11 If there are excessive clearances between the gears and body it may be caused by play in the driveshaft in the pump body, in which case it would be advisable to renew the complete pump with an exchange unit.
12 Check the condition of the pump plate. Ensure that the surface which comes into contact with the pump gears is not worn and also check the relief valve seat for wear.

14 Oil filter and housing – general

1 When work is of a routine maintenance nature and the filter is due for renewal, a useful tool to loosen an unobliging filter cartridge is a small chain or strap wrench. This type of wrench will grip the old filter firmly and enable the most obstinate of cartridges to be freed. Do not use such a wrench to tighten the new filter into position.
2 If the engine has been completely dismantled the opportunity of cleaning the oil filter support fitting interior and exterior should not be missed. Both oil pressure and low oil pressure indicator transducers are mounted on the support fitting and in later models the oil level/temperature sender as well; these should be renewed or checked if possible at your local electrical workshop if there is any doubt as to their operation.

15 Cylinder bores – examination and remedial action

1 The cylinder bores must be examined for taper, ovality, scoring and scratches. Start by carefully examining the top of the cylinder bores. If they are at all worn a very slight ridge will be found on the thrust side. This marks the top of the piston travel. The owner will have a good indication of the bore wear prior to dismantling the engine, or removing the cylinder head. Excessive oil consumption accompanied by blue smoke from the exhaust is a sure sign of worn cylinder bores and piston rings.
2 Measure the bore diameter just under the ridge with a micrometer and compare it with the diameter at the bottom of the bore, which is not subjected to wear. If the difference between the two measurements is more than 0.006 in (0.15 mm) then it will be necessary to fit special rings or to have the cylinders rebored and fit oversize pistons and rings. If no micrometer is available remove the rings from a piston and place the piston in each bore in turn about three quarters of an inch below the top of the bore. If an 0.010 in (0.25 mm) feeler gauge can be slid between the piston and the cylinder wall on the thrust side of the bore then remedial action must be taken.
3 Oversize pistons are available in three sizes as detailed in the Specifications, and if the bores are badly worn they must be rebored and new oversize pistons fitted.

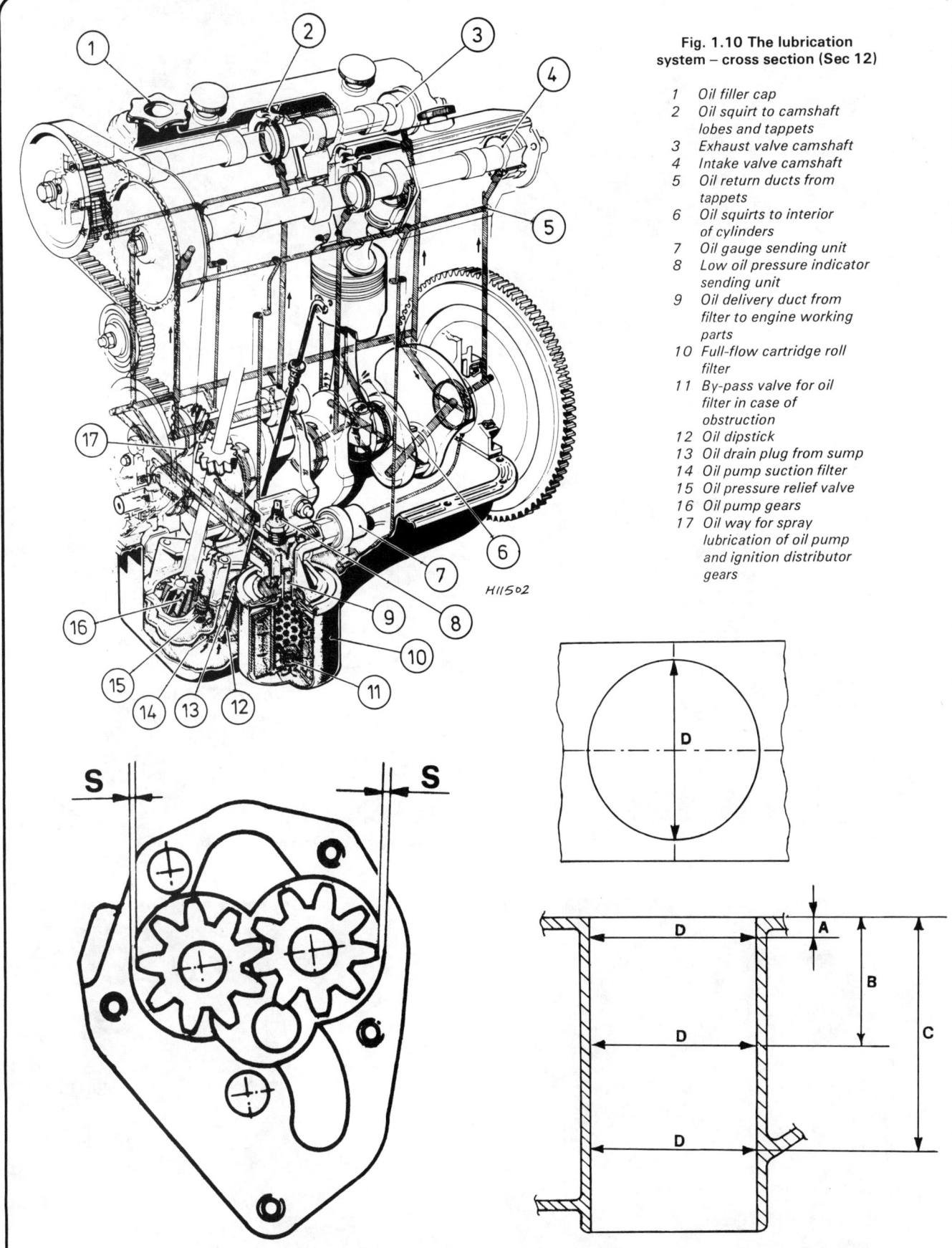

Fig. 1.10 The lubrication
system – cross section (Sec 12)

1. Oil filler cap
2. Oil squirt to camshaft lobes and tappets
3. Exhaust valve camshaft
4. Intake valve camshaft
5. Oil return ducts from tappets
6. Oil squirts to interior of cylinders
7. Oil gauge sending unit
8. Low oil pressure indicator sending unit
9. Oil delivery duct from filter to engine working parts
10. Full-flow cartridge roll filter
11. By-pass valve for oil filter in case of obstruction
12. Oil dipstick
13. Oil drain plug from sump
14. Oil pump suction filter
15. Oil pressure relief valve
16. Oil pump gears
17. Oil way for spray lubrication of oil pump and ignition distributor gears

Fig. 1.11 Checking the oil pump gears mating clearances (Sec 13)

S = Clearance between gears and body

Fig. 1.12 Cylinder bores – checking the bores should be done at points D (Sec 15). For A, B and C see the Specifications

13.4a With the bolts removed separate the two halves of the pump – the lower half contains the relief valve and spring ...

13.4b ... and beneath the centre plate are the pump gears

13.5 Remove the relief valve and spring and check its operation and condition

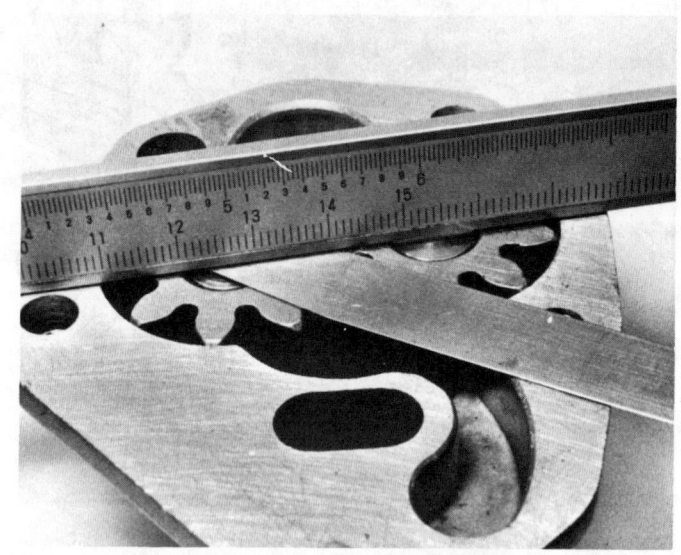

13.7 Checking the oil pump gear endfloat

13.8 Checking the backlash between the two gears

13.9 Checking the clearance between the gears and body

13.10 Remove the driven gear and check its condition

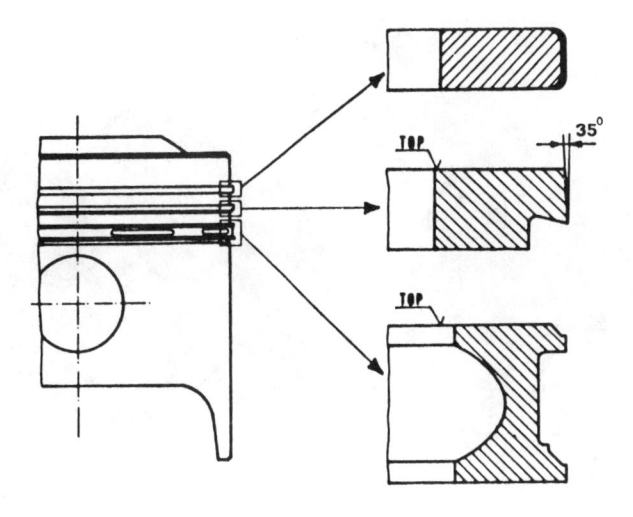

Fig. 1.13 The piston rings – correct fitting detail (Sec 16)

16 Pistons and piston rings – examination and renewal

1 Once the piston/connecting rod has been removed from the engine the piston rings may be slid off the piston.
2 Slide the piston rings carefully over the surface of the piston taking care not to scratch the aluminium alloy from which the piston is made. It is all too easy to break the cast iron piston ring if they are pulled off roughly, so this operation must be done with extreme care. It is helpful to make use of an old feeler gauge.
3 Lift one end of the piston ring to be removed out of its groove and insert the end of a feeler gauge under it.
4 Move the feeler gauge slowly around the piston and apply slight upwards pressure to the ring as it comes out of the groove so that it rests on the land above. It can then be eased off the piston and the feeler gauge used to prevent it slipping into other grooves as necessary.
5 If the old pistons are to be refitted having removed the piston rings, thoroughly clean them. Take particular care to clean out the piston ring grooves. At the same time do not scratch the aluminium. If new rings are to be fitted to the old pistons, then the top ring should be stepped to clear the ridge left above the previous top ring. If a normal but oversize new ring is fitted, it will hit the ridge and break, because the new ring will not have worn in the same way as the old, which will have worn in unison with the ridge.
6 Before fitting the rings on the pistons each should be inserted into the cylinder bore and the gap measured with a feeler gauge (photo). This should be as detailed in the Specifications. It is essential that the gap is measured at the bottom of the ring travel. If it is measured at the top of a worn bore and gives a perfect fit, it could easily seize at the bottom. If the ring gap is too small rub down the ends of the ring with a fine file, until the gap, when fitted, is correct. To keep the rings square in the bore for measurement, line each up in turn with an old piston in the bore upside down, and use the piston to push the ring down about 3 in (75 mm). Remove the piston and measure the piston ring gap.
7 Refit the rings to the piston using the old feeler gauge to help the process, as with removal.
8 Check the clearance of the piston rings in the grooves. These should be as detailed in the Specifications (photo).
9 If the clearances are too great check whether the rings or grooves are worn. If it is the grooves then the piston will have to be renewed.
10 When fitting new pistons and rings to a rebored engine the ring gap can be measured at the top of the bore as the bore will now not taper. It is not necessary to measure the side clearance in the piston ring groove with rings fitted, as the groove dimensions are accurately machined during manufacture. When fitting new oil control rings it may be necessary to have the groove widened by machining to accept the new wider rings. In this instance the manufacturer's representative will make this quite clear and will supply the address to which the pistons must be sent for machining.
11 When new pistons are fitted, take great care to fit the exact size best suited to the particular bore of your engine. Lancia go one stage further than merely specifying one size of piston for all standard bores. A range of different sizes is available either from the piston manufacturers or from the dealer for the particular model of car being repaired. The different classes are all given in the Specifications.
12 On the second and third piston rings, the top side of the ring is marked with Top, always fit this side uppermost and carefully examine all rings for this mark before fitting.
13 Before refitting the piston assemblies stagger the piston ring gaps evenly around the piston.

17 Gudgeon pin – renewal

1 The gudgeon pin connects the piston to the connecting rod at the small-end. In all engine types, except the 1400, the gudgeon pin is held in place by a circlip at either end and can be driven or pushed out quite easily once the circlips have been removed. The gudgeon pin in the 1400 engine is a press fit and has to be pressed out of the piston and connecting rod. This must be carried out by a competent engineering works or Lancia dealer. The refitting too requires the connecting rod to be heated up to ensure that it fits correctly on the pin.

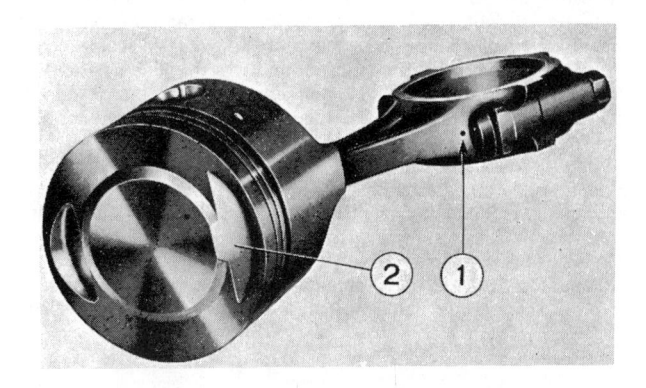

Fig. 1.14 Connecting rod and piston alignment – for refitting purposes (Sec 17)

1 Connecting rod oil hole   2 Inlet valve groove

16.6 Checking the piston ring gap in the bore

16.8 Checking the piston ring fit in the grooves

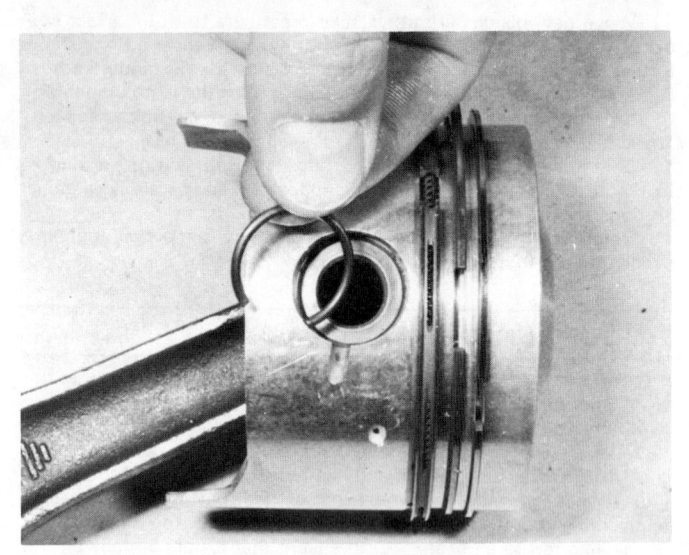

17.2 Removing a gudgeon pin retaining circlip

17.4 Make sure that the valve grooves in the piston are the right way round

2   To renew the push fit gudgeon pin proceed as follows. First remove the circlips from either end of the gudgeon pin (photo).
3   Then push out the gudgeon pin and separate the piston and connecting rod. Examine the gudgeon pin for wear or scoring.
4   To refit the piston and connecting rod together line up the holes, ensuring that the exhaust valve groove (small one) in the piston head is on the same side as the numbered side of the connecting rod (photo).
5   Push the gudgeon pin home. It should be a good push fit when it has been lightly smeared with engine oil. Hold the assembly up with the gudgeon pin in the vertical plane and it should not fall out. If it does fall out it is too sloppy a fit, and is worn, or the small-end bush is worn and needs renewing.
6   Check the fit of the gudgeon pin in the small-end bush and check both parts according to the specifications. Renew the parts if necessary.
7   If the fit is alright, refit the circlips to hold the gudgeon pin in place.

18 Small-end bush – renewal (except 1400 models)

1   The small-end bush is fitted into the upper end of the connecting rod and connects the rod to the piston by way of the gudgeon pin.
2   To renew the bush first remove the gudgeon pin as described in the previous Section, and separate the piston from the connecting rod.

3   To renew the bush requires the use of a press and special reaming equipment. Take the connecting rod to your local Lancia dealer and get him to renew the bush for you, and also ream it to the required size.
4   Refit the connecting rod to the piston as described in the previous Section.

19 Ancillary driveshaft and oil pump/distributor drive gear – examination

1   The ancillary driveshaft runs in bushes which are interference fits in bores in the cylinder block. There are not any undersize bushes supplied and therefore if the bush internal bores are found to be worn the cylinder block should be taken to your local Lancia dealer who will have the necessary equipment to fit new bushes into the cylinder block.
2   The diameters of the shaft journals should also be checked for wear, and if found the only course of action is to renew the shaft.
3   The fitted clearance between the bushes and shaft journals should be as given in the Specifications.
4   Together with the inspection of the shaft bearings the gears for the distributor drive and oil pump drive, and the cam lobe for actuating the fuel pump should also be examined for wear. As stated earlier if wear is found the only recourse is to renew the shaft.
5   Finally the oil pump and distributor drive pinion bearing journal

and gear teeth should be examined for wear. The pinion remains in the bush – which is an interference fit in the cylinder block – by the end thrust from the helical gear arrangement. It is necessary therefore that the thrust faces on the pinion and bush are not scored or worn. Like the driveshaft if wear is found, renewal of the relevant component is the only action possible.

20 Flywheel – examination and renewal of the starter ring gear

1 Check the flywheel for scoring marks or damage. If necessary it will have to be refaced professionally. Check first whether it has already been refaced and therefore whether refacing it again will reduce the thickness below the specified minimum. If the flywheel is too thin to be refaced then it will have to be renewed.
2 Check the starter ring gear teeth for wear or broken teeth and renew it if necessary.
3 To renew the flywheel starter ring gear is easily done with the aid of a fly press. The flywheel is placed with the clutch face downwards and the starter ring resting on blocks and round its outer edge. The flywheel is then pressed out of the ring gear.
4 Refitting is carried out after thoroughly cleaning the ring gear seating on the flywheel.
5 Start by heating the new starter ring gear in an oil bath or oven to 80°C (176°F). Then place it on the blocks with the inside diameter chamfered edge facing upwards and place the flywheel in position and press it home.
6 With the new ring gear fitted to the flywheel clean off all the oil left over from the bath (if it was heated up that way) using paraffin.

21 Crankshaft – examination and renovation

1 Examine the crankpin and main journal surfaces for signs of scoring or scratches and check the ovality and taper of the crankpins and main journals. If the bearing surface dimensions do not fall within the tolerance ranges given in the specifications, the crankpins and/or main journals will have to be reground (photo).
  If the crankshaft is reground the workshop should supply the necessary undersize bearing shells.

22 Big-end and main bearing shells – examination and renovation

1 Big-end bearing failure is accomplished by a noisy knocking from the crankcase and a slight drop in oil pressure. Main bearing failure is accompanied by vibration which can be quite severe as the engine speed rises and falls, and a drop in oil pressure.
2 Bearings which have not broken up, but are badly worn will give rise to low oil pressure and some vibration. Inspect the big-ends, main bearings and thrust washers for signs of general wear, scoring, pitting and scratches. The bearings should be matt grey in colour. With lead-indium bearings, should a trace of copper colour be noticed the bearings are badly worn as the lead bearing material has worn away to expose the indium underlay. Renew the bearings if they are in this condition or if there is any sign of scoring or pitting.
3 The undersizes available are designed to correspond with the regrind sizes, ie 0.02 in (0.508 mm) bearings are correct for a crankshaft reground – 0.02 in (0.508 mm) undersize. The bearings are in fact, slightly more than the stated undersize as running clearances have been allowed for during their manufacture.

23 Engine reassembly – general

1 To ensure maximum life with minimum trouble from a rebuilt engine, not only must every part be correctly assembled, but everything must be spotlessly clean, all the oil-ways must be clear, locking washers and spring washers must always be fitted where indicated and all bearings and other working surfaces must be thoroughly lubricated during assembly. Before assembly begins renew any bolts or studs whose threads are in any way damaged; whenever possible use new spring washers.
2 Apart from your normal tools, a supply of non-fluffy rag, an oil can filled with engine oil (an empty washing-up fluid plastic bottle thoroughly clean and washed out will invariably do just as well), a supply of new spring washers, a set of new gaskets and a torque wrench should be collected together.

24 Crankshaft – refitting

Ensure that the crankcase is thoroughly clean and that all oil ways are clear. A thin twist drill is useful for cleaning them out. If possible blow them out with compressed air. Treat the crankshaft in the same fashion and then inject engine oil into the crankshaft oil-ways.
Commence work on rebuilding the engine by renewing the crankshaft and main bearings.
1 Never refit old main bearing shells unless they are almost new (a false economy to do so). Fit the upper halves of the main bearing shells to their location in the crankcase, after wiping the location clean (photo). The centre bearing shells are different as they have no oil channel and are wider (photo).
2 Note that on the back of each bearing is a tab which engages in locating grooves in either the crankcase or the main bearing cap housings.
3 If new bearings are being fitted, carefully wipe away all traces of protective grease with which they are coated.
4 With the five upper bearing shells securely in place, wipe the lower bearing cap housings and fit the three lower shell bearings to their caps ensuring that the right shell goes into the right cap if the old bearings are being refitted (photo).
5 Wipe the recesses either side of the rear main bearing which locates the upper halves of the thrust washers.
6 Smear some grease onto the plain sides of the upper halves of the thrust washers and carefully place them in their recesses.
7 Generously lubricate the crankshaft journals and the upper and lower main bearing shells and carefully lower the crankshaft into position. Make sure that it is the right way round (photo).
8 Lubricate the crankshaft journals, injecting oil into the oil ways to ensure adequate lubrication upon the initial start of the engine.
9 Fit the main bearing caps into position ensuring that they locate properly. The mating surfaces must be spotlessly clean or the caps will not seat correctly. Make sure that the bearing caps are in the correct positions. The front one is stepped to allow for the oil pump fitting and the remainder are marked with notches (I-II-III-IIII) from the second cap from the front of the engine to the rear. Also make sure that the caps are the right way round. The tags in the bearing cap and seat must be on the same side (away from the oil pump) (photo).
10 When refitting the rear main bearing cap ensure that the thrust washers, generously lubricated, are fitted with their oil grooves facing outwards and the locating tab of each washer in the slot in the bearing cap.
11 Refit the main bearing cap bolts and screw them up finger tight.
12 Test the crankshaft for freedom of rotation. Should it be very stiff

21.1 Checking the crankpin using a micrometer

24.1a Refitting the upper main bearing shells ...

24.1b ... noting that the centre bearing shells are different

24.4 Fitting the lower bearing shell to the bearing cap ensuring that the tab engages in the groove

24.7 Generously lubricate the bearing shells with engine oil before refitting the crankshaft

24.9 Refitting the main bearing caps – note that the front one (already in place) is different

24.13a Tightening the main bearing caps ...

24.13b ... but note that the front cap is tightened to a different torque from the others

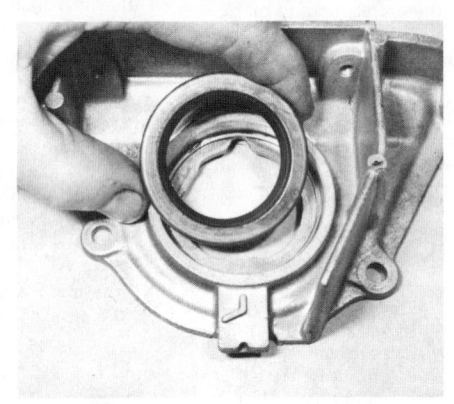

24.16a Offering up the front seal to the front housing ...

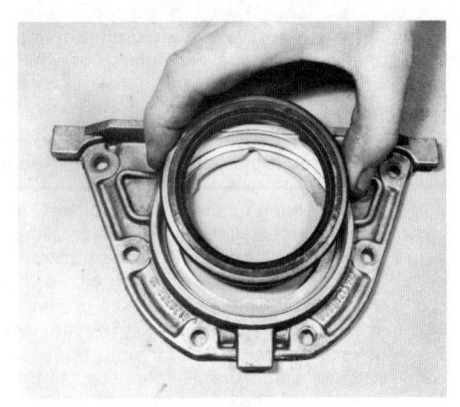

24.16b ... and the rear seal to the rear housing

24.18a With the gasket in place refit the front ...

24.18b ... and rear housings to the crankcase

25.5 Check the oil hole in the base of the connecting rod

to turn, or possess high spots, a most careful inspection must be made, preferably by a skilled mechanic with a micrometer to trace the cause of the trouble. It is very seldom that any trouble of this nature will be experienced when fitting the crankshaft.

13 Tighten the main bearing cap bolts to the torque specified in the Specifications. Note that the front cap bolts are tightened to a different torque from the rest. The 1400 engine differs also (photos).

14 Recheck that the crankshaft rotates freely.

15 Use a large screwdriver to lever the crankshaft forwards and backwards to check the amount of endfloat. Use a feeler gauge to check the clearance which should not exceed that quoted in the Specifications. If the clearance is too much then oversize thrust washers will have to be fitted to reduce the clearance so that it agrees with the specified figure.

16 Refit the new oil seals to the front and rear oil seal housings. Drive out the old seals, place the housing face down and tap the new seal into place with the lips facing inwards. Make sure the seal is fitted and tapped home keeping it level (photos).

17 Smear the new gaskets for the front and rear housings with grease and place them in position on the crankcase.

18 Check that the mating faces of the housings are clean, lightly oil the lips of the seals and grease the front and rear ends of the crankshaft and fit the housings. Secure them with the appropriate bolts (photos).

25 Pistons, connecting rods and big-end bearings – refitting

1 Lay the piston and connecting rod assembly in the correct order ready for refitting into their respective bores. There are numbers stamped on the big-end bearing bosses and on the bearing caps. The connecting rods and caps are numbered 1 to 4 from front to rear.

2 With a wad of clean lint free rag wipe the cylinder bores clean, and lay the block on its side.

3 Position the piston rings so that their gaps are 120° apart and then lubricate the rings.

4 Fit the piston ring compressor to the top of the piston, making sure it is tight enough to compress the piston rings. If you haven't got a ring compressor it is possible to use a large jubilee clip and a thin sheet of soft metal. Whatever you use don't scratch the piston.

5 Using a piece of fine wire double check that the little jet hole in the connecting rod is clean (photo).

6 The pistons, complete with connecting rods, are fitted to their bores from above. Before fitting them check that the upper bearing shell is properly located (photo).

7 As each piston is inserted into its bore ensure that it is the correct piston – connecting rod assembly for the particular bore and that the connecting rod is the right way round, also that the front of the piston is towards the front of the bore, ie towards the front of the engine. Lubricate the piston with clean engine oil.

8 The piston will slide into the bore only as far as the bottom of the piston ring compressor. Gently tap the top of the piston with a wooden or plastic hammer whilst the connecting rod is guided into position on the crankshaft (photo).

9 Generously lubricate the crank pin journals with engine oil and turn the crankshaft so that the crank pin is in the most advantageous position for the connecting rod to be drawn onto it.

10 Wipe clean the connecting rod bearing cap and back of the shell bearing and fit the shell bearing in position ensuring that the locating tongue at the back of the bearing engages the locating groove in the connecting rod cap (photo).

11 Generously lubricate the shell bearing and offer up the connecting rod bearing cap to the connecting rod. Ensure that the numbers on the connecting rod and cap match up.

12 Fit the connecting rod nuts and tighten them but not completely.

13 Carry out the same procedure for the other three piston connecting rod assemblies.

14 When all the big-end bearing caps have been fitted tighten the nuts to the torque wrench setting given in the Specifications at the beginning of the Chapter (photo). Rotate the crankshaft each time so that the bearing cap is in the easiest place to tighten the nuts.

26 Ancilliary driveshaft – refitting

1 First of all insert the oil pump and distributor drive pinion into the bush in the crankcase. It is advisable to liberally coat the bearing surfaces of the pinion gear with engine oil before fitting, and don't forget to oil the shaft bearings and journals.

2 Carefully insert the ancillary driveshaft into the cylinder block taking care not to scratch the bearing bores in the bushes with the gears and cam lobes machined onto the shaft (photos).

3 With the driveshaft in place, it is retained by the thrust plate which is secured by two bolts onto the cylinder block (photo).

4 The shaft oil seal should be renewed at any major overhaul of the engine. Press out the old seal from the carrier fitting and fit a new one. Smear the carrier gasket with grease, and offer the seal carrier assembly onto the engine (photo).

5 Secure the shaft oil seal carrier and the five bolts and tighten them progressively and evenly. Then fit the auxiliary shaft wheel and lightly tightening the centre bolt at this time.

27 Oil pump – refitting

1 Fit a new gasket or O-ring to the oil pump body to crankcase oil channel and hold it in place with a dab of grease. Prime the pump with engine oil and then offer it up to the lower crankcase (photo).

2 Be careful not to push the drive pinion out of its bush when trying to engage the pinions internal spline with the pump shaft spline.

3 Once the pump is in place it is secured by two long bolts which pass through the pump body into the crankcase. Tighten these bolts progressively to the specified torque.

4 If it was removed, refit the oil return pipe from the breather housing. It is secured by two bolts into the edge of the crankcase. Make sure it is located properly at the upper end.

28 Oil sump – refitting

1 Before refitting the sump ensure that the following items have been assembled into and onto the crankcase, and that all seals and

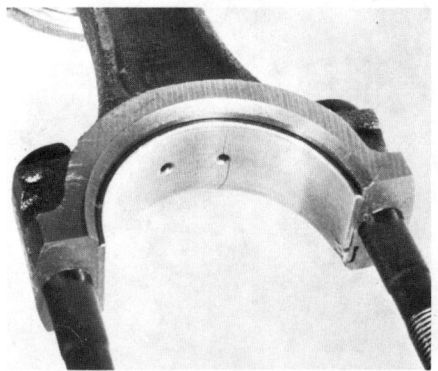

25.6 Check that the upper shells are correctly located

25.8 Refitting No 1 cylinder piston assembly using the ring compressor tool

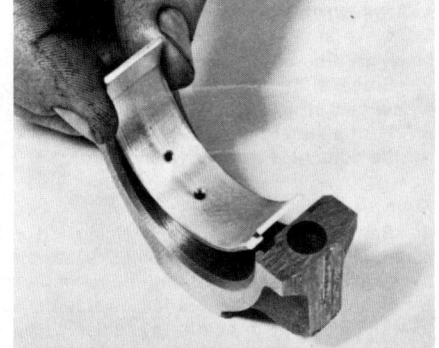

25.10 Refit the lower bearing shell to the bearing cap

gaskets have been refitted and the securing bolts have been tightened to the specified torques.

(a) Crankshaft – end float
(b) Main bearing caps
(c) Big end caps, con-rods and piston assembly
(d) Crankshaft freedom to turn
(e) Rear crankshaft oil seal and housing
(f) Front crankshaft oil seal and housing
(g) Auxiliary drive shaft
(h) Distributor and oil pump drive pinion
(i) Oil pump assembly
(j) Oil return pipe

2 Clean mating sump and crankcase surfaces of old gasket, and place the new gasket – smeared with grease – onto the crankcase face. Ensure that you have the correct gasket, as there are three different possibilities offered in the engine overhaul gasket set. The holes must all be correctly aligned.
3 Offer the sump onto the crankcase taking care not to disturb the new gasket resting on the joint face.
4 Secure the sump with the bolts and tighten progressively
5 Check that the sump plug has been refitted and tightened up.

29 Flywheel and clutch refitting

1 Stand the engine the right way up on its sump, and use wooden blocks to help support it.
2 Turn the crankshaft so that No 1 and No 4 pistons are at top dead centre and clean the mating surfaces of the flywheel and crankshaft.
3 Offer up the flywheel onto the crankshaft mounting flange so that the small white dot near the periphery of the wheel is vertically above the centre of the flywheel.
4 Refit the bolts and plate through which they are bolted. Tighten the bolts progressively to the specified torque. To prevent the crankshaft and flywheel from rotating make up a bracket and bolt it to the crankcase, or screw a sufficiently long bolt into the rear of the flywheel using one of the clutch cover mounting bolt screw holes so that it locks the flywheel against the crankcase (photo). Make sure however that the bolt which is used for this simple method of locking the flywheel has the correct thread and does not damage the hole.
5 If the clutch pressure plate assembly is being refitted then its correct position on the flywheel is given by the mark made when it was removed.
6 Locate the driven plate in the central position using a mandrel, and tighten the clutch cover bolts as described in Chapter 5.

30 Ancillary components – refitting (Stage one)

With the main crankcase assembly refitted it is now time to refit the external components which would have been removed during the dismantling procedure in Section 10.
If necessary they would have been overhauled according to the instructions given in the appropriate Chapters.

Water pump
1 Make sure the mating surfaces of the water pump and cylinder block are free of old gasket or jointing compound.
2 Smear a little grease onto the water pump joint face and place on a new gasket.
3 Fit the water pump to the cylinder block securing it with 4 bolts. All the bolts have copper washers.

Belt tensioner idler assembly
4 Refit the plate and spring retaining bolt and secure it, then refit the spring and secure it with the bolt, 2 spacers and washers. Fit the idler to the stud and secure it with the spacer, washer and nut.
5 Push the idler plate as far towards the spring as it can go. Then lock the idler nut to hold it in position.
6 Lever the long end of the spring into the hole in the plate using a screwdriver (photo).
7 Then unlock the idler unit so that it is ready for refitting the timing belt.

Oil filter and mounting block
8 Refit the four retaining bolts and their respective washers to the oil filter housing, then fit a new gasket over the ends of the bolts. Offer up the assembly to the crankcase and tighten the bolts to the specified torque.
9 Fit a new oil filter. Lightly smear the oil seal with oil and screw the oil filter on by hand, until it just touches the mounting block. Then screw it another half turn, again by hand. Never do up an oil filter using any form of tool (photo).

Crankcase breather
10 Fit the rubber breather tubes to the breather housing and clean the mating faces of the housing and cylinder block.
11 Fit a new gasket onto the breather housing and position the assembled breather system onto the cylinder block immediately above the fuel pump.
12 Tighten the single bolt which secures the breather housing to the block and fit the short angled breather tube to the union beside the distributor position. Tighten the pipe clips.

Fuel pump
13 The fuel pump may only be refitted now if the cylinder head is being refitted with the inlet manifold attached. Otherwise it will have to wait, until the inlet manifold has been refitted first.
14 Assemble the fuel pump, and spacer fitting new gaskets.
15 Fit the fuel pump onto the engine ensuring that as the pump is offered up to the cylinder block the pump actuating arm is positioned over the auxiliary shaft cam lobe. The pump will be damaged if the lever arm is allowed to pass beneath the cam lobe.
16 Secure the pump with the two bolts and spring washers and tighten to the specified torque.

31 Cylinder head – reassembly

1 Section 11 concerned itself with the overhaul of the cylinder head, valve gear and camshafts. This Section covers the rebuilding and refitting of those components so that the cylinder head can be refitted as a complete unit.
2 Start by refitting the valve gear to the cylinder head, as follows.
3 Rest the cylinder head on its side, or if the manifold studs are fitted with the gasket surface downwards.
4 Fit the valve into the same guide from which it was removed. Slip the oil seal over the protruding stem onto the top of the guide.
5 Place the lower spring seat and washer in position.
6 Fit both inner and outer valve springs.
7 Fit the springs cap onto the springs and position the spring compression tool onto the valve assembly.
8 Compress the springs sufficiently for the collets to be slipped into place in the groove in the top of the valve stem.

25.14 Torque tighten the big-end bearing cap nuts (note the connecting rod and cap numbers on the shoulders)

26.2a Oil the bearing shells and ...

26.2b ... insert the ancillary driveshaft

26.3 Fit the thrust plate and tighten the bolts

26.4 Offering up the ancillary shaft front seal housing to the crankcase

27.1 Refitting the oil pump to the crankcase (note the O-ring in the pump body held in place with grease)

29.4 Lock the flywheel and tighten the retaining bolts

30.6 Lever the spring using a strong screwdriver blade to engage it in the idler plate hole

30.9 Fitting a new oil filter

31.15 Refit the camshaft drive wheel to the camshaft

31.17 Refit the tappets to the bores (note the earth wire and bracket on this rear camshaft end cover)

31.19 Invert the camshaft housing and lower it into position with the tappets over the valves

31.21 Checking the tappet clearances (cam lobe vertical)

Chapter 1 Engine

9 Remove the valve spring compressor and repeat this procedure until all eight valves have been assembled into the cylinder head.
10 The next task is to refit the camshafts into the camshaft housings as follows.
11 Wipe the camshaft bearing journals and cam lobes and liberally lubricate with engine oil.
12 Wipe the inside bearing bores of each camshaft housing and lubricate with engine oil.
13 Carefully insert the camshafts into the housings so that the surfaces of the bearings in the housing are not scratched by cam lobes as they pass through.
14 Fit the camshaft end plate and gasket and tighten the three securing nuts to the torque specified. Note that the exhaust camshaft rear plate has the earth wire and a bracket attached to it.
15 Refit the respective camshaft drive wheel onto the shaft projecting through the front of the housing. The wheel is located by a dowel pin and secured by a single bolt. A tag washer is fitted beneath the bolt and this serves not only to lock the bolt when it is tightened later, but also to retain the dowel pin in its bore (photo).
16 The next stage is to refit the camshaft housings to the cylinder head, taking one assembly at a time.
17 Start by refitting the tappets, with the shims fitted in the tops, to the bores in the housing from which they were removed, unless of course new tappets are being fitted. Lubricate them internally and grease the exteriors to hold them in position when the housing is turned over (photo).
18 Fit a new gasket in position on the cylinder head and ensure that all the holes are lined up. If the engine involved is one of the early ones which had studs instead of bolts then there is no problem.
19 Invert the housing and lower it into position on the cylinder head. Now it will become obvious as to why the tappet buckets were greased earlier (photo). With the early type engine this task is made harder by having to locate the housing on the studs before lowering it into place. An assistant is a very useful person for this task.
20 Refit the bolts or nuts and washers and tighten them to 16 lbf ft (21 Nm).
21 Then check the tappet clearances (refer to Section 32 if necessary). They should be as given in the Specifications (photo).
22 If the clearances are not correct then the shims must be changed as described in Section 32.
23 Now repeat the procedure for the other camshaft housing. Do not refit the camshaft covers until the cylinder head has been refitted and the timing set up.
24 Refit the spark plugs and thermal transducers to the cylinder head. Tighten them to the specified torques.
25 The exhaust manifold is fitted to the cylinder head next. As many times before, ensure that the mating surfaces are clean and free from fragments of old gaskets.
26 Fit a new gasket – dry – over the studs and then offer up the exhaust manifold. Secure with the nuts tightened to the specified torque.
27 The inlet manifold may or may not be fitted complete with the carburettor assembly, but in any event use new gaskets and ensure that the mating faces are free from fragments of old gaskets. Air leaks on the inlet manifold system can cause a lot of bother! Tighten the nuts and bolts to the specified torque.
28 The whole cylinder head assembly is now ready for refitting to the cylinder block.

32 Valve/tappet clearance setting (engine in or out of car)

1 Remove the camshaft housing covers.
2 Rotate the camshaft; (by rotating the crankshaft – if this task is performed with the engine in the car) until the cam which moves the tappet to be checked is perpendicular to the tappet face plate and the valve is closed.
3 Measure the existing tappet gap and record the measurement so that the required thickness of the shims can be calculated.
4 The clearances should be as given in the Specifications. If the clearances are different from those specified when the engine is cold, proceed as follows.
5 Rotate the camshaft until the valve is fully open then insert Lancia Tool 88013036 or alternatively a bent screwdriver to hold the tappet down (photo).
6 Rotate the camshaft again so that the shim may be extracted from the top of the tappet with a thin small screwdriver inserted into two grooves in the side of the tappet barrel.
7 Measure the thickness of the tappet shim with a micrometer (photo). Add or subtract as appropriate the difference between the measured gap (paragraph 3) and the desired gap (paragraph 4) from the thickness of the shim to derive the thickness which should produce the desired gap.
8 The Specifications list the range of thicknesses of shims available. There are 30 different thicknesses and the thickness is marked on one face of the plate. This face must be on the tappet barrel side when fitted, and it is well to check that the plate is the thickness marked (photo).
9 Insert the new shim into the recess in the top of the tappet. Turn the camshaft again so that the spacing tool or bent screwdriver (paragraph 5) can be extracted. Turn the camshaft again to the position indicated in paragraph 2 and check that the desired tappet/cam gap has been achieved.
10 Repeat this procedure on each of the eight valve tappets. Refit the camshaft housing covers.

33 Cylinder head – refitting to cylinder block

1 It is essential to position the camshafts in the cylinder head and the crankshaft in the engine block in their correct relative positions before the head is lowered onto the engine. Once the head has been fitted it is equally important that the camshafts are not turned until they have been coupled by the timing belt to the crankshaft. There will be severe – possibly damaging mechanical interference of pistons and valves if these precautions are not taken.
2 Turn the camshafts so that the reference marks, ie the small holes in the faces of the camshaft wheels, are adjacent to the lugs on the camshaft housings. The cylinder head is now in the correct condition for fitment to the engine.
3 Turn the engine so that the No 1 and 4 pistons are at top dead centre. The engine is now in the correct condition to accept the cylinder head.
4 Place a new cylinder head gasket on the engine block. The gasket is marked as to which way up it should be. It can only fit one way or the oil and waterways won't line up. Also ensure that the dowels are in position (photo).
5 Lower the cylinder head carefully onto the engine. Take great care not to knock the open valves against the engine block face, they may bend (photo).
6 A couple of dummy studs screwed into the cylinder block will aid the task of lowering the cylinder head safely onto the block. However the dowels should be sufficient.
7 Once the head is in place the securing bolts may be screwed in and tightened to the specified torque in two stages (photo). The bolts must on no account be tightened fully in one go. The correct tightening sequence is shown in Fig. 1.17.

34 Timing wheels and belt – examination

1 The belt has a rubber surface and is reinforced for strength, and as a consequence it is most unlikely that the wheels will need anything more than a clean in an oil grease solvent and wiping dry.
2 The idler wheel which maintains the tension in the timing belt runs on a prepacked ball bearing race. Whenever the timing belt is being changed or attended to, it is advisable to check the excessive play of the wheel and bearing. The wheel bearing assembly should be renewed if discernable play is felt.
3 The belt however does not wear but fatigues, and because failure of this belt would be catastrophic for the engine, it must be renewed at regular intervals even though on the surface it might appear serviceable.
4 The timing belt must be renewed at intervals *not* exceeding 37 000 miles (60 000 km) – Lancia justifiably recommend that it should be renewed every 25 000 miles (40 000 km). It must be checked anyway every 12 000 miles (19 000 km).
5 It is important to remember that when handling a new belt – when fitting – to avoid bending it to a sharp angle or the fibres which reinforce the belt will be seriously weakened.

32.5 Removing the shim while holding the tappet down with a bent screwdriver (note that the cam lobe points directly upwards)

32.7 Measure the shim thickness using a micrometer

32.8 The shim size is marked on the side which faces the tappet

33.4 Set Nos 1 and 4 cylinders to TDC and fit a new head gasket, ensuring that the dowels are in place

33.5 Lowering the cylinder head onto the block

33.7 Torque tightening the cylinder head bolts in sequence

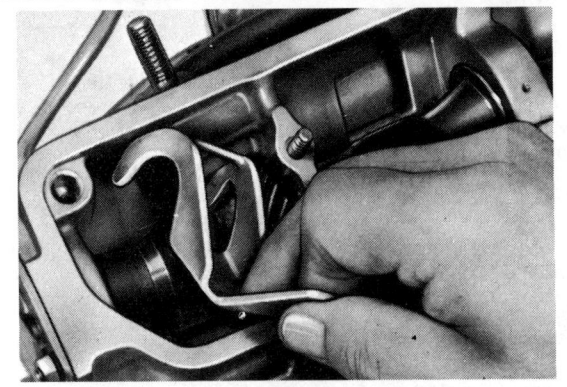

Fig. 1.15 Inserting the tappet locking tool over the camshaft (Sec 32)

Fig. 1.16 The tappet locking tool locks in position with the tappet held down so the shim can be removed (Sec 32)

1 Shim

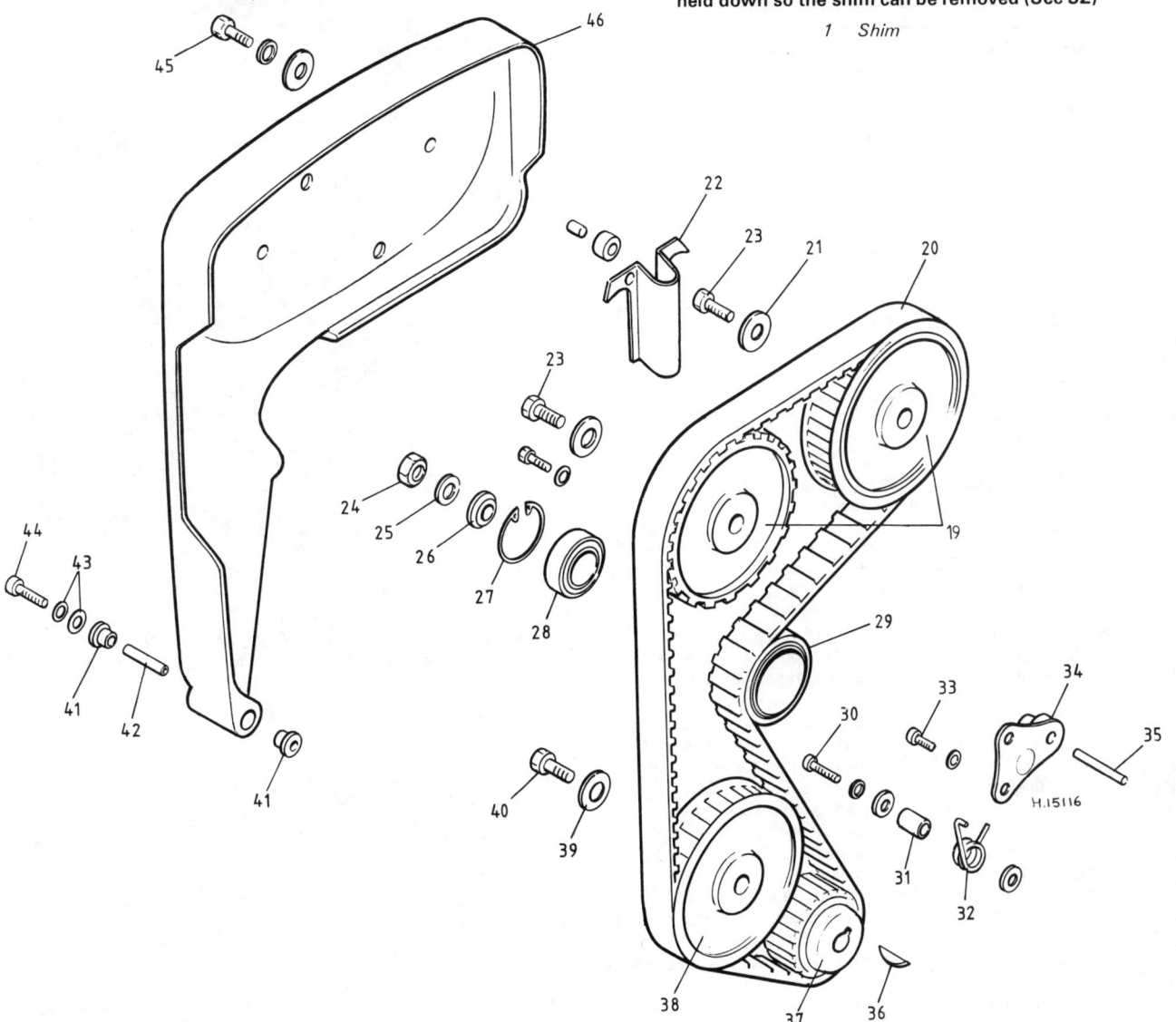

Fig. 1.18 Timing belt, cogs and cover – (Sec 34 and 35)

19 Camshaft drive wheels
20 Timing belt
21 Lock plate
22 Timing pointers and cover mounting bracket
23 Drive wheel retaining bolt
24 Belt tensioner locking nut
25 Lock washer
26 Spacer
27 Circlip
28 Bearing
29 Tensioner idler
30 Spring mounting bolt
31 Spring sleeve
32 Spring
33 Tensioner plate mounting bolt
34 Tensioner mounting plate
35 Stud
36 Woodruff key
37 Crankshaft timing belt drive cog
38 Ancillary shaft drive wheel
39 Lock plate
41 Bushes
42 Spacers
43 Washers
44 Timing cover lower bolt
45 Timing cover upper bolt
46 Timing cover

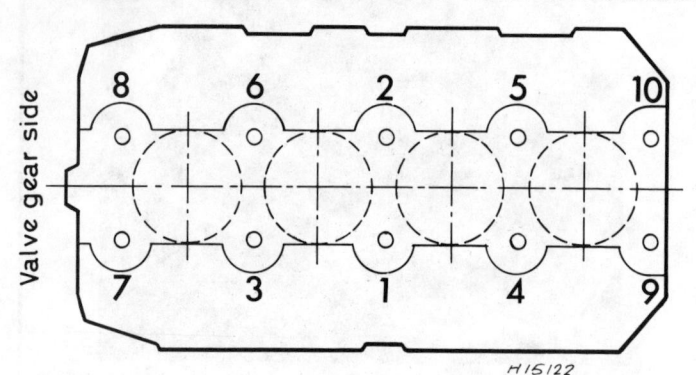

Fig. 1.17 Cylinder head retaining bolt tightening sequence (Sec 33)

35 Timing belt – refitting, alignment and tensioning

1 Tighten the ancillary driveshaft drive wheel if it has been removed. In order to lock the wheel so that the bolt could be tightened to the correct torque we found that a short length of flat steel bar could be neatly wedged in the front oil seal housing and between the notches on the drive wheel (photo). Be careful that it stays in position and does not fly out or damage the teeth of the drive wheel should it slip.

2 Refit the crankshaft timing belt drive cog. The groove on the inside fits over the key which fits in the crankshaft keyway. There is a timing mark or notch in one face of the cog; this faces outwards.

3 Refit the timing pointers and cover mounting bracket to the front of the cylinder head, if you can't get the proper Lancia tool for refitting the timing belt.

4 If it is possible to obtain the special Lancia tool for aligning all the timing marks on the various pulleys now is the time to fit it (Fig. 1.19). There are two versions of this tool. One has the markings for 1300, 1600, 1400 and 1800 engines which is the one shown in the figure. The other is solely for 2000 engines.

5 Align the various pulleys and drive wheels to match up with the marks or holes in the tool. In the case of the crankshaft pulleys and drive wheels to match up with the marks or holes in the tool. In the case of the crankshaft pulley, it cannot be refitted before the timing belt has been fitted and we found that the crankshaft is correctly aligned if the mark in the outer face of the crankshaft timing belt drive cog is lined up with the mark on the front oil seal housing directly above it.

6 It is possible to set up the pulleys and drive wheels correctly without this tool. With the pointers bracket refitted ensure that the timing holes in the drive wheels for the camshafts are lined up with the pointers. Then ensure that the crankshaft drive cog mark is lined up with the mark above it on the front seal housing. Lining up the ancillary driveshaft drive wheel is more difficult, but the timing hole needs to be lined up with the tensioner spring mounting bolt on a line between the drive wheel retaining bolt and the spring mounting bolt (photo).

7 Now refit the belt. With the Lancia tool this should present no problems. Make sure that the belt tensioner is slackened off (ie pushed as far as it can go towards the water pump).

8 Without the tool great care needs to be taken. Fit the belt over the crankshaft drive cog to start with and then the ancillary shaft. Feed it as shown in photo 35.6 up over the tensioner and the exhaust camshaft drive wheel and then lever it finally on to the inlet camshaft drive wheel. The belt must locate in the notches in the drive wheels as it is fitted.

9 When we fitted the timing belt we found great difficulty in fitting the belt to the notches without having to move the camshaft slightly one way or the other.

10 The nearest fit we would achieve was with the drive wheels both set slightly to the left of the camshaft housing legs when viewed from the rear (photo).

11 If the camshaft drive wheels were both rotated the other way to accept the next full section then the gaps were considerably greater between the timing holes and the lugs, especially on the inlet camshaft side.

12 By rotating the inlet camshaft along by 1 notch of the belt (back towards the lug) and leaving the exhaust camshaft stationary the set-up was then correct. This was found by trial and error (photo).

13 Never move the camshaft drive wheels more than 1 notch out of step with each other or the likelihood of series damage being done to the engine by the valves lifting the pistons is very high indeed.

14 When the belt has been refitted satisfactorily release the tensioner locknut and tension the belt. If the setting up tool has been used remove it and then refit the crankshaft pulley and nut.

15 Rotate the crankshaft several turns to check the belt is running smoothly then slacken the belt tensioner and then relock it to the correct torque as specified (photo).

16 Lock the flywheel so that it cannot turn and then tighten up the crankshaft pulley nut to the correct torque. Also check that the camshaft and ancillary shaft drive wheels are tightened to the correct torque.

17 Refit the pointer bracket if the aligning tool was used.

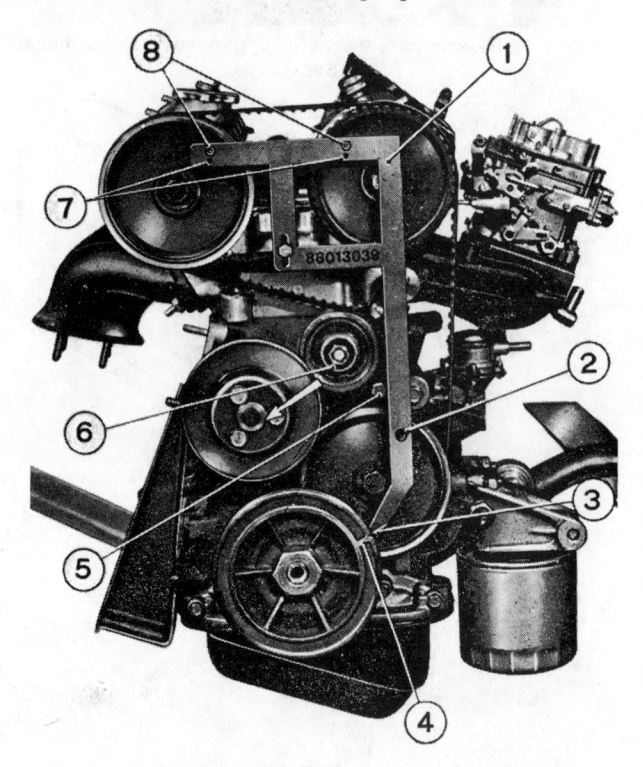

Fig. 1.19 Lancia Tool No 88013039 for aligning the timing markings and refitting the timing belt (Sec 35)

1 Tool
2 Ancillary driveshaft aligning hole
3 TDC mark on tool
4 TDC mark on crankshaft pulley
5 Tensioner plate clamp bolt
6 Tensioner idler locknut
7 1400 engine – camshaft drive wheel aligning holes
8 1300, 1600 and 1800 engines – camshaft drive wheel aligning holes

(There is a separate tool for the 2000 engine which is similar – Lancia No 88013042)

36 Ancillary components – refitting (Stage Two)

1 With the cylinder head refitted and the tappets and timing set up the remainder of the ancillary components can be refitted to the assembly.

2 Begin by refitting the inlet manifold complete with the carburettor, as described in Chapter 3, if it was not refitted with the cylinder head.

3 Then refit the fuel pump to the crankcase below the inlet manifold, as described in Chapter 3. Refit also the fuel lines from the pump to the carburettor.

4 Refit the crankcase breather upper hose to the PCV valve above the fuel pump.

5 Refit the distributor next. Set the engine to TDC on No 4 cylinder, as for removing the distributor. This can be verified by inspecting the cam lobe positions. They should be pointing upwards with both valves closed. Then refit the distributor as described in Chapter 4.

Chapter 1 Engine

35.1 Refitting the ancillary drive wheel and tightening the bolt

35.6 The timing belt refitted with the cogs and drive wheels all lined up correctly

35.10 The timing belt refitted as near the marks as we could get it the first time – but not yet right

35.12 With the camshaft drive wheel timing holes lined up with the lugs on the housings like this the engine then ran perfectly

35.15 Relocking the belt tensioner idler

6 Refit the pulley to the water pump and secure it with the three retaining bolts, as described in Chapter 2.
7 Refit the exhaust manifold as described in Chapter 3.
8 Refit the alternator. Secure the alternator to the oil filter housing with the long pivot bolt and nut and attach it to the adjusting bracket with bolt, washer and nut. The main pivot bolt fits from front to rear.
9 Refit the camshaft housing covers using new gaskets. Ensure that the cover with the oil filler aperture is on the exhaust camshaft housing. Before you secure each cover with the two knurled bolts prime the housing with engine oil as the camshafts run effectively in an oil bath. Note that the front bolt on the inlet cover has a bracket fitted beneath it which carries the ignition cables.
10 Refit the top end rear water rails. Renew the O-ring seals where the top rail meets the cylinder head and where the rear rail meets the water pump (photo). Don't forget that the top rail front mounting also has a bolt which goes through the camshaft timing pointers bracket.
11 Secure both water rail rear end mounting brackets to the engine stabiliser bar mounting bracket by two bolts and nuts.
12 Refit the thermostat and by-pass hose to the top and rear water rail, as described in Chapter 2.
13 On later models with automatic chokes refit the choke thermostat housing water pipe to the front end of the rear water rail by the pump inlet, and reconnect the other water pipe from the choke thermostat housing to the inlet manifold.
14 Refit the oil level dispstick/pipe to the sump. A new rubber

36.10 Fit a new O-ring to the top water rail joint

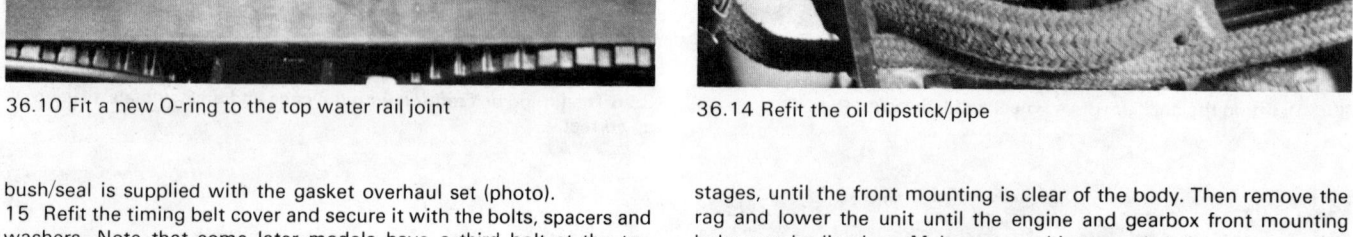

36.14 Refit the oil dipstick/pipe

bush/seal is supplied with the gasket overhaul set (photo).
15 Refit the timing belt cover and secure it with the bolts, spacers and washers. Note that some later models have a third bolt at the top which secures a bracket, which carries the water hose from the choke thermostat housing to the rear water rail.
16 Refit the alternator and water pump drivebelt. Check its condition and renew it if necessary. Then adjust the belt tension as described in Chapter 10.
17 Refit the driveshaft centre section and bearing housing to the crankcase as a complete unit, as it was removed. This of course only applies to later models with the three-piece driveshaft layout, which is covered in Chapter 8. Tighten the two nuts at the bottom and single bolt at the top to the specified torque.
18 Refit the gearbox to the engine as described in Chapter 6. This is a two man job as the gearbox/differential unit is quite heavy and awkward and the two assemblies need to be lined up first. Don't forget to refit the crankcase end plate first. Remember that as the two units are mated the primary shaft has to engage in the splines in the clutch disc at the same time as the driveshaft centre section (where fitted) splined end has to engage in the engine side differential unit driveshaft.
19 Secure the bellhousing to the crankcase and refit the starter motor. Don't forget the clutch return spring bracket.
20 The engine/gearbox assembly is now ready for refitting to the car.
21 Make one final check to ensure that all the components are fitted correctly in the right positions and that all mounting bolts are tight.

37 Engine/gearbox assembly – refitting

1 Refitting of the engine and gearbox is the reverse of the removal sequence. Make sure that all loose leads, hoses or pipes are tucked out of the way where they won't get caught up as the engine is lowered into position.
2 Ensure that the chain between the lifting tackle or crane and the engine has enough length so that the assembly can be twisted and tilted as it is refitted.
3 Fasten the lifting chains in the same positions that were used for removing the unit.
4 Position the lifting gear above the engine bay and slowly lower the unit. The engine mounting will easily foul the right-hand suspension housing so cover the engine mounting with a thick rag to prevent any damage to the bodywork. This will in fact also help the engine slide into position.
5 An assistant needs to work first of all at the gearbox end and needs to push the end down to guide it between the battery tray mounting arms. Then work on the right to keep the engine mounting etc free of the body work.
6 The engine/gearbox unit needs to be lowered very carefully in stages, until the front mounting is clear of the body. Then remove the rag and lower the unit until the engine and gearbox front mounting holes can be lined up. Make sure at this stage that the rear mounting bracket is not trapped underneath the differential unit. It may be necessary to raise the engine slightly and lever it slightly to line up the rear mounting. Refit the mounting bolts as the holes line up. Then lower the engine down and remove the lifting tackle.
7 When the engine has been refitted check that all the items removed or disconnected at removal have been refitted and all connections made.
8 Check that the drain tap is closed, and drain plugs tightened. Fill the cooling system with coolant and the engine oil level and fill as necessary. Make sure that all tools, rags, etc have been removed from the engine compartment.
9 With the help of an assistant refit the bonnet. Line up the marks previously scribed or refit the pins and circlips.

38 Engine – initial start-up after overhaul or major repair

1 Make sure that the battery is fully charged and that all lubricants, coolant and fuel are replenished. Check that the camshafts have been primed with oil.
2 If the fuel system has been dismantled it will require several revolutions of the engine on the starter motor to pump the petrol up to the carburettor. An initial prime of about 1/3 of a cup full of petrol poured down the air intake of the carburettor will help the engine to fire quickly, thus relieving the load on the battery. Do not overdo this, however, as flooding or fire may result.
3 As soon as the engine fires and runs keep it going at a fast tick over only, (no faster) and bring it up to the normal working temperature.
4 As the engine warms up there will be odd smells and some smoke from parts getting hot and burning off oil deposits. The signs to look for are leaks of water or oil which will be obvious if serious. Check also the exhaust pipe and manifold connections, as these do not always 'find' their exact gas tight position until the warmth and vibration have acted on them, and it is almost certain that they will need tightening further. This should be done, of course, with the engine stopped.
5 When normal running temperature has been reached adjust the engine idling speed as described in Chapter 3.
6 Stop the engine and wait a few minutes to see if any lubricant or coolant is dripping out when the engine is stationary.
7 Road test the car to check that the timing is correct and that the engine is giving the necessary smoothness and power. Do not race the engine – if new bearings and/or pistons have been fitted it should be treated as a new engine and run in at a reduced speed for the first 500 miles (800 km).
8 If the engine runs roughly check that the timing belt is fitted correctly.

39 Cylinder head, camshafts and housings – removal and refitting (in situ)

1 Disconnect the battery cable, and remove the right-hand body bracer.
2 Drain the cooling system, as described in Chapter 2.
3 Undo the front exhaust pipe to manifold bolts and remove them, disconnect the front stay and remove the gasket.
4 Remove the complete air cleaner assembly as described in Chapter 3.
5 Disconnect the choke control cable if fitted and remove the fuel feed and return pipes from the carburettor.
6 Disconnect the throttle link arm from the carburettor and the linkage bracket from the camshaft housing.
7 Disconnect the inlet manifold to servo pipe.
8 Remove the distributor cap and detach the plug leads and tuck them away in the right-hand inner wing.
9 Disconnect the thermal sender wires from the units in the cylinder head, and the carburettor solenoid wire at the connector.
10 Remove the engine splash guard in order to remove next the timing belt cover. Release the automatic choke thermostat water feed pipe from the top of the cover first, if it is a later model car. Also remove the alternator drivebelt.
11 Turn the engine over with a spanner on the crankshaft pulley nut to set the engine at TDC on No 4 cylinder (ie line up the holes in the camshaft drive wheels with the lugs on the camshaft housings, (refer to Section 33).
12 Remove the top water rail.
13 Undo the bolts and remove the engine main stabiliser rod.
14 Remove the lower water rail, and detach the choke thermostat hose from the thermostat (if fitted).
15 Remove the alternator adjusting bolt and nut.
16 Release the timing belt tensioner and remove the timing belt from the camshaft drive wheels.
17 Remove the camshaft housing covers. Check that the settings have not changed. The camshaft and housings can be removed separately from the head at this stage although it is not easy to remove the exhaust housing because of its location. This would apply if only the camshafts and/or tappets needed renewal.
18 To remove the cylinder head unto the mounting bolts in the correct order and lift the complete assembly off the cylinder block with inlet and exhaust manifolds still attached.
19 Overhauling the cylinder head assembly is covered in Section 11.
20 Refitting the assembly is the reverse procedure. Always use a new cylinder head gasket. Refer to the appropriate Sections for setting up the camshafts and checking the clearances and refitting the timing belt.

40 Oil sump – removal and refitting (in situ)

1 Drain the cooling system as described in Chapter 2.
2 Drain the engine oil by removing the sump plug. Place a container underneath it. Then remove the air cleaner assembly.
3 Undo the mounting nuts and bolts and remove the front exhaust pipe stay.
4 Undo the inner CV joint on the right-hand driveshaft and move the driveshaft aside, as described in Chapter 8.
5 Undo the exhaust pipe to manifold nuts and remove them. Lower the pipe.
6 Undo the main engine mounting bolts and nuts and remove the bolts. Then undo the nut on the front gearbox mounting bolt in the underside of the subframe.
7 Disconnect the front gear linkage control shaft from the connector rod and idler arm as described in Chapter 6.
8 Remove the battery and tray.
9 Disconnect the clutch cable from the operating arm and undo the cable bracket mounting nuts and bolts and place it out of the way.
10 Disconnect the radiator top hose, and the bottom hose.
11 Undo the mounting bolts and remove the engine stabiliser bar.
12 Disconnect the heater hoses from the rear water rail.
13 Remove the flywheel lower cover plate.
14 Attach lifting chains to the engine lifting eyes as for removing the engine from the car. Position the hoist and lift the engine up slowly in order to be able to reach all the sump mounting bolts. Check as the engine is raised that nothing catches or snags.

Fig. 1.20 Removal of the sump in situ (Sec 40)

1 Engine mounting bolts
2 Sump
3 Gearbox front mounting bolt and nut
4 Front exhaust pipe stay
5 Right-hand inner CV joint (early model)
6 Gear linkage idler arm
7 Connector rod
8 Exhaust downpipe

15 Undo the sump bolts and prise the sump away from the crankcase.
16 Refit the sump having cleaned off both mating faces. Grease the new gasket and fit it onto the sump. Offer the sump and gasket up to the crankcase and secure it with the retaining bolts. Fit all the bolts finger tight to start with and then tighten them evenly to the correct torque (including the drain plug).
17 Lower the engine. Refit the mounting bolts and nuts and then refit the rest of the components in the reverse order. Tighten all bolts and nuts to their specified torques.

41 Big-end bearings, pistons, connecting rods and oil pump – removal and refitting in situ

1 With the sump removed as described in the previous Section the oil pump can be removed as can the big-end bearing shells.
2 With the sump removed and the cylinder head removed as described in the two previous sections the pistons and connecting rod assemblies can be removed.
3 Refer to the appropriate sections of the engine dismantling procedure for the detailed instructions of the various operations.
4 Refitting is the reverse procedure to removal in all cases.

42 Main engine mounting – renewal

1 Jack the car up and support it on stands.
2 Undo the engine mounting bolts and nuts and remove them.
3 Remove the front exhaust pipe stay.
4 Disconnect the inner CV joint from the drive flange on the right-hand side of the differential unit on early cars.
5 Place a jack under the engine with a block of wood interposed between the engine and the jack.
6 Raise the car gently and slowly. It only needs to be raised enough to remove the engine mounting from the subframe. If you are lucky there will be enough flexibility in the exhaust pipe and it will not be necessary to undo the exhaust pipe to manifold joint. This is not always the case, especially on the early models.
7 Refit the new mounting to the subframe, lower the jack and refit the engine mounting bolts. Refit the nuts and tighten them to the specified torque.
8 Refit the rest of the components in the reverse order to removal.

43 Engine shock absorbers – renewal

1 There are two shock absorbers attached to the engine and gearbox assemblies. One is mounted between the right-hand side of the body and the engine main mounting bracket. The other is mounted between the left-hand side of the body and the rear end of the gearbox. Both units are identical (photo).
2 To remove the right-hand unit first remove the right-hand roadwheel having jacked the car up at the front.
3 Undo the necessary attachments, remove the alternator drivebelt and the timing belt cover.
4 This should create enough space to reach the upper and lower mounting bolts for the shock absorber. Remove the bolts and lift the unit away.
5 Fit a new unit in the reverse order. Make sure that all the components, hoses, etc, removed or detached are properly fitted. Don't forget to tension the alternator drivebelt.
6 In order to renew the left-hand engine shock absorber it will be necessary to remove the left hand front roadwheel having jacked the car up.
7 Then remove the battery and battery tray to reach the upper mounting bolt and nut in the body section.
8 Undo the nut and bolt at the bottom mounting point on the gearbox and remove the unit (photo 5.20).
9 Fit the new unit in the reverse order and tighten all the mounting bolts to that specified torques.

44 Notes for cars fitted with air conditioning systems

These notes should be used in conjunction with the appropriate Section for the part of the car on which you are working.

Let's start by saying that the basic engine bay is overfull with components anyway and no task is all that simple, so with the addition of the air conditioning system, which is described in some detail in Chapter 12, the problem grows more acute.

Basically to carry out any major task requires the removal of the compressor or its drivebelt. This can be done without detaching the refrigerant pipes.

In major engine operations the condenser radiator will also have to be removed, and the system evacuated and later recharged.

If jobs such as cylinder head or sump removal are carried out with the engine in the car, the compressor and/or drivebelt will have to be removed to allow the space required either to get at the components or jack the engine up to remove the sump. The same is true for setting the engine time as renewing the timing belt.

In the case of removing and refitting the sump in situ it is unlikely that this can be done without removing the refrigeration pipes from the condenser.

If the system has to be disconnected, always take the car to a refrigeration engineer. *It is dangerous to disconnect the system without specialist equipment.*

45 Notes for USA models

The USA and California version of the Lancia Beta contain much more additional equipment than those cars supplied for the UK or European markets. This includes the anti-pollution components in addition to the air conditioning system, all of which add to the already crowded engine bay.

Therefore when it comes to carrying out any major work on the engine there is always extra work to be done before any major dismantling or removal of major assemblies can take place.

The anti-pollution equipment is covered in Chapters 3 and 4, but these notes are intended to be read in conjunction with the procedures given in the rest of the Sections of this Chapter.

The North American models have a different distributor arrangement, which is covered in greater detail in Chapter 4. This means that the engine removal and dismantling procedures will be slightly different, because the unit is mounted on the exhaust valve camshaft housing, although the setting up of the engine to remove and refit it is basically the same.

The fuel pump is of the electro mechanical type and is fitted to the right-hand inner wing not to the engine (Fig. 1.21).

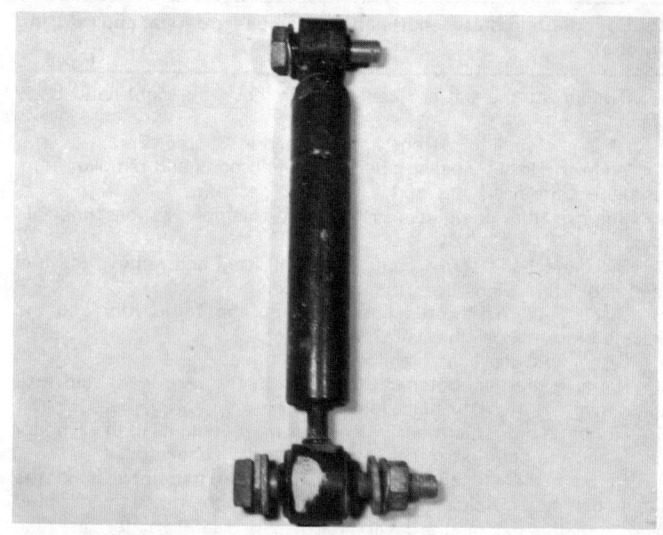

43.1 Engine shock absorber

The petrol fume venting system requires more pipes from and to the carburettor.

The air pump fro the EGR system is mounted on the left-hand end of the engine and driven by a drivebelt off a pulley driven by an extension of the inlet camshaft.

The oil filter housing and filter are mounted differently. The filter is mounted horizontally and not vertically and the housing is set more in the centre of the front of the engine beneath the PCV valve.

The additional work required for the major operations are listed below in sub-sections:-

Cylinder head removal and refitting – in situ

In addition to the procedures detailed in Section 39 of this Chapter the following work will also need to be carried out.

Detach the following from the inlet manifold (refer to Fig. 1.22)

(a) Diverter valve vacuum pipe
(b) Diverter valve solenoid switch wiring
(c) CB points change over switch wiring
(d) Fast idle device vacuum pipe
(e) Petrol fume separation and feed pipes to the carburettor

Remove the exhaust gas recirculation system air feed pipe to the cylinder head (Fig. 1.23).

Remove the exhaust duct piping from the cylinder head (Fig. 1.24).

Remove the distributor from the rear camshaft housing, as described in Chapter 4.

Remove the air pump drivebelt.

Remove the air pump pipe bracket from the stabiliser bar mounting bolt.

Undo the front single exhaust pipe at the clamp below the exhaust gas separator which is integral with the exhaust manifold (Fig. 1.25)

Engine and gearbox assembly – removal and refitting

Carry out all the operations described in Section 5 and also those described in the first two paragraphs of the previous sub-section. In addition the following extra tasks will have to be carried out before the engine and gearbox assembly can be removed from the car.

Remove the air conditioner compressor and drivebelt (see Chapter 12).

Remove the condenser for the air conditioning unit.

Remove the oil filter and housing (Fig. 1.22).

Disconnect the pipe from the air pump to the diverter valve, and the pipe from the diverter valve to cylinder head (air feed pipe).

Remove the air pump drivebelt.

Remove the air pump and its mounting bracket.

Disconnect all the wires from the gearbox switches and note which fits where. Draw a diagram to help.

Remove the distributor as described in Chapter 4.

Note that the method of attaching the rear lifting chain will need

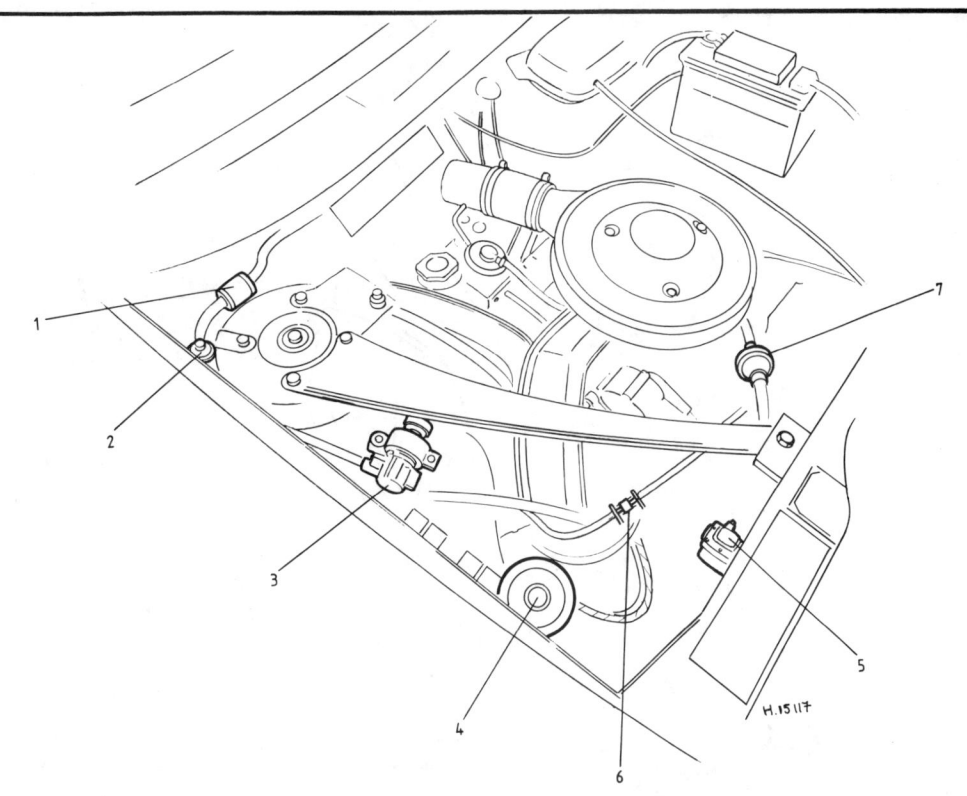

Fig. 1.21 USA version fuel pump and filter location (Sec 45)

1 Fuel filter
2 One-way fuel stop valve (if fitted)
3 Fuel pump
4 Power steering reservoir
5 Fast idling electrovalve
6 One-way valve
7 Fuel pressure regulator

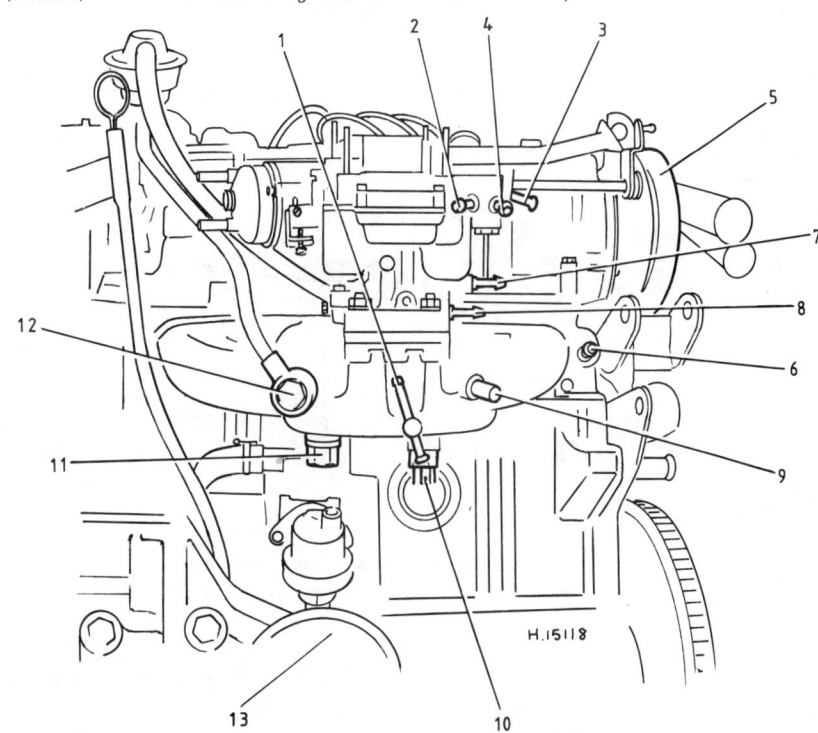

Fig. 1.22 Inlet manifold and carburettor component fittings (Sec 45)

1 Brake servo and fast idle device solenoid valve vacuum union
2 Fuel filler
3 Petrol fumes filter line union
4 Fuel recovery line outlet
5 Air pump drive pulley
6 Semi-automatic choke line union
7 Union for air cleaner
8 Union for petrol fumes filter
9 Diverter valve vacuum union
10 Diverter valve control solenoid valve temperature switch
11 Breakers change-over temperature switch
12 Exhaust gas recirculation front pipe mounting plug
13 Oil filter and housing

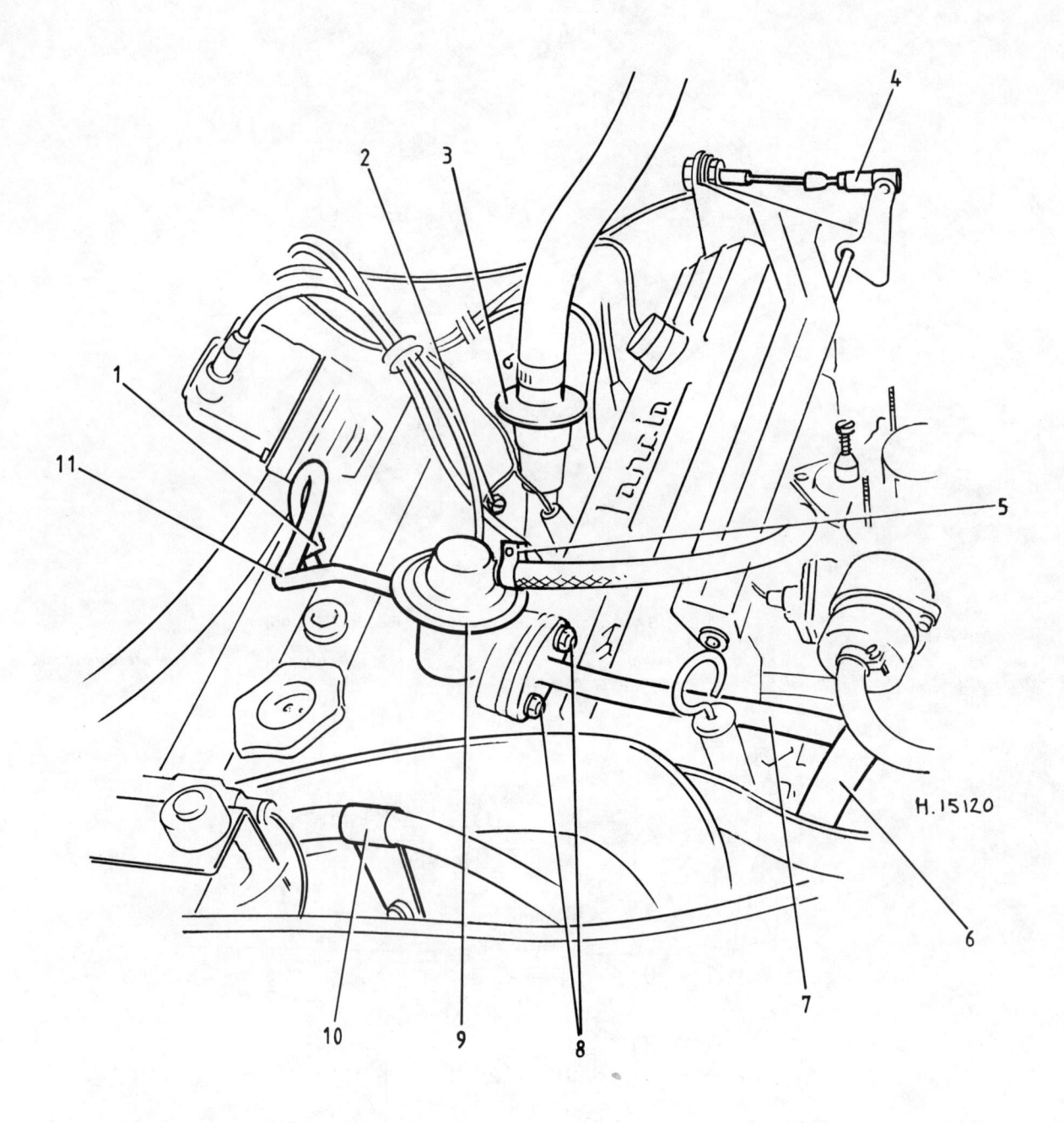

Fig. 1.23 EGR system pipework (Sec 45)

1. Exhaust gas recirculation lower pipe mounting union
2. Exhaust ducts air delivery pipe mounting screw
3. Exhaust ducts air delivery pipe and clamp
4. Accelerator control cable
5. Clamp and pipe between recirculation valve and thermostat valve
6. Semi-automatic choke water outlet hose
7. Exhaust gas recirculation upper pipe
8. Pipe and valves mounting bolts
9. Exhaust gas recirculation valve
10. Hose (6) mounting bracket
11. Exhaust gas recirculation intermediate pipe

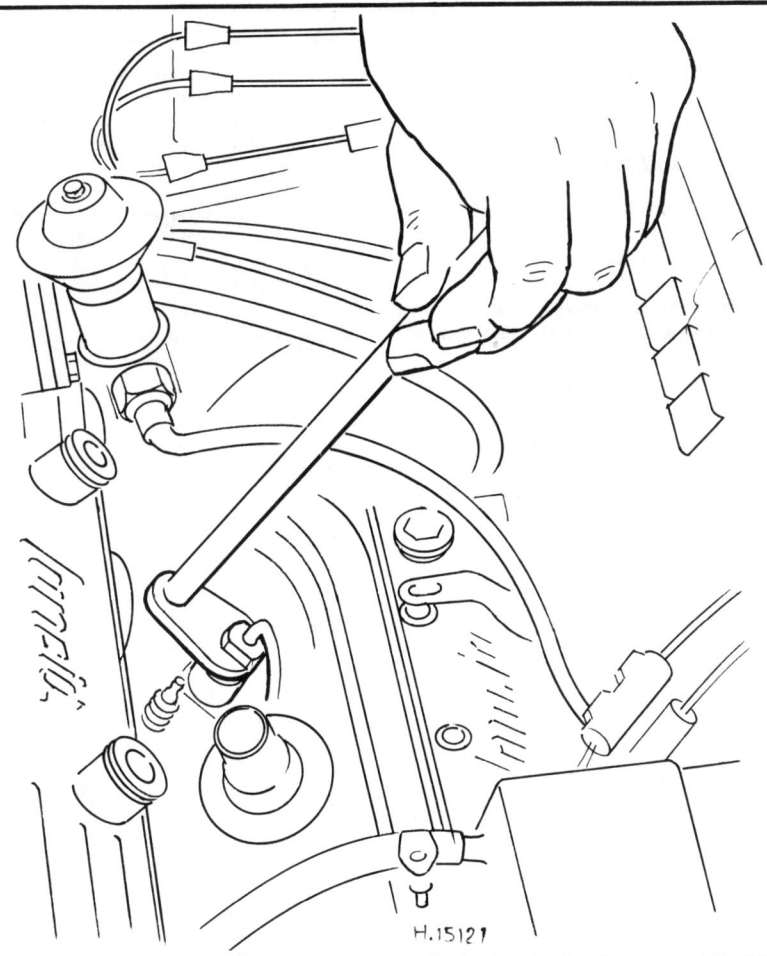

Fig. 1.24 Undoing the exhaust piping delivery duct union on the cylinder head using Lancia tool No 88011461 (Sec 45)

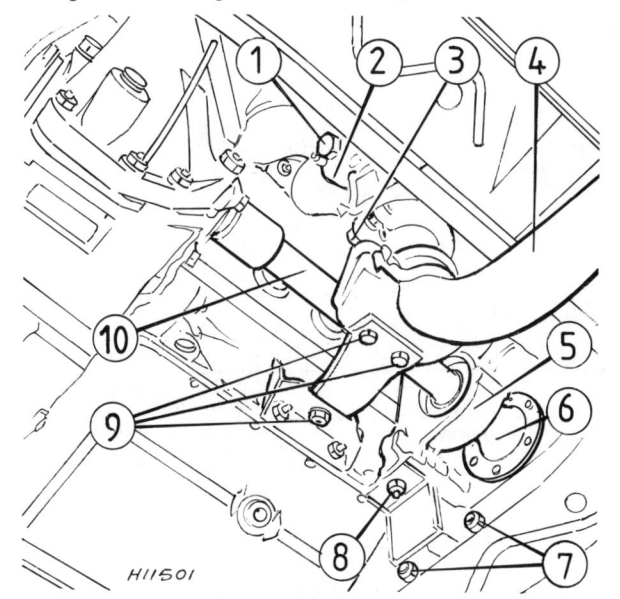

Fig. 1.25 The exhaust pipe to exhaust gas separator connections (Sec 45)

1. Union to secure exhaust gas recirculation lower pipe
2. Exhaust gas separator body
3. Exhaust manifold and pipe upper mounting bracket bolts
4. Front exhaust pipe
5. Right-hand driveshaft extension support
6. Drive flange
7. Buffer mounting bolts
8. Support-to-buffer fixing nut
9. Exhaust pipe lower bracket mounting bolt and screws
10. Right-hand driveshaft extension

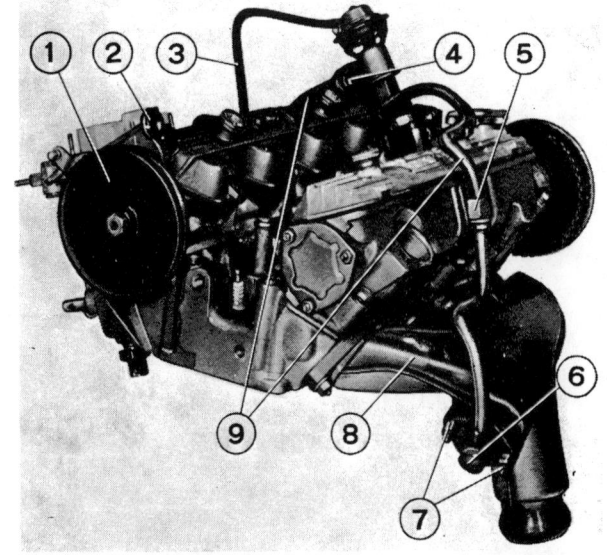

Fig. 1.26 The cylinder head assembly (USA) (Sec 45)

1. Air pump pulley
2. Accelerator control/support bracket
3. Pipe between recirculation valve and thermostatic valve
4. Intermediate pipe and exhaust gas recycling valve attachment bolts
5. Union
6. Connector
7. Exhaust gas separator/body attachment nuts
8. Exhaust manifold
9. Exhaust recycling pipe

Chapter 1 Engine

to be modified. Lancia use a special hook under the right-hand end of the exhaust manifold with an eye at the top.

Cylinder head assembly – overhaul procedure

In addition to the procedure detailed in Section 11 of this Chapter, there are major differences between the two assemblies. The inlet camshaft is extended through the rear cover plate via an oil seal in a housing and a pulley is attached to the end of the camshaft. This drives the air pump via a V-drivebelt. The right-hand (rear/exhaust) camshaft drives the distributor via a spur gear and the distributor is mounted into the rear camshaft housing. Also there is the EGR system piping and exhaust gas separator (Fig. 1.26).

With the assembly removed from the cylinder block, begin by removing the air pump drive pulley from the rear end of the inlet camshaft. Lock up the camshaft front drive wheels so that the nut can be undone.

Remove the EGR system pipes and valve.

Undo the retaining nut and remove the throttle linkage bracket from the inlet camshaft housing.

When removing the inlet camshaft note that there is a seal and housing in the rear end as well as the front.

When checking the rear camshaft for wear check also the distributor drivegear teeth for wear and chipping.

46 Fault diagnosis – engine

Symptom	Reason(s)
Engine will not turn over when starter switch is operated	Flat battery Bad battery connections Bad connections at solenoid switch and/or starter motor Starter motor jammed Defective solenoid Starter motor defective
Engines turns over normally but fails to fire and run	No spark at plugs No fuel reaching engine Too much fuel reaching the engine (flooding)
Engine starts but runs unevenly and misfires	Ignition and/or fuel system faults Incorrect valve clearances Burnt out valves
Lack of power	Ignition and/or fuel system faults Incorrect valve clearances Burnt out valves Worn out piston or cylinder bores
Excessive oil consumption	Oil leaks from crankshaft oil seal, camshaft cover gasket, camshaft seals or housing joints Worn piston rings or cylinder bores resulting in oil being burnt by engine Smoky exhaust is an indication Worn valve guides and/or defective valve stem seals
Excessive mechanical noise from engine	Wrong valve clearances Worn crankshaft bearings Worn cylinders (piston slap)
Unusual vibration	Misfiring on one or more cylinders Loose mounting bolts

NOTE: When investigating starting and uneven running faults do not be tempted into snap diagnosis. Start from the beginning of the check procedure and follow it through. It will take less time in the long run. Poor performance from an engine in terms of power and economy is not normally diagnosed quickly. In any event the ignition and fuel system must be checked first before assuming any further investigation needs to be made.

Chapter 2 Cooling system

Contents

Cooling system – draining 2	Expansion tank – general 9
Cooling system – filling 4	Fault diagnosis – cooling system 15
Cooling system – flushing 3	General description 1
Drivebelt – adjustment 13	Radiator – inspection and cleaning 8
Drivebelt – removal and refitting 12	Radiator – removal and refitting 5
Electric fan operation – checking, testing and renewal of the thermal switch 7	Temperature transducers, switches and gauges 14
	Thermostat – removal, testing and refitting 10
Electric fan and motor – removal and refitting 6	Water pump – removal and refitting 11

Specifications

Type .... Pressurized, assisted by pump and fan

Capacity (including heater) .... 1.8 gals (8.30 litres/2.20 US gals)

Thermostat
Type .... Behr-Thomson (double valve type)
Opening temperature .... 78° to 82°C (172° to 179°F)
Fully open .... 95°C (203°F)
Travel when open .... 7 mm (0.275 ins)

Electric coolant fan
Fan cut-in temperature .... 90° to 94°C (194° to 201°F)
Fan cut-out temperature .... 85° to 89°C (185° to 192°F)
Engine overheating warning dashboard light comes on at .... 112° to 118°C (233° to 244°F)

Water pump type .... Impeller type – mounted on right-hand side of engine and belt driven off the crankshaft pulley

Torque wrench settings

	lbf ft	Nm
Transmitters in cylinder head	36	49
Water pump pulley bolts	88	119
Water pump retaining bolts	17	23
Cylinder head water elbow bolts	8	11

1 General description

The engine cooling water is circulated by a thermo/syphon, water pump assisted system, and the coolant is pressurised. This is primarily to prevent premature boiling in adverse conditions and to allow the engine to operate at its most efficient running temperature; this being just under the boiling point of water. The overflow pipe from the radiator is connected to an expansion tank which makes topping-up virtually unnecessary. The coolant expands when hot, and instead of being forced down an overflow pipe and lost, it flows into the expansion tank. As the engine cools the coolant contracts and because of the pressure differential flows back into the radiator.

The cap on the expansion tank is set to a pressure of 15 lbf/in^2 (1.05 kgf/cm^2) which increases the boiling point of the coolant to 230°F (110°C). If the water temperature exceeds this figure and the water boils, the pressure in the system forces the internal valve of the cap off its seat thus exposing the expansion tank overflow pipe down which the steam from the boiling water escapes and so relieves the pressure. It is therefore important to check that the expansion tank cap is in good condition and that the spring behind the sealing washers has not weakened. Check that the rubber seal has not perished and its seating in the neck is clean to ensure a good seal. A special tool which enables a cap to be pressure tested is available at some garages.

The system functions as follows: The water flow round the system is controlled by a special double valve type thermostat, located in the junction of the lower radiator, bypass and water pump hoses. This valve governs the flow of water from the cylinder head bypass hose and the lower radiator hose into the water pump. If the engine is cool the pump forces the water through the cylinder head and into the bypass hose to the thermostat. As the engine temperature reaches its proper running level, the thermostat bellows move to close the bypass hose and open the lower radiator hose. The hot water from the cylinder head will then flow down through the radiator up through the lower radiator hose and the thermostat and upwards to the pump.

The method of supplying hot water to the car interior heating system is common to all engines; a union in the left-hand end of the rear water rail enables hot water to flow to the heater and it is returned to the cooling system at the water pump inlet, again on the rear water rail.

2 Cooling system – draining

1 With the cooling system cold unscrew and remove the radiator drain plug. Do not remove the plug whilst the engine is hot (photo).
2 If antifreeze is being used in the cooling system it should be collected in a bowl located under the bottom hose. Do not carry out the instructions in paragraph 5.
3 Turn the car heater control to HOT to ensure that the heater system will drain and fill without air locks forming.
4 Release the expansion tank pressure cap.
5 Undo and remove the cylinder block drain plug. This is located to the rear of the right-hand side of the cylinder block, and is rather difficult to reach.
6 If no radiator drain plug is fitted, slacken the hose clip and carefully ease the bottom hose off at its connection on the radiator.
7 When the coolant has finished running out of the cylinder block drain hole, probe the orifice with a short piece of wire to dislodge any particles of rust or sediment which may be causing a blockage preventing complete draining.

3 Cooling system – flushing

1 Generally even with proper use, the cooling system will gradually lose its efficiency as the radiator becomes choked with rust scale, deposits from the water and other sediment. To clean the system out, remove the radiator drain plug, cylinder block plug and bottom hose and leave a hose running in the expansion tank filler hole for fifteen minutes.
2 Reconnect the bottom hose, refit the cylinder block plug and refill the cooling system as described in Section 4, adding a proprietary cleaning compound. Run the engine for fifteen minutes. All sediment and sludge should now have been loosened and may be removed by then draining the system and refilling again.
3 In very bad cases the radiator should be reverse flushed. To do this, remove the radiator, invert it, and insert a hose in the bottom outlet until the water runs clear.

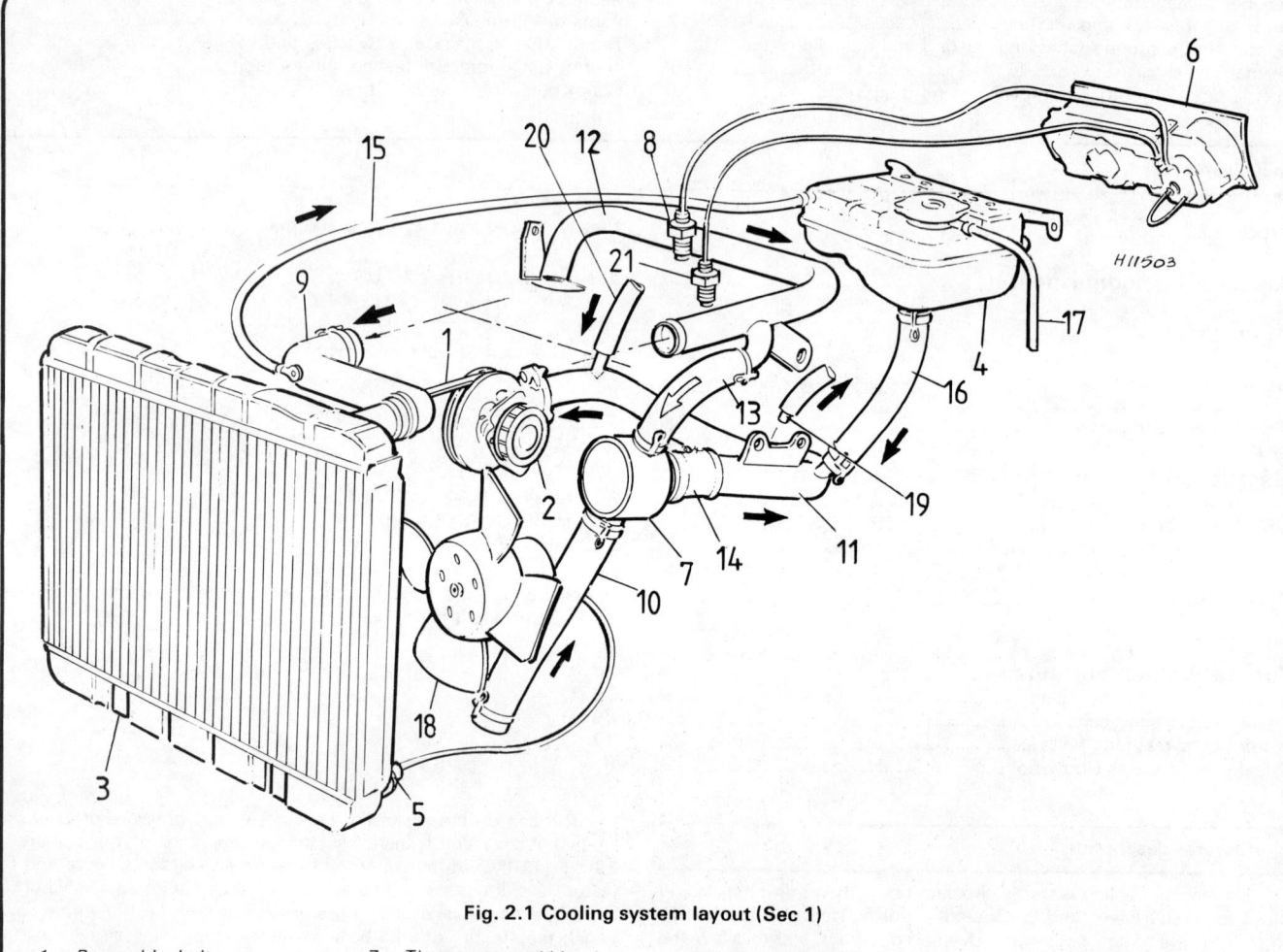

Fig. 2.1 Cooling system layout (Sec 1)

1 Pump drivebelt
2 Coolant pump
3 Radiator
4 Expansion tank
5 Electric fan thermal switch
6 Gauges
7 Thermostat and blender
8 Temperature gauge transmitter
9 Top hose
10 Bottom hose
11 Rear water rail
12 Top water rail
13 Bypass hose
14 Thermostat to water rail hose
15 Expansion pipe
16 Expansion tank return pipe
17 Overflow pipe
18 Electrically driven fan
19 Heater hose feed
20 Heater hose return
21 Engine overheating transmitter

Chapter 2 Cooling system

2.1 The radiator drain plug is fitted in the bottom right-hand side of the radiator and can be reached from underneath

4 Cooling system – filling

1 Fit the cylinder block drain plug and if the bottom hose has been removed it should be reconnected.
2 Fill the system slowly to ensure that no air locks develop, Check that the valve to the heater unit is open, otherwise an air lock may form in the heater. The best type of water to use in the cooling system is rain water, mixed 50/50 with a proprietary brand of antifreeze with a glycol base.
3 Fill up the cooling system through the expansion tank to the level indicated, just below the centre join. Refit the expansion tank cap.
4 Start the engine and run at a fast idle speed for 30 seconds.
5 Stop the engine and top up the system through the expansion tank cap.
6 Run the engine until it has reached its normal operating temperature. Stop the engine and allow to cool.
7 Top up the expansion tank to the level marked.
8 The system should be drained and refilled at least every two years. The system needs to have a 50/50 water and antifreeze mixture or it must have a rust inhibitor added.

5 Radiator – removal and refitting

1 Drain the cooling system.
2 Slacken the clip securing the expansion tank hose to the radiator. Carefully ease the hose from the union pipe on the radiator (photo). Remove the air cleaner assembly on cars fitted with air conditioning
3 Slacken the clips securing the radiator top and bottom hoses to the radiator pipes and carefully ease the two hoses from them.

4 Undo and remove the nuts and bolts which secure the main engine stabiliser bar to the engine bracket and the front body section. Lift the stabiliser bar out (photo).
5 Undo the single bolt which secures the top radiator mounting bracket.
6 Disconnect the three spade connectors from the thermal switch on the bottom of the left-hand side of the radiator. Disconnect the battery negative terminal.
7 Disconnect the electric fan motor wiring as the radiator and fan assembly is lifted out. There are two wires. The blue is connected to the red one at the battery and the black one is connected to the blue one.
8 With the assembly removed the fan and drive motor which is mounted on a bracket can be removed from the radiator if required. It is mounted by three bolts and is removed as described in Section 6.
9 Refitting is the reverse sequence to removal. Check before lowering the radiator into position that the rubber mountings at the bottom are in place and in good condition (photo).
10 Refill the cooling system as described in section 4.

6 Electric fan and motor – removal and refitting

1 Remove the radiator and fan assembly complete, as described in Section 5.
2 Undo the three bolts which secure the fan motor and fan mounting bracket to the radiator frame.
3 Remove the assembly from the radiator (photo).
4 To remove the fan from the mounting bracket undo the 3 nuts and remove the spring washers, bushes and spacers. Withdraw the fan and motor unit from the bracket.
5 To remove the fan from the motor hold the fan blades and undo the nut on the front end of the shaft. Then remove the washer and tap the fan off the motor shaft. Do not lose the key from the key way.
6 Refit the components in the reverse order to removal and then check that the fan operates correctly.

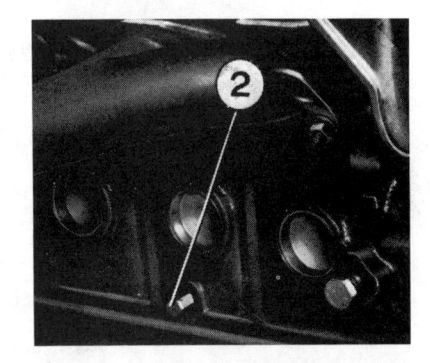

Fig. 2.2 The cylinder block drain plug (Sec 2)

2 Drain plug

5.2 The top hose and expansion tank hose are next to each other

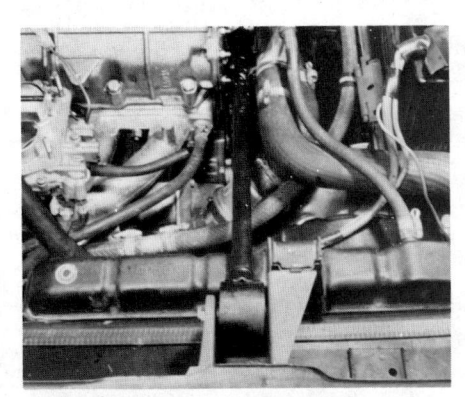

5.4 Remove the main engine stabiliser bar and radiator top clamp next to it

5.9 Replacing the radiator left-hand rubber mounting insert

Chapter 2 Cooling system

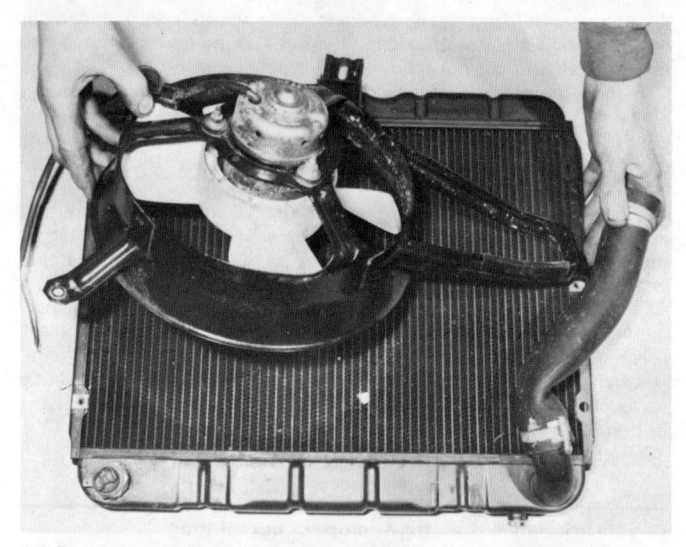

6.3 Removing the fan and motor assembly on the mounting bracket from the radiator

7.5 The thermal switch for the electric fan is located in the bottom left-hand corner of the radiator

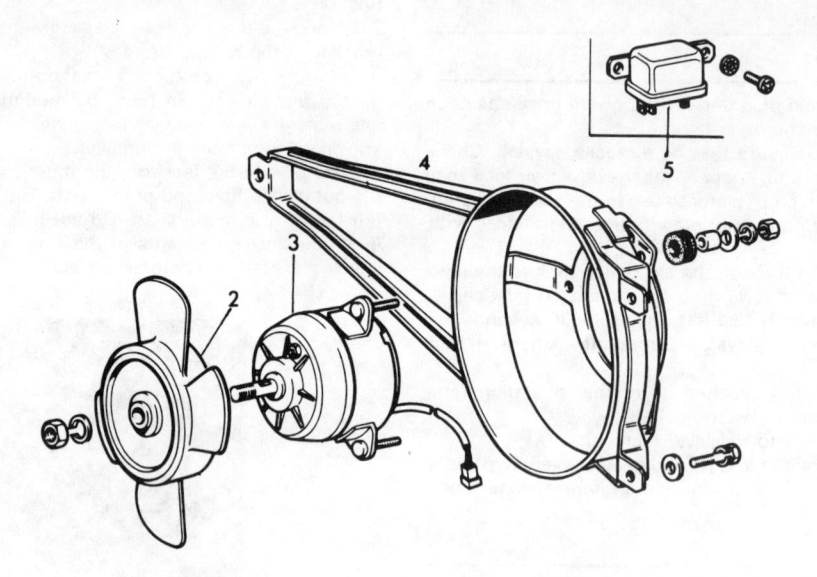

Fig. 2.3 Electric fan and motor assembly layout (Sec 6)

2 Plastic fan
3 Motor
4 Mounting bracket
5 Fan relay

7 Electric fan operation – checking, testing and renewal of the thermal switch

1 If the fan is suspected of incorrect operation then first of all check the main fuse in the fuse box.
2 Assuming that the fuse is all right and the fan still does not work, then disconnect the wires from the thermal switch in the bottom of the radiator and connect them together. Make sure before doing so that the ignition switch is turned off.
3 Switch the ignition on and the fan should now operate. If it works then the thermal switch is faulty and must be renewed. If it does not work then either the electric fan motor or the solenoid, mounted on the left-hand wing, is faulty. To check where the fault lies reconnect the wires to the thermal switch and connect a spare wire direct from the positive terminal of the battery to the fan electrical supply, at the 16 amp fuse in the fusebox.
4 If the fan now works then the fault lies in the relay, which must be renewed. If however the fan still doesn't work then the fault lies in the fan motor, which will have to be taken out and renewed as described in Section 6.

5 To renew the thermal switch drain the cooling system as described in Section 2 and remove the wires from the thermal switch. Undo the switch using a suitably sized spanner (photo). Fit the new switch using a new washer. Refill the system and connect the wiring. Finally test the operation of the fan.

8 Radiator – inspection and cleaning

1 With the radiator out of the car, any leaks can be soldered up or repaired. Clean out the inside of the radiator by flushing as described in Section 3.
2 When the radiator is out of the car, it is advantageous to turn it upside down for reverse flushing. Clean the exterior of the radiator by hosing the radiator matrix with a strong jet of water to clean away road dirt, dead flies etc.
3 Inspect the radiator hoses for cracks, internal and external perishing and damage caused by overtightening of the hose clips. Renew the hoses as necessary. Examine the radiator hose clips and renew them if they are rusted or distorted. The drain plugs and

Chapter 2 Cooling system

washers should be renewed if leaking.
4   With the type of hose clips that are fitted to most of the hoses it is sometimes very difficult to get them to refit properly. Therefore it is advisable to renew them if they are at all doubtful.

9 Expansion tank – general

1   The coolant expansion tank is mounted on the left-hand side of the rear bulkhead of the engine compartment, and does not require any maintenance. It is important that the expansion tank filler cap is not removed whilst the engine is hot, or scalding may result.
2   Should it be found necessary to remove the expansion tank, partly drain the cooling system to empty the tank, then disconnect the radiator to expansion tank hose and the expansion tank to rear water rail hose at the tank (photo).
3   Remove the bracket screws and carefully lift away the tank.
4   Refitting is the reverse sequence to removal. Add a mixture of 50% water and 50% antifreeze solution until it is up to the level mark.
5   Then refit the pressure cap.

10 Thermostat – removal, testing and refitting

1   To remove the thermostat drain the cooling system as described in Section 2 of this Chapter.
2   With sufficient water drained from the cooling system, the clips securing the water hoses to the thermostat housing are loosened and the hoses eased off the housing (photo).
3   The thermostat can then be removed from the car (photo).
4   As can now be seen, the thermostat is unlike any normal type as it is a complete one-piece housing with a double valve system inside (Fig. 2.4).
5   Test the thermostat for correct functioning by suspending it together wth a thermometer on a string in a container of cold water. Heat the water and note the temperature at which the thermostat begins to open. The opening temperature should be as per the Specifications at the beginning of this Chapter, and it should be fully open also as specified.
6   Discard the thermostat if it opens too early, or does not close when the water cools down. If the thermostat is stuck open when cold it will be apparent when it is first exposed.
7   Refitting the thermostat is the reverse procedure to removal.

9.2 The expansion tank is mounted on the rear bulkhead

10.2 Removing the hoses from the thermostat housing

10.3 The thermostat removed

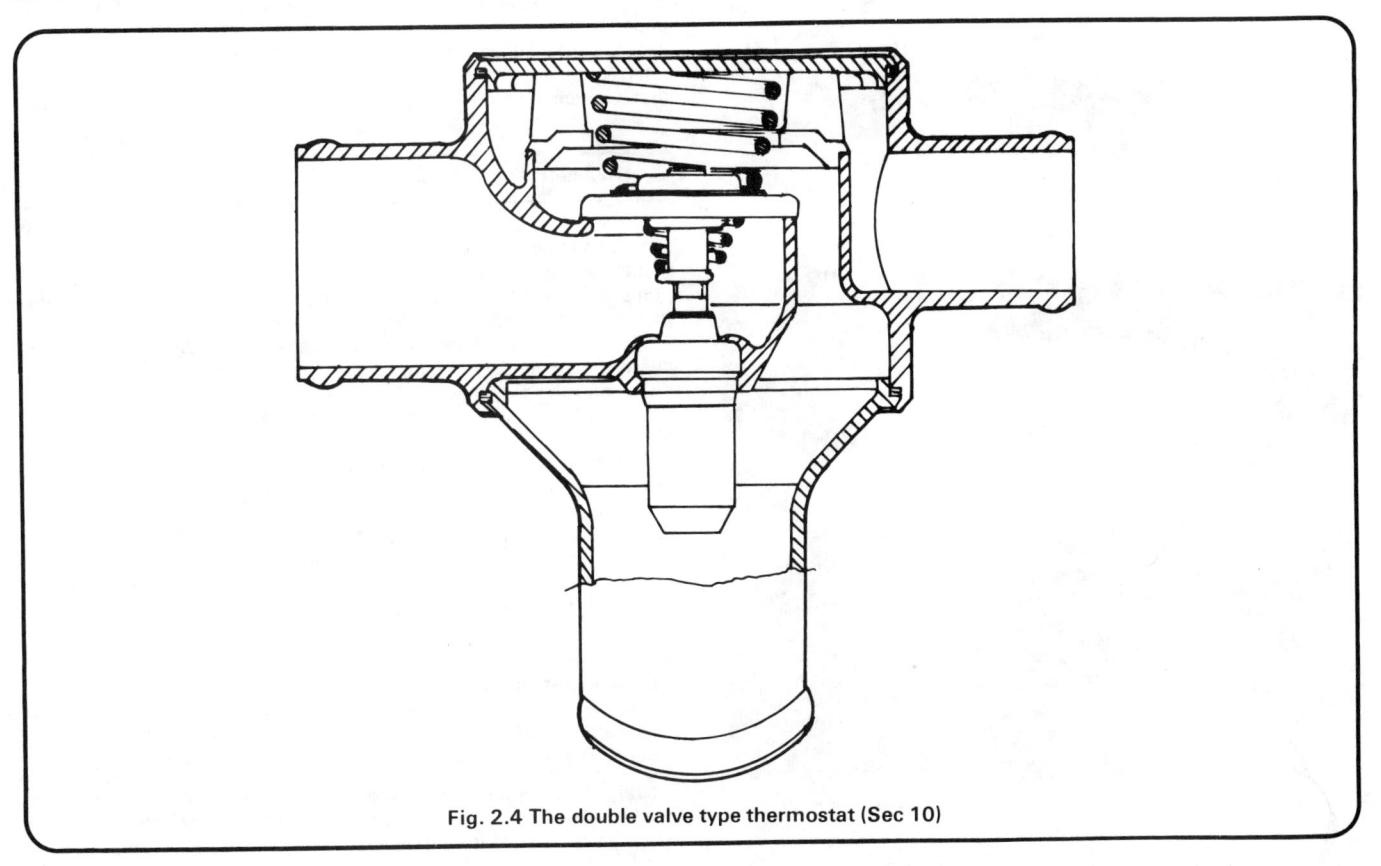

Fig. 2.4 The double valve type thermostat (Sec 10)

11.5 Lift the pump drive pulley away (engine out)

11.9 Fit a new gasket and offer up the water pump (engine out)

11.11 Always use a new O-ring when refitting the rear water rail to the pump

11 Water pump – removal and refitting

1   This operation is easy to carry out with the engine out of the car, but it can also be done (with a little difficulty) with the engine in situ.
2   Jack the front of the car up and support it on axle stands. Remove the right-hand front roadwheel and the engine splash guard.
3   Drain the cooling system as described in Section 2.
4   Slacken the alternator mounting bolt and remove the drivebelt from the pump pulley. It will be easier if the pulley mounting bolts have been slackened off with the belt still tight.
5   Undo the pump pulley bolts and remove the pulley (photo).
6   Undo the two nuts which secure the rear water rail to the pump inlet pipe, and slacken the rear water rail mounting bolts at the left-hand end of the engine so that the water rail can be separated from the pump inlet. Remove the timing belt cover.
7   Undo the four pump mounting bolts and remove the pump unit from the cylinder block.
8   If the pump is worn out then renew it. It is far easier and wiser in the long run than trying to strip and renew the bearings or impeller, if indeed they can be obtained. Most motor factors and garages will only carry a complete replacement unit.
9   Fit a new gasket to the cylinder block face using a smear of grease and offer up the pump (photo).
10 Refit the four bolts and tighten them up. Refit the timing belt cover.
11 Then refit the rear water rail, using a new O-ring which fits in the groove in the water rail flange (photo).
12 Refit the pump pulley, the alternator and the drivebelt.
13 Tighten the pulley mounting nuts and adjust the alternator so that the drivebelt tension is correct, as described in Section 13.
14 Refit the rest of the components and refill the cooling system as described in Section 4.

12 Drivebelt – removal and refitting

1   Loosen the alternator mounting and adjuster nuts and bolts.
2   Tip the alternator towards the engine and slip the belt from the pulleys.
3   Refitting of the drivebelt is the reverse of removal, but it is important to tension the belt correctly (see Section 13).
4   If a new drivebelt has been fitted then it will need to be retensioned after approximately 250 miles (400 km) owing to the fact that a certain amount of stretching occurs during the bedding-in stage.

13 Drivebelt – adjustment

It is important to keep the drivebelt correctly adjusted and it should be checked every 6000 miles (10 000 km) or 6 months. If the belt is loose it will slip, wear rapidly and cause the alternator and water pump to malfunction. If the belt is too tight the alternator and water pump bearings will wear rapidly and cause premature failure.

The drivebelt tension is correct when there is 0.5 in (13 mm) of lateral movement at the mid point position between the alternator pulley and the pump pulley, under a load of 11 lbf (5 kgf).

To adjust the belt, slacken the securing bolts and move the alternator in or out until the correct tension is obtained. It is easier if the alternator bolts are only slackened a little so it requires some effort to move the unit. In this way the tension of the belt can be arrived at more quickly than by making frequent adjustments. If difficulty is experienced in moving the unit away from the engine, a tyre lever placed behind the unit and resting against the block gives a good control so that it can be held in position whilst the securing bolts are tightened. Be careful of the alternator cover – it is fragile.

14 Temperature transducers, switches and gauges

1   There are several electrical systems associated with the cooling system. These are:

   (a) The thermal switch which operates the electric fan motor
   (b) The thermal sender unit which gives a continuous signal to the temperature gauge

Chapter 2 Cooling system

(c) The thermal sender unit which indicates high temperature and operates the temperature warning light

2   The sender units for the gauge and warning light are screwed into the top of the cylinder head – between the spark plugs. The left-hand sender unit operates the warning light.
3   The thermal switch that operates the fan drive is screwed into the bottom of the left-hand side of the radiator, and is covered in detail in Section 7.
4   If unsatisfactory gauge readings are being obtained or the temperature warning system is suspected, the respective thermal sender units may be tested by removing the cable connection on the appropriate unit and placing the metal cable end on a good earthing point, for example a cleaned point on the cylinder head. Disconnect the lead from the low oil pressure switch.

5   Switch on the ignition. If the temperature gauge sender unit is being tested and the gauge needle quickly moves to the HOT sector of the dial, a new temperature sender should be fitted. Alternatively should the temperature gauge not respond, check the continuity of the wire to the gauge, and finally if necessary check the gauge itself by substituting another.
6   The test procedure is similar for the warning system. Having directly bypassed the sender unit and the light glows immediately the ignition is switched on, the sender unit should be renewed. If the warning light fails to glow then the continuity of the wire and bulb should be checked.
7   Before removing any of the thermal devices from the engine or radiator drain the cooling system as described in Section 2.
8   Once new devices have been fitted the cooling system may be refilled as described in Section 4 of this Chapter.

15 Fault diagnosis – cooling system

Symptom	Reason(s)
Loss of coolant	Leak in system Defective pressure cap on expansion tank Overheating causing too rapid evaporation due to excessive pressure in system Blown cylinder head gasket causing excess pressure in cooling system forcing coolant out Cracked block or head due to freezing
Overheating	Insufficient coolant in system Water pump not turning properly due to slack drivebelt Kinked or collapsed water hoses causing restriction to circulation of coolant Faulty thermostat (not opening properly) Engine out of tune Blocked radiator either internally or externally Cylinder head gaskets blown forcing coolant out of system New engine not run-in
Engine running too cool	Thermostat jammed open Thermostat missing

Chapter 3 Fuel, emission control and exhaust systems

Contents

Air cleaner and element – removal and refitting	2
Automatic choke – removal and refitting	10
Automatic choke diaphragm control unit – removal and refitting	11
Automatic choke operation on Weber DATR and DATRA series carburettors – description	12
Carburettor – description and principles	5
Carburettor – overhaul	14
Carburettors – removal and refitting	6
Carburettors – setting and adjustment for slow running	7
Carburettors – setting the fast idling speed	8
Carburettor – Weber DATRA Series (USA) – setting the engine idling and fast idling and checking the CO content	9
Choke cable – renewal	4
Emission control systems (UK models) – general description and maintenance	20
Emission control systems (USA versions) – general description and maintenance	21
Exhaust manifold gasket – renewal	24
Exhaust system – general	22
Exhaust system – renewal	23
Fast idling device diaphragm – renewal	13
Fault diagnosis – fuel, carburation, and exhaust systems	25
Fuel pump – mechanical – removal and refitting	17
Fuel tank – removal and refitting	19
Fuel tank sender unit – testing, removal and refitting	18
General description	1
In-line fuel filter – removal and refitting	15
Inlet manifold – removal and refitting	16
Throttle cables – removal and refitting	3

Specifications

Fuel pump type .... Mechanical or electric depending on model

Fuel tank
Capacity (including reserve) .... 10.77 Imp gals (12.94 US gals,/49 litres)
Reserve capacity .... 1.76 Imp gals, (2.11 US gals, 8 litres)

Air cleaner .... Renewable paper element type mounted in canister on top of carburettor

Carburettor application (UK models)

	Weber	Solex
1300 models		
Up to engine No 7800	32 DMTR 31	C32 CIC-3
From engine No 7801	32 DATR 3/250	C32 CIC-3
1400 models	32 DMTR 24	C32 CIC-1
1600 models:		
Series A	34 DMTR 21	C34 CIC-1
Series B	34 DATR 1/200 or 34 DATR 2/250*	C34 TCIC-1
1800 models	34 DMTR 21	C34 CIC-1
2000 models	34 DATR 2/200 or 34 DATR 2/250**	C34 TCIC-1

Note: *The Weber 34 DATR 2.250* fitted to the 1600 'B' series is different in specifications from the Weber 34 DATR 2.250** fitted to the 2000 models*

Carburettor application (USA models)
Non-California versions:
   1975 models .... Weber 32 DATRA 3/100
   1976/1977/1978 models .... Weber 32 DATRA 13/100
California version
   1976 models .... Weber 32 DATRA 12/100
   1977/1978 models .... Weber 32 DATRA 12/101

Chapter 3 Fuel, emission control and exhaust systems

Weber carburettor data

	32 DMTR 31		32 DMTR 21	
	Primary	Secondary	Primary	Secondary
Choke/barrel size mm (in)	22 (0.867)	22 (0.867)	22 (0.867)	22 (0.867)
Venturi diameter mm (in)	4 (0.157)	4 (0.157)	4 (0.157)	4 (0.157)
Main jet size mm (in)	1.10 (0.0433)	1.10 (0.0433)	1.10 (0.0433)	1.15 (0.452)
Air corrector jet size mm (in)	1.90 (0.0748)	2.60 (0.102)	1.90 (0.0748)	2.40 (0.0944)
Emulsion tube type	F.30		F.30	
Idling jet size mm (in)	0.50 (0.0196)	0.70 (0.0275)	0.50 (0.0196)	0.70 (0.0275)
Idling air jet size mm (in)	1.10 (0.0433)	0.70 (0.0275)	1.10 (0.0433)	0.70 (0.0275)
Pump jet size mm (in)	0.45 (0.0175)	–	0.45 (0.0175)	–
Pump discharge jet size mm (in)	0.40 (0.0157)	–	0.40 (0.0157)	–
Pump diaphragm stroke mm (in)	3.5 (0.138)		3.5 (0.138)	
Super feed jet size mm (in)	1.25 (0.0492)	–	1.10 (0.0433)	–
Super feed mixture bore mm (in)	–	–	–	2.00 (0.0787)
Needle valve size mm (in)	1.50 (0.0590)		1.50 (0.0590)	

	34 DMTR 21		32 DATR 3/250	
	Primary	Secondary	Primary	Secondary
Choke/barrel size mm (in)	25 (0.985)	26 (1.023)	22 (0.867)	22 (0.867)
Venturi diameter mm (in)	4 (0.157)	4 (0.157)	4 (0.157)	4 (0.157)
Main jet size mm (in)	1.20 (0.0472)	1.50 (0.0590)	1.10 (0.0433)	1.15 (0.0452)
Air corrector jet size mm (in)	1.60 (0.0629)	2.40 (0.0629)	1.90 (0.0748)	2.30 (0.0905)
Emulsion tube type	F.30		F.30	
Idling jet size mm (in)	0.50 (0.0196)	1.00 (0.0393)	0.50 (0.0196)	0.70 (0.0275)
Idling air jet size mm (in)	1.10 (0.0433)	0.70 (0.0275)	0.10 (0.0433)	0.70 (0.0275)
Pump jet size mm (in)	0.45 (0.0175)	–	0.50 (0.0197)	–
Pump discharge jet size mm (in)	0.40 (0.0157)	–	0.40 (0.0157)	–
Pump diaphragm stroke mm (in)	3.5 (0.138)		3.0 (0.0118)	
Super feed jet size mm (in)	–	1.10 (0.0433)	–	1.10 (0.0433)
Super feed mixture bore mm (in)	–	2.00 (0.0787)	–	–
Needle valve size mm (in)	1.75 (0.0689)		1.50 (0.0590)	

	34 DATR 1/200 and 34 DATR 2/250*		34 DATR 2/200 and 34 DATR 2/250**	
	Primary	Secondary	Primary	Secondary
Choke/barrel size mm (in)	25 (0.985)	27 (1.0629)	25 (0.985)	27 (1.0629)
Venturi diameter mm (in)	4 (0.157)	4 (0.157)	4 (0.157)	4 (0.157)
Main jet size mm (in)	1.20 (0.0472)	1.50 (0.0590)	1.20 (0.0472)	1.50 (0.0590)
Air corrector jet size mm (in)	1.70 (0.0669)	2.10 (0.0826)	1.70 (0.0669)	2.40 (0.0944)
Emulsion tube type mm (in)	F.30		F.30	
Idling jet size mm (in)	0.50 (0.0196)	0.80 (0.0314)	0.50 (0.0196)	0.80 (0.0314)
Idling air jet size mm (in)	1.10 (0.0433)	0.70 (0.0275)	1.10 (0.0433)	0.70 (0.0275)
Pump jet size mm (in)	0.50 (0.0197)	–	0.50 (0.0197)	–
Pump discharge jet size mm (in)	–	–	–	–
Pump diaphragm stroke mm (in)	–	–	–	3.0 (0.118)
Super feed jet size mm (in)	–	1.00 (0.0393)	–	1.10 (0.0433)
Super feed mixture bore mm (in)	–	2.00 (0.0787)	–	2.00 (0.0787)

Chapter 3 Fuel, emission control and exhaust systems

Needle valve size mm (in)	1.75 (0.0689)	1.75 (0.0689)

Note: *The Weber 34 DATR 2/250* fitted to the 1600B series is different in specification to the Weber 34 DATR 2/250** fitted to the 2000 models*

	32 DATRA 12/100 and 32 DATRA 12/101		32 DATRA 3/100 and 32 DATRA 13/100	
	Primary	Secondary	Primary	Secondary
Choke/barrel size mm (in)	22 (0.867)	25 (0.985)	22 (0.867)	25 (0.985)
Venturi diameter mm (in)	4 (0.157)	4 (0.157)	4 (0.157)	4 (0.157)
Main jet size mm (in)	1.10 (0.0433)	1.25 (0.0493)	1.15 (0.0452)	1.25 (0.0493)
Air corrector jet size mm (in)	1.85 (0.0728)	2.20 (0.0866)	1.95 (0.0767)	2.20 (0.0866)
Emulsion tube type mm (in)	F.30		F.30	
Idling jet size mm (in)	0.50 (0.0196)	0.70 (0.0275)	0.50 (0.0196)	0.70 (0.0275)
Idling air jet size mm (in)	1.10 (0.0433)	0.70 (0.0275)	1.10 (0.0433)	0.70 (0.0275)
Pump jet size mm (in)	0.55 (0.0216)	–	0.55 (0.0216)	–
Pump discharge jet size mm (in)	0.40 (0.0157)	–	0.40 (0.0157)	–
Pump diaphragm stroke mm (in)	3.5 (0.138)	–	–	–
Super feed jet size mm (in)	–	0.85 (0.0334)	–	0.85 (0.0334)
Super feed mixture bore mm (in)	–	2.00 (0.0787)	–	2.00 (0.0787)
Needle valve size mm (in)	1.75 (0.0689)		1.75 (0.0689)	

Solex carburettor data

	C32 CIC-1		C34 CIC-1	
	Primary	Secondary	Primary	Secondary
Barrel/choke size mm (in)	22 (0.867)	22 (0.867)	25 (0.985)	26 (1.013)
Auxiliary venturi mm (in)	4.4 (0.173)	4.4 (0.173)	4.4 (0.173)	4.4 (0.173)
Main jet size mm (in)	1.10 (0.0433)	1.00 (0.0393)	1.25 (0.0443)	1.25 (0.0443)
Air corrector jet size mm (in)	1.80 (0.0708)	2.10 (0.0629)	1.60 (0.0629)	2.70 (0.1062)
Emulsion tube type	C25597	C25602	C25517	C25518
Idling jet size mm (in)	0.45 (0.0175)	0.40 (0.0157)	0.45 (0.0175)	0.70 (0.0275)
Idling air jet size mm (in)	1.00 (0.0393)	1.00 (0.0393)	1.10 (0.0433)	1.00 (0.0393)
Needle valve size mm (in)	1.8 (0.0708)		1.8 (0.0708)	

	C32 CIC-3		C34 TCIC-1	
Barrel/choke size mm (in)	22 (0.867)	22 (0.867)	25 (0.985)	26 (1.023)
Auxiliary venturi mm (in)	4.4 (0.173)	4.4 (0.173)	4.4 (0.173)	4.4 (0.173)
Main jet size mm (in)	1.05 (0.0413)	1.05 (0.0413)	1.25 (0.0443)	1.40 (0.0551)
Air corrector jet size mm (in)	1.80 (0.0768)	2.10 (0.0826)	1.60 (0.0629)	2.30 (0.0905)
Emulsion tube type	C25597	C25602	No 46	No 46
Idling jet size mm (in)	0.47 (0.0185)	0.47 (0.0185)	0.45 (0.0175)	0.70 (0.0275)
Idling air jet size mm (in)	1.00 (0.0393)	1.00 (0.0393)	1.10 (0.0433)	1.00 (0.0393)
Needle valve size mm (in)	1.4 (0.0551)		1.8 (0.0708)	

Torque wrench settings

	lbf ft	Nm
Inlet manifold to cylinder head	18	24
Exhaust manifold to cylinder head	18	24
Exhaust pipe to manifold		
Threads dry	17	23
Lubricated	15	20
Front exhaust pipe stay bracket to crankcase	11	14
Front exhaust pipe stay to front bracket	33	46
Front exhaust pipe stay to exhaust pipe clamp bracket	11	14
Front exhaust pipe clamp	11	14
Positive crankcase ventilation valve to crankcase	16	22

Chapter 3 Fuel, emission control and exhaust systems

	lbf ft	Nm
USA Emission control system		
Air pump to stay	18	24
Air pump drive pulley to camshaft	40	54
Inlet manifold to EGR pipe	40	54
Recirculation pipe to exhaust gas separator	40	54
Exhaust gas separator to manifold	11	14

1 General description

The Lancia Beta series have a conventional fuel and carburation system. The fuel tank is positioned at the rear of the car and a fuel pump, either mechanical, driven off the auxiliary driveshaft at the front of the engine, or electric, mounted at the rear of the body on the right-hand side, feeds the carburettor.

The carburettor, which may be either of the Weber or Solex type, is of the twin choke design and is mounted on the front of the engine on a four branch inlet manifold.

The amount of fuel in the tank is measured by the tank unit and this information is transmitted by a sender unit to the gauge on the instrument panel.

Emission control on UK models is of the normal type with air intake pre-heating devices and positive crankcase ventilation systems. There is also a petrol recirculation device.

In models restricted for the USA much more complicated emission control systems are involved. These include exhaust gas recirculation (EGR), an air pump, with a diverter valve and fuel evaporation control to ensure the most efficient burning of the fuel and thereby the best possible reduction in noxious gas output. These are all covered in greater detail in Section 21.

The exhaust system consists of an exhaust manifold on the rear of the engine and a three section exhaust pipe system with 2 silencer boxes. The front section is a dual pipe siamesed into a single pipe near the rear end. The rest of the system is a single pipe with the tail pipe

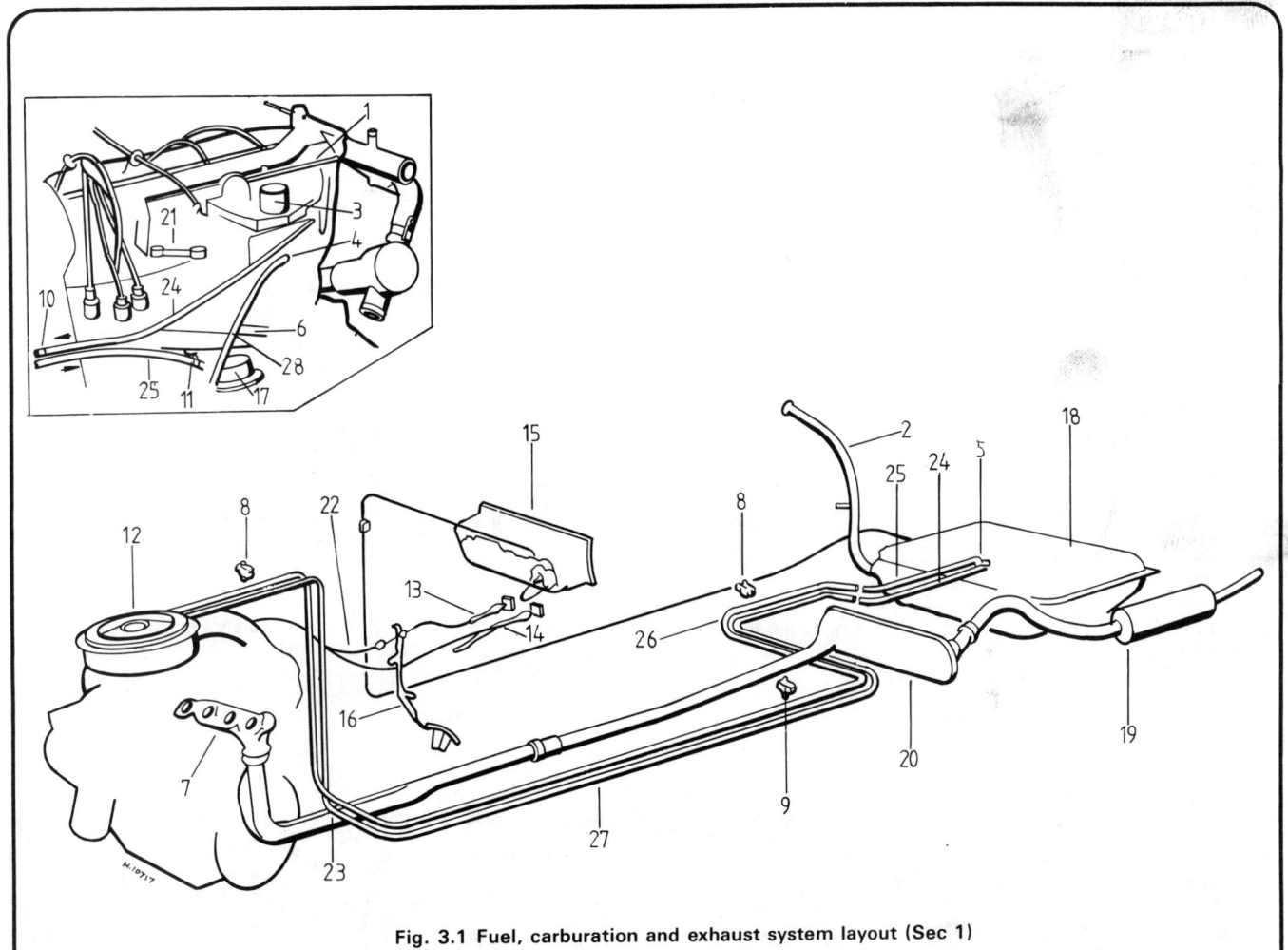

Fig. 3.1 Fuel, carburation and exhaust system layout (Sec 1)

1 Throttle control shaft (fitted to some models)
2 Fuel tank filler pipe
3 Carburettor
4 Fuel inlet to carburettor
5 Fuel tank transmitter unit
6 Inlet manifold
7 Exhaust manifold
8 Hose clip
9 Hose clip
10 Hose clip
11 Hose clip
12 Air cleaner
13 Hand throttle
14 Choke cable
15 Fuel gauge
16 Throttle pedal
17 Fuel pump (mechanical)
18 Fuel tank
19 Tail silencer
20 Main silencer
21 Throttle linkage
22 Throttle cable
23 Front exhaust pipe
24 Fuel return hose
25 Fuel supply line
26 Fuel supply line
27 Fuel return hose
28 Fuel feed pipe from pump to carburettor

coming out on the left-hand rear side. The system is supported by a stay from the lower crankcase at the front and rubber mountings at the silencer boxes.

On some of the earlier models a hand operated choke was fitted to the carburettor, but later models all have thermostatically operated semi-automatic chokes.

2 Air cleaner and element – removal and refitting

1 The air cleaner housing is located in the centre of the engine compartment on top of the carburettor, and the element is of the disposable paper cartridge type. Three bolts secure the top of the air cleaner housing to the body (photo). Once these nuts have been removed, together with their washers, the top cover can be lifted off.
2 The cleaner element is now exposed and can simply be lifted out of the housing. Wipe over the interior and fit the new element which should seat snugly over the locating ridges in the bottom of the housing (photo).
3 When refitting the cover ensure that the sealing ring is intact and in place.
4 On some patterns of cleaner the intake pipe may be positioned for 'Summer' or 'Winter' conditions by turning the cover so that the open end front intake collects cool air for 'Summer' driving or the rear intake feeds warm air from the exhaust manifold for 'Winter' conditions. This can be seen in photos 2.1 and 2.2. There is an arrow on the front intake of the air cleaner and an I (for winter) or E (for summer) setting on the top cover. A plate blanks off the appropriate intake pipe.
5 To remove the air cleaner housing first remove the top cover and filter element as already described.
6 The housing body is secured to the top of the carburettor by a plate and four nuts (photo).
7 Undo the nuts and remove the plate.
8 The air cleaner can now be lifted up so that the pipe from the carburettor and PCV hose can be disconnected.
9 On some early models the air cleaner assembly can now be removed complete with intake pipes. In later models however, especially those with thermostatic controls for the hot/cold air intake the hot air flexible hoses will have to be disconnected from the air cleaner intake (Fig. 3.2).
10 On some models the rigid hot air intake pipe has to be disconnected from the mounting bracket on the cam cover.
11 Lift the air cleaner away and remove the gasket.
12 Renew the gasket and refit the assembly in the reverse order to removal.

3 Throttle cables – removal and refitting

1 There are two throttle cables fitted. One is a hand throttle cable which is mounted on the dashboard and the other is the main throttle cable from the accelerator pedal to the carburettor.

Hand throttle
2 The hand throttle cable runs from the knob on the dashboard panel to the accelerator pedal upper end.
3 The cable should first be disconnected from the pedal, by undoing the cable retaining nut.
4 Remove the hand throttle control knob from the front of the dashboard. Its seating has two lugs which hold it in place and they have to be released from behind the panel.
5 With the control end removed the cable can be withdrawn completely through the hole in the dashboard and be renewed.
6 Refitting is the reverse procedure to removal. Check that the cable is properly adjusted so that the accelerator pedal can operate freely and correctly.

Main throttle cable
7 Disconnect the throttle cable from the linkage at the carburettor end (photo).
8 Undo the inner cable adjuster locknut at the cable end bracket behind the carburettor (photo).
9 Withdraw the inner and outer cable through the mounting bracket and recover the washers and bush, and other locknut.
10 Disconnect the throttle cable from the accelerator pedal.
11 Withdraw the inner and outer cable complete through the bulkhead into the engine bay.
12 Refit the cable in the reverse order to removal and then check and adjust it as necessary so that with the accelerator pedal fully depressed the throttle valve of the primary choke is fully open.

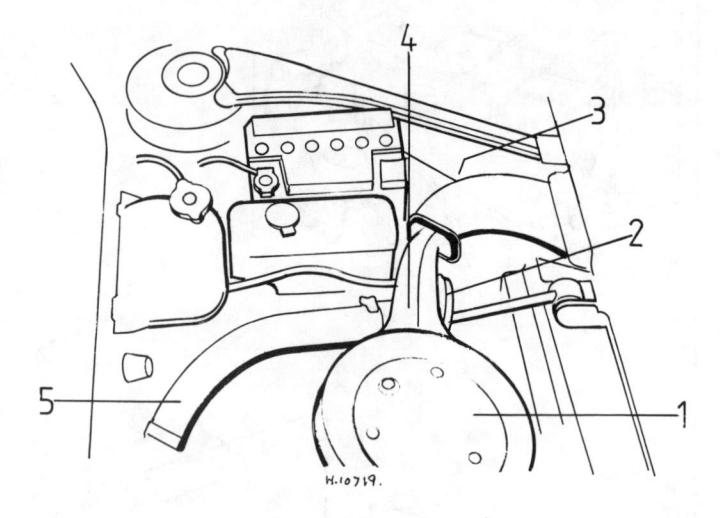

Fig. 3.2 Under bonnet view of later air cleaner with thermostatic heat control (Sec 2)

1 Air cleaner
2 Thermostatic control unit
3 Cold air feed duct
4 Cold air intake
5 Hot air intake hose

2.1 The air cleaner cover is secured by three bolts

2.2 Fitting a new element

2.6 The air cleaner body is secured to the carburettor by a plate and four nuts

Chapter 3 Fuel, emission control and exhaust systems

3.7 The throttle cable front end connection

3.8 The throttle cable adjusting nuts and mounting bracket

4 Choke cable – renewal

1 The choke cable is only fitted to earlier models which do not have a semi-automatic choke arrangement.
2 To remove the choke cable start by disconnecting it at the carburettor from the operating lever. Undo the nuts which hold the outer cable in position.
3 Separate the choke and throttle cables which are held together by clips.
4 Tape a length of wire to the front end of the choke cable to make refitting the new cable easier.
5 If it is not possible to reach the choke control knob from behind the dashboard without removing the instrument panel then this will have to be done as described in Chapter 10.
6 Disconnect the wires from the choke warning light.
7 Release the lugs on the choke control knob mounting plate with a screwdriver and withdraw it into the car.
8 Feed the complete cable through the dashboard until the front end is clear. Remove the tape and leave the piece of guide wire in place.
9 Refit the new choke cable assembly and tape it onto the guide wire. Then gently pull the other end of the wire inside the engine bay and feed the cable through the bulkhead and into position. Secure the control knob to the dashboard panel and reconnect the warning light wires.
10 Reconnect the carburettor end of the cable.
11 Check that with the choke knob pushed fully home the operating lever on the carburettor is completely disengaged and the panel warning light is off. Also check that with the knob pulled out the light is on and the choke is operating correctly.

5 Carburettor – description and principles

1 All the carburettors fitted are of 'fixed choke' design and are manufactured by Weber or in some instances Solex. The term 'fixed choke' relates to the throat (venturi) into which the main petrol jet sprays the fuel, and because on these carburettors this venturi is a fixed size, several other petrol jets are incorporated in the carburettor to enrich the fuel/air mixture passing into the engine as and when necessary. All fuel jets take the form of inserts screwed into the carburettor body.
2 The carburettors function as follows: Petrol is pumped into the float chamber and is regulated by the needle valve actuated by the float. As air is sucked into the engine a slight vacuum – proportional to air flow – is created in the venturi of the carburettor; this vacuum draws a corresponding amount of fuel from the float chamber through the emulsion tube where it mixes with the small amount of air coming from the air correction jet. The fuel air emulsion passes on through the main jet and into the main air stream in the venturi of the carburettor.

3 The flow of air through the carburettor is controlled by the butterfly valve operated by the accelerator linkage. The engine develops power in proportion to the amount of air and fuel drawn into the engine, and the economy is dependant on the relative proportions of fuel and air taken into the engine. On these fixed choke carburettors the relative proportions of the fuel and air are controlled by the sizes of the main jet, carbutettor throat – 'venturi', and the other minor enrichment jets. All these sizes are fixed and decided by the engine designers. The only adjustment or trim to the fuel/air mixture available is provided by the accelerator's butterfly valve stop and the 'idling mixture control screw'. Sections 7 and 8 set down the carburettor adjustment procedure.
4 As mentioned earlier the fixed jet carburettors require additional fuel jets to enrich the air/fuel mixture when necessary. When power is required for acceleration the engine needs an enriched mixture of fuel and air. An acceleration jet is provided in the throat of the carburettor and it is supplied with fuel under pressure from the accelerator pump. The pump is actuated by levers connected to the accelerator linkage. There is also a compensating jet which introduces additional fuel into the throat of the carburettor to create the slightly rich mixture necessary to sustain the engine at high speeds.
5 In addition to the accelerator jet and compensating jets, there is the idling jet. The flow of fuel through this jet is adjustable. When the engine is idling and the accelerator butterfly valve almost closed, the slight vacuum in the inlet manifold sucks air and fuel via the pilot jet through the idling jet orifice. The three minor jets therefore provide the mixture enrichment necessary when accelerating, maintaining high speed, and engine idling.
6 Arrangements for cold starting consist of a choke flap above the main jet and venturi. To start the engine when cold, appreciably more fuel is required to overcome the losses due to condensation of the fuel vapour in the inlet manifold ducts. By closing the choke valve an increased vacuum is created in the carburettor barrel and this sucks a greater amount of fuel out of the main jet system. The choke flap is mounted eccentrically on the support spindle so that when the engine demands more air the flap will partially open itself against a control spring to admit the air into the carburettor and engine. The Solex carburettors have a valve which admits fuel directly into the barrel for cold starting.
7 On early models the choke is controlled by manual means, whereas on later models the choke is entirely automatic. The automatic choke is controlled by a bi-metal spring which opens or closes the choke flap dependant on the engine coolant temperature in the spring housing. Two small hoses connect the housing to the cooling system. One is connected to the rear water rail and runs over the camshaft drivebelt cover to the carburettor; the other is attached to the right-hand end of the inlet manifold.
8 Both the Weber and Solex type of carburettor are what is normally referred to as 'Compound Carburettors'. In these carburettors the accelerator linkage does not open the butterfly valves in the two

Chapter 3 Fuel, emission control and exhaust systems

throats simultaneously. The principle employed here is that only one barrel is used when air requirements are low and the second barrel is opened only when extra power and air is required. This arrangement gives a smooth running engine at low speed whilst maintaining efficiency at high engine speeds.

6 Carburettor – removal and refitting

1 The carburettor is one of the easier components to remove from the very crowded engine bay.
2 It can be removed complete with the inlet manifold if it is desired, or it can be removed on its own, which is the procedure described in the following instructions.
3 Remove the air cleaner assembly as described in Section 2.
4 Disconnect the plastic arm which connects the throttle linkage to the carburettor control arm (photo).
5 Disconnect the choke cable from the carburettor operating lever, where the manual type of choke is fitted.
6 In later cars which have automatic chokes unscrew the hose clamps on the water inlet and outlet pipes to the carburettor housing. Have handy two small corks of the size of the water hoses. Pull each hose off in turn and plug the end with a cork. Then tuck the loose ends away. Note which hose fits where. (The bottom hose is connected to the inlet manifold).
7 Disconnect the solenoid cable at the connector (photo).
8 Undo the petrol feed and return pipe clamps and disconnect the pipes from the carburettor (photo).
9 Disconnect the emission control hoses on USA models.
10 Undo the four nuts which secure the carburettor to the inlet manifold and lift it away, recovering the washers as this is done.
11 Retrieve the gasket and plug the inlet manifold with a clean non-fluffy piece of rag.
12 Refitting the carburettor is the exact reversal of removal. You must be sure that the correct number of gaskets have been replaced and that the control cables have been fitted correctly to give the full range of movement the accelerator and choke linkages require.

13 Finally adjust the slow running as described in Section 7.

7 Carburettors – setting and adjustment for slow running

Weber

1 The amount of adjustment that can be carried out on the later Weber carburettors is very limited. However on the earlier 34 DMTR series more adjustment is possible. This is due to anti-pollution regulations.
2 To set the slow running adjustment start the engine and allow it to warm up thoroughly. This can be gauged when the electric fan cuts in.
3 With the DMTR series carburettors allow the cooling fan to cut out before starting work.
4 Turn the bypass idling setting screw until the engine is ticking over at 800 to 850 rpm (Fig. 3.3). The figure shows the carburettor with the air cleaner removed, so as to show the components better. Needless to say the slow-running adjustment has to be carried out with the air cleaner fitted.
5 If this operation is carried out and the engine runs smoothly then it is correctly set as far as the slow running is concerned. If however the engine does not run smoothly then proceed as follows.
6 Screw the bypass idling setting screw right in. Adjust the throttle opening screw and the idling fuel adjusting screw until the engine ticks over at 700 to 750 rpm. Once this has been achieved lock the locknut on the throttle opening screw. Screw the bypass idling screw out until the engine tickover rises to 800 to 850 rpm.
7 With the later DATR series carburettors the only external adjustment is via the throttle opening setting screw and the automatic choke setting screw (photo).
8 With the automatic choke on this model set the air cleaner to the 'Summer' position, as described in Section 2, and allow the engine to get really well warmed up before attempting to adjust the slow running setting, which can only be done legally by altering the throttle opening screw.

6.4 Disconnect the plastic link arm to the carburettor

6.7 Carburettor fuel cut-off solenoid wire connector

6.8 Disconnect the petrol feed and return hoses (arrowed)

7.7 The only external adjusting screws on the DATR Series Weber carburettors are arrowed

7.8 The slow-running adjusting screw (arrowed) with plug removed on DATR Series carburettor

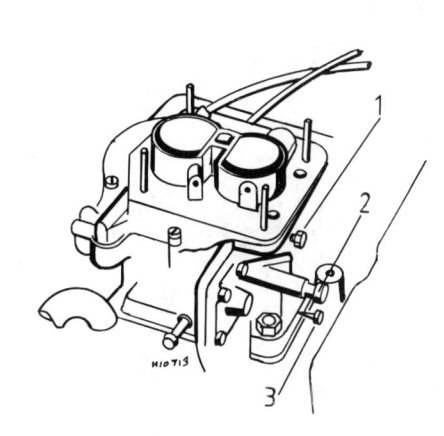

Fig. 3.3 Weber 34 DMTR type carburettor adjusting screws (Sec 7)

1. Throttle opening setting screw
2. Bypass idling setting screw
3. Idling fuel setting screw

Fig. 3.4 Solex 34 CIC-1 type carburettor (Sec 7)

1. Slow-running fuel adjusting screw
2. Primary choke throttle opening setting screw
3. Choke strangler control rod cotter key
4. Accelerating pump injector plug
5. Choke strangler control rod.
6. Secondary choke progression jet
7. Choke cut-off device
8. Choke control cable securing screw
9. Secondary choke throttle control intermediate lever
10. Throttle control lever

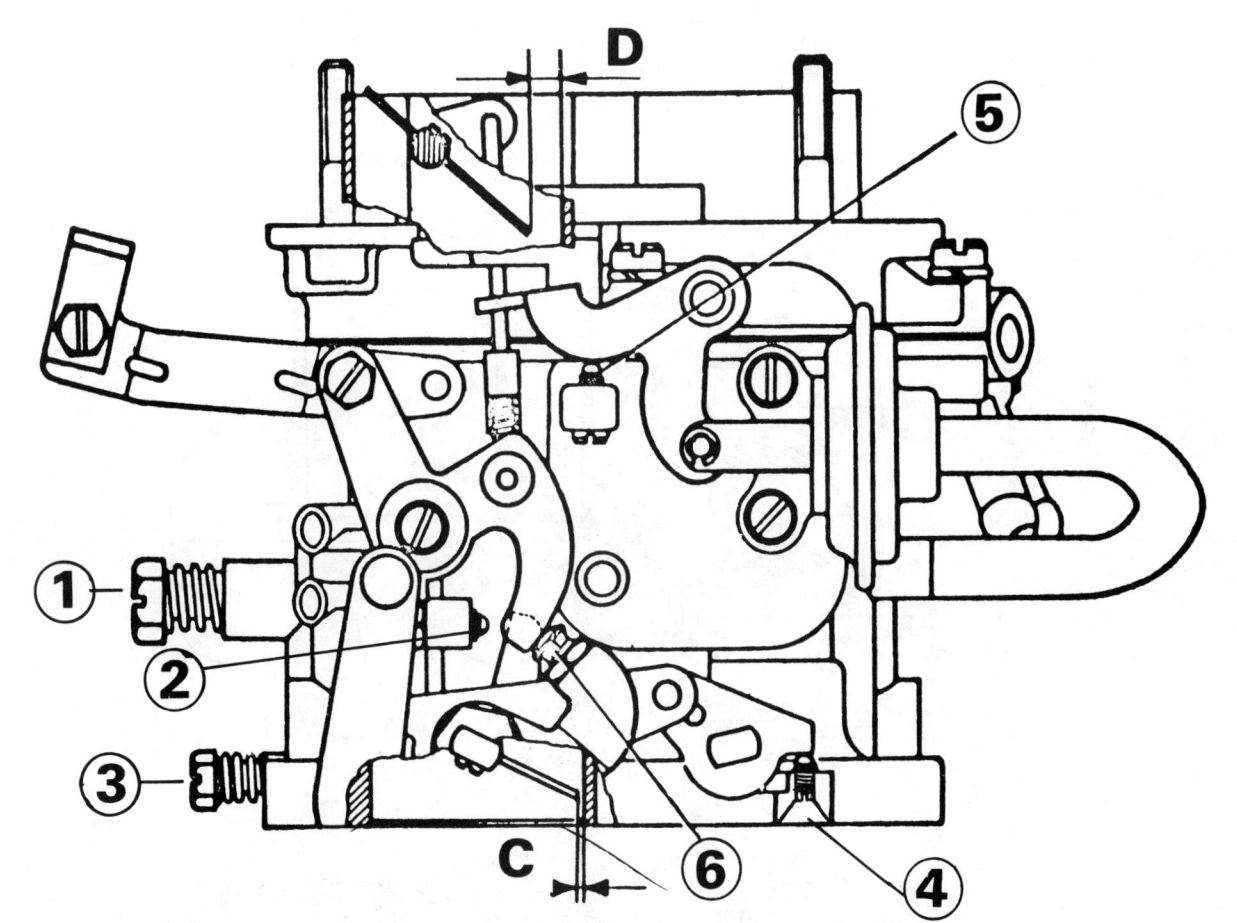

Fig. 3.5 Fast idling adjustment – Weber DMTR carburettors (Sec 7)

1. Bypass idling setting screw
2. Primary choke butterfly opening setting screw
3. Fuel idling setting screw
4. Secondary choke butterfly opening setting screw
5. Vacuum-controlled choke opening setting screw
6. Fast idling setting screw (with choke on)

C = 1600 and 1800 engines 0.9 to 0.95 mm (0.0357 to 0.0374 in)
1400 engines 0.85 to 0.9 mm (0.0334 to 0.0357 in)
D = 5.75 to 6.25 mm (0.226 to 0.246 in)

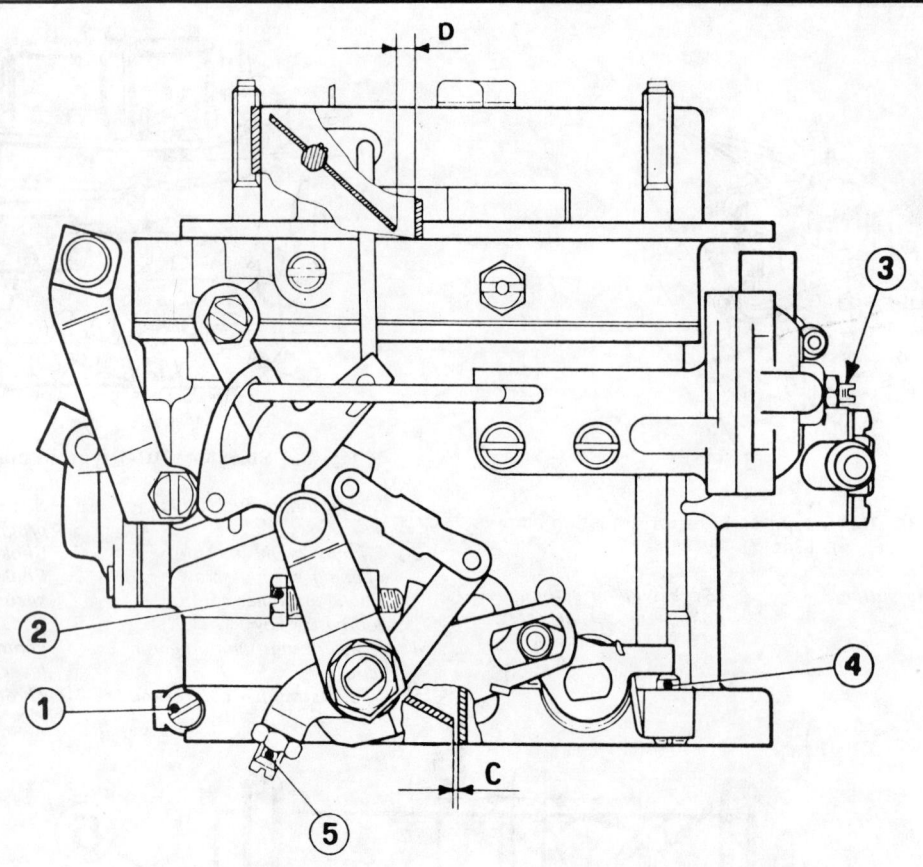

Fig. 3.6 Fast idling speed adjustment – Solex carburettors (Sec 8)

1 Fuel idling setting screw
2 Primary choke butterfly opening setting screw
3 Choke butterfly opening setting screw
4 Secondary choke butterfly opening setting screw
5 Fast idling setting screw (with choke on)

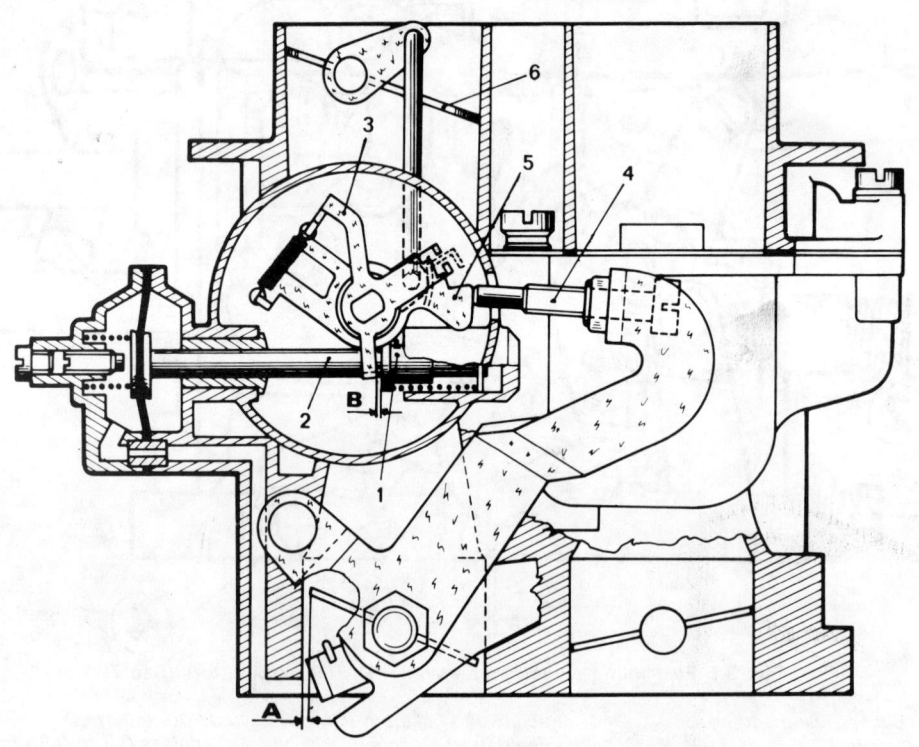

Fig. 3.7 Fast idling speed adjustment – Weber DATR and DATRA carburettors (Sec 8)

1 Bush
2 Rod
3 Control lever
4 Fast idling adjusting screw with choke on
5 Control cam
6 Choke butterfly

Chapter 3 Fuel, emission control and exhaust systems

9 There is a slow running adjusting screw in the base of the carburettor, on the right in the DATR series and on the left in the DATRA series carburettors (photo). This is covered by a plug and is almost impossible to find. By prising the plug out the screw can be adjusted. The tickover speed should be set to 800 to 850 rpm. It may be necessary to adjust the throttle opening screw as well.

10 Due to modern anti-pollution legislation modern carburettors have become less and less adjustable and more and more complicated with solenoid valves to cut off the fuel supply when the ignition is switched off, and systems to reduce pressure in the float chamber at tickover speed and bleed pipes to run-off excess fuel. In certain countries it is illegal to tamper with a carburettor once it has been set, as this can increase the CO content of the exhaust gases.

Solex

11 Warm the engine up just as for the Weber carburettor.

12 Adjust the primary choke throttle adjusting screw so that the engine ticks over at 800 to 850 rpm (Fig. 3.4).

13 Screw in the idling fuel adjusting screw until the engine speed drops due to lack of fuel, then slowly open the screw until the fastest engine speed is encountered, but don't open it further than necessary. At this point the fuel/air mixture is in the correct proportions.

14 If the engine tickover speed has dropped below 800 rpm, increase it by adjusting the primary choke throttle adjusting screw as necessary.

8 Carburettors – settiing the fast idling speed

Weber DMTR series carburettor

1 To set the fast idling speed with the choke on is achieved by altering the choke butterfly opening setting screw and the primary choke butterfly setting screw as necessary (Fig. 3.5).

2 The gap C between the primary choke butterfly and the barrel wall should be 0.90 to 0.95 mm (0.0357 to 0.0374 ins) when the gap D, the measurement between the choke butterfly and the wall is 5.75 to 6.25 mm (0.226 to 0.246 ins). In the case of 1400 engines the gap C should measure 0.85 to 0.90 mm (0.0334 to 0.0357 ins)

Solex carburettor

3 The fast idling speed settings for the Solex carburettors are similar to the Weber (Fig. 3.6).

4 Adjusting screw 3 for the choke butterfly will alter gap D and adjusting screw 5 will alter gap C.

5 The clearance of gap C should be 1.00 to 1.1 mm (0.0393 to 0.0433 ins) on all engines and gap D 4.5 mm (0.177 ins).

Weber DATR and DATRA series carburettors

6 To adjust the fast idling speed on these carburettors first remove the choke thermostat control housing from the side of the carburettor, as described in Section 10.

7 Then set up the linkage as shown in Fig. 3.7. This is with the engine cold, having stalled. In these circumstances gap A should be as follows:-

　1300 engines 1.00 to 1.05 mm (0.0393 to 0.0413 ins)
　1600 engines 1.00 to 1.05 mm (0.0393 to 0.0413 ins)
　2000 engines 1.15 to 1.20 mm (0.0452 to 0.0472 ins)
　1800 engines (USA) 1.05 to 1.15 mm (0.0413 to 0.0452 ins)

8 Gap B, measured on the rod, should in all cases be 0.3 to 1.00 mm (0.0118 to 0.0393 ins). This measurement must be made with the choke butterfly closed.

9 To adjust the clearance alter the setting of the fast idling screw 4, if necessary.

10 If the gap A is less than specified the fuel mixture will be too weak and if larger the fuel mixture will then be too rich.

9 Carburettors Weber DATR series (USA) – setting the engine idling and fast idling and checking the CO content

1 Set the slow running or tickover as described in Section 7.

2 To check the carbon monoxide (CO) content in the exhaust gas, insert the probe of an exhaust gas analyser into the rear end of the exhaust pipe. This must be done with the air cleaner set in the Summer position.

3 Adjust the idling setting screw if necessary to ensure that the CO content of the exhaust does not exceed the maximum limit on the plate which is located in the engine compartment (Fig. 3.8).

4 When this has been checked make sure that the tickover speed has not altered, and that the engine runs smoothly.

5 Push down the fast idling check button on the solenoid valve. Keep the button down and operate the throttle control lever. The engine speed recorded on the tachometer should be within the limits set out on the plate in the engine compartment.

6 If the engine speed is not correct adjust the fast idling screw as necessary (see Fig. 3.8).

10 Automatic choke – removal and refitting

1 This applies to Weber DATR and DATRA model carburettors, which have a thermostatically controlled choke.

2 Remove the carburettor as described in Section 6.

3 Disconnect the rod from the choke butterfly operating lever by removing the small split pin.

4 Scribe the relationship of the housing to the choke unit body. Undo and lift the thermostat housing retaining collar screws and lift the thermostat housing away (photo).

5 Remove the plastic disc to reach the choke unit retaining screws (photo).

6 Remove the two retaining screws and release the choke adjusting screw from the choke mechanism and lift the choke unit away from the carburettor body (photo).

7 Refitting is the reverse process to removal. Make sure that the choke adjusting screw is aligned correctly to the choke operating cam and that the thermostat spring hooked end locates in the fork of the choke operating arm. This ensures that the thermostat housing is fitted the right way round. Also line up the scribed marks.

8 Refit the carburettor as described in Section 6 and check that it operates correctly. Set and adjust it as described in the foregoing Sections.

11 Automatic choke diaphragm control unit – removal and refitting

1 With the thermostat housing and plastic disc removed from the choke unit as described in Section 10, undo the three screws which retain the choke diaphragm control assembly cover.

Fig. 3.8 Weber DATR carburettor (cover removed) (Sec 9)

1　Idling adjusting screw
2　Fast idling adjusting screw
3　Throttle opening adjusting screw

10.4 The thermostat housing is retained by a collar and three screws (one arrowed)

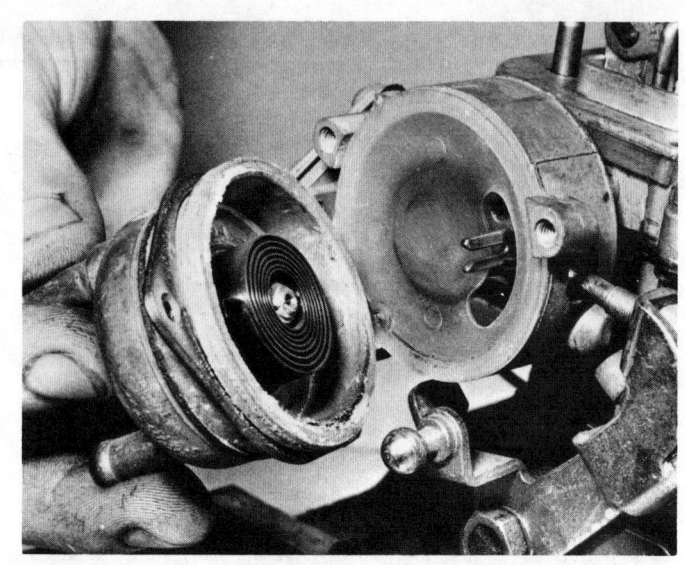

10.5 With the thermostat housing removed the plastic disc can be removed

10.6 The choke unit is retained to the carburettor by two screws in the upper section

11.2 Removing the choke diaphragm control cover

11.3 Removing the diaphragm and control rod assembly

14.4 Lifting off the carburettor top cover (Weber DATR Series) complete with floats and gasket

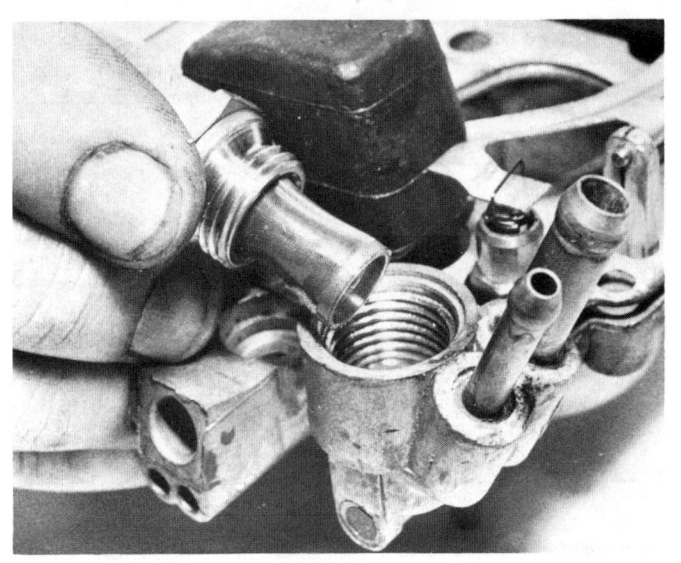

14.5 Removing the fuel inlet filter (Weber DATR Series)

14.6 To remove the gasket the float pin must be removed to release the float assembly. Note the clip which retains the needle to the float arm

14.8 The slow running jets (outer pair) and emulsion tubes (inner pair) on Weber DATR Series

14.9 Removing the accelerator pump diaphragm (Weber DATR Series)

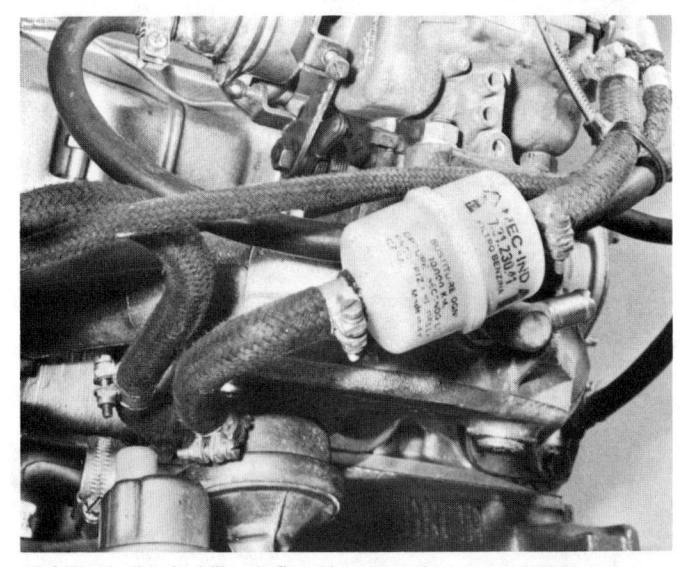

15.2 The in-line fuel filter is fitted between the mechanical fuel pump and the carburettor

16.5 Removing the inlet manifold from the engine (photo taken with engine out of car)

2 Carefully remove the cover and spring taking care not to damage the diaphragm if it is stuck to the cover (photo).
3 Rotate the choke mechanism inside the unit and ease the diaphragm control rod outwards. Check that the diaphragm is separated from the inner flange and note that there is a tiny vacuum pipe in the upper edge over which the diaphragm fits. Compress the plastic lugs on the collar and carefully withdraw the complete assembly (photo).
4 This assembly has to be renewed as a complete unit.
5 Refit the assembly carefully in the reverse order to removal. Make sure that the diaphragm locates correctly and that the choke control lever engages correctly on the rod behind the bush. This can be seen in photo 10.6.
6 Refit the rest of the assemblies as described in Section 10.

12 Automatic choke operation on Weber DATR and DATRA carburettors – description

1 The choke in the DATR and DATRA series Weber carburettors is of the semi-automatic type. This means that it has to be engaged, when the engine is cold, by fully depressing the accelerator pedal. Having done this the accelerator pedal can then be slowly released to its normal position.
2 When the engine is started the bi-metallic spring, which controls the choke butterfly via the operating lever, spindle and rod, keeps the choke closed.
3 At the same time the cam lever will be in the position shown in Fig. 3.7. This gives the fast idling setting. It can be adjusted by the fast idling adjusting screw.
4 This setting delivers a suitably enriched mixture to help cold starting.
5 Once the engine has started the vacuum created in the main venturi beneath the primary choke butterfly operates the diaphragm in the choke control assembly via a channel in the carburettor body and the small pipe which runs through the diaphragm, as described in Section 11.
6 The diaphragm moves the rod and balanced by the return spring the bush pushes on the operating lever and operates against the bi-metallic spring to open the choke butterfly slightly and thus weaken the mixture being drawn into the engine, which is more suitable for the running conditions now prevalent.
7 If the accelerator is depressed during this stage the fast idling adjuster will be disengaged from the choke operating cam and the cam spring will reduce the opening of the primary choke.
8 As the engine coolant warms up and circulates in the system so the water in the thermostat housing increases in temperature and causes the bi-metallic spring to turn and act on the choke operating lever and thus, by the linkage, open the butterfly gradually. This will happen until the choke is fully open.

13 Fast idling diaphragm – renewal

1 This Section applies to Weber DATR and DATRA carburettors which have a fast idling diaphragm assembly fitted to them.
2 Remove the carburettor as described in Section 6.
3 Undo the four screws and remove the housing cover complete with diaphragm.
4 Remove the diaphragm from the cover and renew it.
5 Refit in the reverse order to removal.

14 Carburettors – overhaul

1 Do not dismantle the carburettor unless it is absolutely necessary, when systematic diagnosis indicates that there is a fault with it. The internal mechanism is delicate and finely balanced, and unnecessary tinkering will probably do more harm than good.
2 Although certain parts may be removed with the carburettor still

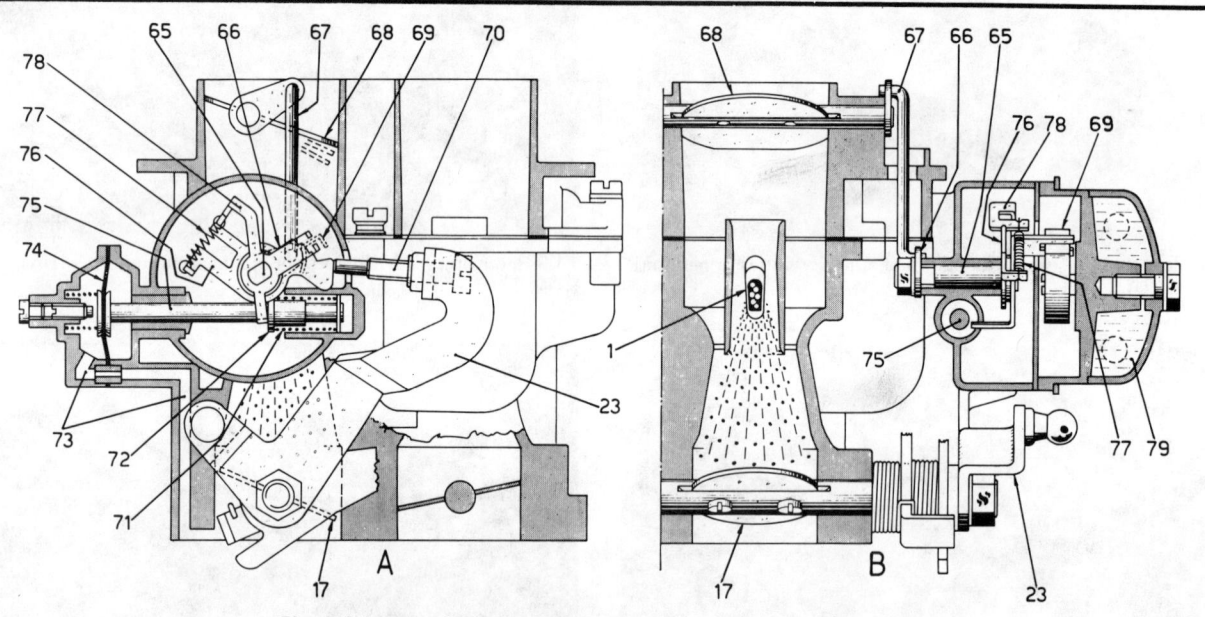

Fig. 3.9 Automatic choke – Weber DATR and DATRA carburettors (Sec 12)

A End on view

B Carburettor through section

17 Primary choke butterfly
23 Throttle operating lever
65 Spindle
66 Bi-metallic spring operating lever
67 Rod
68 Choke butterfly
69 Bi-metallic spring end
70 Fast idling adjusting screw
71 Return spring
72 Bush
73 Vacuum channel to diaphragm
74 Diaphragm
75 Operating rod
76 Choke operating lever from cam
77 Spring

1 Nozzle
17 Primary choke butterfly
23 Throttle operating lever
65 Spindle
66 Lever
67 Rod
68 Choke butterfly
69 Bi-metallic spring
75 Rod from diaphragm
76 Cam
77 Spring
78 Lever
79 Water jacket

Chapter 3 Fuel, emission control and exhaust systems

attached to the engine, it is considered safer to remove it and work over a bench.
3 With the carburettor off the car disassembly is as follows. (This covers all carburettors fitted with slight variations, which, because of their simple and obvious nature, are not specifically mentioned each time, but it does not cover the automatic choke working of these carburettors so fitted. The automatic choke is covered fully in Sections 10, 11 and 12).
4 Remove the top of the carburettor by undoing its fixing screws with the correct size screwdriver. These screws are made of comparatively soft metal and will damage very easily. Lift off the top cover complete with floats and gasket (photo). The fast idling adjustment screw has first to be released from the choke cam, on Weber carburettors fitted with automatic choke systems.

Weber carburettors

5 To remove the floats it is first necessary to remove the large filter retaining nut and filter on the underside of the top cover (photo).
6 Now the float pivot pin can be removed and the float assembly and gasket can be lifted away from the top cover (photo).
7 Note that the fuel inlet control needle is attached to the float arm by a clip and will come out at the same time. Note how it is attached and take care not to lose the clip and needle.
8 Unscrew and remove the 2 slow running jets and 2 emulsion tubes from the body of the carburettor (photo).
The emulsion tubes are the inner pair and the slow running jets the outer pair. The main jets lie directly beneath the emulsion tubes.
9 Where they are fitted, on Weber DMTR carburettors, unscrew all external adjusting screws and float chamber drain plugs. If necessary remove the accelerator pump diaphragm (4 screws) (photo).

Solex carburettors

10 Unscrew and remove the progression jet (6) from the float chamber cover and also remove the slow running fuel cut-off device, which incorporates the slow running jet for the primary choke (Fig. 3.4).
11 Unscrew and remove the main emulsion tubes and main jets from the body of the carburettor. The main jets are located directly beneath the emulsion tubes (Fig. 3.10).
12 Undo the float pivot screw and remove the pivot and the float.
13 Remove the choke tube from the body of the carburettor.

Weber and Solex carburettors

14 There is no point under any circumstances in removing any more parts from the carburettor. If any of these parts are in need of attention, then a complete new carburettor is needed. It is safer and more efficient if you have reached this stage of need of repair to renew the complete unit.
15 All the parts which have been separated should be thoroughly cleaned in methylated spirit or clean petrol. Do not use any scrapers, emery paper, wire wool, or hard projections such as a pin, on these parts. All have been machined to extremely fine tolerances. Blow all parts dry.
16 Inspect for blockages and scoring, the float for a puncture and the needle valve for easy operation. On plastic and metal floats make sure the metal tag is not bent or distorted. Renew any parts which are obviously worn or damaged but also do so to any which you even suspect. (Make sure that parts are still available though before throwing away – you may need a temporary repair).
17 Refit all parts in the reverse of their removal using new copper washers and gaskets all the way through. Do not overtighten anything, and do not use any gasket cement. If the top of the carburettor does not fit flat it needs renewing. No mention has been made of throttle, choke and butterfly spindles. As these spindles run directly in the body or top of the carburettor they are likely over a period of usage to wear. It is impractical to renew parts of these and any wear or failure in these parts must mean complete renewal.
18 When the carburettor has been reassembled, and refitted to the car, it must be checked and set up correctly as covered in Sections 7 and 8.

15 In-line filter – removal and refitting

1 The in-line fuel filters are normally only fitted to later models. However, it is possible to fit a fuel filter of this type to any fuel system.
2 The in-line filter is fitted in the fuel supply hose between the fuel pump and carburettor (photo). On models with electric fuel pumps the filter is located next to the pump under the right-hand side of the body at the rear.
3 The fuel filter is a plastic container with a paper filter in the middle through which the fuel flows.
4 To remove it undo the pipe clamps at the top and bottom of the filter and pull off the hoses.
5 The filter is of the throw-away variety and should be checked every 6000 miles (10 000 kms) and renewed every 12 000 miles (20 000 km).
6 Fit the filter in the reverse order to removal but ensure that the arrow on the body of the filter points in the direction of the petrol flow (ie towards the carburettor).

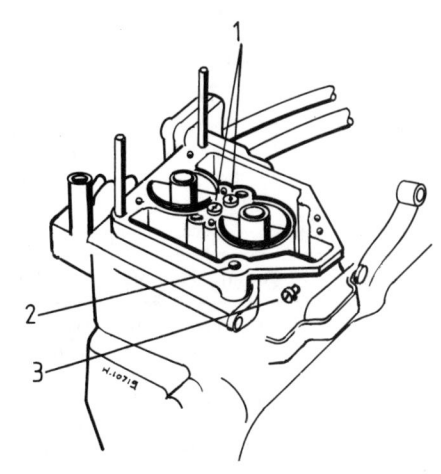

Fig. 3.10 Solex CIC carburettor with top cover removed (Sec 14)

1 Emulsion tubes
2 Slow running jet bore
3 Float pivot screw

16 Inlet manifold – removal and refitting

1 The inlet manifold can be removed with or without the carburettor fitted to it.
2 Before the inlet manifold can be removed it is first necessary to remove the air cleaner, the distributor on cars which have it fitted to the front of the engine (see Chapter 4), and also the fuel pump (see Section 17). The reason for this is that the inlet manifold is mounted on studs which protrude at an angle from the cylinder head, and the manifold when removed has to be drawn slightly downwards and will foul both the distributor and fuel pump. It does however just clear the crankcase ventilation valve.
3 Undo and remove the four bolts along the bottom edge of the manifold and slacken the nuts on the two studs which sit between the inlet manifold branches.
4 Disconnect the carburettor and choke linkage and the automatic choke water inlet and outlet pipes where fitted. Also disconnect the carburettor solenoid wire at the connector.
5 Remove the remaining 2 nuts and lift the inlet manifold away from the cylinder head, and place it to one side (photo).
6 Remove the inlet manifold gasket, and clean off all traces of the old gasket from the face of the manifold and cylinder head.
7 Fit a new gasket and position it on the studs. Make sure that you have the correct gasket for the particular engine type. Inlet manifold gaskets are shaped for the specific cylinder head and the centre elongated hole is the one that really matters, or you will have a water leak.
8 Offer up the inlet manifold to the cylinder head and refit it in the reverse order to removal. Tighten up the mounting nuts and bolts to the specified torque.
9 Finally refit all the ancillary components and check the operation of the engine.

Chapter 3 Fuel, emission control and exhaust systems

17 Fuel pump – mechanical – removal and refitting

This type of fuel pump cannot be stripped but must be renewed if it malfunctions.

1 The fuel pump can be removed in situ once the air cleaner has been removed, as described in Section 2. On 1400 models the inlet manifold must also be removed to reach the pump. On models fitted with air conditioning, remove the alternator and air conditioning pressure switch.
2 On Saloon models the pump can be readily reached when the radiator grille has been removed. This cannot be done with the Coupe, Spider or HPE models.
3 Disconnect the inlet and outlet hoses from the petrol pump and plug the ends of the pipes to prevent the ingress of dirt.
4 Undo the two bolts which secure the pump and spacer to the crankcase and remove the pump and gasket between the pump and spacer. Then remove the spacer block and gasket (photo).
5 Refitting is the reverse procedure to removal but remember to use new gaskets both between the block and spacer and between the spacer and pump.
6 Start the engine up and check that the pump operates correctly and that there are no fuel leaks from the pipe connections. The top pipe on the pump is the inlet and the bottom the outlet to the carburettor.

17.4 Removing the fuel pump from the engine (with spacer behind it)

18 Fuel tank sender unit – testing, removal and refitting

1 On early models where it is not possible to reach the fuel tank sender unit from the luggage area, the tank will have to be removed first, as described in Section 19.
2 On later models remove the carpet or mat in the luggage area, then remove the circular wooden panel which is held in place by three cross-head screws. Beneath this panel is a plastic lid with a seal. Prise it out to reveal the top of the fuel tank sender unit (photo).
3 With the tank lowered or with the cover removed, the next job is to check that the tank sender unit is actually faulty. To do this switch on the ignition having removed the feed and earth wires from the sender unit and then earth the feed wire. (Note which wire fits on to which connector). Check that the needle on the fuel gauge goes from the empty to the full position. If this does not happen then the wiring or the fuel gauge itself is faulty. If it does happen then the fuel tank sender unit needs to be renewed.
4 Undo the clamps on the fuel feed and fuel return hoses and pull them off, but disconnect the battery negative terminal first.
5 On early models the tank must be removed from the car in order to remove the tank unit. Undo the nuts which retain the sender unit to the tank.
6 In later models, working through the hole in the luggage compartment floor, undo the nuts which retain the sender unit to the fuel tank.
7 Ease the sender unit free from the tank and lift it out ensuring that the old gasket does not break up and fall into the tank.
8 Clean off and remove all traces of the old gasket from the tank top.
9 Refitting the tank unit is the reverse procedure to removal, but ensure that a new gasket is used. When the wiring has been reconnected check that the fuel gauge records the level in the tank correctly, and that the fuel warning light operates correctly.

18.2 Removing the plastic bung from the floor of the HPE luggage compartment to reach the sender unit

19 Fuel tank – removal and refitting

1 Drain the fuel tank.
2 Loosen the hose clamp on the filler pipe where it meets the neck of the fuel tank and remove the hose from the tank pipe. In some cases this is done from underneath and in others from inside the boot, behind the spare wheel position.
3 On later models lift up the floor mat in the boot, remove the wooden panel and round plastic plug and disconnect the wiring to the fuel tank sender unit. Undo the fuel pipe clamps and pull off the hoses.
4 On models with emission control systems which include a fuel recovery tank, disconnect the hose from the fuel tank to the recovery tank.
5 Unscrew the nuts which retain the fuel tank to the rear floor pan and lower the tank.
6 On early models the wiring and fuel hoses can now be discon-

20.7 Removing the positive crankcase ventilation (PCV) valve and oil separator (photo shows engine removed)

Chapter 3 Fuel, emission control and exhaust systems

nected from the tank sender unit before the tank is removed from the car.

7 Refit the fuel tank in the reverse order to removal. Fill it up and test the operation of the fuel gauge and warning lights.

20 Emission control systems (UK models) – general description and maintenance

1 On UK models the emission control systems are divided into four sections. Firstly there is the air intake control to the carburettor itself. This has already been covered in Section 2 to a certain extent. Secondly there is the positive crankcase ventilation (PCV) system which feeds the blow-by gases produced in the crankcase to an oil separator and then to the air cleaner and thus back into the engine to be burnt (Fig. 3.13). Thirdly there is a fuel return system which feeds excess fuel from the carburettor via a small return pipe to the fuel tank. There is also an ignition controlled solenoid valve in most carburettors which shuts off the fuel flow to the engine via the idle duct.

2 On late models there is a fourth control system which is a fuel evaporative emission control system. This means that there is a trap between the fuel tank and atmosphere venting pipe (Fig. 3.12).

3 The whole idea behind these systems is that they combine to reduce both the amount of noxious gases and CO (carbon monoxide) being discharged into the atmosphere.

4 To consider the various control systems let us start with the air intake control system. The air is drawn into the air cleaner via the intake pipe over the exhaust manifold or the ram pipe at the front of the engine. In early models this is controlled by altering the air cleaner top cover to the summer or winter position, thereby blanking off one or other of the intake pipes (see Section 2).

5 As emission control regulations have developed, along with more sophisticated carburettors, so the temperature of the incoming air has been more carefully regulated. As can be seen from Section 2 of this Chapter, the later air intake systems have a bi-metallic spring controlled flap valve, which, when the engine is cold, blocks off the cold air intake completely. When the engine is started warm air is drawn from the intake pipe above the exhaust to help cold starting and warm the engine up faster. This in turn cuts down the amount of time that the choke is required thus reducing the amount of fuel consumed under the rich running conditions.

6 As the engine warms up so the bi-metallic spring rotates and gradually opens the flap valve to allow a mixture of warm and cold air to enter the carburettor. When the engine is fully warmed up the warm air intake is blocked off allowing only cold air to enter.

7 The PCV valve which is mounted above and to the rear of the mechanical fuel pump can be removed after the inlet manifold has been removed. It is held in place by a single bolt and has an O-ring on which it seats (photo). The gases from the crankcase are fed up into the valve which also acts as an oil separator. The oil is returned to the engine by the long pipe running into the sump and the fumes are fed up to the air cleaner by the right-hand pipe. As a safety measure there is a wire mesh flame trap inserted in the end of the pipe.

8 To prevent vapour build up in the carburettor and to maintain the temperature of the fuel at a constant level a fuel return line is fitted to all Weber carburettors to feed excess fuel from the carburettor to the fuel tank.

9 Most modern carburettors have built in to them a separate idling fuel supply channel which is controlled by a solenoid valve. This applies to all DATR and DATRA series Webers and the Solex CIC type carburettors. The solenoid valve is connected to the ignition system so that when the ignition is switched off the solenoid valve closes and the fuel supply to the engine is immediately stopped. This prevents the engine running-on (Fig. 3.11).

10 The purpose of the fuel evaporation control system is to prevent the vapour in the fuel tank venting direct to the atmosphere. To achieve this a carbon filled recovery tank is fitted to the rear of the left-hand side of the car. A pipe is connected from the fuel tank to the bottom of the recovery tank, and a vent pipe is fitted to the top of it, so that the vapour can vent to the atmosphere having been purged of petrol fumes. Any petrol which gathers in the carbon filler recovery tank will drain back down the pipe into the fuel tank (Fig. 3.12).

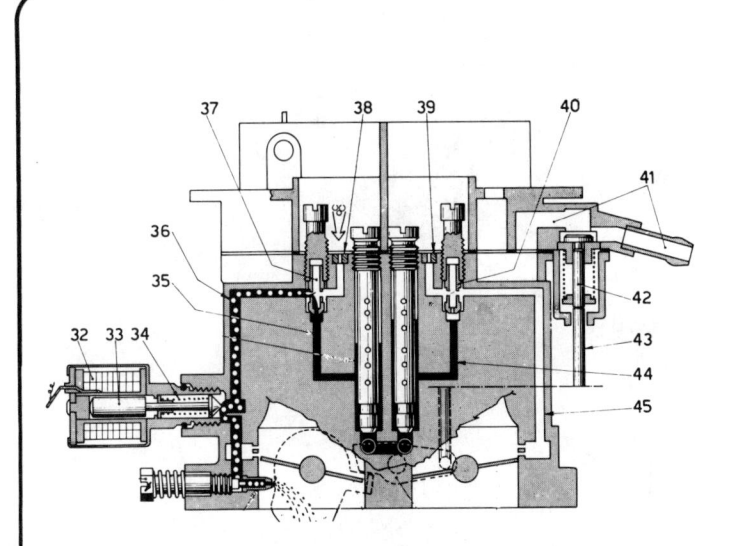

Fig. 3.11 Sectional view of Weber DATRA series carburettor (Sec 20)

- 32 Solenoid
- 33 Piston
- 34 Spring
- 35 Duct
- 36 Duct
- 37 Slow-running jets
- 38 Calibrated bush
- 39 Calibrated bush
- 40 Slow running jets
- 41 Fuel bleed off duct
- 42 Valve
- 43 Rod
- 44 Duct
- 45 Duct

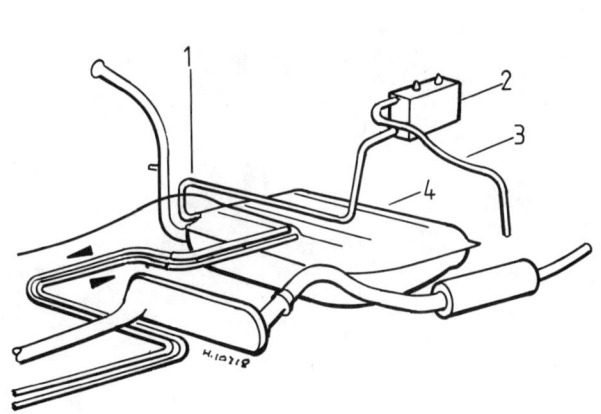

Fig. 3.12 Fuel tank evaporation system (Sec 20)

- 1 Recovering tank-to-fuel tank hose
- 2 Recovering tank
- 3 Breather hose
- 4 Fuel tank

Chapter 3 Fuel, emission control and exhaust systems

21 Emission control systems (USA versions) – general description and maintenance

1 The USA versions have a more complicated system of emission control than most other versions due to the very stringent American anti-pollution laws (Fig. 3.14).
2 Apart from the emission control systems which the UK models have fitted, the USA models have several more. Firstly they are fitted with an exhaust gas recirculation (EGR) system (Fig. 3.15). This feeds exhaust gases back to the inlet manifold via the EGR valve to be reburnt, thus reducing the output of nitrogen oxides by reducing the combustion temperatures. Secondly they are fitted with an air pump (Fig. 3.16). This is mounted above the gearbox and is belt driven by a pulley operated off the left-hand end of the front camshaft. It pumps clean air into the exhaust system via a diverter valve which is controlled by the vacuum in the inlet manifold (Fig. 3.17). This air injection system basically promotes prolonged burning of exhaust gases before they are expelled into the atmosphere thus reducing the CO content of the exhaust gases.
Thirdly the USA versions are fitted with a slightly more complicated fuel tank vapour recirculation unit (Fig. 3.18).
The purpose of this system is to vent the petrol vapour in the fuel tank through a carbon packed canister into the inlet manifold of the engine so that it is burnt. The petrol filler cap is a completely sealed type.
The carbon filled canister is located in the engine compartment. There are three pipe connections, one from the fuel tank vent system, one from the inlet manifold, and the last one from the air cleaner (Fig. 3.19).
The fuel vapours 'soak' into the carbon, which is purged of the vapour when the engine is running. The vacuum in the inlet manifold draws warm air from the air cleaner through the carbon taking the vapour into the engine through the inlet manifold pipe. Between the petrol tank and carbon canister there is a three way valve. The valve allows air into the vent system to compensate for consumption of fuel.
The valve also allows vapours to pass along the vent line to the carbon canister and engine (Fig. 3.20).
The third mode of operation of the valve allows vapour to vent directly to the atmosphere in the event of a blockage in the vent pipe to the canister.
Lastly the USA versions are also fitted with a distributor type which has two sets of contact breaker points. This is to ensure the best ignition advance curve to help prevent wastage of fuel and thus reduce the nitrogen oxide output. This is covered in Chapter 4.

22 Exhaust system – general

1 The exhaust system is conventional in its arrangement and working. It is not too difficult to repair, having two silencers and supplied in three sections. The forward section consists of the twin pipes from the manifold and a convergent section. The middle section includes the main silencer and the rear section comprises the rear silencer and tail pipe (Fig. 3.21).
2 It is wise only to use the original type exhaust clamps and proprietary made system. When any one section of the exhaust system needs renewal it often follows that the whole system is best renewed.
3 It is most important when fitting exhaust systems that the twists and contours are carefully followed and that each connecting joint overlaps the correct distance. Any stresses or strains imparted in order to force the system to fit the hanger rubbers will result in early fractures and failures.
4 When fitting a new part or a complete system it is well worth removing *all* the systems from the car and cleaning up all the joints so that they will fit together easily. The time spent struggling with obstinate joints whilst flat on your back under the car is eliminated and the likelihood of distorting or even breaking a section is greatly reduced. Do not waste time trying to undo rusted clamps and bolts. Cut them off. New ones will be required anyway if they are bad.

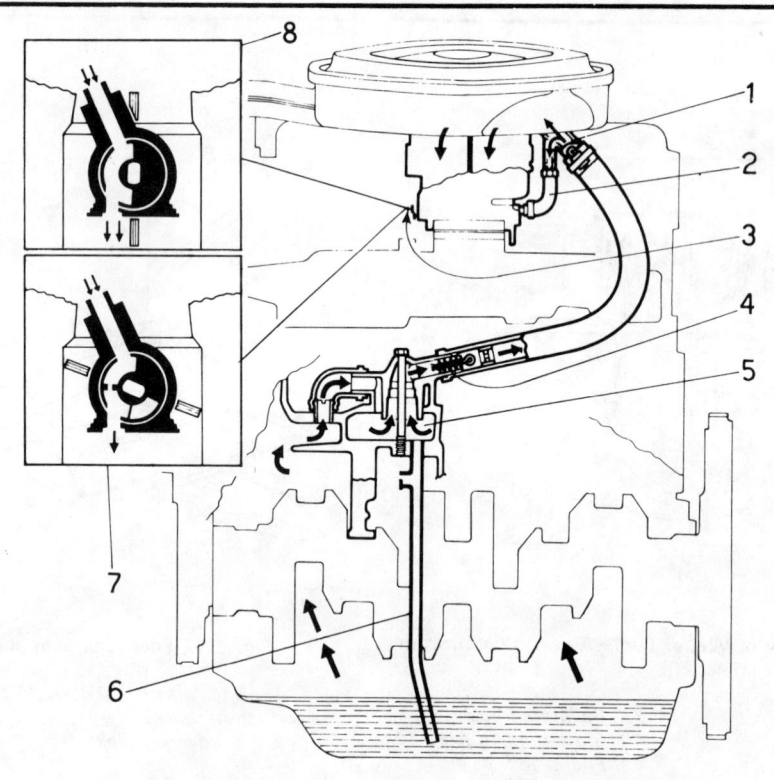

Fig. 3.13 The crankcase emission control system (Secs 20 and 21)

1 Emission feedback tubing to air cleaner
2 Air cleaner-to-carburettor tubing
3 Control valve
4 Flame trap
5 Liquid/vapour separator
6 Oil drain tubing into sump
7 Control valve with engine idling
8 Control valve with engine beyond idling

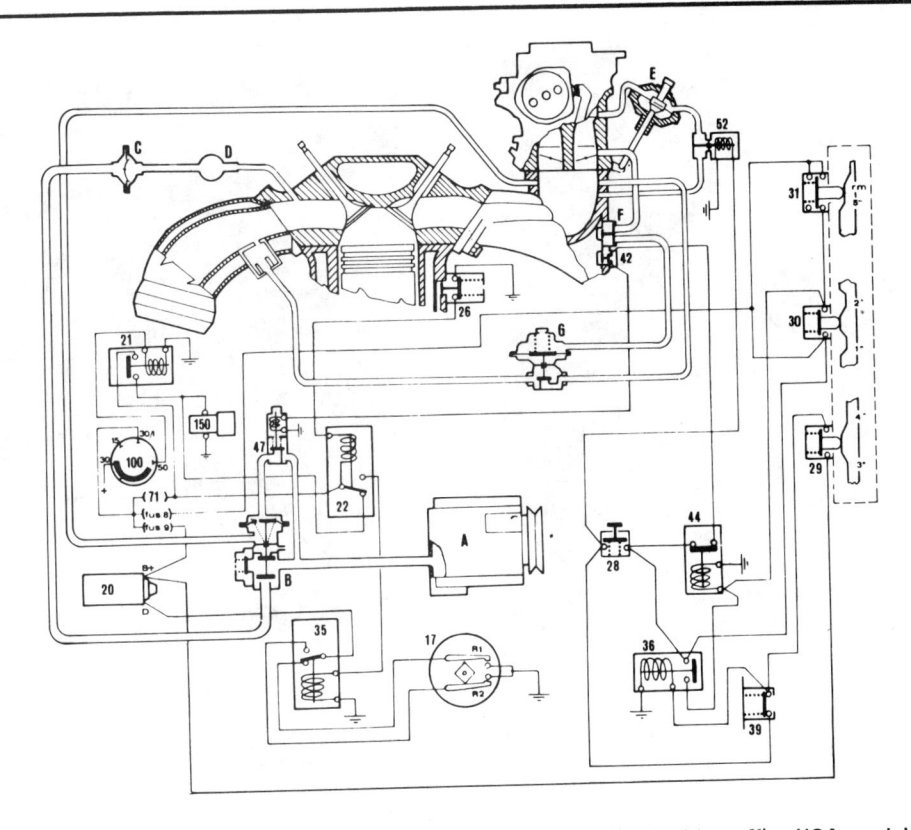

Fig. 3.14 Lay-out of exhaust emission control system at rest (engine cold; ignition off) – USA models (Sec 21)

- A Air pump
- B Diverter valve
- C Non-return valve
- D Exhaust valves air flow chamber
- E Fast idling chamber
- F EGR valve control thermostat valve
- G EGR valve
- 17 Ignition distributor
- 20 Ignition coil
- 21 Fuel pump solenoid switch
- 22 Pump and ignition remote control switch
- 26 Engine oil low pressure switch
- 28 Fast idling control switch
- 29 3rd and 4th speed switch
- 30 1st and 2nd speed switch
- 31 5th speed switch
- 35 Ignition solenoid switch
- 36 Emission control system solenoid switch
- 39 Switch on clutch pedal
- 42 Solenoid valve thermoswitch
- 44 Emission control system solenoid switch
- 47 Diverter valve control solenoid valve
- 52 Carburettor fast idling control solenoid valve
- 71 Fuel pump fuse
- 100 Ignition switch
- 150 Fuel pump

Fig. 3.15 EGR system components layout (Sec 21)

1. Union for securing exhaust gas recirculation bottom tubing
2. Exhaust ducts air delivery tubing screw
3. Clamp and exhaust ducts air delivery tubing
4. Accelerator control cable
5. Clamp and tubing between recirculation valve and thermostat valve
6. Semi-automatic choke water outlet tubing
7. Exhaust gas recirculation top tubing
8. Tubing and valve mounting bolts
9. Exhaust gas recirculation valve
10. Tubing mounting bracket
11. Exhaust gas recirculation intermediate tubing

Fig. 3.16 The air injection system air pump (Sec 21)

1 Rear fixing screws
2 Pump support
3 1st and 2nd gear switch
4 3rd and 4th gear switch
5 5th gear and reverse commutator
6 Pump through bolt
7 Air outlet tubing
8 Air pump body

Fig. 3.17 Air pump and diverter valve (Sec 21)

1 Air pump
2 Belt stretcher nut
3 Recovery tank
4 Brake fluid reservoir
5 Breather tubing
6 Brake servo unit
7 Diverter valve solenoid valve
8 Diverter valve
9 Bolt
10 Brake master cylinder
11 Valve tubing clip
12 Clip and tubing from valve to exhaust ducts air delivery tubing
13 Ignition coil

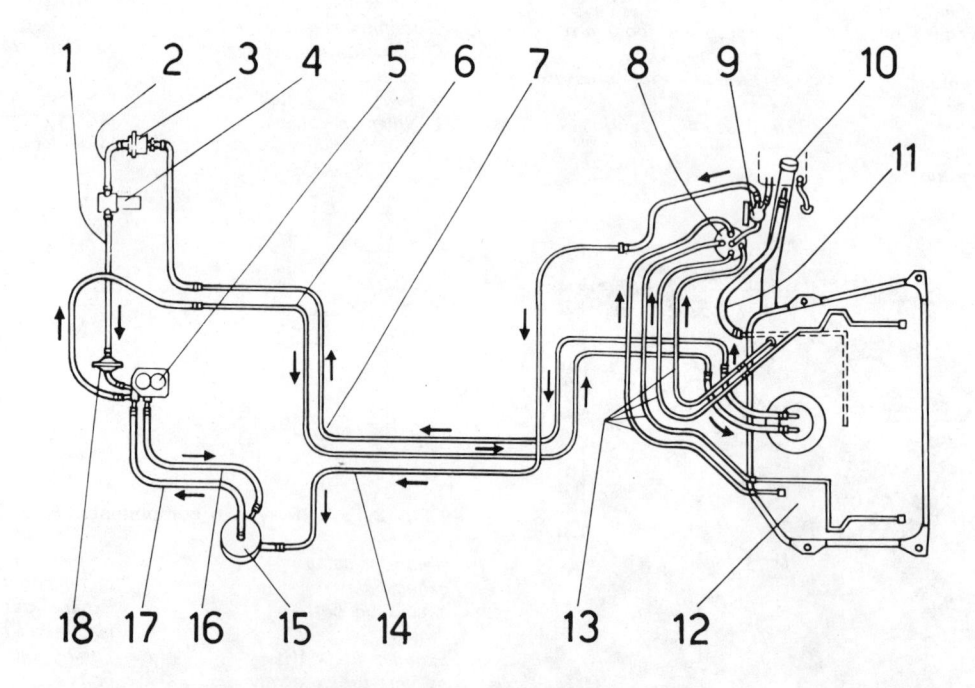

Fig. 3.18 Fuel supply and evaporative emission control system (Sec 21)

1 Pump-to-carburettor supply tubing
2 Filter-to-pump tubing
3 Filter
4 Electric pump
5 Carburettor
6 Carburettor-to-tank leak-off tubing
7 Tank-to-filter tubing
8 Vapour/liquid separator
9 Three-way valve
10 Filler with cap
11 Tank breather tubing
12 Tank
13 Tank vapour outlet tubing
14 Three-way valve-to-activated carbon trap vapour outlet
15 Activated carbon trap
16 Carburettor float chamber vapour vent tubing
17 Activated carbon trap-to-carburettor suction tubing
18 Pressure regulator

Chapter 3 Fuel, emission control and exhaust systems

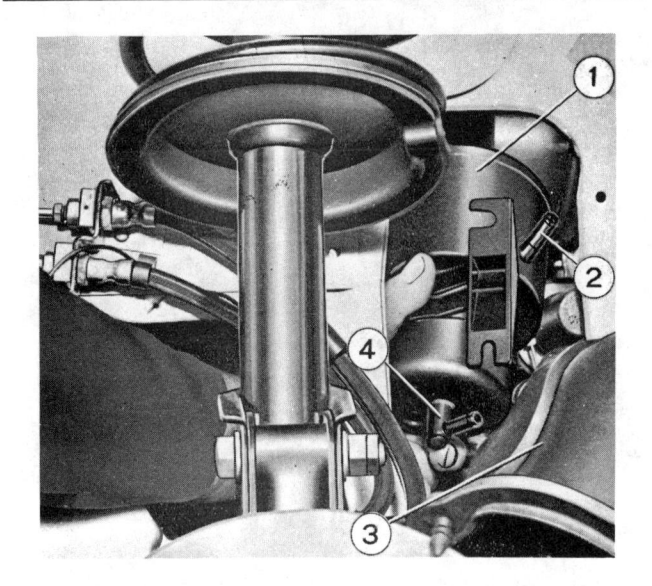

Fig. 3.19 Removing the petrol vapour trap (Sec 21)

1 Petrol vapour trap
2 Union for petrol vapour tubing from carburettor
3 Rubber guard
4 Union for petrol vapour tubing from tank

23 Exhaust system – renewal

1 Liberally apply penetrating fluid to all the nuts and bolts on the section(s) that need(s) to be renewed.

Front pipe
2 Slacken off the clamp bolt and nut on the joint between the front and centre sections of the exhaust system.
3 Undo and remove the four nuts which secure the front pipe to the exhaust manifold. Note that there are two lock tab plates between the pairs of studs that need to be knocked flat first of all (photo).
4 Undo the front exhaust pipe clamp nuts, which are again locked by lock tabs, and remove the upper part of the clamp.
5 Undo and remove the single nut and bolt, which secure the front supporting stay to the crankcase bracket.
6 The front pipe section can now be separated from the centre section and then be lowered from the manifold.
7 Refitting is the reverse procedure to removal. Remember to tighten all the clamp and mounting bolts to the correct torques. Also renew the front gasket between the front pipes and the manifold.

Centre section
8 Slacken the front section to centre section clamp nut and bolt and slide the clamp forwards.
9 Undo the centre section to rear section clamp nut and bolt behind the main silencer, in front of the rear crossmember (photo).

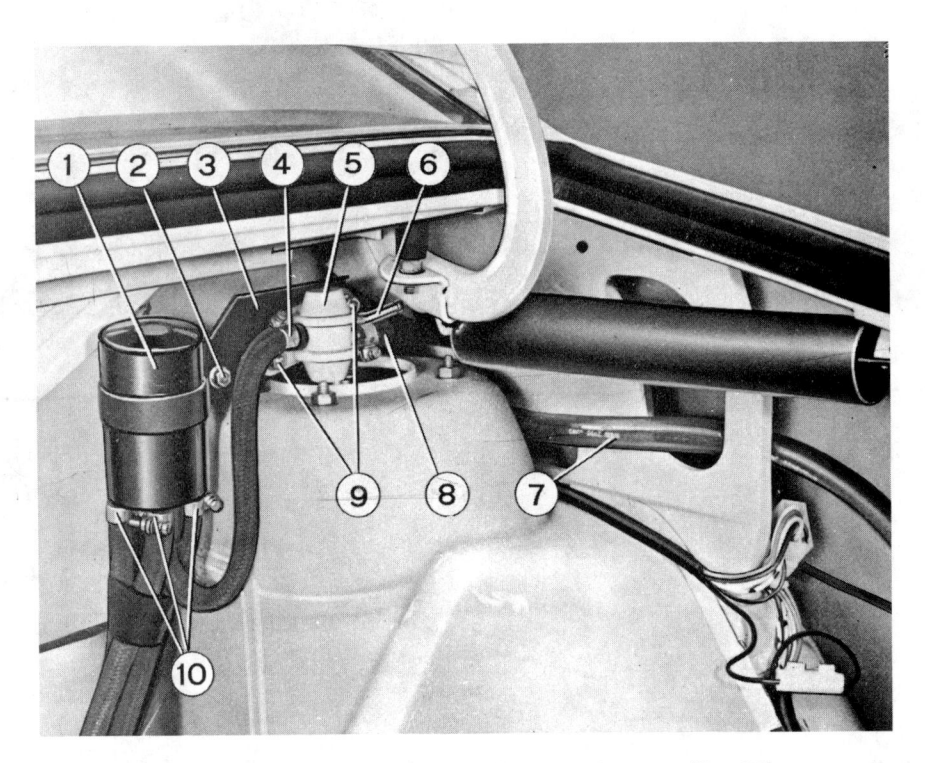

Fig. 3.20 Fuel evaporative emission control system (Sec 21)

1 Petrol vapour separator
2 Separator clamp mounting bolt
3 Support bracket
4 Clip and tubing from separator to valve
5 Tank vent valve
6 Vapour safety outlet tubing
7 Filler-to-tank tubing
8 Clip and tubing from valve to petrol vapour trap
9 Valve mounting bolts
10 Clips and tubing from tank to separator

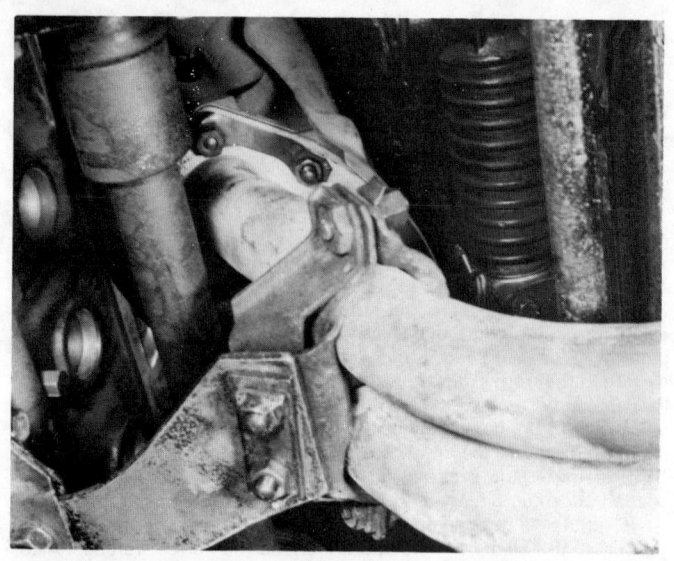

23.3 The front exhaust pipe is attached to the manifold and supported by a stay

23.9 The centre to rear section joint is just to the rear of the main silencer

23.10 There is a hanger like this on either side of the main silencer

23.14 The tail pipe silencer has a hanger mounted onto it

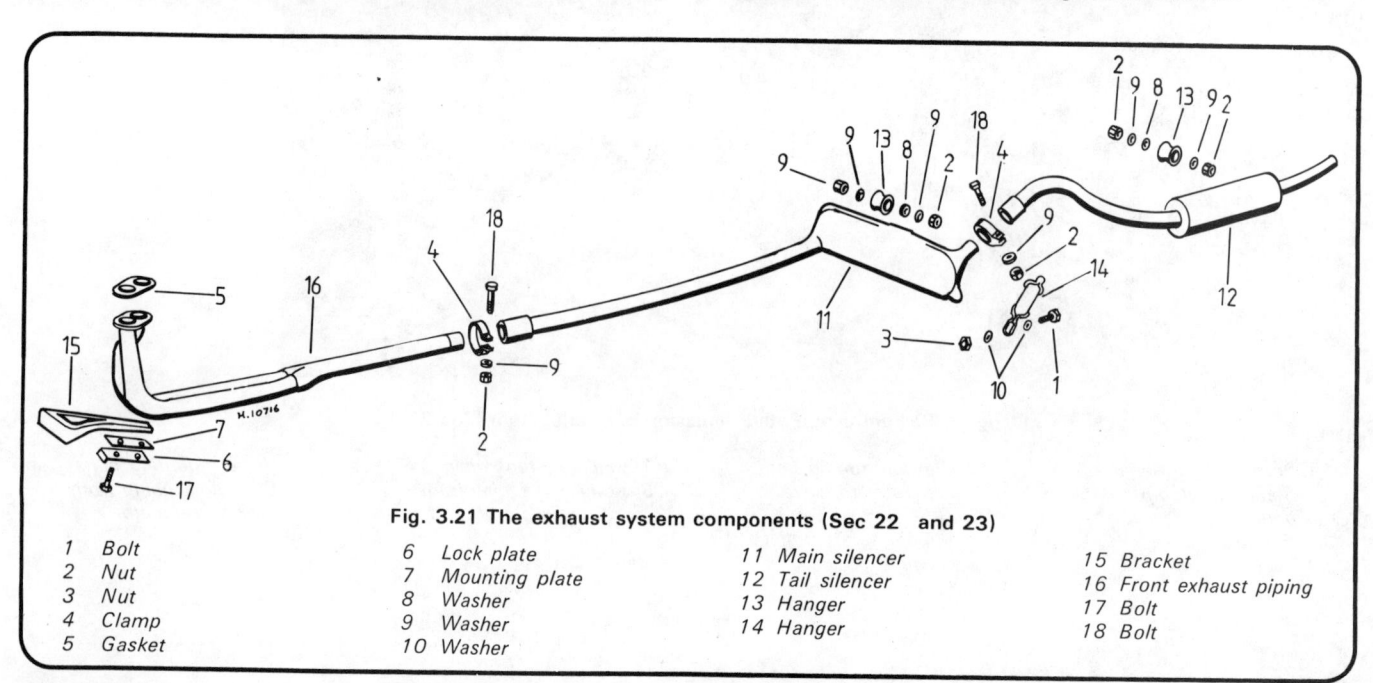

Fig. 3.21 The exhaust system components (Sec 22 and 23)

1	Bolt	6	Lock plate	11	Main silencer
2	Nut	7	Mounting plate	12	Tail silencer
3	Nut	8	Washer	13	Hanger
4	Clamp	9	Washer	14	Hanger
5	Gasket	10	Washer		
15	Bracket				
16	Front exhaust piping				
17	Bolt				
18	Bolt				

Chapter 3 Fuel, emission control and exhaust systems

10 Undo the nuts and bolts which secure the main silencer to the rubber hangers on either side (photo).
11 The centre section can now be separated from the front and rear sections and removed from the car.
12 Refitting is the reverse procedure to removal. It is wiser to use new nuts and bolts for the pipe clamps if the old ones are badly rusted. Also inspect the rubber hangers carefully and renew them if necessary.

Rear section
13 Slacken the clamp nut and bolt between the centre and rear sections of the exhaust pipe.
14 Undo and remove the nut and bolt, or both nuts, depending on the type of hanger which is fitted, from the hanger on the tail silencer (photo).
15 Remove the tail section.
16 Check the condition of the hanger and then fit the new section in the reverse order to removal and tighten the clamp nut and bolt.

24 Exhaust manifold gasket – renewal

1 Should the exhaust manifold gasket blow then it is just possible to renew it with the engine in the car. However this only applies to UK models. Cars fitted with EGR systems can only have their exhaust manifold removed with the engine out of the car.
2 With a suitable cranked spanner undo the five manifold retaining nuts, noting that the right-hand one has an engine lifting eye attached to it. The heat guard may tend to get in the way while the nuts are being removed.
3 Lift the manifold off the studs and lower it down.
4 The old gasket can now be removed.
5 Clean up the cylinder head flange face and scrape any old bits of gasket or carbon deposit from the manifold as well.
6 Fit a new gasket to the studs and refit the exhaust manifold. Refit the nuts and lifting eye and tighten the nuts to the specified torque.

25 Fault diagnosis – fuel, carburation, and exhaust systems

Symptom	Reason(s)
Smell of petrol when engine is stopped	Leaking fuel lines or unions Leaking fuel tank
Smell of petrol when engine is idling	Leaking fuel line unions between pump and carburettor Overflow of fuel from float chamber due to wrong level setting, ineffective needle valve or punctured float
Excessive fuel consumption for reasons not covered by leaks or float chamber faults	Worn jets Over-rich setting Sticking mechanism
Difficult starting, uneven running, lack of power, cutting out	One or more jets blocked or restricted Float chamber fuel level too low or needle valve sticking Fuel pump not delivering sufficient fuel

Unsatisfactory engine performance and excessive fuel consumption are not necessarily the fault of the fuel system or carburettor. In fact they more commonly occur as a result of ignition and timing faults. Before acting on the following it is necessary to check the ignition system first. Even though a fault may lie in the fuel system it will be difficult to trace unless the ignition is correct. The faults below, therefore, assume that this has been attended to first (where appropriate).

Chapter 4 Ignition system

Contents

Coil – removal and refitting	12
Condenser – testing, removal and refitting	5
Distributor (single points) – removal and refitting	6
Distributor (Marelli type S147HX) – overhaul	7
Distributor contact points (single set) – gap adjustment	3
Distributor contact points (single set) – removal and replacement	4
Distributor contact breaker points (Marelli type S144 CBY) – adjustment, removal and refitting	8
Distributor (Marelli type S144 CBY) – overhaul	10
Distributor (Marelli type S144 CBY) – removal and refitting	9
Electronic ignition – general and maintenance	11
Fault diagnosis – engine fails to start	16
Fault diagnosis – engine misfires	17
General description	1
Ignition timing – static	13
Ignition system – fault finding	15
Routine maintenance – ignition components	2
Spark plugs and HT leads	14

Specifications

Contact breaker type ignition system

Distributor application (UK models)

	Standard models	Models with air conditioner
1300 engines	Marelli S147 CX	S144 Q
1400 engines	Marelli S147 CX	–
1600, 1800 and 2000 engines	Marelli S147 HX or Bosch JFR 4 (R)	Marelli S144 BA or S144 E

Distributor application (USA models)

1800 models	Marelli S144 CBY with 2 contact breaker points system
1800 California models	Marelli S144 M

Adjustment data

Distributor rotation	Clockwise
Contact breaker points gap:	
All main points	0.37 to 0.43 mm (0.014 to 0.016 in)
Auxiliary points (California models)	0.31 to 0.49 mm (0.012 to 0.019 in)
Dwell angle (all distributors)	55° ± 3°
Ignition timing (static):	
All models except 1800 Coupe, Spider and 1800 USA models	10° BTDC
1800 Coupe and Spider models	7° 30′ BTDC
1800 USA models	0° (TDC)

Firing order ... 1-3-4-2. No 1 cylinder on right-hand end (front of engine)

Ignition coil

Type	Bosch K12V, Marelli Bes 200A or K litz G37 SU
Location of coil	Front right-hand wing

Spark plugs

	UK	USA
Type:		
Bosch	W6D	W7D
Champion	N77	N97
Marelli	CW78LP	CW7LP
Gap:		
Bosch	0.5 to 0.7 mm (0.019 to 0.027 in)	
Champion	0.5 to 0.6 mm (0.019 to 0.024 in)	
Marelli	0.6 to 0.7 mm (0.024 to 0.027 in)	

Torque wrench settings

	lbf ft	Nm
Spark plugs	27	37

Chapter 4 Ignition system

Electronic type ignition system

Distributor type .... Bosch JGFU 4 – Z44 – 14483 – A2

Distributor rotation .... Clockwise

Ignition timing (static) .... 10° BTDC

Ignition coil
Type .... Bosch KW12V
Location of coil .... Front right-hand inner wing
Location of electronic ignition control unit .... Mounted on plate attached to right front suspension mounting

Spark plugs
Type .... Bosch W6D, Champion N7Y or Marelli CW78LP
Spark plug electrode gap .... 0.6 to 0.7 mm (0.023 to 0.027 in)

Torque wrench settings | lbf ft | Nm
Spark plugs .... 27 | 37

1 General description

Up to November 1978 all models were fitted with a conventional ignition system employing a distributor with contact breaker points, a coil and spark plugs. From that point onwards all production models have an electronic ignition system fitted as standard equipment. This system employs a special Bosch distributor without contact breaker points, a special Bosch coil and an electronic control unit, plus, of course, spark plugs.

The main reason for fitting an electronic ignition system is to provide a more regular spark and therefore more efficient ignition.

In order that the internal combustion engine can operate properly, the spark which ignites the air-fuel charge in the combustion chamber must be delivered at precisely the correct moment. This correct moment is that which will allow time for the charge to burn sufficiently to create the highest pressure and temperature possible in the combustion chamber as the piston passes top dead centre and commences its power stroke. The distributor and ignition coil are the main devices which ensure that the spark plug ignites the charge as required.

Very high voltages need to be generated in the ignition system in order to produce the spark across the plug gap which ignites the fuel/air charge. The device in which these high voltages – several thousand volts – are generated is the coil (or electronic ignition pack which is fitted to more recent models). The coil contains two sets of windings – the primary and the secondary windings. A current at 12 volts is fed through the primary windings via the contact breaker mechanism in the distributor. It is precisely when the flow is interrupted by the contact breaker that the huge voltage is momentarily induced in the secondary windings and that voltage is conveyed via HT leads and the rotor arm in the distributor cap to the appropriate spark plug.

It follows therefore that the contact breakers must part the instant a spark is required and the rotor arm must be aligned to the appropriate stud in the distributor cap which is connected to the spark plug which 'needs' the spark. The distributor shaft revolves at half crankshaft speed, and there are four rises on the distributor cam and four studs in the distributor cap, to cater for the four sparks the engine requires each two revolutions of the crankshaft.

The timing of the ignition is set by two means. The first is referred to as static timing and the second, which is fully automatic, is the centrifugal advance mechanism. The mechanism comprises two weights on an arm on the distributor shaft. As the shaft speed increases the weights move outwards against their restraining springs. The contact breaker cam is attached to the weights so that as they move out it is rotated relaitve to the distributor shaft, and therefore the contact breaker will open earlier relative to the distributor shaft and engine crankshaft as required.

The static ignition advance is that nominal amount which corresponds to the time for combustion at the idling speed of the engine. It is necessary therefore when carrying out ignition and carburation adjustments to ensure that the engine is turning at the speed appropriate to that test or adjustment.

Models sold in the USA (not California) are fitted with special distributors with two sets of contact breaker points (see Specifications). When electronic ignition is fitted, the distributor also incorporates a vacuum advance control.

2 Routine maintenance – ignition components

a) **Spark plugs**
Remove the plugs and thoroughly clean away all traces of carbon. Examine the porcelain insulation round the central electrode inside the plug; if damaged discard the plug. Reset the gap between the electrodes. Do not use a set of plugs for more than 12,000 miles. It is false economy. Check them every 6,000 miles.

b) **Distributor**
Every 6,000 miles remove the cap and put one or two drops of engine oil on the centrifugal weight pivots. Smear the surfaces of the cam itself with petroleum jelly. Do not over-lubricate as any excess could get onto the contact point surfaces and cause ignition difficulties.

Every 6,000 miles examine the contact point surfaces. If there is a build up of deposits on one face and a pit in the other it will be impossible to set the gap correctly and they should be refaced or renewed. Set the gap when the contact surfaces are in order.

c) **General**
Examine all leads and terminals for signs of broken or cracked insulation. Also check all terminal connections for slackness or signs of fracturing of some strands of wire. Partly broken wire should be renewed.

The HT leads are particularly important as any insulation faults will cause the high voltage to 'jump' to the nearest earth and this will prevent a spark at the plug. Check that no HT leads are loose or in a position where the insulation could wear due to rubbing against part of the engine.

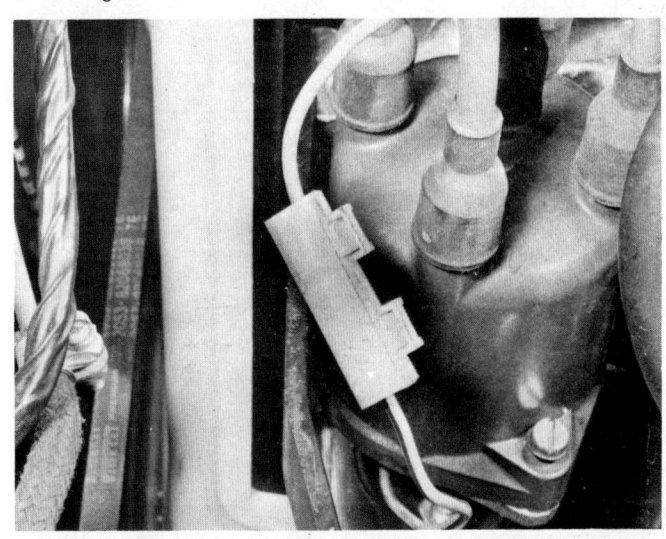

3.1 The Marelli type of distributor cap is held in place by two screws

3 Distributor contact points (single set) – gap adjustment

1 To adjust the contact breaker points so that the correct gap is obtained, first release the two screws securing the cap to the distributor body and lift the cap away. Clean the inside and outside of the cap with a dry cloth. It is unlikely that the four segments will be badly burned or scored, but if they are, the cap must be renewed. If only a small deposit is on the segments it may be scraped away using a small screwdriver (photo).
2 Push in the carbon brush located in the top of the cap several times to ensure that it moves freely. The brush should protrude by at least a quarter of an inch.
3 Gently prise open the contact breaker points to examine the condition of their faces. If they are rough, pitted or dirty, it will be necessary to remove them for resurfacing, or for new points to be fitted.
4 Presuming the points are satisfactory, or they have been cleaned or renewed, measure the gap between the points by turning the engine over until the contact breaker arm is on the peak of one of the four cam lobes. Insert a feeler gauge of the specified size to measure the gap.
5 A feeler gauge should now just fit between the points.
6 If the gap varies from the specified amount, slacken the contact plate securing screw and adjust the contact gap by inserting a screwdriver in the notched hole of the stationary point and table and turn in the required direction to increase or decrease the gap.
7 Tighten the securing screw and check the gap again. Refit the distributor cap and secure it with the two screws.

4 Distributor contact points (single set) – removal and refitting

1 If the contact breaker points are burned, pitted or badly worn, they must be removed and either renewed or their faces must be filed smooth.
2 To remove the points, first unscrew the terminal nut and remove it together with the washer under its head. Remove the low tension cable from the terminal.
3 Unscrew and remove the contact breaker locking screw.
4 Detach the low tension and condenser cables and lift away the contact breaker points.
5 To reface the points, rub the faces on a fine carborundum stone or on fine emery paper. It is important that the faces are rubbed flat and parallel to each other so that there will be complete face to face contact when the points are closed. One of the points will be pitted and the other will have deposits on it.
6 It is necessary to remove completely the built up deposits, but not necessary to rub the pitted point right to the stage where all the pitting has disappeared, though obviously if this is done it will prolong the time before the operation of refacing the points has to be repeated.
7 Refitting the contact breaker points is the reverse sequence to removal. It will be necessary to adjust the points gap as described in Section 3.

5 Condenser – testing, removal and refitting

1 The purpose of the condenser (sometimes known as a capacitor), is to ensure that when the contact breaker points open there is no sparking across them which would waste voltage and cause wear.
2 The condenser is fitted in parallel with the contact breaker points. If it develops a short circuit, it will cause ignition failure as the points will be prevented from interrupting the low tension circuit. In the USA model with 2 sets of contact breaker points there are 2 condensers; 1 for each set of contact breaker points.
3 If the engine becomes very difficult to start or begins to miss after several miles running, and the breaker points show signs of excessive burning, then the condition of the condenser must be suspect. A further test can be made by separating the points by hand with the ignition switched on. If this is accompanied by a flash it is indicative that the condenser has failed.
4 Without special test equipment the only sure way to diagnose condenser trouble is to replace a suspected unit with a new one and note if there is any improvement.
5 To remove the condenser from the distributor, first remove the distributor cap, and rotor arm.
6 Detach the condenser cable from the terminal block and remove the condenser fixing to the body. Lift away the condenser.
7 Refitting the condenser is a reversal of the removal procedure.

6 Distributor (single points) – removal and refitting

1 To remove the distributor from the engine, start by removing all the HT leads from the spark plugs and remove the battery connections – a safety precaution.
2 Remove the HT leads from the coil, and disconnect the LT lead from the snap connector (photo). Remove the cap.
3 Turn the crankshaft until the rotor arm is pointing to the insert in the distributor cap which is connected to the number 4 spark plug lead. Apply the handbrake, and select a gear to ensure that the engine does not turn subsequently. Mark the distributor body and cylinder block in relation to each other.
4 Undo and remove the nut, washer and plate securing the distributor to the cylinder block (photo).
5 Lift away the distributor. Refitting the unit is simply the reversal of the removal procedure. Remember to hold the rotor arm and distributor shaft aligned to the No 4 cylinder insert in the cap, when refitting the distributor into the engine (photo).
6 If the crankshaft has been rotated, before the distributor was refitted, it may be necessary to retime the ignition as described in Section 11 of this Chapter.

7 Distributor (Marelli type S147 HX) – overhaul

1 Remove the distributor from the car as described in Section 6. Give the outside a good clean in a petrol bath, and dry it off.
2 Fit the distributor body in a vice with protected jaws.
3 Undo the two screws which secure the rotor arm to the shaft and lift it away.
4 Beneath the rotor arm on the top of the rotor shaft is a mounting plate with the centrifugal advance weights attached to it. This is different from most distributors where they are fitted to the bottom of the shaft inside the distributor body.
5 Remove the contact breaker set and the condenser.
6 Using a parallel pin punch drive the pin out of the distributor shaft at the bottom of the body. Then slide off the oil thrower and washer.

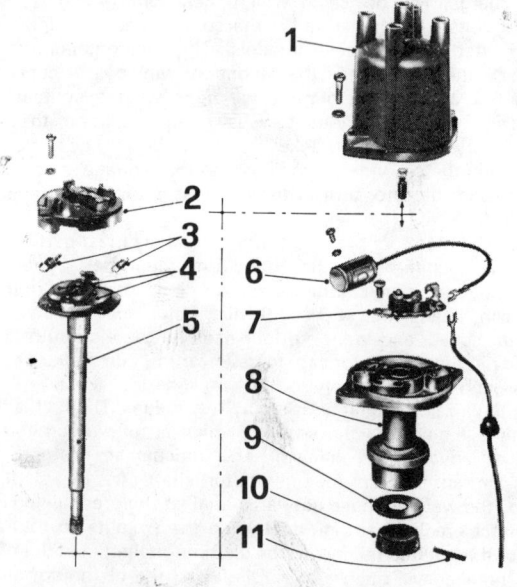

Fig. 4.1 Marelli S147 HX distributor exploded view (Sec 7)

1 Distributor cap
2 Rotor
3 Balance weight springs
4 Centrifugal weights
5 Rotor shaft
6 Condenser
7 Contact breaker
8 Distributor body
9 Washer
10 Oil thrower
11 Parallel pin

Chapter 4 Ignition system

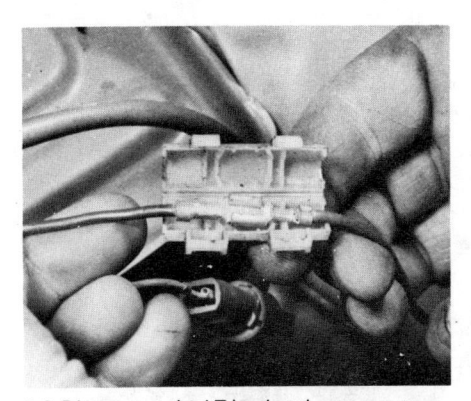

6.2 Disconnect the LT lead at the connector

6.4 Undoing the distributor clamp plate locking nut

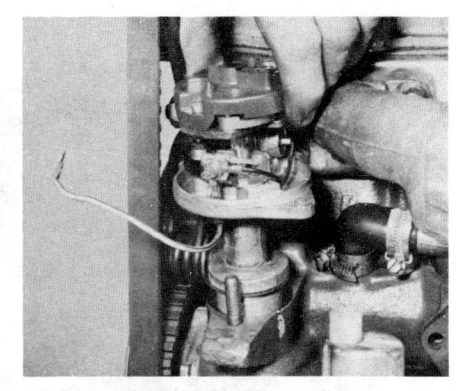

6.5 Ensure that the distributor is lined up correctly before inserting it

7 Check the movement of the shaft in the body. If there is too much play then a new distributor is required.
8 Withdraw the shaft from the body and check the centrifugal advance weights and springs.
9 If the centrifugal advance mechanism and/or the distributor shaft in the body, are found to be worn and sloppy, you should check whether spares are available and consider an exchange distributor. Very often in fine mechanisms such as distributors, it will not be a case of needing to renew just one component.
10 Examine the balance weights and pivot pins for wear, and renew the weights or centre shaft if a degree of wear is found.
11 Examine the centre shaft and the fit of the cam assembly on the shaft. If the clearance is excessive compare the items with new units and renew either, or both, if they show excessive wear.
12 If the shaft is a loose fit in the distributor bushes and can be seen to be worn it will be necessary to fit a new shaft and bushes.
13 Examine the length of the balance weight springs and compare them with new springs. If they have stretched, they should be renewed.
14 Check the contact breaker points as described in Section 2. Check the distributor cap for signs of tracking, indicated by a thin black line between the segments. Renew the cap if any signs of tracking are found.
15 If the metal portion of the rotor arm is badly burnt or loose renew the arm. If slightly burnt clean the arm with a fine file. Check that the carbon brush moves freely in the centre of the distributor cover.
16 Reassembly is a straightforward reversal of the dismantling process, but there are several points which should be noted.
17 Lubricate the balance weights and other parts of the mechanical advance mechanism, and the distributor centre shaft, with oil, during assembly. Do not oil excessively but ensure these parts are adequately lubricated.
18 Check the action of the weights in the fully advanced and fully retarded positions and ensure they are not binding.
19 Finally, set the contact breaker gap to the correct clearance when the distributor has been refitted to the car.

8 Distributor contact breaker points (Marelli type S144 CBY) – adjustment, removal and refitting

The procedure for adjusting or renewing the contact breaker points on one of the Marelli distributors fitted with 2 sets of contact breaker points is basically the same as for a standard distributor as described in Sections 3 and 4. Ensure that each set of contact breaker points is set to the correct gap (see the Specifications).

9 Distributor (Marelli type S144 CBY) – removal and refitting

1 The USA versions fitted with this type of distributor have the unit mounted on the rear of the engine driven off the camshaft (Figs 4.2 and 4.3).
2 The procedure for removing and refitting this type of unit is exactly

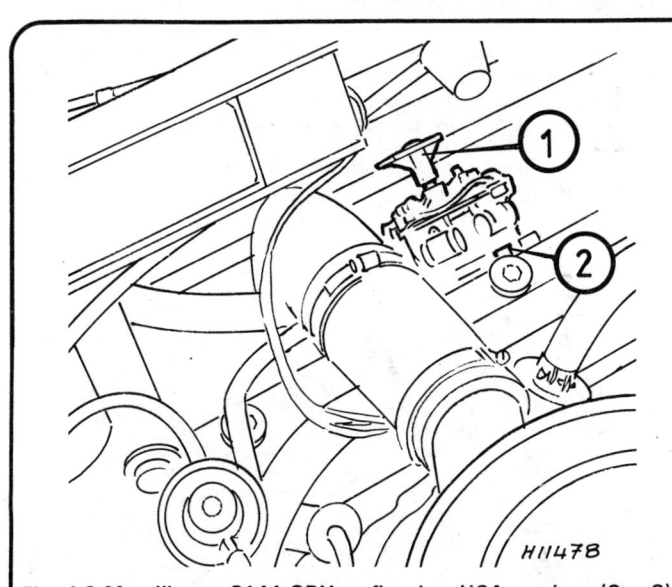

Fig. 4.2 Marelli type S144 CBY as fitted to USA versions (Sec 8)

1 Rotor arm
2 Locking nut for distributor clamp plate

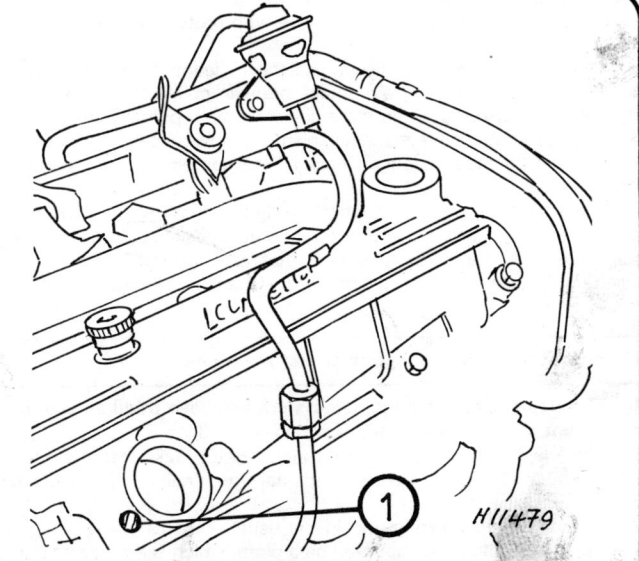

Fig. 4.3 The USA version distributor mounting viewed from the rear (distributor removed) (Sec 8)

1 Distributor mounting bolt

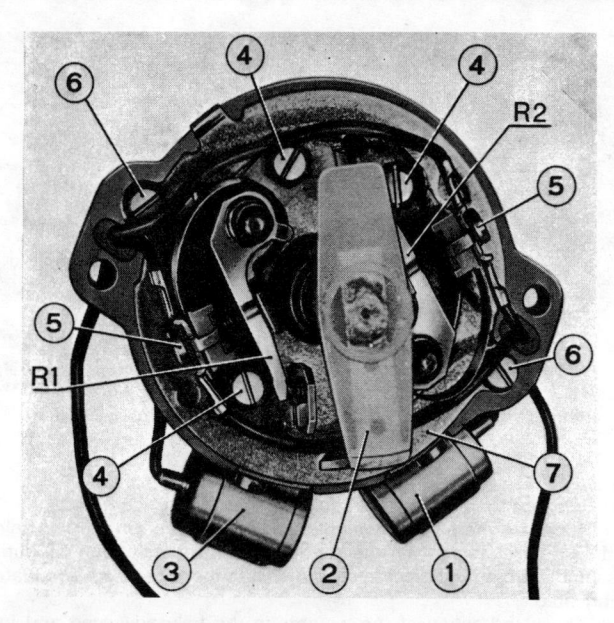

Fig. 4.4 Marelli S144 CBY distributor as fitted to USA models (Sec 8)

1. Condenser
2. Rotor arm
3. Condenser
4. Contact breaker points attachment screws
5. Leads retaining screws
6. Contact breaker plate attachment screws
7. Contact breaker plate
R1 Main contact breaker points
R2 Second contact breaker points

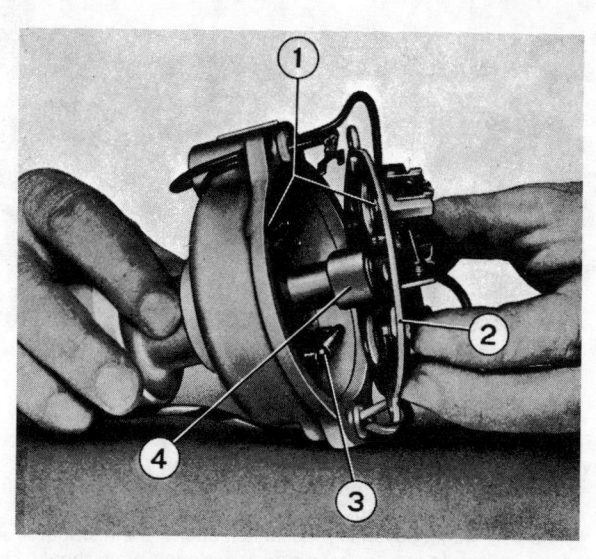

Fig. 4.5 The S144 CBY distributor with the mounting plate lifted away (Sec 8)

1. Plate assembly references
2. Plate
3. Automatic advance weights return spring
4. Cam

the same as for the standard distributor as described in Section 6, except that there are two LT leads, one for each set of points.
3   When the distributor is refitted the meshing of the gear on the end of the distributor shaft with the camshaft drivegear will alter the distributor setting. Do not lock the distributor clamp nut until the timing has been checked, as described in Section 11.

10 Distributor (Marelli type S144 CBY) – overhaul

1   With the distributor removed from the car, give the outside of the unit a good clean and then dry it off.
2   Place the distributor body in a vice with protective jaws. Remove the rotor arm.
3   Undo the screws and disconnect the LT leads from the points.
4   Undo and remove the contact breaker points mounting plate screws (6), and remove the plate complete with points (Fig 4.5).
5   The centrifugal advance mechanism can now be seen. Remove the springs and balance weights. Check all the components as described in Section 7.
6   Clean all the parts thoroughly and lightly oil all the moving parts before refitting them. Note that the contact breaker points mounting plate has a locating lug to ensure that it is refitted in the correct position. Reassembly is a reversal of the dismantling procedure.

11 Electronic ignition – general and maintenance

1   The Bosch electronic ignition system has been fitted as standard equipment to all models since November 1978.
2   The electronic ignition system includes a breakerless distributor with a vacuum advance system, a special Bosch coil and an electronic control unit.
3   The distributor is mounted in the usual place as is the coil. The electronic control unit is mounted on a plate, which is attached by two of the right-hand front suspension upper mounting nuts to the suspension housing (Fig 4.6).
4   The purpose of electronic ignition is to provide a more powerful and therefore more efficient spark.

5   The amount of maintenance for the electronic ignition system is minimal. Apart from checking the plugs regularly, the distributor cap should be removed every year (12 000 miles (20 000 km)) and the reluctor extracted.
6   Check the cap and reluctor for cracks and the terminals for carbon deposits, or wear. Clean them both using a petrol dampened cloth and scrape off any carbon deposits. If the cap or reluctor are worn then they must be renewed.

12 Coil – removal and refitting

1   The coil is located on the right-hand front inner wing and is secured in place by a strap and two nuts.
2   Undo one nut and slacken off the other.
3   Disconnect the low tension leads from the coil. Note which fit where.
4   Disconnect the HT lead from the centre of the coil.
5   Lift off the coil and mounting bracket.
6   Refit the assembly in the reverse order and reconnect the HT and LT leads.

Fig. 4.6 The electronic ignition control unit is mounted on the right-hand suspension mounting (Sec 11)

Measuring plug gap. A feeler gauge of the correct size (see ignition system specifications) should have a slight 'drag' when slid between the electrodes. Adjust gap if necessary

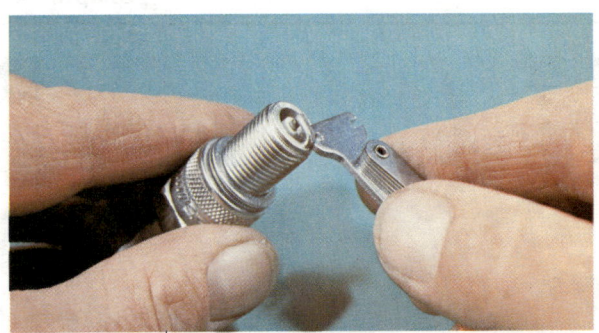

Adjusting plug gap. The plug gap is adjusted by bending the earth electrode inwards, or outwards, as necessary until the correct clearance is obtained. Note the use of the correct tool

Normal. Grey-brown deposits, lightly coated core nose. Gap increasing by around 0.001 in (0.025 mm) per 1000 miles (1600 km). Plugs ideally suited to engine, and engine in good condition

Carbon fouling. Dry, black, sooty deposits. Will cause weak spark and eventually misfire. Fault: over-rich fuel mixture. Check: carburettor mixture settings, float level and jet sizes; choke operation and cleanliness of air filter. Plugs can be re-used after cleaning

Oil fouling. Wet, oily deposits. Will cause weak spark and eventually misfire. Fault: worn bores/piston rings or valve guides; sometimes occurs (temporarily) during running-in period. Plugs can be re-used after thorough cleaning

Overheating. Electrodes have glazed appearance, core nose very white – few deposits. Fault: plug overheating. Check: plug value, ignition timing, fuel octane rating (too low) and fuel mixture (too weak). Discard plugs and cure fault immediately

Electrode damage. Electrodes burned away; core nose has burned, glazed appearance. Fault: pre-ignition. Check: as for 'Overheating' but may be more severe. Discard plugs and remedy fault before piston or valve damage occurs

Split core nose (may appear initially as a crack). Damage is self-evident, but cracks will only show after cleaning. Fault: pre-ignition or wrong gap-setting technique. Check: ignition timing, cooling system, fuel octane rating (too low) and fuel mixture (too weak). Discard plugs, rectify fault immediately

13.5A Engine at TDC – Check specifications for correct timing

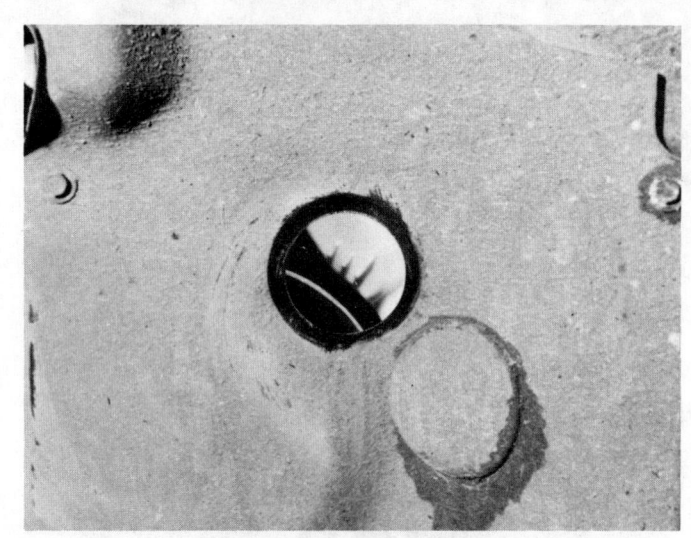

13.5B The edge of the crankshaft pulley and the timing marks can be viewed through the hole

13 Ignition timing – static

1   If the clamp plate pinch nut has been loosened on the distributor and the static timing lost or if for any other reason, it is wished to set the ignition timing proceed as follows:
2   Refer to Section 3 and check the contact breaker points (if fitted). Reset as necessary.
3   Assemble the clamp plate to the distributor body but do not tighten the pinch nut fully.
4   It will be as well to remove the camshaft covers so that the exact position of the engine assembly can be determined.
5   Slowly turn the crankshaft until the groove in the crankshaft pulley lines up with the static ignition point on the timing belt cover (see Fig 4.7), and the crankshaft lobes operating the No 4 cylinder valves are pointing upwards and inclined towards the centre line of the engine. The piston in the No 4 cylinder should then be just approaching Top Dead Centre and just about to commence the power stroke (photos).
6   Provided the distributor shaft has not been disturbed or has been refitted correctly the rotor arm should now point to the position of the insert in the distributor cap which is connected to the spark plug in the No 4 cylinder.
7   With the distributor body lightly clamped with the retaining plate, slowly rotate the distributor body anti-clockwise until the contact breaker points are just beginning to open. Tighten the distributor clamp.
8   Difficulty is sometimes experienced in determining exactly when the contact breaker points open. This can be ascertained most accurately by connecting a 12v bulb in parallel with the contact breaker points (one lead to earth and the other from the distributor low tension terminal). Switch on the ignition and turn the distributor body until the bulb lights up, indicating the points have just opened.
9   If it was not found possible to align the rotor arm correctly one of two things is wrong. Either the distributor driveshaft has been incorrectly fitted in which case the distributor must be removed and refitted as described in Section 6 of this Chapter and Chapter 1; or the distributor cam assembly has been incorrectly fitted on the driveshaft. To rectify this, it will be necessary to partially dismantle the distributor and check the position of the cam assembly on the centrifugal advance mechanism: it may be 180° out of position.
10 It should be noted that this adjustment is nominal and the final adjustment should be made under running conditions.
11 First start the engine and allow to warm up to normal running temperature, then in a road test accelerate the car in top gear from 30 to 50 mph (50 to 80 kmph) whilst listening for heavy pinking of the engine. If pinking is heard, the ignition needs to be slightly retarded until only the faintest trace of pinking can be heard under these operating conditions.
12 Since the ignition advance adjustment enables the firing point to be related correctly to the grade of fuel used, the fullest advantage of any change of fuel will only be obtained by re-adjustment of the ignition settings as described in paragraph 11.

14 Spark plugs and HT leads

1   The correct functioning of the spark plugs is vital for the correct running and efficiency of the engine.
2   At intervals of 6000 miles (10 000 km), the plugs should be removed, examined, cleaned and if worn excessively, renewed. The condition of the spark plugs will also tell much about the overall condition of the engine.
3   If the insulator nose of the spark plug is clean and white, with no deposits, this is indicative of a weak mixture, or too hot a plug (a hot plug transfers heat away from the electrode slowly – a cold plug transfers it away quickly.
4   If the top of the insulator nose is covered with hard black-looking deposits, then this is indicative that the mixture is too rich. Should the plug be black and oily, then it is likely that the engine is fairly worn, as well as the mixture being too rich.
5   If the insulator nose is covered with light tan to greyish brown deposits, then the mixture is correct and it is likely that the engine is in good condition.
6   If there are any traces of long brown tapering stains on the outside of the white portion of the plug, then the plug will have to be renewed, as this shows that there is a faulty joint between the plug body, and the insulator, and compression is being allowed to leak away.
7   Plugs should be cleaned by a sand blasting machine which will free them from carbon more thoroughly than cleaning by hand. The machine will also test the condition of the plugs under compression.

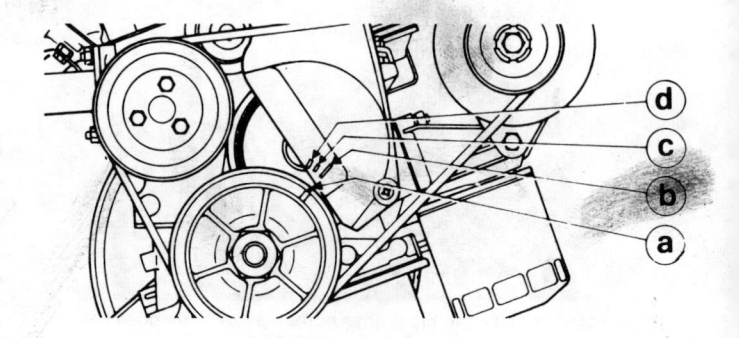

Fig. 4.7 The timing marks (Sec 13)

a – Mark on crankshaft pulley
b – 0° (TDC)
c – 5° BTDC
d – 10° BTDC

Any plug that fails to spark at the recommended pressure should be renewed.

8 The spark plug gap is of considerable importance, as if it is too large or too small, the size of the spark and its efficiency will be seriously impaired. The spark plug should be set to the figure given in the Specification at the beginning of this Chapter.

9 To set it, measure the gap with a feeler gauge, and then bend open, or close, the outer plug electrode until the correct gap is achieved. The centre electrode should never be bent as this may crack the insulation and cause plug failure if nothing worse.

10 Refit the plugs, and refit the leads from the distributor in the correct firing order, which is given in the Specifications.

11 The plug leads require no routine attention other than being kept clean and wiped over regularly.

12 At intervals of 6000 miles (10 000 km) or 6 months, pull the leads off the plugs and distributor one at a time and make sure no water has found its way onto the connections. Remove any corrosion from the brass ends, wipe the collars on top of the distributor, and refit the leads.

15 Ignition system – fault finding

By far the majority of breakdown and running troubles are caused by faults in the ignition system either in the low tension or high tension circuits.

There are two main symptoms indicating ignition faults. Either the engine will not start or fire, or the engine is difficult to start and misfires. If it is a regular misfire, ie the engine is running on only two or three cylinders, the fault is almost sure to be in the secondary or high tension circuit. If the misfiring is intermittent, the fault could be in either the high or low tension circuits. If the car stops suddenly, or will not start at all, it is likely that the fault is in the low tension circuit. Loss of power and overheating, apart from faulty carburation settings, are normally due to faults in the distributor or to incorrect ignition timing.

16 Fault diagnosis – engine fails to start

1 If the engine fails to start and the car was running normally when it was last used, first check there is fuel in the petrol tank. If the engine turns over normally on the starter motor and the battery is evidently well charged, then the fault may be in either the high or low tension circuits. First check the HT circuit. **Note**: *If the battery is known to be fully charged, the ignition light comes on, and the starter motor fails to turn the engine* **check the tightness of the leads on the battery terminals** *and also the secureness of the earth lead to its connection to the body. It is quite common for the leads to have worked loose, even if they look and feel secure. If one of the battery terminal posts gets very hot when trying to work the starter motor this is a sure indication of a faulty connection to that terminal.*

2 One of the commonest reasons for bad starting is wet or damp spark plug leads and distributor. Remove the distributor cap. If condensation is visible internally dry the cap with a rag and also wipe over the leads. Refit the cap.

3 If the engine still fails to start, check that current is reaching the plugs, by disconnecting each plug lead in turn at the spark plug end, and holding the end of the cable about $\frac{3}{16}$ inch (5 mm) away from the cylinder block. Spin the engine on the starter motor.

4 Sparking between the end of the cable and the block should be fairly strong with a strong regular blue spark. (Hold the lead with rubber to avoid electric shocks). If current is reaching the plugs, then remove them and clean and regap them to the specified clearance. The engine should now start.

5 If there is no spark at the plug leads take off the HT lead from the centre of the distributor cap and hold it to the block as before. Spin the engine on the starter once more. A rapid succession of blue sparks between the end of the lead and the block indicate that the coil is in order and that the distributor cap is cracked, the rotor arm faulty, or the carbon brush in the top of the distributor cap is not making good contact with the rotor arm. Possibly, the points are in bad condition (mechanical contact breaker). Renew them as described in this Chapter.

6 If there are no sparks from the end of the lead from the coil check the connections at the coil end of the lead. If it is in order start checking the low tension circuit.

7 Use a 12V voltmeter or a 12V bulb and two lengths of wire. On conventional distributors switch on the ignition and ensure that the points are open. On breakerless distributors ensure that the segments on the rotor are not adjacent to the trigger coil. Make a test between the low tension wire to the coil (+) terminal and earth. A reading of 7 to 8 volts should be obtained. No reading indicates a break in the supply from the ignition switch. A correct reading indicates a faulty coil or condenser, or a broken lead between the coil and the distributor. On breakerless ignition systems have it checked by a Lancia dealer.

8 Take the condenser wire off the points assembly and with the points open test between the moving point and earth. If there is now a reading then the fault is in the condenser. Fit a new one as described in this Chapter.

9 With no reading from the moving point to earth, take a reading between earth and the CB or negative (-) terminal of the coil. A reading here shows a broken wire which will need to be renewed between the coil and distributor. No reading confirms that the coil has failed and must be renewed, after which the engine will run once more. Remember to refit the condenser wire to the points assembly.

17 Fault diagnosis – engine misfires

1 If the engine misfires regularly run it at a fast idling speed. Pull off each of the plug caps in turn and listen to the note of the engine. Hold the plug cap in a dry cloth or with a rubber glove as additional protection against a shock from the HT supply.

2 No difference in engine running will be noticed when the lead from the defective plug is removed. Removing the lead from one of the good cylinders will accentuate the misfire.

3 Remove the plug lead from the end of the defective plug and hold it about $\frac{3}{16}$ inch (5 mm) away from the block. Start the engine. If the sparking is fairly strong and regular the fault must lie in the spark plug.

4 The plug may be loose, the insulation may be cracked, or the electrodes may have burnt away giving too wide a gap for the spark to jump. Worse still, one of the electrodes may have broken off.

5 If there is no spark at the end of the plug lead, or if it is weak and intermittent, check the ignition lead from the distributor to the plug. If the insulation is cracked or perished, renew the lead. Check the connections at the distributor cap.

6 If there is still no spark, examine the distributor cap very carefully for tracking. This can be recognised by a very thin black line running between two or more electrodes, or between an electrode and some other part of the distributor. These lines are paths which now conduct electricity across the cap thus letting it run to earth. The only answer is a new distributor cap.

7 Apart from the ignition timing being incorrect, other causes of misfiring have already been dealt with under the section dealing with the failure of the engine to start. To recap – these are that:

  (a) *The coil may be faulty giving an intermittent misfire*
  (b) *There may be a damaged wire or loose connection in the low tension circuit.*
  (c) *The condenser may be short circuiting (if fitted)*
  (d) *There may be a mechanical fault in the distributor (broken driving spindle or contact breaker spring, if fitted).*

8 If the ignition timing is too far retarded, it should be noted that the engine will tend to overheat, and there will be a quite noticeable drop in power. If the engine is overheating and the power is down, and the ignition timing is correct, then the carburettor should be checked, as it is likely that this is where the fault lies.

Chapter 5 Clutch

Contents

Clutch assembly – inspection and renewal in situ	5
Clutch assembly – inspection and renewal on the bench	6
Clutch cable – removal and refitting	4
Clutch pedal – removal, refitting and adjustment	3
Clutch pedal and release lever free travel – adjustment	2
Clutch release mechanism – removal, inspection and refitting	7
Fault diagnosis – clutch	8
General description	1

Specifications

Type .... Single dry plate, diaphragm spring

Make .... Valeo or Fichtel and Sachs (F & S)
1300, 1400 and 1600 models .... 200 mm (7.874 in)
1800 and 2000 models .... 215 mm (8.464 in)

Disc thickness
1300, 1400 and 1600 models .... 7.5 to 8.0 mm (0.295 to 0.315 in)
1800 and 2000 models .... 7.40 to 7.85 mm (0.291 to 0.309 in)

Minimum thickness .... 6.5 mm (0.256 in)

Clutch pedal free travel .... 15 mm approximately (0.69 in approximately)

Clutch release lever free travel ...... 3 to 5 mm (0.118 to 0.196 in)

Clutch release lever wear stroke
1300, 1400 and 1600 models .... 9 mm (0.354 in)
1800 and 2000 models .... 12 mm (0.472 in)

Cable length between bracket and release lever
Clutch pedal released .... 100 mm (3.93 in)
Clutch pedal depressed .... 64 mm (2.52 in)

Torque wrench settings

	lbf ft	Nm
Bellhousing to crankcase	62	83
Clutch release bearing sleeve retaining bolt	6	8
Clutch to flywheel bolts	18	24

1 General description

The reason for the clutch unit being fitted between the engine and the transmission is so that there can be some form of connection between the engine and the transmission. It enables the engine torque to be progressively applied to the transmission so enabling the car to move off gradually from rest, and then for the gear to be changed easily as the speed increases or decreases.

The main parts of the clutch assembly are; the clutch driven plate assembly, the cover assembly and the release bearing assembly. When the clutch is in use the driven plate assembly, being splined to the clutch shaft, is sandwiched between the flywheel and pressure plate by the diaphragm spring. Engine torque is therefore transferred from the flywheel to the clutch driven plate assembly and then to the transmission unit clutch shaft.

By depressing the clutch pedal, the clutch release bearing assembly is drawn against the diaphragm spring by a cable connecting the pedal to the release bearing throwout yoke. The pressure on the driven plate assembly by the diaphragm spring is released and therefore the drive between the engine and transmission is broken.

When the clutch pedal is released, the diaphragm spring forces the pressure plate into contact with the high friction linings on the clutch drive plate, at the same time forcing the clutch driven plate assembly against the flywheel and so taking the drive up.

As the friction linings on the clutch driven plate wear, the pressure plate automatically moves closer to the driven plate to compensate. This makes the centre of the diaphragm spring move nearer to the release bearing, so decreasing the release bearing clearance and therefore the clutch free pedal travel. The clearance will have to be adjusted as described in Section 2.

2 Clutch pedal and release lever free travel – adjustment

1 When it is noticed that the pedal free travel is appreciably less than 15 mm (0.6 inch) the clutch cable needs to be adjusted.
2 Adjustment of the free travel both at the pedal and at the release lever on top of the bellhousing is made by tightening up the large

Chapter 5 Clutch

butterfly nut on the end of the cable where it passes through the release arm (Fig. 5.1).

3 The clutch cable will require periodic adjustment at the friction linings on the driven plate wear down and as the cable stretches slightly. This should not require attention more than every 12 500 miles (20 000 km), unless the conditions are very arduous or the car is used for a lot of towing.

4 The pedal free travel will be correct once the release lever clearance has been set. Tighten up the butterfly nut until the release lever has a free travel of 3 to 5 mm (0.118 to 0.196 in) (Fig. 5.2).

3 Clutch pedal – removal, refitting and adjustment

1 Slacken off the adjuster nut and locknut on the release arm end of the clutch cable.
2 As mentioned earlier the clutch is actuated by a wholly mechanical system. The clutch and brake pedals are mounted on a common fulcrum, attached to a bracket on the toe board. The procedure for removing the clutch pedal is identical to that for the brake pedal, details of which are given in Chapter 9.
3 Refitting is the reverse procedure but remember to adjust the cable tension and therefore the pedal and release lever free travel when the cable has been refitted to the pedal.

Fig. 5.1 Clutch release lever and cable end adjuster (Sec 2)

1 Release lever
2 Large butterfly adjusting nut

4 Clutch cable – removal and refitting

1 Remove the battery and battery tray. On later models it is better to remove the windscreen washer reservoir as well.
2 Undo and remove the large butterfly unit from the end of the cable and slip the cable off the release lever. Remove the locknut from the cable.
3 Disconnect the cable from the clutch pedal upper end. It is held in place by a pin with a circlip on its end.
4 Withdraw the clutch cable through the bulkhead into the engine bay and release the front end of the cable from the bracket on the gearbox casing.
5 The inner and outer cables cannot be separated. They are supplied as a complete replacement unit.
6 Refit the cable in the reverse order to removal.
7 When the cable has been refitted, adjust the pedal and release lever free travel, as described in Section 2.

5 Clutch assembly – inspection and renewal in situ

Except on very early models, renewing the clutch assembly, the release bearing or the bearing support sleeve and oil seal can be done without removing the engine and gearbox assembly from the car. However, to do this requires special equipment which will have to be borrowed or made up. To fabricate the necessary cradle which is needed to support and move the gearbox away from the engine is not difficult, and adequate arrangements can be made to support the engine assembly. However, before attempting to carry out this operation read through the following instructions and assess the complete task before starting, as the necessary equipment will have to be ready when needed (Figs. 5.4 and 5.5).

Obviously it is much easier to change the clutch or its components with the engine and gearbox assemblies removed from the car and if the gearbox has to be removed to investigate or repair some defect, then it is advisable to remove the clutch assembly as described in this Chapter and check its condition. It can then be renewed if necessary and fitted quite easily.

1 Remove the hub caps, where fitted, and slacken the front roadwheel bolts.
2 Jack up the front of the car and support it firmly on axle stands positioned on either side of the subframe side-members where they meet the rear crossmember.
3 Drain the cooling system and then remove the radiator as described in Chapter 2.
4 Disconnect the battery and remove it and the tray. Remove the large plastic windscreen washer reservoir fitted to later models.
5 On models fitted with air conditioning, remove the alternator

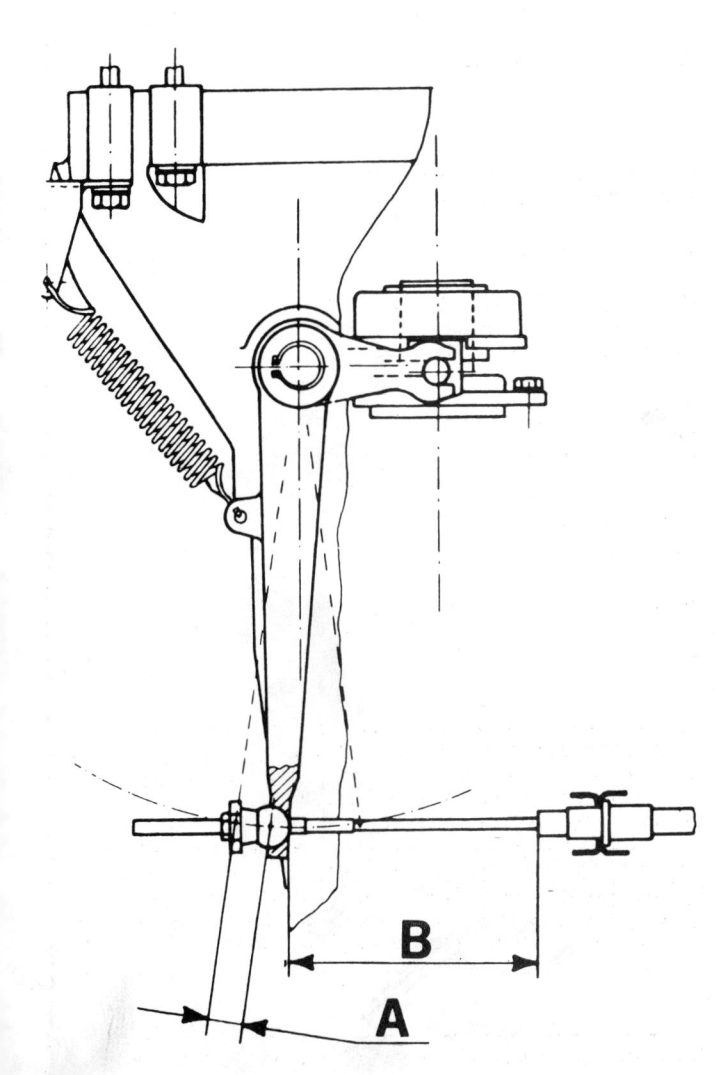

Fig. 5.2 Clutch release lever and cable adjustment (Sec 2)

A = Wear stroke
B = Length of cable between bracket and release lever

Chapter 5 Clutch

(Chapter 10) and carburettor hoses and slacken the compressor fastenings in order to remove the drivebelt. Disconnect the electrical leads, remove the compressor and place it on one side taking care not to invert it as this action could result in oil entering the conditioner circuit. Do not remove any of the air conditioning hoses. See Chapter 1, Section 44.

6   On USA models remove the coolant expansion tank (Chapter 2), the air pump, and detach the activated carbon canister from the left-hand wheel arch without disconnecting the inlet and outlet hoses. Also disconnect the leads from the gear switches on the gearbox.

7   Remove the gear linkage front control shaft, as described in Chapter 6. Disconnect the speedometer cable from the driven gear output shaft in the differential cover.

8   Unhook the clutch release lever return spring from its bracket and then release the forward end of the clutch cable from the lever.

9   Remove the circlip and lift off the release lever from the shaft. There is no need to scribe the position of the lever to the shaft as there is a master spline to ensure that it fits in the correct position (photo).

10   Undo the reversing light switch wires and tuck them out of the way.

11   Undo the clutch cable mounting bracket retaining bolts and lift it away from the gearbox casing.

12   Undo the bolts and remove the engine splash guard.

13   Remove the two front roadwheels, and the steering arm boots.

14   Undo the inner CV joints where they meet the drive flanges, and move the driveshafts as far to the rear as is possible.

15   Remove the driveshaft centre section where fitted (see Chapter 8).

16   On models fitted with the self-levelling headlamp system, disconnect the front sensor link from the left-hand wishbone.

17   Undo the sensor mounting bracket bolts and place the unit to one side. Take extreme care not to pull the pipes off the unit, or the whole self-levelling system will be ruined.

18   Undo the exhaust pipe to manifold mounting nuts and also the support bracket to crankcase nuts, as described in Chapter 3.

19   Undo the nuts and remove the bolts from the rear engine mounting.

20   Undo the retaining bolts and remove the flywheel cover plate.

21   Undo the nut and bolt which secure the gearbox to the front mounting and withdraw the bolt.

22   Undo the rear bellhousing to crankcase mounting bolt and nut. Then remove the front bellhousing to crankcase mounting bolt. It is to these bolt holes in the bellhousing that the mounting cradle is attached.

23   Undo and remove the gearbox end damper. This is attached to the end cover of the gearbox and to the body. It is the lower mounting point on the gearbox end cover to which the other end of the cradle is attached.

24   Attach the cradle, that has been borrowed or fabricated, to the bellhousing and gearbox end cover. Then place a trolley jack underneath the cradle in line with the engine and gearbox and attach it to it.

25   Take the weight of the gearbox on the cradle and jack.

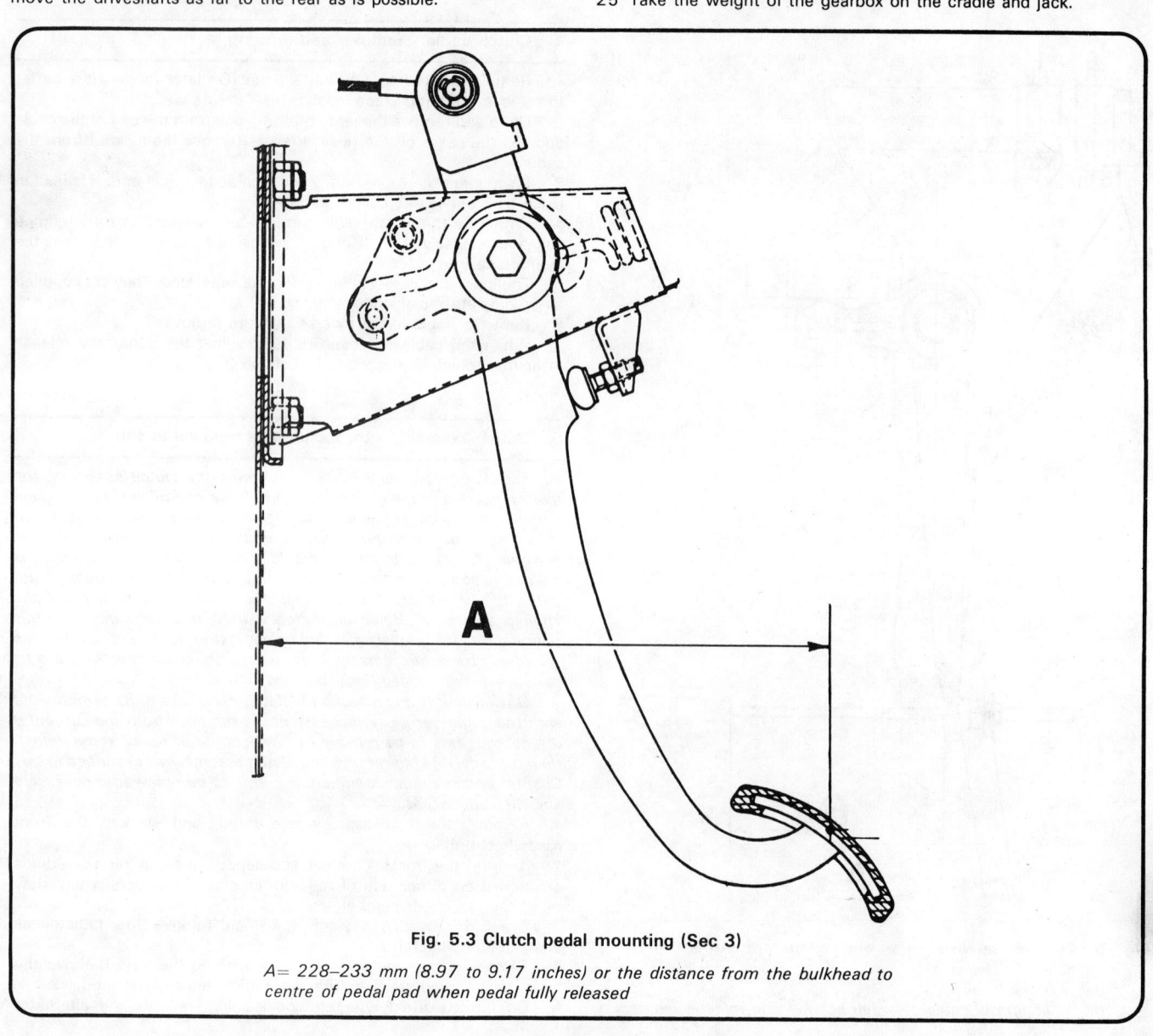

Fig. 5.3 Clutch pedal mounting (Sec 3)

A= 228–233 mm (8.97 to 9.17 inches) or the distance from the bulkhead to centre of pedal pad when pedal fully released

Chapter 5 Clutch

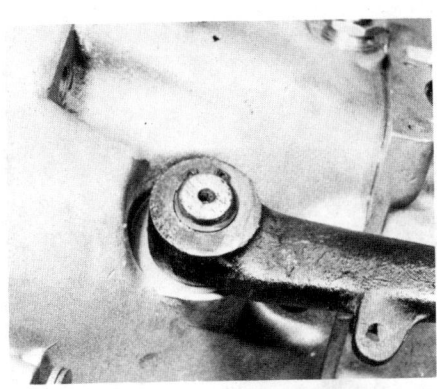

5.9 The circlip which retains the release lever has to be removed first

5.33 Remove the 6 bolts which secure the cover to the flywheel (engine out of car)

5.38 Refit the clutch disc to the flywheel using a mandrel (engine out of car)

5.39a Refitting the pressure plate and cover assembly to the flywheel (engine out of car)

5.39b Pressure plate assembly correctly located on dowels (engine out of car)

26 Remove the starter motor, as described in Chapter 10.
27 Remove the gearbox rear mounting and its bracket from the subframe and the gearbox front mounting bracket from the gearbox as described in Chapter 6.
28 Place another jack under the engine/gearbox assembly and jack up the engine until the timing gear cover touches the body. Then support the engine in that position. This can be done either by making up a special bracket as shown in Fig. 5.5 or by using a wooden baulk and axle stands. Alternatively a hoist or crane could be attached to the engine lifting eyes to support the weight.
29 With the engine firmly supported and the gearbox cradle and jack taking the weight of the gearbox/differential unit remove the two upper mounting bolts which secure the bellhousing to the crankcase.
30 Now pull the trolley jack slowly away and the gearbox, on its cradle, will be separated from the engine assembly. Pull the gearbox as far to the left-hand side of the engine bay as it will go.
31 When the gearbox has been moved away to its fullest extent, lower the jack and allow the gearbox to rest on the subframe. This will allow greater access to the clutch assembly.
32 Scribe the position of the clutch pressure plate cover to the flywheel.
33 Undo and remove, in a diagonal and progressive manner, the six bolts and spring washers which secure the clutch cover to the flywheel. This will prevent distortion of the cover and the cover suddenly flying off. Remove the cover and note which way the clutch disc is fitted (photo).
34 Examine the clutch disc friction linings for wear and loose rivets, and the disc for rim distortion, cracks, broken hub springs and worn splines. The surface of the friction linings may be highly glazed, but as long as the clutch material pattern can be clearly seen this is satisfactory. Ensure that the lining thickness is well within the minimum thickness given in the Specifications. If the linings are more than 75% worn renew the disc.
35 Always renew the clutch driven plate as an assembly to preclude further trouble, but, if it is wished to merely renew the linings, the rivets should be drilled out and not knocked out with a punch. The manufacturers do not advise that only the linings are renewed and personal experience dictates that it is far more satisfactory to renew the driven plate complete than try to economise by fitting only new friction linings.
36 Check the machined faces of the flywheel and the pressure plate. If either is grooved they should be machined until smooth or they should be renewed.
37 If the pressure plate is cracked or split or if the pressure of the diaphragm spring is suspect it is essential that an exchange unit is fitted.
38 Fit the clutch disc to the flywheel and use a mandrel to hold it in position and centralise it (photo). Ensure that the clutch disc is fitted the correct way round.
39 Refit the pressure plate and cover assembly lining up the previously made scribe marks (photo). If a new pressure plate assembly is being fitted refer to Fig. 5.6 for the correct fitting position.
40 Tighten the clutch cover securing bolts firmly in a diagonal sequence to ensure that the cover plate is pulled down evenly and without distortion of the flange. Finally tighten the bolts down to the specified torque.
41 Remove the mandrel from the centre of the clutch disc.
42 Refit the gearbox/differential unit in the reverse order to that in which it was removed. Remember to take care when locating the front end of the primary shaft into the splined centre section of the clutch disc. Lightly grease the splines before starting. When it is correctly aligned slide the gearbox unit in to mate with the engine assembly.
43 Tighten all nuts and bolts, as they are refitted, to their specified torques, and refill the cooling system.

6 Clutch assembly – inspection, renewal – on the bench

1 With the engine and gearbox assembly removed from the car as described in Chapter 1, first remove the starter from the bellhousing. It is held in place by 3 bolts through the bellhousing. Note that the lower bolt also retains the clutch release lever return spring bracket.
2 Undo the 4 large bolts which secure the bellhousing to the

Fig. 5.4 Gearbox unit support cradle (Sec 5)

Fig. 5.5 Engine assembly support bracket B with gearbox cradle A in background
(Sec 5)

Chapter 5 Clutch

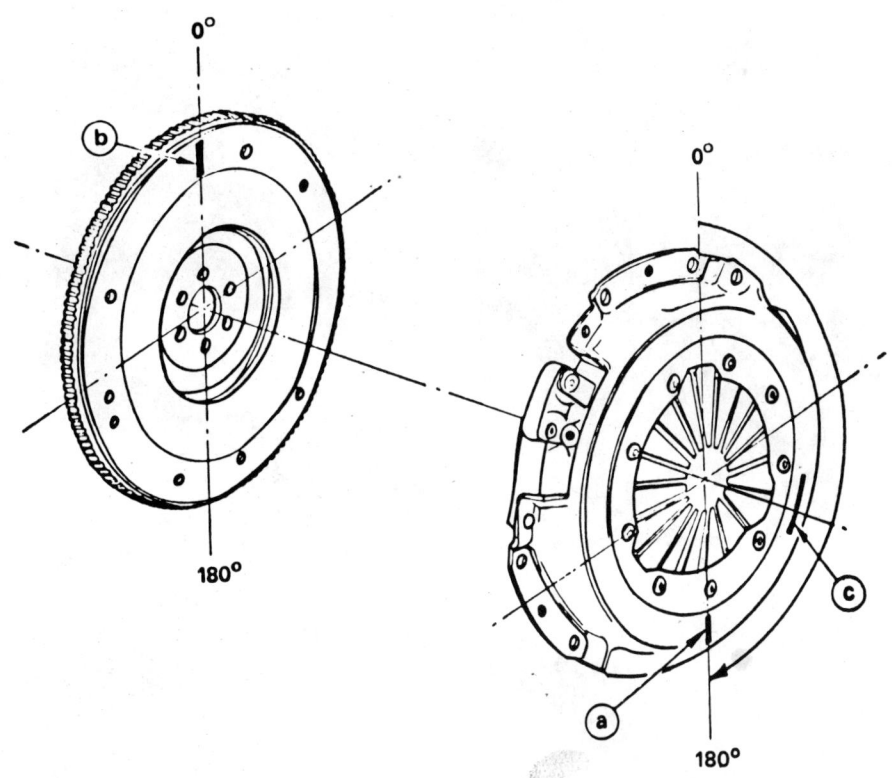

Fig. 5.6 Correct alignment of paint markings on flywheel and pressure plate cover (Sec 5)

a – Yellow paint mark showing clutch radial unbalance
b – Red paint mark showing flywheel radial unbalance
c – Yellow paint mark showing clutch circumferential unbalance
Location of a as to b – 180° away from b
Location of c as to b – any position

crankcase. There are two on top and one at each side. The one on the rear side has a nut on the front end. Remove the flywheel lower cover plate.
3 Withdraw the bellhousing and gearbox assembly from the engine assembly.
4 The clutch assembly can now be removed and refitted as described in Section 5 paragraphs 32 to 41. This also includes the inspection of the clutch assembly components.
5 Refit the bellhousing and gearbox assembly to the engine unit in the reverse order to removal. Refit the starter motor. The engine gearbox assembly is now ready to be refitted as described in Chapter 1.

7 Clutch release mechanism – removal, inspection and refitting

1 The clutch release mechanism consists of the release bearing, support sleeve, sleeve oil seal and release arm and bush (Fig. 5.7).
2 Except on very early models the clutch release mechanism can be removed either with the engine/gearbox assembly in the car or with the units removed. On early models the engine/gearbox must be removed. If the task is to be carried out in situ refer to Section 5 which describes the clutch assembly renewal procedure with the engine/gearbox in situ. Follow the instructions from the beginning to paragraph 31.
3 If the task is to be done with the engine and gearbox assemblies removed from the car then refer to Section 6 and follow the instructions from the beginning to paragraph 3.
4 With the bellhousing separated from the engine the release bearing can be pulled off the sleeve (photo).
5 Remove the three support sleeve retaining bolts and tap the sleeve off the end of the primary shaft (photo).
6 To remove the release arm first remove the circlip and pull off the release lever from the outer end of the release arm. Remove the plastic bush in the outside of the bellhousing (photo).
7 With the bush removed hold the inner release arm lever and push the arm outwards through the bellhousing so that the inner end is freed from its seat (photo).
8 Pull the release arm lever back into the bellhousing at an angle thus releasing the outer end as well.
9 Inspect the components for wear. There should be no harshness in the release bearing and it should spin reasonably freely, allowing for the fact that it has been pre-packed with grease. Check that it slides freely on the support sleeve, and check that the lugs are in good condition.
10 Check the support sleeve oil seal. This stops the oil coming out of the gearbox via the primary shaft bearing. If there is any trace of oil on the clutch or the splined end of the primary shaft then the seal may be faulty.
11 To renew the oil seal, tap out the old seal from the seat in the sleeve and fit a new seal with the lips to the rear. Tap it into position squarely.
12 Check the release arm for wear in the end pivots and for fitting in their bushes. There should not be too much clearance. Check the hooked ends of the levers, which hold and operate the release bearing, for wear.
13 Inspect the bush that fits on the outer splined end of the release arm and ensure that it is a good fit and not worn.
14 Renew any badly worn components.
15 Refit the release arm, bush and release lever – note the master spline to guide it on. Fit a new gasket on the rear end of the release bearing support sleeve and grease the inner lips of the oil seal before the sleeve is fitted to the primary shaft. Line up the bolt holes and tap the sleeve into position. Refit the three bolts and tighten them to the specified torque.
16 Re-engaging the release bearing in the release arm levers is simply a matter of twisting the arm so that the lugs on the bearing casing engage in the ends of the levers as the bearing is slid onto the sleeve. Lightly grease the sleeve first.

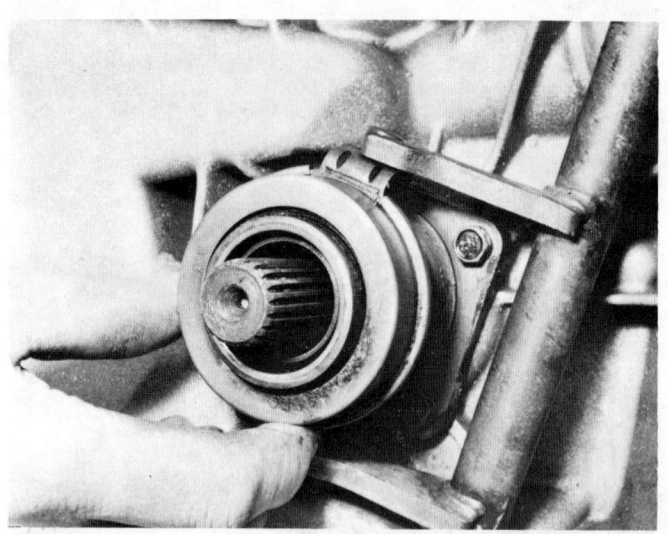

7.4 Pull the release bearing off the sleeve and lugs will disengage from the levers

7.5 Sliding the release bearing support sleeve off the primary shaft

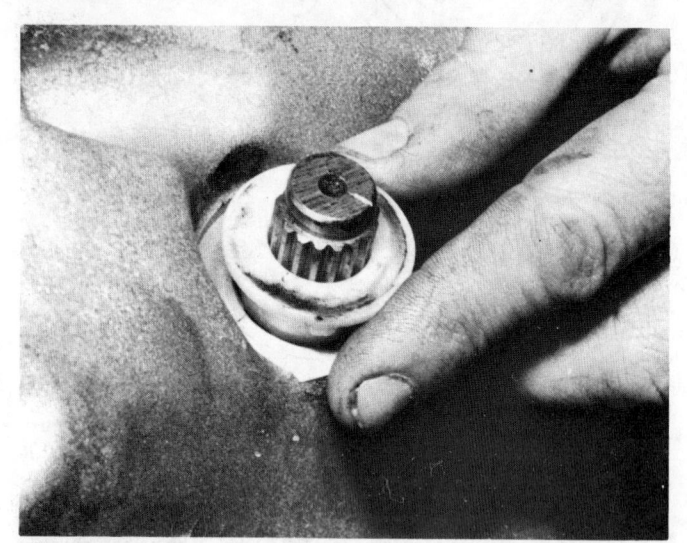

7.6 Withdrawing the plastic bush from the outside end of the release arm

7.7 Push the release arm outwards through the casing to free the inner end from its seat

17 Lightly grease the splined end of the primary shaft and the gearbox assembly is ready to be joined up to the engine.

18 Refit the gearbox assembly as described in either Section 5 or 6 depending on whether the operation was carried out in situ or on the workbench.

8 Fault diagnosis – clutch

There are four main faults to which the clutch and release mechanism is prone. They may occur by themselves or in conjunction with any of the other faults. They are clutch squeal, slip, spin and judder.

Clutch squeal – diagnosis and remedy

If, on taking up the drive or when changing gear, the clutch squeals, this is indicative of a badly worn clutch release bearing.

As well as regular wear due to normal use, wear of the clutch release bearing is much accentuated if the clutch is ridden or held down for long periods in gear, with the engine running. To minimise wear of this component the car should always be taken out of gear at traffic lights and for similar hold-ups.

The clutch release bearing is not an expensive item, but it is difficult to get at.

Clutch slip – diagnosis and remedy

Clutch slip is a self-evident condition which occurs when the clutch driven plate is badly worn, oil or grease have got onto the flywheel or pressure plate faces, or the pressure plate itself is faulty.

The reason for clutch slip is that due to one of the faults above, there is either insufficient pressure from the pressure plate, or insufficient friction from the driven plate, to ensure solid drive.

If small amounts of oil get onto the clutch, they will be burnt off under the heat of the clutch engagement, and in the process, gradually darken the linings. Excessive oil on the clutch will burn off leaving a carbon deposit which can cause quite bad slip, or fierceness, spin and judder.

If clutch slip is suspected, and confirmation of this condition is required, there are several tests which can be made.

With the engine in second or third gear and pulling lightly, sudden depression of the accelerator pedal may cause the engine to increase its speed without any increase in road speed. Easing off on the accelerator will then give a definite drop in engine speed without the car slowing.

In extreme cases of clutch slip the engine will race under normal acceleration conditions.

If slip is due to oil or grease on the linings a temporary cure can sometimes be effected by squirting carbon tetrachloride into the clutch. The permanent cure is, of course, to renew the clutch driven plate and trace and rectify the oil leak.

Fig. 5.7 Clutch release mechanism (Sec 7)

1 Clutch release bearing
2 Release arm
3 Plastic bush
4 Support sleeve
5 Sleeve retaining bolt

Clutch spin – diagnosis and remedy

Clutch spin is a condition which occurs when there is an obstruction in the clutch, either in the gearbox input shaft or in the operating lever itself, or oil may have partially burnt off the clutch lining and have left a resinous deposit which is causing the clutch disc to stick to the pressure plate or flywheel.

The reason for clutch spin is that due to any, or a combination of the faults just listed, the clutch pressure plate is not completely freeing from the driven plate even with the clutch pedal fully depressed.

If clutch spin is suspected, the condition can be confirmed by extreme difficulty in engaging first gear from rest, difficulty in changing gear, and sudden take up of the clutch drive at the fully depressed end of the clutch pedal travel as the clutch is released.

Check the clutch cable adjustment.

If these points are checked and found to be in order then the fault lies internally in the clutch, and it will be necessary to remove the clutch for examination.

Clutch judder – diagnosis and remedy

Clutch judder is a self-evident condition which occurs when the gearbox or engine mounts are loose or too flexible, when there is oil on the face of the clutch friction plate, or when the clutch pressure plate has been incorrectly adjusted.

The reason for clutch judder is that due to one of the faults just listed, the clutch pressure plate is not freeing smoothly from the driven plate and is snatching.

Clutch judder normally occurs when the clutch pedal is released in first or reverse gears, and the whole car shudders as it moves backwards or forward.

Chapter 6 Manual gearbox and automatic transmission

Contents

Automatic transmission – separation from engine	14
Automatic transmission fluid – level checking and renewal	13
Automatic transmission oil seals – renewal	16
Automatic transmission selector linkage – adjustment	15
Automatic transmission unit – general description	12
Fault diagnosis – automatic transmission	18
Fault diagnosis – manual gearbox	17
Gear linkage – description and layout	9
Gear linkage assemblies – removal, checking, refitting and adjusting	10
Gearbox – dismantling	3
Gearbox – reassembly	8
Gearbox components – examination and renewal	4
Gearbox/differential unit mountings – removal and refitting	11
General description – manual gearbox	1
Mainshaft – dismantling and reassembly	5
Manual gearbox/automatic transmission unit – removal and refitting	2
Primary shaft – bearing removal	6
Synchromesh assemblies – inspection, checking and reassembly	7

Specifications

Manual gearbox
Type .... 5 forward speeds, all synchromesh and one reverse speed

Gear ratios
1st:
    Early models and all USA models .... 3.50 : 1
    Later models .... 3.75 : 1
2nd .... 2.23 : 1
3rd .... 1.52 : 1
4th .... 1.15 : 1
5th .... 0.925 : 1
Reverse .... 3.07 : 1

Primary shaft
Endfloat adjustment method .... Shims of different thickness fitted to front and rear of bearing housing
Front shim thickness .... 2.0 mm (0.0787 in)
Rear shim thickness .... 5.5 mm (0.216 in)

Synchro hub clearance
Gap between baulk ring and hub .... 1 mm ± 0.13 mm (0.3937 in ± 0.0051 in)

Driven gear endfloat
1st gear .... 0.04 to 0.365 mm (0.00157 to 0.0143 in)
2nd gear .... 0.04 to 0.271 mm (0.00157 to 0.0107 in)
3rd gear .... 0.041 to 0.318 mm (0.00164 to 0.0125 in)
4th gear .... 0.043 to 0.226 mm (0.00193 to 0.00889 in)
5th gear .... 0.070 to 0.170 mm (0.00275 to 0.00669 in)

Driven gears radial clearance
.... 0.040 to 0.075 mm (0.00157 to 0.00295 in)

Automatic transmission
General
Type .... 3 speed with torque converter
Quadrant positions .... P – R – N – D – 2 – 1
Type of gearing .... Epicyclic
Type of actuation .... 2 hydraulic clutches – 'Forward' and 'Direct drive and reverse'
Locking device .... Brake bands – three sets
Oil cooling method .... External heat exchanger

Gear ratios
1st .... 2.346 : 1
2nd .... 1.402 : 1
3rd .... 1 : 1
Reverse .... 2.346 : 1

Chapter 6 Manual gearbox and automatic transmission

Converter ratios
- 1600 models .... 2.05 : 1
- 2000 models .... 1.95 : 1

Transfer gears reduction
- 1600 models .... 1.17 : 1
- 2000 models .... 1.24 : 1

Kickdown speeds
- Top to second .... Below 116 km/h (72 mph)
- Second to first .... Below 65 km/h (40 mph)

Change up speeds at full throttle (Drive setting)
- First to second gear .... 80 km/h (50 mph) at 6000 rpm
- Second to top gear .... 130 km/h (81 mph) at 5800 rpm

All models

Torque wrench settings

	lbf ft	Nm
Gear remote control idler to subframe	51	70
Gear remote control shafts to idler	29	40
Control rod to shaft	7	10
Gear lever to control rod	22	30
Selector arm to selector rod	24	36
Selector forks to selector rods	21	39
Main and primary shaft locking ring nuts	99	135
Bearing plate to casing	11	15
Gearbox end cover bolts	11	15
Differential cover to carrier:		
Small nuts	18	25
Large nuts	36	50
Gearbox casing to bellhousing	18	25
Differential side gear cover plates	11	15
Clutch cable bracket to gearbox	11	15
Clutch release guide to bellhousing	6	8
Gearbox assembly to engine	62	85
Clutch cover plate to bellhousing	11	15
Selector shaft bush to casing	73	100
Speedometer driven gear locknut	7	10
Front gearbox mounting to subframe	4	6
Front gearbox mounting bracket to gearbox	14	20
Front gearbox mounting bolt to mounting	41	56
Rear gearbox mounting bracket to subframe	21	30
Gearbox to rear gearbox mounting	19	26
Gearbox shock absorber:		
Nut and bolt to gearbox	43	60
Nut and bolt to body	18	25

1 General description – manual gearbox

Most models are fitted with the standard five-speed all synchromesh manual gearbox. The automatic gearbox, which is now available as an option in the 1600 and 2000 models of the Saloon, Coupe and HPE, was only introduced in September 1979. In the UK only a very limited number of automatic units have been fitted. This option has not been made available for the Spider model.

Both the manual and automatic transmission units are mounted in line with the engine with the differential unit mounted integrally to the rear side of the gearbox. The engine occupies the right-hand side of the engine bay and the transmission the left.

The gearbox and differential unit share a common oil bath and the dipstick for both is in the differential housing cover.

The manual gearbox has two shafts. The primary or input shaft has the appearance of a layshaft that is to say the gears are fixed in it. The mainshaft, which lies to the rear, is of a conventional pattern with the final drive pinion on the right-hand, or front end, of the shaft, meshing directly with the crownwheel of the differential unit.

The manual gearboxes tend to be notchy especially the engagement of first or second gear, this being more noticeable when the unit is cold. In terms of wear second gear synchromesh usually disappears first, this being an indication of general wear in the gearbox. Early models were sometimes found to have rather rattly gearboxes, but better machining and closer tolerances seem to have improved this in later models.

2 Gearbox and automatic transmission unit – removal and refitting

1 There is no way in which the transmission can be removed without taking out the complete engine and transmission assembly.

2 Full details of this operation will be found in Chapter 1. Refitting the transmission is the reverse operation.

3 Gearbox – dismantling

1 With the engine and gearbox assembly removed from the car the first task is to separate the assemblies.

2 Begin by removing the starter motor. This is held into the bellhousing by three bolts from the rear, through the bellhousing itself. Note that the lower bolt also retains the clutch control arm return spring bracket. Also remove the forward section of the remote control assembly from the selector shaft lever (photo).

3 Remove the bolts which attach the bellhousing to the crankcase. There are 2 on top and one at each side. The bolt on the rear side of the bellhousing has a nut on the front end; the others bolt directly into the crankcase.

4 The bellhousing and gearbox can now be separated from the engine. If the unit is of a later model with the 'three-piece' driveshaft layout there is no need to remove the centre section; the splined shaft

Chapter 6 Manual gearbox and automatic transmission

end will separate from the differential unit side gear as the two assemblies are parted.

5 Drain the oil from the gearbox/differential unit if this has not already been done (photo).

6 From inside the bellhousing remove the release bearing and then undo and remove the three bolts which hold the bearing support sleeve into the casing. It is held into the casing and will need to be tapped out (photo). Turn the gearbox up on end so that the whole assembly sits on the bellhousing.

7 Undo and remove the seven bolts which secure the gearbox end cover to the main casing and remove the cover (photo).

8 The gear and synchro assembly which are contained in this rear compartment are the fifth speed assembly. Undo the selector fork locking bolt and engage 5th gear by hand, by pushing the synchro sleeve downwards (photo).

9 With a screwdriver or bar, inserted through the locking bolt hole at the outer end of the main selector shaft, which protrudes through the gearbox casing, engage another gear simultaneously so that the shafts are locked.

10 Use a centre punch to open up the crimped sections of the locking ring nuts on the rear ends of both the primary and main shafts.

11 To remove the ring nuts will require the use of a four legged peg spanner. If it is not possible to borrow one, then one will have to be made up using a piece of tube of the appropriate size (photo).

12 Remove the nut from the end of the primary shaft first. Disengage 5th gear, that is to say slide the synchro sleeve upwards, so that the springs, balls and pins do not jump out of the synchro hub.

13 Now it is safe to remove the locking ring unit from the end of the mainshaft.

14 Lift off the 5th gear synchro hub and sleeve and synchro fork from the mainshaft (photo).

15 Lift off the 5th gear from the mainshaft and the end gear from the

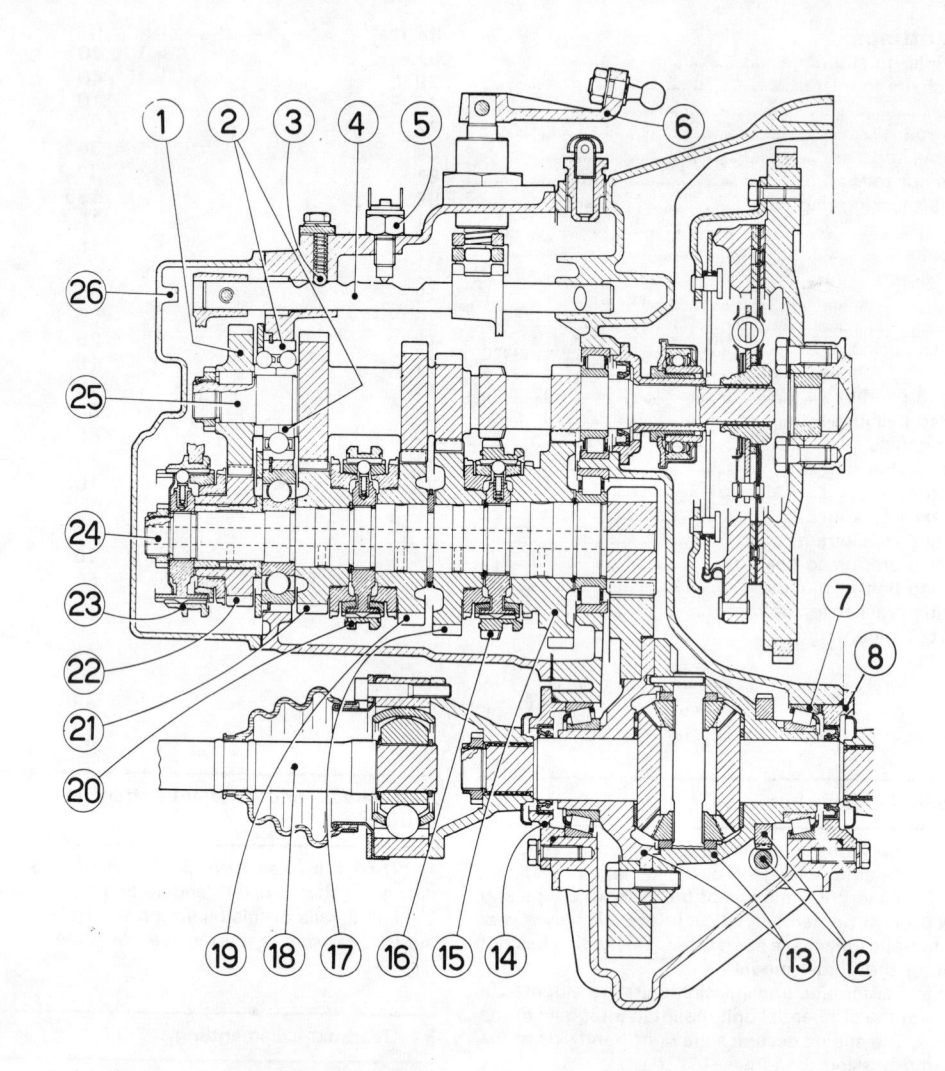

Fig. 6.1 Manual 5-speed gearbox layout (Sec 1)

1 Fifth speed primary gear
2 Single or double row ball bearing
3 Selector rod locking ball
4 Fifth-speed and reverse selector rod
5 Reversing light switch
6 Gear selector shaft outer lever
7 Taper roller thrust bearing
8 Engine side cover plate
12 Speedometer drive and driven gear
13 Differential half-casings
14 Gearbox side cover plate
15 First-speed driven gear
16 First, second and reverse gear synchromesh hub and sleeve
17 Second speed driven gear
18 Left-hand drive shaft
19 Third-speed driven gear
20 Third and fourth gear synchro
21 Fourth-speed driven gear
22 Fifth-speed driven gear
23 Fifth gear synchro
24 Main shaft
25 Primary shaft
26 Gearbox rear cover

Chapter 6 Manual gearbox and automatic transmission

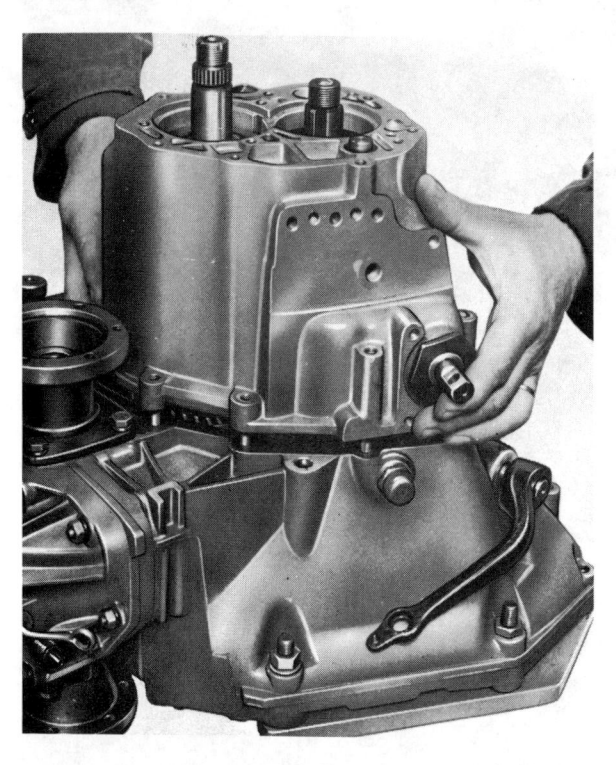

Fig. 6.2 Lifting off the gearbox casing (Sec 3)

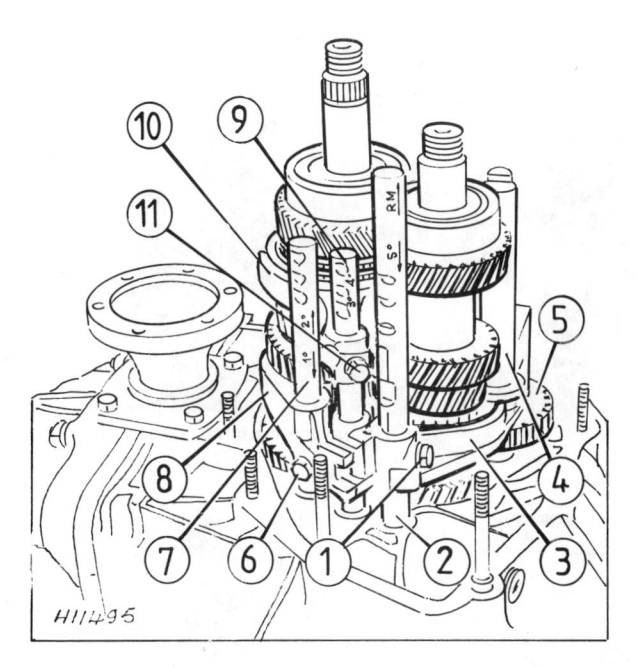

Fig. 6.3 Gearbox selectors and forks (Sec 3)

1 Reverse gear fork locking bolt
2 Reverse and fifth gear selector rod
3 Reverse gear selector fork
4 Reverse gear shaft
5 Reverse gear
6 First and second gear selector fork lock bolt
7 First and second gear selector rod
8 First and second gear selector fork
9 Third and fourth gear selector rod
10 Third and fourth gear selector fork
11 Third and fourth gear selector fork lock bolt

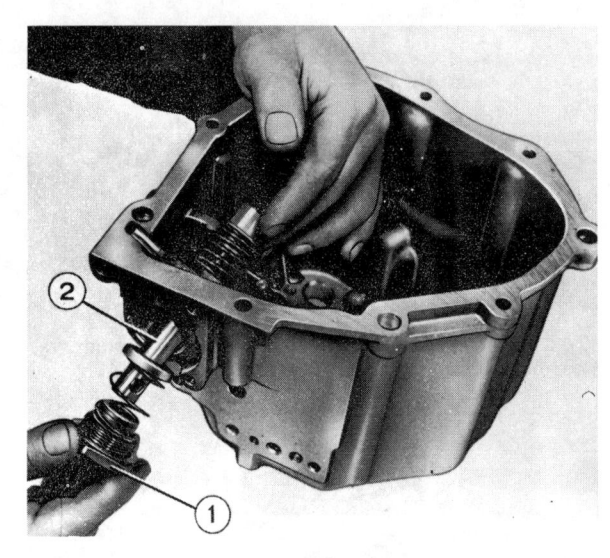

Fig. 6.4 Removing the selector shaft and bush assembly (Sec 3)

1 Selector shaft bush   2 Selector shaft

primary shaft. Mark the gears so that they are refitted the right way round. Remove the key from the primary shaft keyway. Remove the spacer from the primary shaft where one is fitted (photo).

16 Undo and remove the five bolts and lift off the bearing retaining plate.

17 Remove the cover plate on the side of the gearbox casing and carefully remove the three springs, which are all the same length and the interlocking balls. It may be found easier if the gearbox is tipped over onto its side to remove them (photo).

18 Remove the reversing light switch from the top side of the gearbox casing (photo).

19 The next stage is to remove the bearing retaining circlips. To do this the circlips must be rotated so that the gaps face each other in the middle (photo).

20 Unscrew the selector shaft bush, in the outside of the main gearbox casing, about half way. Pull the selector shaft out and turn it so that you can check that the selector arm has disengaged from the selector forks in the gearbox. The gearbox casing cannot be removed until this has been done (photo).

21 Undo the nine nuts and remove the casing straight upwards (Fig. 6.2).

22 Undo and remove the reverse gear selector fork locking bolt. Ensure that the other two rods are in neutral and then withdraw the rod. With the rod out the selector fork itself can be removed. So that there is no chance of confusion later refit the fork on the rod, and place them both to one side (photo).

23 The reverse gear and shaft may now be freed off and lifted out and placed to one side (photo).

24 Undo and remove the first and second gear selector fork locking bolt (photo). Pull the rod out, and lift up the gears to withdraw the selector fork. Refit the fork to the rod and place it to one side.

25 Undo and remove the bolts which secure the selector fork and dog to the third and fourth gear selector rod (photo). Pull the rod up to withdraw it but take care that the retaining pin in the lower section does not become lost (photo). Then remove the rod with the selector fork and dog still attached.

26 Using a magnet, remove the retainer pins from the selector rod seats (photo).

27 The two shafts, primary and main, can now be lifted out as complete assemblies together (photo).

28 With the main gear assemblies removed from the gearbox casing the selector shaft and bush can be removed. Remove the bush completely by unscrewing it, remembering that it has already been half undone. There is a spring and seal beneath it (Fig. 6.4).

29 Remove next the selector shaft and arm complete with its spring and cap (Fig. 6.4).

30 With the gearbox dismantled the two shaft assemblies can now be examined and overhauled as necessary, as can the various ancillary components. These subjects are covered in the following Sections.

3.2 With the engine/gearbox assembly removed from the car the starter, front remote control shaft and clutch return spring can be removed

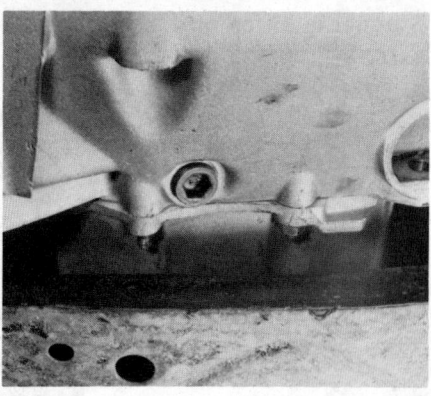

3.5 Gearbox drain plug location

3.6 Removing the bearing support sleeve

3.7 Removing the gearbox end cover

3.8 Undoing the fifth gear selector fork locking bolt

3.11 Removing the primary shaft nut using a made up peg spanner

3.14 Removing the fifth gear synchro mechanism and fork from the mainshaft

3.15 Removing the fifth-speed drivegear from the end of the primary shaft. Note the key which must not be lost

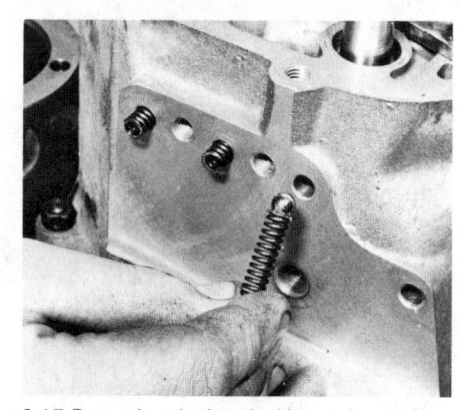

3.17 Removing the interlocking springs and balls

3.18 The reversing light switch loosened

3.19 With the circlips aligned like this the end of one can be released and levered out ...

3.20 Unscrew the selector shaft bush half-way out

Chapter 6 Manual gearbox and automatic transmission

4 Gearbox components – examination and renewal

1 With the basic components removed from the gearbox the examination of parts can commence once they have all been thoroughly cleaned in paraffin and dried.
2 The gearbox has been stripped probably because of wear or malfunction, possibly excessive noise, ineffective synchromesh or failure to stay in a selected gear. The most common fault in this gearbox is the early failure of 2nd gear synchromesh.
3 The cause of most gearbox problems is bad wear in the mainshaft or primary shaft bearings and wear on the synchro rings.
4 Examine the teeth of all gears for signs of uneven or excessive wear and, of course, chipping. If a gear on the mainshaft requires renewal then check that the corresponding gear on the primary shaft is not equally damaged. If it is, then the complete primary shaft will have to be renewed.
5 All gears should be a good running fit on the shaft with no signs of rocking. The hubs should be a good fit on the splines. Check the endfloat of all the gears with the measurements given in the Specifications.
6 Selector forks should be examined for signs of wear or ridging on the faces which are in contact with the operating sleeve.
7 Check the selector rod pins, balls and springs and ensure that they are in good condition.
8 The bearings may not be obviously worn, but if the gearbox has had to be dismantled, which is a long job anyway it would be wise to consider their renewal.
9 The same applies to the synchromesh assemblies, which are covered in Section 7.
10 Check carefully the drive pinion on the front end of the mainshaft. This meshes with the crownwheel in the differential unit and if it is worn, then the crownwheel must also be checked. The two gears are supplied as a complete set and must be renewed together. Dismantling, reassembly and overhaul of the differential unit is covered in Chapter 7.
11 Strip the selector shaft and inspect its parts. The shaft should be held in a vice with protective jaws and the inner lever, spring and cap

3.22 Unlocking the reverse gear selector fork bolt

3.23 The reverse gear and shaft are now free to be removed

3.24 Undoing the first and second gear selector fork lock bolt

3.25a Unlocking the third and fourth gear selector fork lock bolt

3.25b The third and fourth gear selector rod with dog selector fork and lower retaining pin

3.26 Removing a retaining pin using a magnet

3.27 Lifting out the two shaft assemblies together

4.12 Renewing the release bearing sleeve oil seal. The lips face towards the gearbox

Chapter 6 Manual gearbox and automatic transmission

can be removed. The O-ring in the selector shaft bush should be renewed as well.

12 Inspect the oil seal in the clutch release bearing support sleeve. It is wise to renew it anyway while the assembly is dismantled as it is not worth the risk of having it fail only perhaps a few thousand miles later on. Tap out the old seal and fit a new one (photo).

13 Before deciding to dismantle the mainshaft consider after thorough inspection whether the general state of the gearbox is going to make this a worthwhile operation. If the general condition is very poor and there is a lot of wear evident then it would be much wiser to consider purchasing a replacement gearbox, even at this late stage. However, it will be necessary to reassemble the old gearbox for exchange purposes.

5 Mainshaft – dismantling and reassembly

Read the instructions through completely before attempting to strip the mainshaft gears. In practice the gears should be a push fit onto the mainshaft, that is to say they should be able to be fitted and removed by hand. However, in practice, we have not found this to be so and it is quite possible that you may have to find someone with a press, or a local engineering firm, who can do the job for you. The front bearing behind the final drive pinion has to be pressed off anyway.

1 Clamp the mainshaft in a vice with protective jaws, with the rear end of the shaft pointing upwards.

2 Pull off the rear bearing. To do this it will be necessary to make up some form of bracket for the legs of the puller to grip on. We used a large (3 in) exhaust U-bolt clamp quite successfully. This was fitted round the bearing with the retaining circlip refitted. The bracket part of the clamp was half cut away so that the face was located under the bearing. Fit the nuts tightly and use the puller to extract the bearing and the 5th gear bush if it is still on the shaft (photos). The circlip will have to be renewed afterwards.

3 Remove the 4th speed gear and synchro cone (photo). Then remove the baulk ring (photo).

4 Remove the circlip, which must be renewed (Fig. 6.6).

5 Lift up the third and fourth gear synchro sleeve and release the balls or rollers, depending on which type of synchro hubs are fitted. Hopefully it is the ball type as they are much easier to fit back in again.

6 Remove the hub, baulk ring and 3rd gear. This is where we found the problems beginning, as the gear was not a push fit on the shaft and and had to be pressed out.

7 The second speed gear is held onto the shaft by two half thrust washers which are kept in place by a circlip. Remove the circlip and the two half washers (Fig. 6.7).

8 The second speed gear should now pull off the hub. Again it may have to be pressed off.

9 Lift up the synchro sleeve for first, second and reverse and extract the balls or rollers as before.

10 Remove the synchro hub retaining circlip.

11 The first speed synchro hub, sleeve and baulk ring and gear should be free to be lifted off the mainshaft. Here again a press may be required.

12 Remove the pinion (front) bearing retaining circlip.

13 The front bearing will have to be pressed off the shaft anyway, so whether or not you have experienced difficulty in stripping the other components from the mainshaft this will have to be undertaken by someone with a suitable press.

14 With the mainshaft completely stripped the components can all be thoroughly cleaned for examination and inspection as described in Section 4.

15 Having checked all the components and renewed those which are worn or damaged the reassembly of the mainshaft and its components can commence.

16 Start by refitting the front bearing to the shaft. To do this the bearing needs to be heated up in an oil bath to a temperature of 80°C (176°F). Then it can be pressed into place on the shaft.

17 Next refit the retaining circlips. To do this use a piece of tube slightly larger than the mainshaft, and fit the circlip over it so that the ends of the circlip do not damage the mainshaft when pushing it on (Fig. 6.8).

18 Refit first gear to the shaft complete with synchro hub and baulk

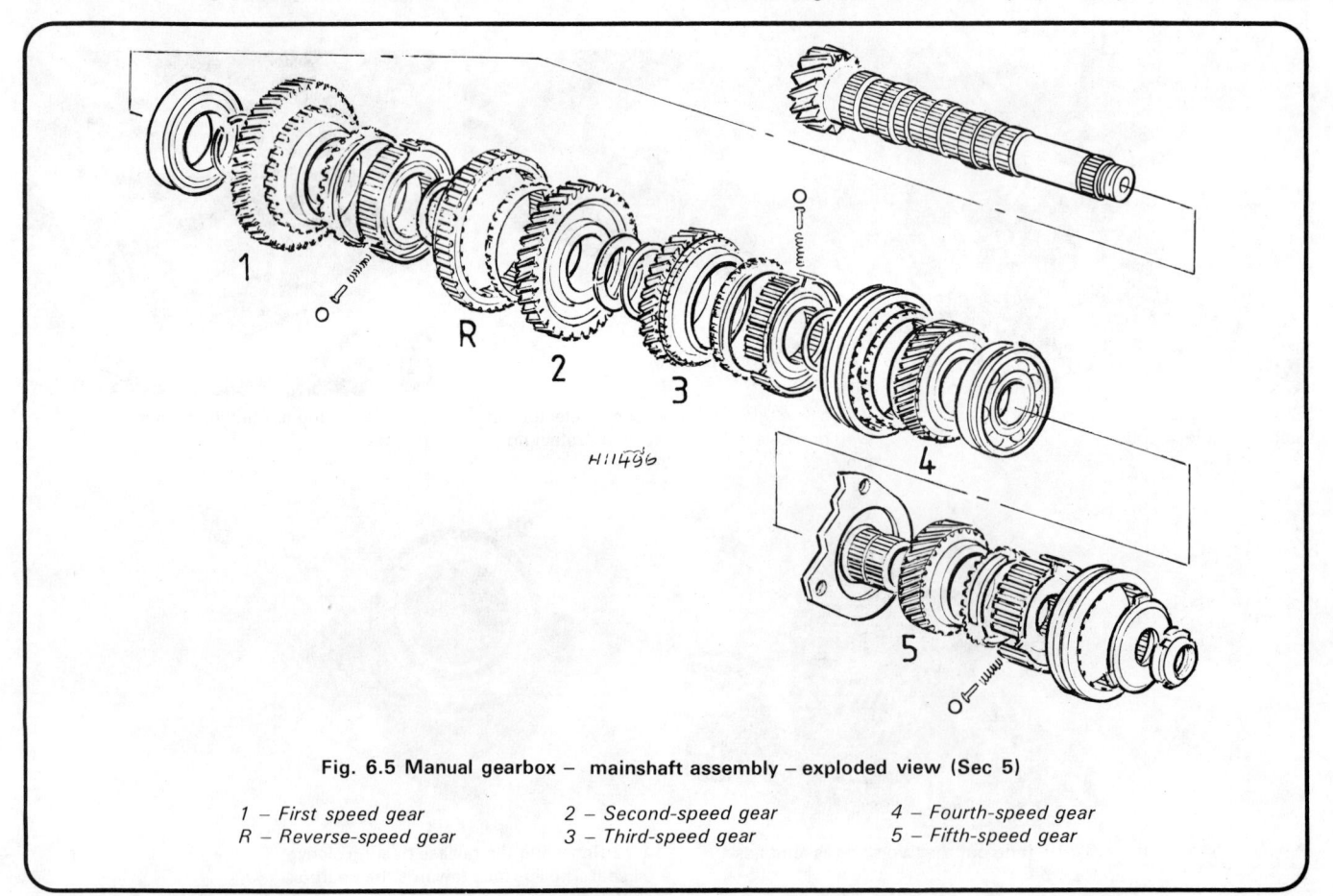

Fig. 6.5 Manual gearbox – mainshaft assembly – exploded view (Sec 5)

1 – First speed gear
R – Reverse-speed gear
2 – Second-speed gear
3 – Third-speed gear
4 – Fourth-speed gear
5 – Fifth-speed gear

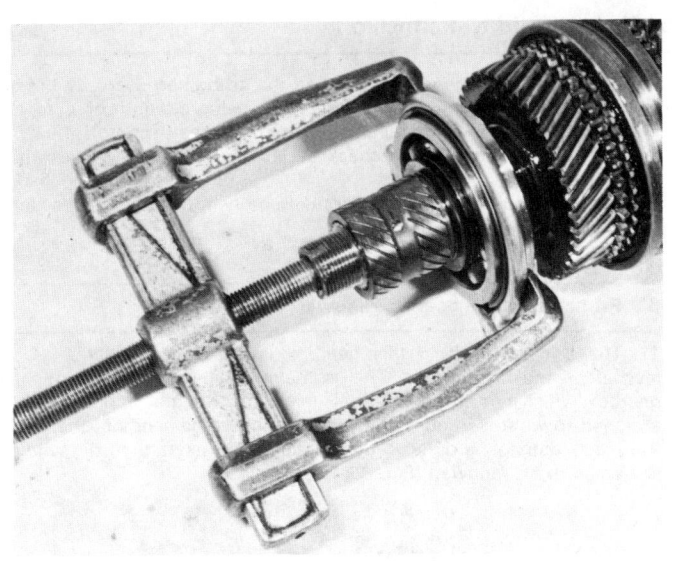

5.2a Pulling the rear bearing off the mainshaft using a home made bracket for the puller to grip on

5.2b The home made bracket in close-up view with the bearing removed

5.3a Lift off the fourth-speed gear ...

5.3b ... and then the baulk ring

5.24 The mainshaft partly rebuilt with final drive pinion on right-hand end, then first gear, hub, reverse, second, third, and third and fourth gear synchro

5.26 Refitting the mainshaft rear bearing and fifth gear sleeve

Chapter 6 Manual gearbox and automatic transmission

the front one, although we did not find this to be necessary. The rear bearing was tapped into place quite easily using a length of suitably sized tube. Refit the 5th gear bush to the mainshaft (photo).

27 With the bearing refitted check the 4th gear endfloat as with the other gears.

28 The mainshaft is now ready for reassembly into the gearbox which is covered in Section 8 (photo).

6 Primary shaft – bearing renewal

1 The primary shaft in this gearbox is basically the same as a layshaft in most gearboxes. This is mainly due to the layout of the gearbox which has the differential unit as an integral part of it, but attached to what is in effect the side of the gearbox and not the rear.

2 If any damage is done to the primary shaft gears then the whole shaft has to be renewed (Fig. 6.11).

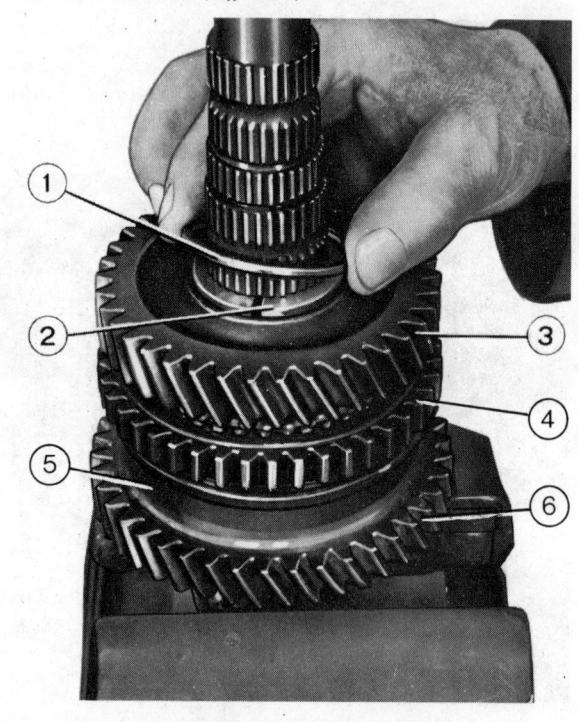

5.28 View of the pinion end of the mainshaft

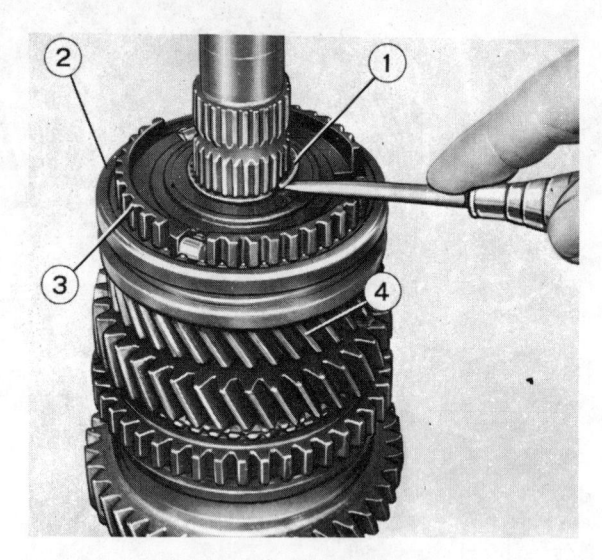

Fig. 6.6 Removing the retaining circlip for the third and fourth gear synchro hub (Sec 5)

1 Circlip
2 Synchro sleeve
3 Synchro hub
4 Third gear

ring. Refit the hub retaining circlip using the piece of tube as described in paragraph 17.

19 Check that the first gear endfloat is within the limits prescribed in the Specifications (Fig. 6.9).

20 Refit the first gear synchro sleeve to the hub with the selector fork groove facing downwards (Fig. 6.10).

21 Then refit the springs, pins and balls or rollers to the synchro hub. This is a very fiddly job if the hubs have rollers, whereas refitting the balls is much easier. It is simply a matter of time and patience before all three are back in place. Then slip the sleeve into the neutral position to stop them escaping again, which they are only too willing to do.

22 Refit the second speed gear to the shaft with the synchro ring underneath. Refit the two half thrust washers and their retaining circlip. Then again check the second gear endfloat as given in the Specifications.

23 Refit the third speed gear with synchro hub and baulk ring to the shaft. Then refit a new circlip to hold the hub in position, using once more the piece of tube to help you. Check that the endfloat is correct.

24 Refit the 3rd gear synchro sleeve to the hub and refit the springs, pin and balls or rollers as before (photo).

25 Refit the 4th gear to the mainshaft complete with its baulk ring.

26 If necessary, heat the rear bearing in an oil bath, as was done for

Fig. 6.7 Removing the second gear half thrust washers and retaining circlip (Sec 5)

1 Circlip
2 Half thrust washers
3 Second gear
4 Reverse gear
5 Synchro sleeve for first, second and reverse gears
6 First gear

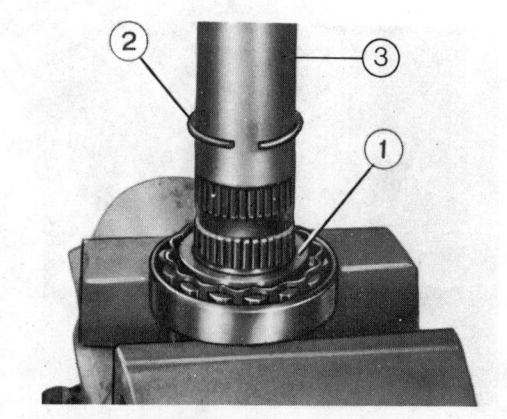

Fig. 6.8 Refitting the front bearing retaining circlip (Sec 5)

1 Front bearing
2 Circlip
3 Tube

Chapter 6 Manual gearbox and automatic transmission

Note that there is a shim between the rear bearing and the 5th gear (Fig. 6.11).

5 While you are there, hopefully having taken the new bearings with you, get the press operator to fit the new ones as well.

6 The primary shaft is now ready for reassembly into the gearbox as described in Section 8.

7 Synchromesh assemblies – inspection, checking and reassembly

1 Check the synchro hubs, sleeves, cones, springs, pins and rollers or balls and baulk rings when they have been dismantled and cleaned.

2 The synchro sleeves should not have excessive wear patterns in their grooves, and the splined centres of the hubs must be a good fit on the mainshaft splines.

3 The teeth in the sleeves and hubs must not show undue signs of wear. If they do then they must be renewed (photo).

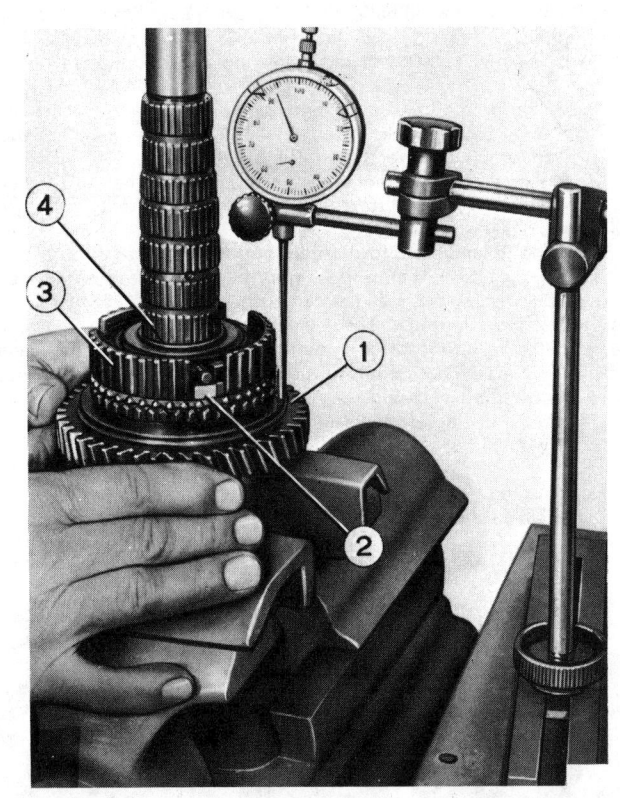

Fig. 6.9 Using a dial gauge on a stand to check the gear endfloat (Sec 5)

1 First gear
2 Outer circlip
3 Synchro hub
4 Hub retaining circlip

3 The primary shaft runs in roller bearings at the front and a single or double row ball-bearing at the rear end. Because of the way that the bearings are fitted it is almost impossible to remove them with a puller, as there is no room between the bearing races and the gears behind them (photo).

4 The only simple way, and probably the quickest, is to take the shaft to a local engineering firm and have the bearings pressed off.

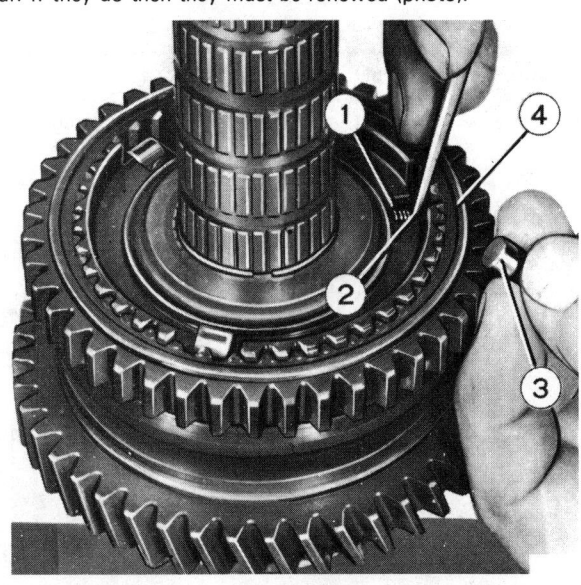

Fig. 6.10 Refitting the synchro sleeve and rollers to the first gear hub (Sec 5)

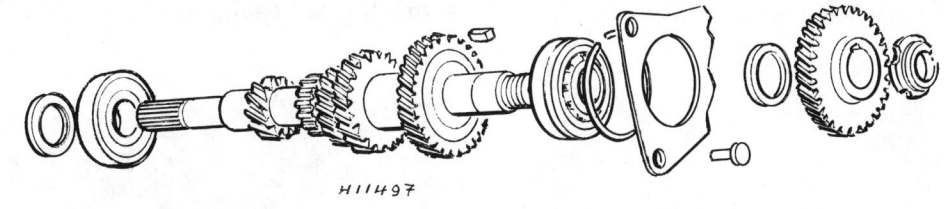

Fig. 6.11 The primary shaft components – exploded view (Sec 6)

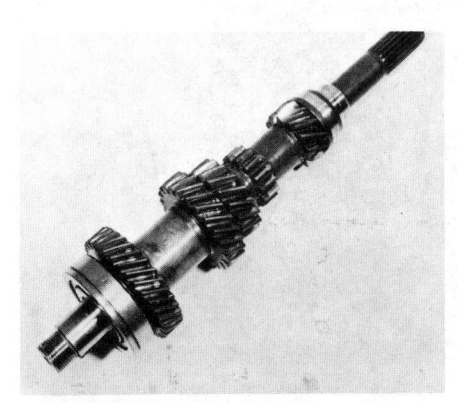

6.3 The primary shaft with front and rear bearings in position

7.3 The synchro hubs and sleeves should be in good condition like these two

7.6 Synchromesh mechanism – springs, pins and roller type assembly

Chapter 6 Manual gearbox and automatic transmission

4 The synchro cones must also be in good condition. The teeth must not be worn and the baulk rings must be circular. There is a tendency for the baulk rings to become slightly oval in shape. They also tend to distort.

5 To check for distortion in the baulk rings fit them to their appropriate cones and check the clearance as shown in Fig. 6.12. This should be as given in the Specifications.

6 When dismantling the mainshaft assembly you will find that in some cases the synchro hubs have balls and in some cases rollers fitted (photo). Refitting the balls, pins and springs is comparatively easy, whereas to get all three rollers with their pins and springs into the hub at the same time can be extremely frustrating. The best way to do it is with two people so that one person can hold the first and second set of balls or rollers, pins and springs in position whilst the third set is fitted. Then the sleeve can be slipped over and into place. Do not attempt to move the sleve until the next gear assembly or the fifth gear retaining plate has been placed in position, or all the rollers or balls, pins and springs will jump out. It is a task that requires a considerable amount of patience, especially with the roller type mechanism.

8 Gearbox – reassembly

1 Refit the selector shaft, spring and inner lever assembly to the gearbox casing and screw the bush into place for reassembly (photo). It should only be screwed half-way home. Don't forget the O-rings and thrust spring.

2 Refit the mainshaft front outer bearing race to the seat in the casing.

3 Mesh the primary and mainshaft together then lower them into the differential casing. Make sure that the final drive pinion engages with the crownwheel correctly (photo).

4 Refit the selector rod retainer pins. Use a magnet to fit them correctly and apply grease to them in position. Then refit the plug to the gearbox casing.

5 Refit the retaining pin to the third and fourth gear selector rod if it has fallen out and then refit the rod with the retaining pin downwards to the casing, at the same time engaging the fork in the 3rd/4th synchro unit sleeve. The selector fork and dog must also be on the rod. This is the reason for placing them back where they came from, or in this case not removing them at all from their selector rods, when the assembly was originally stripped.

6 The third and fourth speed selector rod fits into the centre hole and the locking bolts must be refitted when it is correctly positioned. Then tighten the locking bolts to the correct torque.

7 Grease the reverse gear and refit it to the shaft if it was removed. The chamfer on the gear and the selector fork groove face upwards. Then refit the O-ring to the groove in the base of the shaft (photo).

8 Refit the reverse gear to the casing with the flat face on the top

8.1 The selector shaft and inner lever reassembled and in position

8.2 Refitting the mainshaft front bearing outer race to its seat

8.7 The reverse gear correctly fitted to its shaft with a new O-ring on the bottom end

8.12 This is how the reassembled gearbox should look before the casing is refitted

8.15 The bearing retaining circlips cannot seat properly unless the gaps coincide

8.16 The bearing and reverse shaft retaining plate locates in the groove in the flat face at the end of the reverse gear shaft

8.17 Refit the fifth-speed drivegear with the chamfered edge upwards

8.18 Refitting the fifth-speed driven gear to the mainshaft together with its synchro cone

8.20 The synchro mechanism retaining plate refitted to the fifth-speed assembly

8.22 Crimping the primary shaft locking ring

Chapter 6 Manual gearbox and automatic transmission

end of the shaft facing into the gearbox. When the main casing has been refitted the bearing retaining plate locates against the flat end of the shaft.

9 Refit the reverse gear selector fork to the groove and then refit the selector rod. Feed it through the fork and locate it in position in the casing. Note that all the selector notches face the same way — outwards (Fig. 6.3).

10 Refit the selector fork locking bolt and tighten it to the specified torque.

11 To refit the second gear selector fork it is necessary to engage second gear. This can be done by lifting up the synchro sleeve. Then refit the fork and slide the rod through it. This is the left-hand rod. To locate the rod in the casing the other two selector rods must be in neutral, otherwise their retaining pins will prevent the selector rod from fitting correctly.

12 When the selector rod is fitted correctly, tighten the selector fork locking bolt to the specified torque (photo).

13 Fit a new gasket to the differential carrier and lower the gearbox casing into position. Check that the selector shaft bush has not been screwed in. It should be half undone. If, when lowering the casing, it feels as though it does not want to go fully down, this is because the selector shaft is catching on the selector rods. Pull the selector shaft out as far as possible and the casing will probably slide into position. If it doesn't then the selector shaft bush has been screwed in too much. Unscrew it several turns and try again, until the casing drops right down.

14 Then refit the nuts and tighten them to the specified torque.

15 Refit the bearing retaining circlips; one on the rear bearing outer edge for each shaft. As with removal, it is vital that the gaps in the circlips are lined up or the second one cannot be seated in the groove (photo).

16 Push the reverse shaft downwards to ensure that it has seated properly, then refit the reverse shaft and bearing retaining plate, and secure it with the bolts (photo).

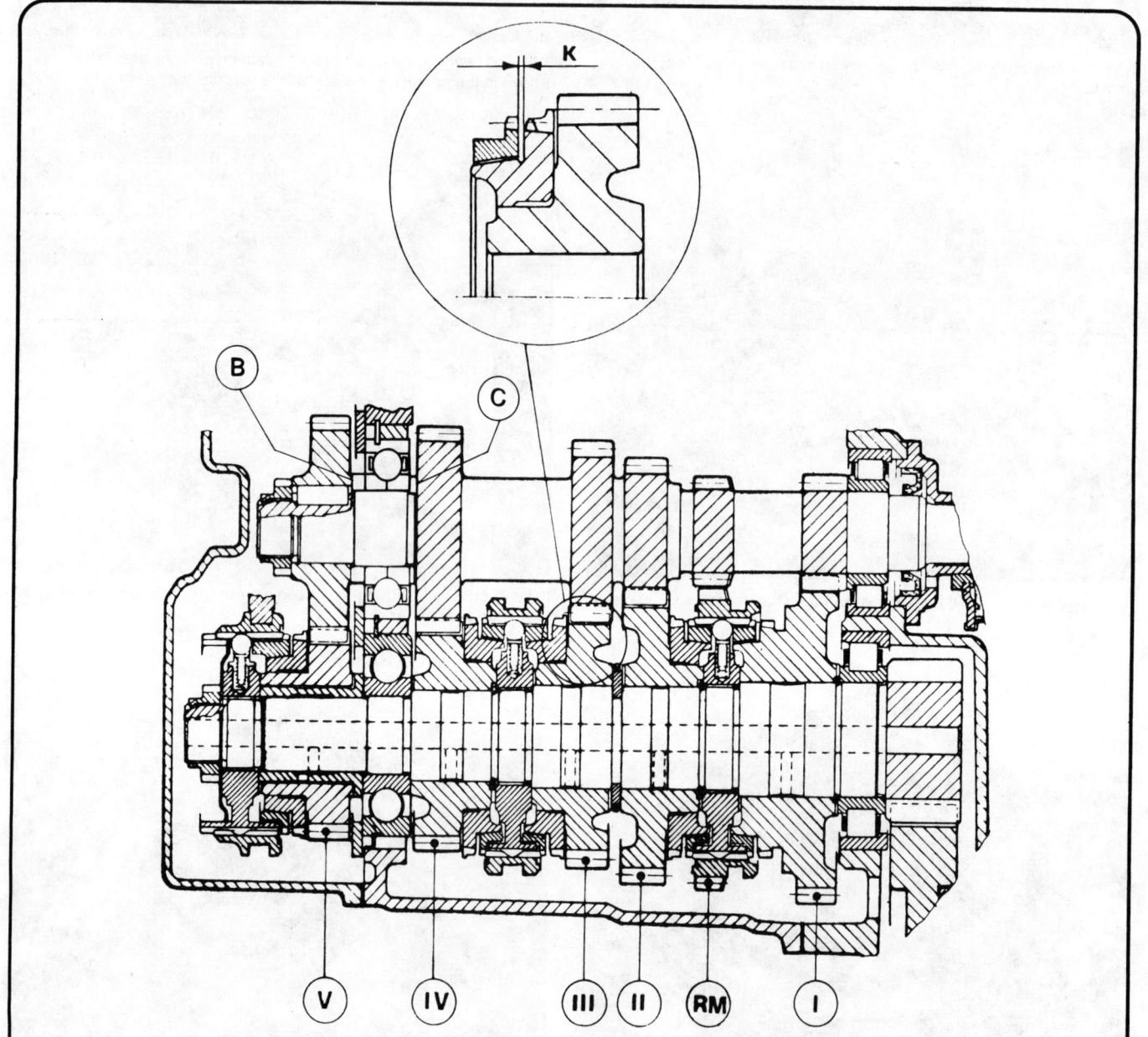

Fig. 6.12 Check the clearance of the baulk rings and synchro cones at gap k. Note that the primary shaft may have two shims at B and C (Secs 6, 7 and 8)

K = specified gap between baulk rings and synchro cones
B = rear shim on primary shaft
C = shim between bearing and fourth-speed drive gear

Chapter 6 Manual gearbox and automatic transmission

17 Refit the key to the keyway and then the spacer and fifth speed gear to the rear end of the primary shaft. Note the marks that were made on dismantling (photo).
18 Refit the 5th gear to the mainshaft, meshing it with the primary shaft gear (photo).
19 Refit the synchro hub, and then refit the pins, springs and balls or rollers. Refit the synchro fork to the sleeve before starting this, so that when the three balls or rollers are in place the sleeve can be slipped over the hub to lock the balls or rollers into position. At the same time as the synchro sleeve is slipped down the selector fork must be engaged on the end of the selector rod.
20 Refit the synchro hub mechanism retaining plate to the mainshaft assembly and screw on the mainshaft locking ring nut (photo). Take great care that the synchro sleeve stays in the neutral position until the locking ring nut has been screwed into the end of the shaft sufficiently far as to hold the retaining plate in place and stop the balls or rollers jumping out yet again.
21 Once the retaining plate is secure slide the synchro sleeve upwards to engage fifth gear and operate one of the selector rods to engage another gear by moving the main selector shaft. Refit the primary shaft rear end locking ring nut and tighten both the shaft locking ring nuts to the specified torque, using the spanner which was made to remove them. By engaging two gears at the same time the shaft will be locked.
22 Lock the rings with a punch (photo).
23 Depress the fifth gear synchro sleeve to put it in neutral and unlock the other gear selected.
24 Refit the fifth speed selector fork locking bolt and tighten it to the specified torque.
25 Lightly grease the end cover gasket and place it in position on the main gearbox casing.
26 Refit the end cover to the casing and refit the seven retaining bolts. Tighten them to the specified torque.
27 Refit the reversing light switch and tighten it to the specified torque, having first refitted the selector rod locking balls and springs into the side of the casing. Refit the cover and gasket and retaining bolts.
28 If the outer lever was refitted temporarily to the selector shaft to lock the gears up to help refitting the gearbox casing then remove it and screw in the selector shaft bush. Tighten the bush to the specified torque.
29 Refit the outer lever to the selector shaft and secure it with the locking bolt.
30 Turn the whole assembly over on its side in order to refit the clutch release bearing sleeve. Fit the new gasket to the gearbox side of the sleeve. Grease the inner lips of the new oil seal and slide the sleeve carefully over the end of the primary shaft. Locate it in position and refit the three retaining bolts and tighten them to the specified torque. Lightly grease the sleeve.
31 Refit the release bearing to the sleeve and engage it in both arms of the operating mechanism, as it slides onto the sleeve.
32 The gearbox and differential unit assembly is now ready for refitting to the engine assembly in the reverse order to its removal. Refit the starter motor and reconnect the front remote control gear linkage to the selector shaft lever and front idler arm.
33 Do not forget to refill the gearbox/differential unit with the correct grade and quantity of oil. Both units have a common oil bath and the unit is filled with oil through the dipstick/filler hole in the top of the differential unit cover. Check that the level is correct according to the dipstick and then refit it.

9 Gear linkage – description and layout

1 The gear linkage between the gear lever inside the car and the gearbox itself is a two section remote control rod system with a central

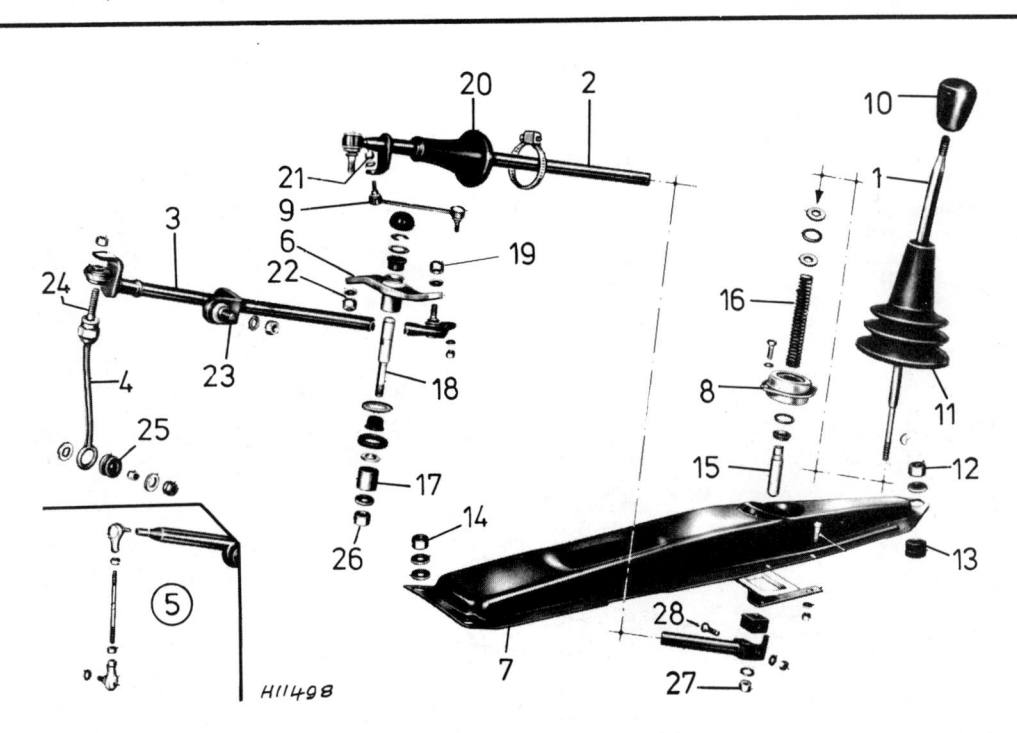

Fig. 6.13 Gear linkage – remote control system – layout (Sec 9)

1 Gear lever
2 Rear control shaft
3 Front control shaft
4 Front stay rod
5 Front stay rod (early type)
6 Control shaft to idler retaining nut
7 Support bracket
8 Pivot
9 Connector rod
10 Gear lever knob
11 Gear lever boot
12 Mounting nut
13 Bush
14 Mounting nut and washers
15 Spacer
16 Gear lever spring
17 Spacer
18 Idler shaft
19 Nut
20 Control shaft boot
21 Connector rod to control shaft nut
22 Control shaft to idler lever nut
23 Front control shaft to selector shaft mounting
24 Front stay rod adjuster
25 Bush
26 Idler shaft nut
27 Gear lever mounting nut
28 Rear control shaft to gear lever lockbolt

idler and a control stay at the front (Fig. 6.13).
2 The gear lever is connected to the rear control shaft and supported by a large bracket, in which it pivots.
3 The gearbox selector shaft is operated by the front control shaft via a short lever and the two shafts are linked together in the centre of the rear engine bay bulkhead by an idler system operating in a horizontal plane.
4 The front end of the front control shaft is located by a thin metal rod or stay which itself is attached to the side of the bellhousing casing.

10 Gear linkage assemblies – removal, checking, refitting and adjusting

1 It is unlikely that it will be necessary to remove the remote control gear linkage system as a complete assembly. It is therefore divided into logical sections.

Gear lever and rear control shaft

2 At the idler on the rear bulkhead of the engine bay remove the nuts which secure the rear control shaft to the idler bar and the small connector rod (photo).
3 Unscrew the gear lever knob and remove it.
4 Lift off the gear lever boot complete with the centre console top plate. By inserting a thin flat bladed screwdriver under the plate it can be carefully prised away from the console, as described in Chapter 12.
5 Remove the centre console, also as described in Chapter 12.
6 Remove the heating and ventilating pipes and ducts to the rear seats. This is also covered in Chapter 12.
7 Remove the front mounting for the centre console from the car floor.
8 Prise the rear control shaft boot from the body panel.
9 Check that the three retaining nuts and relevant washers have been removed from the main support bracket.
10 The complete assembly can now be removed from the car, and the component parts can be stripped out and examined on the bench.
11 With the locking bolt and nut removed from the rear end of the rear control rod, the rod and gear lever shaft can be separated.
12 Remove the two bolts which secure the gear lever pivot joint to the support bracket and lift off the gear lever assembly.
13 Undo the retaining nut on the bottom of the gear lever shaft and the complete assembly on the shaft can be withdrawn, so that any worn components can be renewed. The most likely items to need renewal are the thrust spring and pivot joint.
14 With the rear control rod separated, the large rubber boot or grommet can be removed and a new one fitted if necessary.
15 With the rear part of the remote control gear linkage stripped, inspect all the parts for wear, especially the joints, and renew those which are worn.
16 Reassembly and refitting is the reverse procedure to removal. With the assembly complete refit it to the car, but do not tighten the support bracket mounting nuts at this stage.
17 Reconnect the rear control rod to the central idler and reconnect the connector rod to the rear control rod. Tighten both nuts to their specified torques.
18 Refit the rear control shaft rubber boot to the body panel.
19 Now try all the gears. They should all engage smoothly and without effort. Once engaged there should be a little extra travel on the lever. The gear lever when in neutral should assume a position 15° to the rear of the vertical and aligned centrally (Fig. 6.14).
20 If the position is not correct then it can be altered by moving the support bracket along on its mounting bolts, until it is right. Then tighten up the bracket mounting nuts.
21 Refit the ventilation and heating ducts and centre console. Then refit the gear lever boot and knob.
22 Recheck the gears and if necessary adjust the front stay rod to ensure perfect engagement, and correct upright position of the gear lever in the car, when viewed from the front or rear.

Central idler

23 Undo the nuts and disconnect the central idler small connector rod from both front and rear control shafts.
24 Undo the idler mounting and pivot shaft retaining nut from underneath the bracket (photo).
25 Remove the protective rubber cover from the top of the shaft and remove the shaft and idler lever upwards. As the assembly is withdrawn from the mounting bracket remove the spacers and washers.

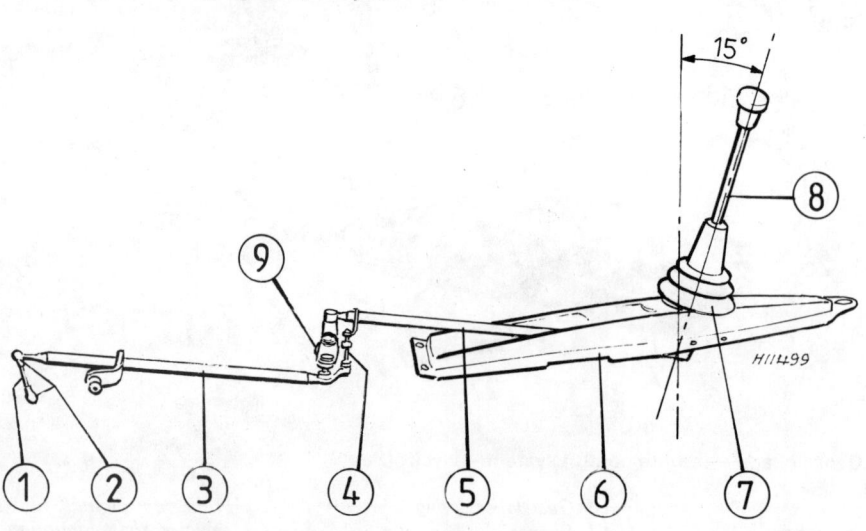

Fig. 6.14 The correct position of the gear lever viewed from the side (Sec 10)

1 Front control shaft stay rod
2 Joints
3 Front control shaft
4 Rod
5 Rear control shaft
6 Support bracket
7 Rubber boot
8 Lever
9 Idler lever

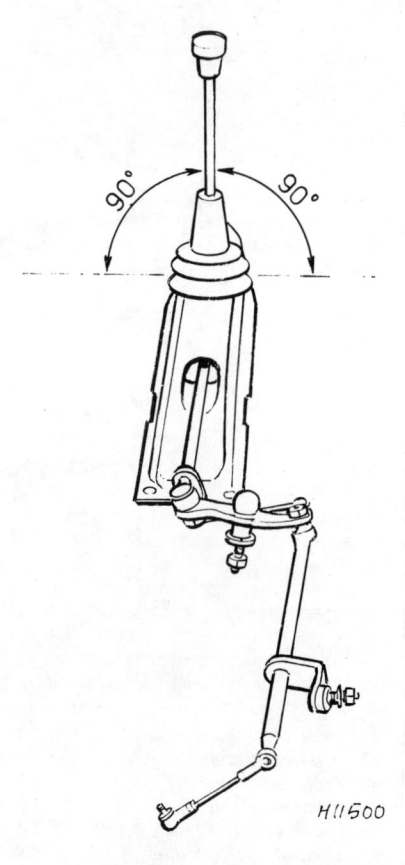

Fig. 6.15 Gear lever – correct alignment – viewed from the front (Sec 10)

Chapter 6 Manual gearbox and automatic transmission

26 With the idler lever and shaft removed from the car the idler pivot bushes can be renewed if necessary.
27 The upper and lower bushes in the idler lever can be driven out and new ones fitted. Pack the inside of the idler lever sleeve with grease.
28 Inspect the idler lever shaft for wear or distortion and renew it if necessary.
29 Reassemble the idler lever and its components and refit the assembly in the reverse order to removal. Finally tighten all the mounting nuts to their specified torque.

Front control shaft

30 Undo the nuts which retain the small connector rod to the rear of the shaft and the shaft to the idler lever.
31 To reach the front control shaft in the engine compartment necessitates a certain amount of 'digging'. The shaft is located beneath the cooling system hoses between the engine and battery. In later models the large windscreen washer reservoir will have to be removed first. If you wish to make the task easier and remove the radiator or hoses, then refer to Chapter 2 for guidance.
32 Once located, disconnect the control shaft bracket from the selector shaft outer lever and release the front end of the control shaft from the top of the stay (photo). In both cases it is matter of undoing one nut.
33 Refitting is the reverse procedure to removal. Check the engagement of the gears when the whole assembly has been refitted and if necessary adjust the front stay rod.

Front stay rod

34 Two types of stay rod have been fitted since production started. The early type has an adjustable centre section with locknuts at the top and at the bottom. The top end is bolted to the front end of the front control shaft and the bottom end is bolted through a bush to the bellhousing. This is shown in Fig. 6.13.
35 The later type of stay is a jointed rod secured to the front control shaft by a locknut on the adjuster end of the stay. The bottom end is attached by a stud and nut through a bush to the bellhousing.
36 In the case of the early type of stay undo the upper and lower bolts and remove the stay. With the later type undo the adjuster locknut above the control shaft end bracket and remove the lower mounting nut and release the stay and bush at the bottom end. Withdraw the top end from the control shaft front bracket.
37 Check that the mounting bushes are in good condition, and if necessary renew them.
38 Check that on the later type of stay there is no excessive play in the stay joint and that the protective rubber cover for the joint is in good condition. If the joint is worn then the stay rod will have to be renewed.
39 Refit the stay rods in the reverse order to removal and then adjust them as follows.

Front stay rod – adjustment

40 If the adjustment is being made as a result of removing the rod for overhaul or renewal of bushes or if the front control shaft has had work carried out on it, then access to the area of the control stay rod will have been cleared. However, if the stay rod needs adjusting after the rear section of the remote control linkage has been overhauled, then it will be necessary to make access to the stay rod. This involves either removing the battery and battery tray or the cooling system hoses and thermostat housing beneath which it lies. In later models the windscreen washer reservoir will need to be removed as well.
41 The stay rod adjustment alters the position of the gear lever in the car from side to side. When correctly adjusted the gear lever when viewed from the front should be vertical. When viewed from the side it should be inclined back 15° from the vertical as well (Fig. 6.15).
42 On early models unlock the two locknuts on the centre section of the stay rod and on later models slacken the top locknut on the adjuster. Rotate the adjuster as necessary.
43 Lengthen or shorten the stay as required so that the gear lever is in the vertical position when viewed from the front or rear. If the gear lever goes further to the left than to the right when moved from side to side then the stay rod needs to be shortened and if the lever goes further to the right than to the left it needs to be lengthened.
44 When the gear lever is correctly aligned tighten up the locknuts and check that the gears engage smoothly and easily. The gear lever should have slight travel after the gear has engaged. If necessary readjust the stay rod.
45 Refit the ancillary components which were removed to reach the stay and refill the cooling system if necessary.

11 Gearbox/differential unit mountings – removal and refitting

1 There are two mountings for the gearbox and differential unit. One is at the front of the gearbox. This comprises a bracket secured to the gearbox casing which is retained by a single nut and bolt to the mounting on the front member of the subframe. The rear mounting is attached to the rear cover of the differential unit by two nuts and bolts and the mounting itself is located on the rear member of the subframe (Fig. 6.16).
2 If one mounting is suspect then both should be renewed, as it is very unlikely that the other will be in good condition.
3 Remove the battery and battery carrier, and the windscreen washer reservoir in later models.
4 Remove the radiator as described in Chapter 2.
5 Remove the gear linkage front control shaft as described in Section 10.
6 Jack up the left-hand front roadwheel and remove it and place an axle stand under the wishbone supported by a wooden block.
7 On cars with self levelling headlamps disconnect the link arm from the front left-hand wishbone and then undo the sensor mounting bracket, and move it to one side in order to reach the gearbox rear mounting (photo).
8 Undo and remove the two nuts from the ends of the rear mounting bolts, which pass through the differential carrier. Then place a jack under the differential carrier/gearbox and take the weight of the unit on the jack so that the two bolts can then be withdrawn.
9 Undo the four bolts which secure the rear mounting to the subframe and remove it through the left-hand wheel arch.
10 To remove the gearbox front mounting on saloon models the task will be made much easier if the radiator grille is removed as described in Chapter 12. This is not possible with the Coupe, Spider/Zagato or HPE and in these models the operation has to be carried out from above and below. However, with the radiator already removed access is possible.
11 Undo and remove the nuts which secure the bracket to the

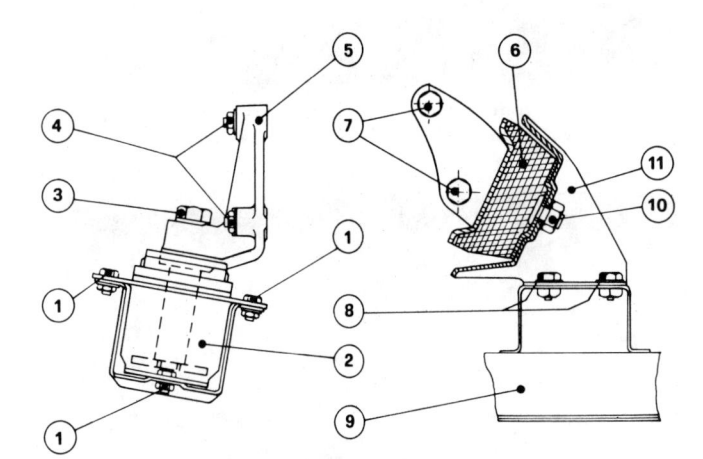

Fig. 6.16 Gearbox/differential unit front and rear mountings (Sec 11)

1 Front mounting to subframe nuts and bolts
2 Front mounting
3 Mounting bolt
4 Mounting bracket to gearbox retaining nuts
5 Mounting bracket
6 Rear mounting
7 Differential to mounting retaining bolts
8 Mounting bracket to subframe nuts and bolts
9 Subframe
10 Mounting to mounting bracket retaining nut
11 Rear mounting bracket

10.2 Disconnect the rear control shaft from the idler and small connector rod

10.24 The idler lever is located by a central shaft, through the mounting bracket, secured by a large nut

10.32 The front control shaft is attached to the gearbox selector shaft by a short lever and supported at the front by the stay rod

11.7 The front headlamp levelling sensor can be seen next to the rear mounting for the gearbox (engine removed for clarity)

11.11 The front gearbox mounting bolt retaining nut is easily reached from underneath the car. The mounting retaining nuts can also be seen

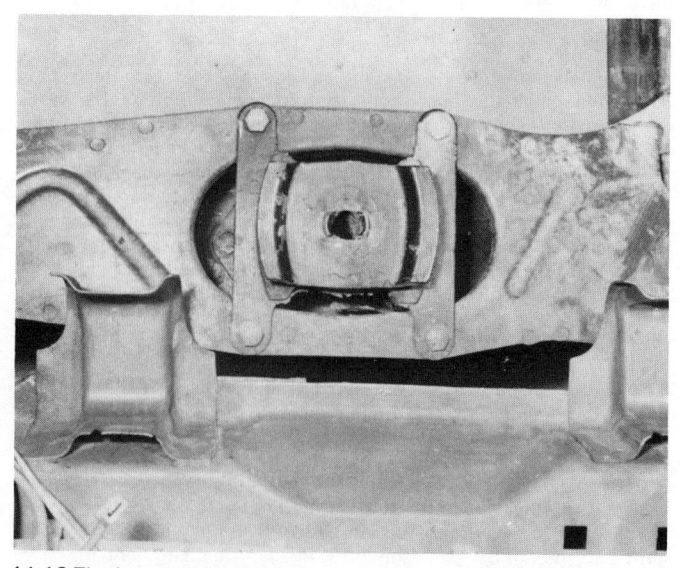

11.13 The front gearbox mounting viewed from above with the engine and gearbox removed from the car

gearbox casing. Then unscrew the long mounting bolt retaining nut which secures the gearbox to the mounting (photo).
12 Move the jack from the rear of the differential unit to the front side of the gearbox and lift it high enough so that the front bracket can be lifted off the studs on the gearbox complete with the mounting bolt. As will have been discovered, the long front mounting bolt cannot be removed immediately even when the retaining nut has been removed.
13 The next task is to undo the six nuts and bolts which secure the gearbox mounting to the subframe. Four are located in the inner and outer upper lips of the subframe and two are recessed and bolted through the bottom of the subframe crossmember (photo).
14 It may be advisable to raise the jack higher to gain sufficient access to remove the nuts and bolts.
15 With the retaining nuts and bolts removed the front mounting can be withdrawn from the subframe crossmember.
16 With the rear mounting, the mounting itself has to be removed from the bracket to which it is secured by a single bolt and nut. The new mounting block can then be attached to it. The front mounting is a complete unit and is renewed as such.
17 Fit the new mountings in the reverse order to their removal. Do not forget to mount the headlamp self-levelling sensor correctly and reconnect the linkage. Tighten all the mounting bolts and nuts to their specified torque.

12 Automatic transmission unit – general description

The Beta is the first ever Lancia to be fitted with automatic transmission and as such it is a great innovation. The transmission was designed for Lancia by the British company, Automotive Products (AP), who have had great success in the past designing and manufacturing automatic transmissions for smaller engined cars, notably the Mini, BL 1100/1300 and Allegro series.

The Lancia unit is now built entirely at a new factory in Italy, specially constructed for the manufacture of automatic transmissions.

The transmission unit is mounted in line with the transverse engine, as is the manual gearbox, and takes up no more room; it is a very compact unit. A torque converter replaces the flywheel and clutch, and directs the drive from the engine to a train of epicyclic bevel gears which are engaged as required for the different transmission ratios.

Engagement of the gears is carried out by two hydraulically operated clutches. The primary or forward clutch engages all the forward speeds. The secondary or direct drive and reverse clutch is engaged together with the primary clutch for direct drive, or top gear, and on its own for reverse gear.

The system of locking devices by engaging the ratios is effected by three sets of brake bands.

Seven helical bevel gears make up the gear train which gives three forward speeds and one reverse speed. The highest speed is in direct drive. This effectively means that there is no relative motion between the gears.

The final drive is by transfer gears to the differential unit mounted on what is effectively the right-hand side of the transmission unit, although in a transverse layout this becomes the rear of the assembly. From the differential unit drive is transmitted to the front roadwheels by driveshafts.

The gearbox and differential unit share a common oil bath and there is an oil pump driven off the rear of the torque converter which supplies the oil both for operation and lubrication. The oil is drawn through a wire mesh filter in the bottom of the unit and is delivered to a valve assembly in the gearbox, located beneath the top cover. A separate heat exchanger is also fitted to keep the transmission fluid within the operating temperature limits, especially necessary if the car is being used for towing.

Lancia have built in a compensating device to prevent the engine slow-running becoming bumpy in traffic, when the car is stationary. A bleed valve from the brake servo system raises the engine revs slightly when the brake pedal is depressed. This system comes into effect when the car is in neutral or in gear and effectively reduces the load imposed by the torque converter.

The green selector which is mounted centrally between the front seats follows the conventional P – R – N – D – 2 – 1 pattern (Fig. 6.17).

There is a detent catch on the top of the gear lever knob to prevent accidental movement of the lever between the gears. The catch must

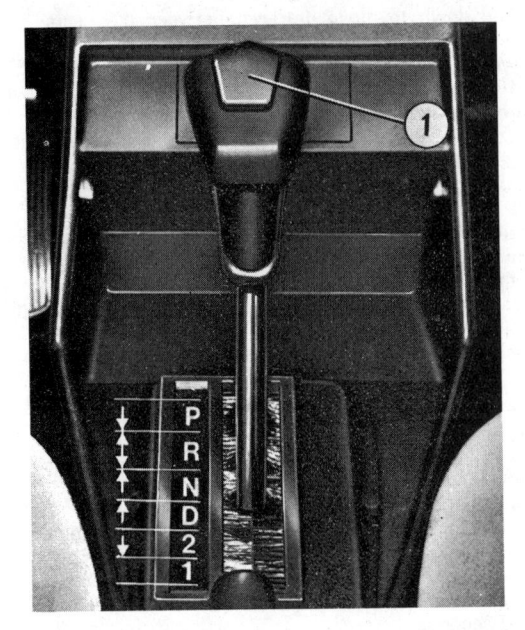

Fig. 6.17 The automatic transmission selector quadrant (Sec 12)

1 Detent catch

Push button down to move the operating lever between the following gears

P to R	R to N	D to N
R to P	N to R	2 to 1

Fig. 6.18 Automatic transmission fluid level dipstick (1) location (Sec 13)

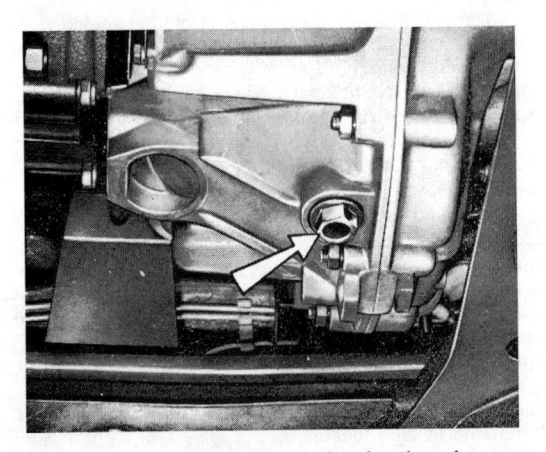

Fig. 6.19 Automatic transmission drain plug location – arrowed (Sec 13)

be pushed down to move the lever between the gears. The catch must be pushed down to move the lever between all gears except D and 2 where it can move freely.

The quadrant has a lighting device which illuminates whichever section of the quadrant the gear lever is opposite.

If 2 or 1 is selected then the car will stay in that gear and will not change up or down. This is very useful in town driving, especially the 2 position.

Due to the complexity of the automatic transmission only the procedures given in the following sections should be attempted by the home mechanic. It must also be remembered that a comprehensive test of the transmission is only possible with it in the car, therefore if a fault develops which cannot be cured by the procedures given, the car should be taken to a Lancia agent with suitable diagnostic equipment before any attempt is made at removing the transmission.

13 Automatic transmission fluid – level checking and renewal

1 Every 6000 miles (10 000 km) check the level of the fluid in the automatic transmission. To do this, run the car on the road until the engine and transmission are at the normal operating temperature; approximately 6 miles (9.5 km) is normal.
2 With the car on level ground move the selector lever slowly through all the positions, then select P and allow the engine to idle for 20 to 30 seconds.
3 With the engine still running, withdraw the dipstick and wipe it with a lint-free cloth. Reinsert it fully, withdraw it, and read off the level which should be between the low and full marks.
4 If necessary top up the level with automatic transmission fluid (ATF) poured through the filler tube.
5 Every 18 000 miles (30 000 km) the fluid must be changed. To do this, run the car on the road for a few miles in order to warm the fluid.
6 Place a suitable container beneath the transmission and clean the surrounding area of the drain plug.
7 Unscrew the drain plug and allow the fluid to drain for a minimum of 10 minutes. Take care to avoid scalding as the fluid may be quite hot.
8 Clean away any metallic particles which may have collected on the magnetic drain plug, then screw it firmly back into the casing. The transmission can now be refilled with fluid and the level checked as described in paragraphs 1 to 4. The exact amount required will depend on how much fluid was left in the torque converter after draining.

14 Automatic transmission – separation from engine

1 The procedure is basically identical to that for the manual gearbox described in Section 3, with the exception of the remote control assembly. However first remove the screws and withdraw the aperture cover from the bottom of the torque converter housing.
2 Unscrew the bolts securing the driveplate to the torque converter. To do this, hold the crankshaft stationary with a sprocket on the pulley nut. It will be necessary to turn the crankshaft to obtain access to all of the bolts.
3 After disconnecting the transmission from the engine as described in Section 3, it can be withdrawn, but make sure that the torque converter is held in engagement with the oil pump to prevent damage to components and loss of fluid. Use a length of wood as a lever if necessary to separate the torque converter from the spigot bearing.
4 Reconnection of the transmission is a reversal of the separation procedure but make sure that the torque converter is fully entered as described in Section 16.

15 Automatic transmission selector linkage – adjustment

1 Disconnect the selector linkage from the lower pivot of the relay arm on the rear of the transmission.
2 Turn the linkage arm at the transmission end fully clockwise, then turn it back three detent position to select D.
3 Have an assistant hold the selector lever in the D position and check that the linkage locates on the relay arm without altering the position of either linkage. If necessary loosen the locknuts and adjust the length of the linkage rod on the transmission. Tighten the locknuts

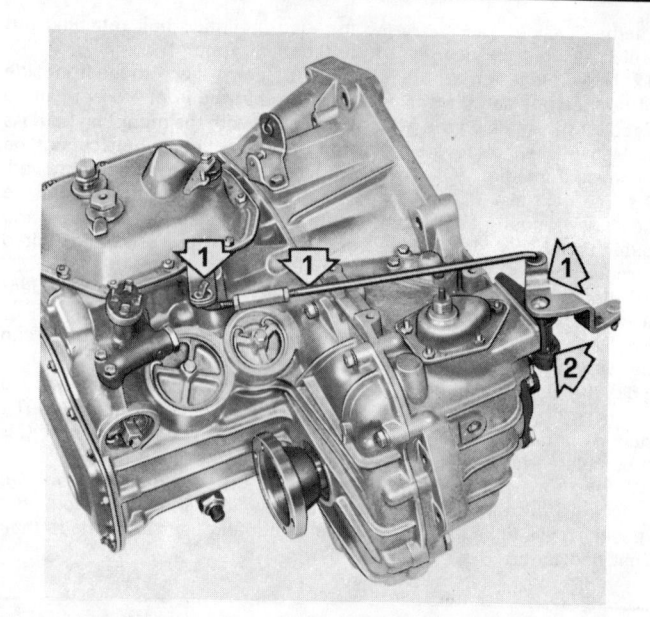

Fig. 6.20 Automatic transmission selector linkage (1) and support bracket (2) (Sec 15)

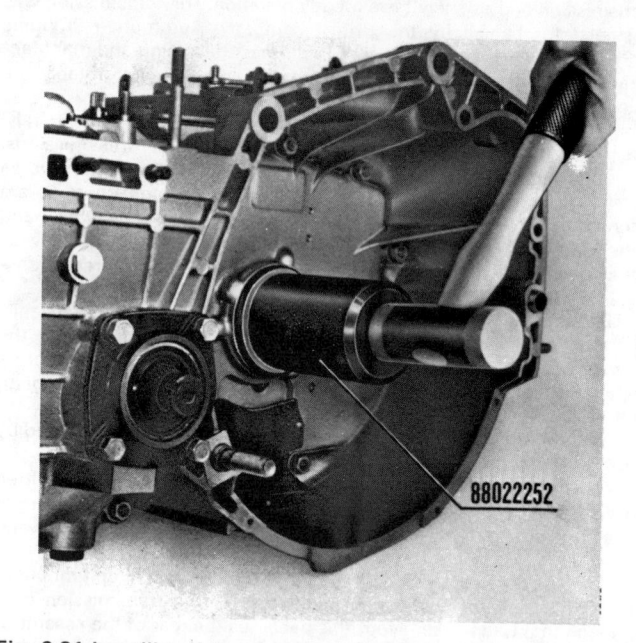

Fig. 6.21 Installing the automatic transmission input oil seal using the special tool (Sec 16)

when the adjustment is correct, and secure the linkage to the pivot on the relay arm.

16 Automatic transmission oil seals – renewal

When renewing the oil seals it is essential to thoroughly clean the surrounding area of the transmission to prevent foreign matter entering the hydraulic system. Failure to observe this could result in malfunction of the unit necessitating extensive repairs.

Differential side gear oil seals
1 The procedure is similar to that described in Chapter 7 for the manual gearbox, but first drain the transmission fluid into a suitable clean receptacle.
2 After fitting the new seal, refill the transmission and top up the level as described in Section 13.

Chapter 6 Manual gearbox and automatic transmission

Input oil seal

3 Remove the engine and transmission from the car as described in Chapter 1, and separate the transmission as described in Section 14 of this Chapter.
4 Place the transmission on a bench and have a container ready to catch the fluid which will escape from the torque converter.
5 Withdraw the torque converter and place it to one side.
6 Using a hooked piece of metal, pull the oil seal from the casing. Note that the oil seal must be renewed every time the torque converter is removed.
7 Clean the oil seal recess in the casing with lint-free cloth, taking care not to allow any foreign matter to enter the transmission.
8 Drive the new seal into position using tubing of suitable diameter. The seal lips must face into the transmission.
9 Carefully slide the torque converter into position whilst turning it as necessary to engage the input shaft splines. When fully installed a click should be heard as the torque converter engages the oil pump, also the front face of the torque converter must be below the mating face of the transmission.
10 Reconnect the transmission to the engine as described in Section 14, and fit the engine/transmission unit to the car as described in Chapter 1. Refill the transmission with fluid to the correct level as described in Section 13.

17 Fault diagnosis – manual gearbox

Symptom	Reason(s)
Weak or ineffective synchromesh	Synchromesh dogs worn, or damaged
Jumps out of gear	Broken gearchange fork rod spring Gearbox coupling dogs badly worn Selector fork rod groove badly worn Remote control linkage incorrectly adjusted
Excessive noise	Incorrect grade of oil in gearbox or oil level too low Bush or bearings worn or damaged Gearteeth excessively worn or damaged Gear remote control linkage front stay rod maladjusted or central idler not lubricated
Gearbox knocks or vibrates when engine is under load	Gearbox and/or engine mountings worn out
Excessive difficulty in engaging gear	Clutch pedal adjustment incorrect Gear linkage remote control maladjusted

18 Fault diagnosis – automatic transmission

Symptom	Reason(s)
Engine will not start in positions N or P	Faulty inhibitor switch Incorrect linkage adjustment
No drive, poor acceleration, low maximum speed	Incorrect fluid level
Severe bump when selecting gear	Idling speed too high

Chapter 7 Differential unit

Contents

Differential side gear oil seals – renewal 4
Differential unit – overhaul 3
Differential unit – removal and refitting 2
Fault diagnosis – differential unit 5
General description 1

Specifications

Final drive ratios	UK	USA
1300 Saloon and Coupe models	4.46 : 1	–
1400 Saloon models	4.46 : 1	–
1600 Saloon and HPE (early models)	4.21 : 1	–
1600 Saloon and HPE (later models and all Coupe and Spider models)	4.07 : 1	–
1800 Coupe and Spider (early models)	3.92 : 1	4.00 : 1
1800 Saloon, Coupe and Spider (later models)	4.07 : 1	4.21 : 1
1800 HPE models	–	4.38 : 1
2000 Saloon, Coupe, Spider and HPE models	3.78 : 1	4.21 : 1
1600 Automatic models	4.38 : 1	–
2000 Automatic models	4.38 : 1	4.84 : 1

Side gear
Endfloat .... 0.20 to 0.30 mm (0.0078 to 0.0118 ins)
Endfloat adjustment shim sizes available 1.8, 1.9, 2.0, 2.1, 2.2
Side gear bearing preload adjusting shims available .... From 1.70 to 2.60 mm in steps of 0.05 mm
Differential side gear bearings revolving torque .... 1 to 1.5 Nm (0.74 to 1.10 lbf ft)

Torque wrench settings

	lbf ft	Nm
Drive flange retaining nut	117	160
Crownwheel gear to half casing	51	70
Differential cover to gearbox casing:		
Small nuts	18	25
Large nuts	36	50
Oil seal covers to casing	10	15
Speedometer gear bush to differential cover	7	10

1 General description

The differential unit is located on the rear or bulkhead side of the gearbox and is part of the gearbox assembly. The differential unit provides the final drive from the gearbox to the driveshafts and thus to the front wheels. The cover of the unit is held on by nuts of different sizes, the four main ones being on either side of the side gear bearings.

The differential unit contains the crownwheel, the planet pinions and side gears (Fig. 7.2). The gearbox and differential unit share a common oil bath and the dipstick is located in the rear topside of the differential housing.

Because the differential unit is part of the gearbox assembly it can only be removed from the car and stripped down as part of the gearbox itself. The gearbox has to be removed complete with the engine anyway, so to carry out any work involving the differential unit is a long and involved task. Only the oil seals can be renewed in situ.

The drive pinion is on the end of the mainshaft in the gearbox and is supplied as a matched set with the crownwheel.

This Chapter only deals with the differential units fitted to manual transmission models.

2 Differential unit – removal and refitting

1 In order to carry out any work on the differential unit the gearbox and engine assembly have to be removed from the car as described in Chapter 1.
2 With the assembly removed from the car, it is possible to remove the differential unit without separating the engine and gearbox assembly. However, it is more than likely that the reason for wanting to remove the differential unit is that the crownwheel and final drive pinion have been damaged or have worn badly. In this case the gearbox and engine assemblies will have to be separated, and the gearbox will have to be stripped down to remove the mainshaft with the final drive pinion at the end of it.
3 To remove the differential unit without stripping the gearbox, first refer to Chapter 8 and remove the driveshaft centre section if your car has a three-piece system fitted. Drain the oil from the differential unit.
4 If however the final drive pinion needs renewing as well then start by removing the gearbox from the engine and strip out the gearbox shaft assemblies as described in Chapter 6.
5 To remove the differential unit itself start by removing the drive

131

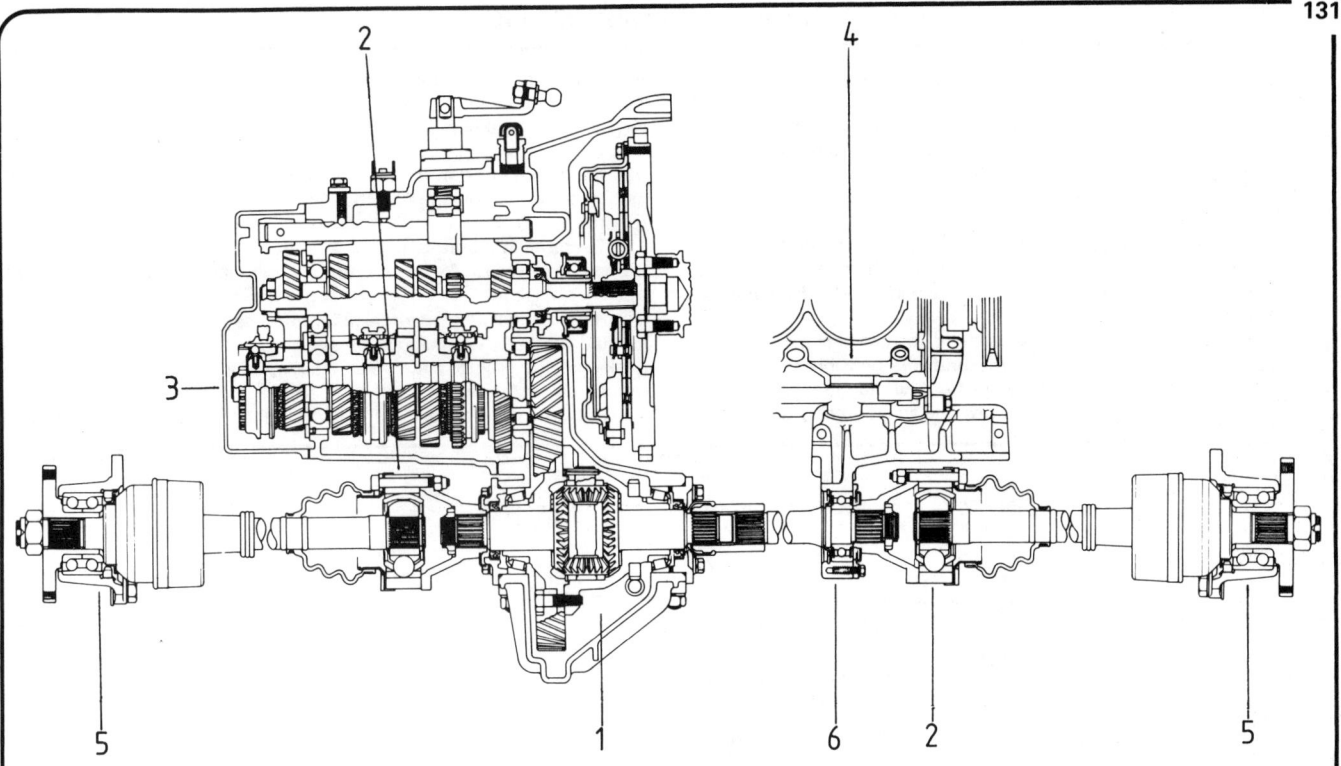

Fig. 7.1 Drive train layout (Sec 1)

1 Differential unit
2 Drive flange/inner CV joint (later model)
3 Gearbox
4 Engine
5 Roadwheel hubs
6 Centre section driveshaft bearing (later model)

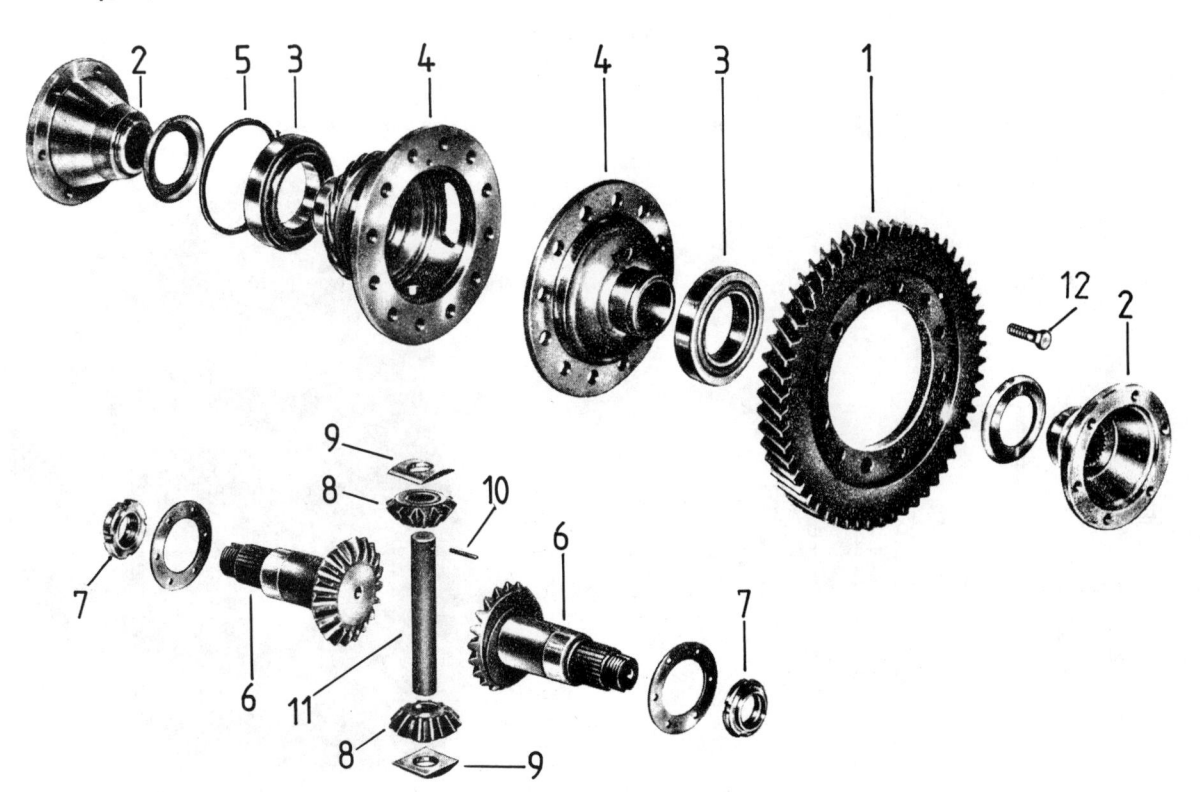

Fig. 7.2 Differential unit components – exploded view (Sec 1)

1 Crownwheel
2 Drive flanges
3 Bearings
4 Half casings
5 Bearing pre-load shim
6 Side gears
7 Locking ring nuts
8 Pinion gears
9 Thrust washers
10 Retaining pin
11 Pinion shaft
12 Bolt

Chapter 7 Differential unit

flange(s) from the end of the side gear shafts. In the early models with the long right-hand driveshaft there are two drive flanges, whereas in the later model with the three piece system there is only one flange on the left-hand side of the unit (photo). Uncrimp the locking ring nuts and remove them. It will be necessary to make up a bracket to prevent the flanges from turning when doing this.

6 Use a puller to remove the drive flange(s).

7 Remove the four bolts on the gearbox side which retain the oil seal and cover plate. Remove the bolts which retain the engine side cover plate. Note that there are shims for the bearing preload adjustment beneath (photo).

8 Since on reassembly there is no need to refit shims to the side gear shaft on the gearbox side of the unit it is possible to strip and reassemble the unit without removing the drive flange on that side.

9 Undo the locking bolt and withdraw the speedometer driven gear from the differential unit cover.

10 Undo the cover mounting nuts. There are six small ones and four larger ones, on either side of the bearing housings.

11 Then lift off the cover (photo).

12 The complete differential unit can now be lifted out. Lift the two side gear shafts and the unit, bearings etc will all come out in one go (photo).

13 To refit the differential unit after overhaul is a step by step procedure, which if followed correctly will leave the unit correctly set up.

14 Refit the assembly to the differential carrier. Make sure that it is located the right way round and seats correctly. Apply a bead of RTV gasket to the face of the carrier, having ensured that there is no dirt, grease or old gasket still adhering to it.

15 Refit the differential cover casing. Refit the nuts and tighten them

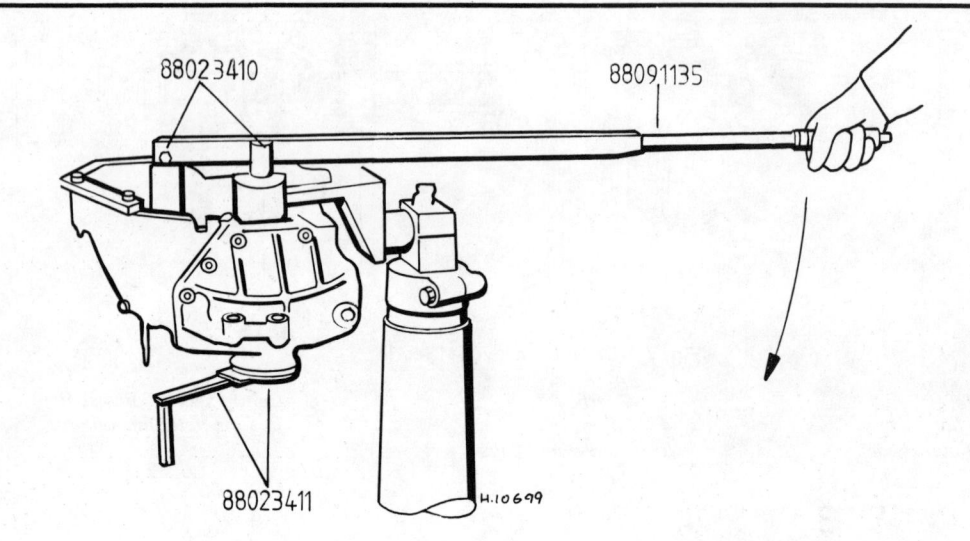

Fig. 7.3 Special tools for bedding-in the outer bearing races (Sec 2)

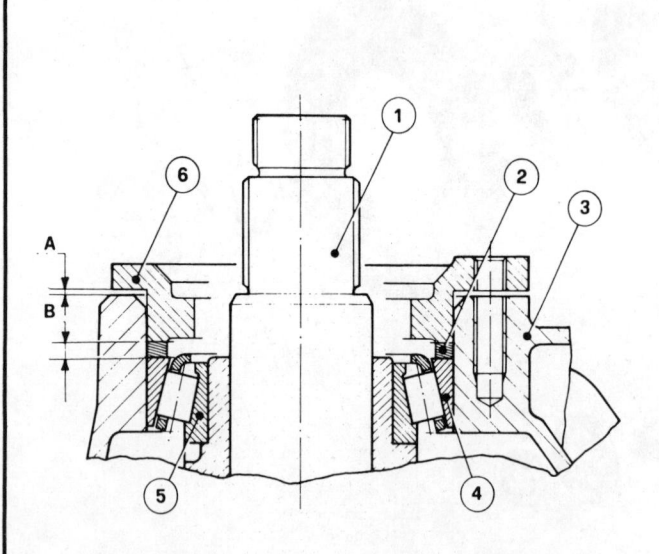

Fig. 7.4 Calculating the pre-load shims for the bearings (Sec 2)

1 Side gear
2 Known thickness shim
3 Differential carrier
4 Bearing outer race
5 Bearing inner race
6 Seal carrying cover

For A and B see text

Fig. 7.5 Checking the rolling torque using a pulley and weight (Sec 2)

Chapter 7 Differential unit

to the correct torques (photo).

16 The next stage is to bed in the taper roller bearing outer races. Lancia use a special piece of equipment which is shown in Fig. 7.3. The basic idea of this apparatus is to apply leverage at the end of the long lever forcing the socket in the bearing housing to press on the outer race of the taper roller bearing. At the same time as pressure is applied, the differential unit must be turned by the handle attached to the other side gear shaft. The load applied at the end of the bar is approximately 100 kg (220 lb). This can be achieved by refitting the gearbox side drive flange to the splined end of the shaft and using that to revolve the differential unit. A handle can be made up quite easily to help in this. Then use a piece of tube the size of the outer race and drive the outer race home, whilst revolving the differential gears.

17 Before commencing this part of the operation the oil seal cover plate must be refitted to the gearbox side of the differential unit, and torque tightened. The oil seal which would have been removed for renewal must not be in position for this bedding in operation.

18 Once the bearings have been bedded in, the next stage is to calculate the thickness of bearing preload shims that need to be fitted to the differential side bearing.

19 To do this fit a shim of a known thickness to the bearing (photo). Refit the seal cover plate, again this side without the seal, and measure the gap between the cover plate and differential housing using a feeler gauge. Subtract the measurement of the space from the

2.5 The left-hand drive flange and crimped retaining ring nut with the seal cover plate behind it

2.7 Remove engine side cover plate – note that there are shims beneath it

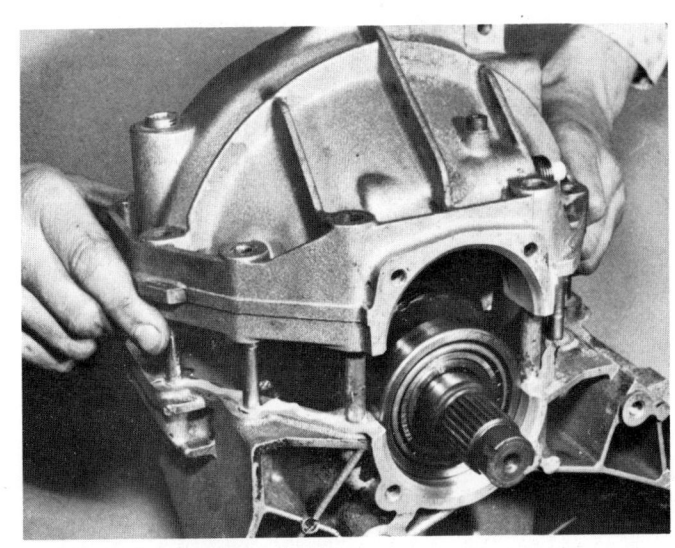

2.11 Removing the differential unit cover

2.12 The whole assembly can now be lifted out

2.15 The differential unit cover showing the location of the dipstick and speedometer driven gear

thickness of the shim which was fitted and add 0.05 mm; this will give you the correct thickness of shim to fit to preload the bearings correctly (Fig. 7.4).

20 Fit the correct shim size and refit the seal cover plate, still without the seal, and torque tighten the four bolts.

21 Now the rolling torque has to be checked and to do this requires the use of a torquemeter, or if one is not available, it can be gauged using weights as shown in Figs. 7.5 and 7.6. The differential unit must be placed vertically as shown and the weight to be used is 2 to 3 kg (4 lbs 7ozs to 6 lbs 10 ozs). Instead of the pulley 88023411 use the drive flange, but the weight must be calculated as follows. Divide the rolling torque value by the radius of the pulley on which the cord runs.

22 If the weight does not turn the differential unit then the shim which has been fitted is too thick and a slightly thinner one must be fitted. Do this and the recheck the rolling torque.

23 When the rolling torque has checked satisfactorily, refit the seal to the seal covers. Refit the seal covers and torque tighten them. Do not forget the O-rings (photo).

24 Refit the drive flange(s) to the side gear shaft(s) and refit and tighten the locking ring nut(s) to the specified torque. Then crimp the ring(s).

25 Reassemble the gearbox as described in Chapter 6 and refit the whole assembly to the engine. The engine and gearbox assembly can be refitted to the car as described in Chapter 1.

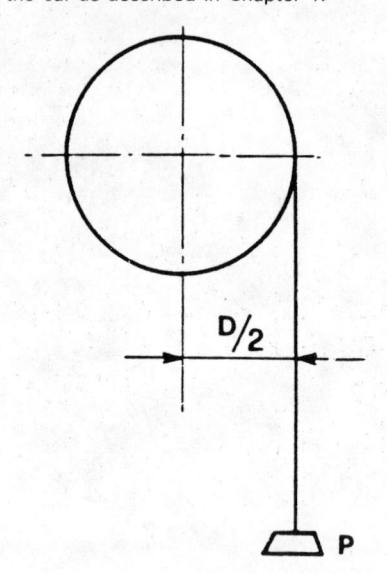

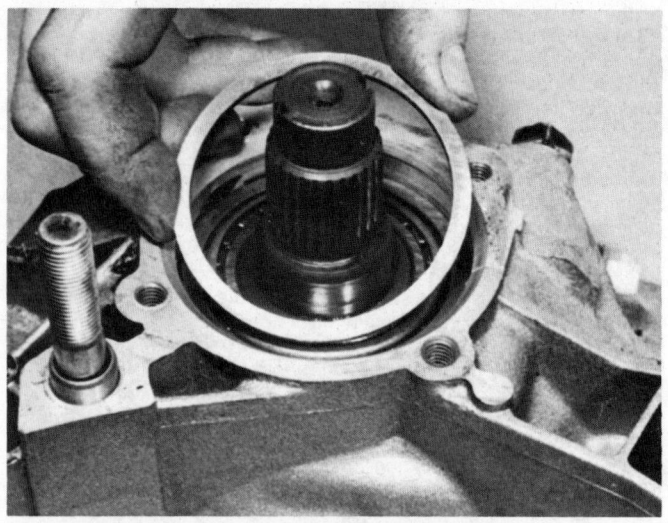

2.19 Fit a shim of known size to the differential side bearing

Fig. 7.6 Calculating the weight to be used to check the rolling torque (Sec 2)

$$\text{Weight }(p) = \frac{\text{Rolling torque value}}{\frac{1}{2}\text{ diameter of the pulley}}$$

2.23 The cover plate with new oil seal and O-ring

3 Differential unit – overhaul

1 With the differential unit removed from the carrier as described in the previous Section proceed as follows. Use a puller to remove the bearings from both side gear shafts (photo).

2 Check the endfloat in the side gears using a dial gauge. This should be as given in the Specifications.

3 Clamp the crownwheel in a vice with protective jaws and remove the bolts that join the two half casings (Fig. 7.8).

4 Push out the pinion shaft retaining pin and remove the shaft. Lift out the pinion gears and thrust washers. Note which thrust washers fit in which positions. The side gears and shims can be removed from the half casings (Fig. 7.9).

5 If the crownwheel needs to be renewed it can be driven off the half casing using a hide hammer.

6 The crownwheel and drive pinion, which is part of the mainshaft, are supplied as a matched set.

7 Clean off all the components in a paraffin bath and dry them thoroughly before inspecting them.

8 Check that the gear teeth are not worn or chipped. Check that there are no hairline cracks in the gear wheels. Ensure that there is no

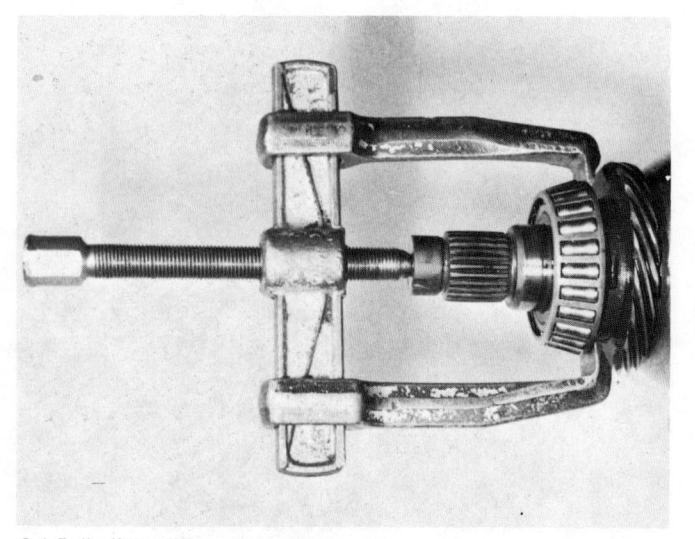

3.1 Pull off the differential bearings

Chapter 7 Differential unit

wear or damage to the side gear shafts or the pinion shaft. Check the bearings for wear or damage.

9 Remove the oil seals from the cover plates and the O-ring from the speedometer driven gear, as they must be renewed.

10 Reassemble the differential half casing in the reverse order to the dismantling process. Make sure all the thrust washers and shims are correctly fitted. Secure the two half casings together, but only use three bolts equally spaced at this stage, in case the shims need to be changed.

11 Check the side gear to pinion endfloat, using a dial gauge. The correct amount of endfloat is 0.20 to 0.30 mm (0.00786 to 0.0118 ins). If the endfloat measurement is incorrect then the shims on either side gear shaft, which lie between the gear end and the half casing, will have to be changed. These adjusting shims are available in various thicknesses as shown in the Specifications (Fig. 7.10). Undo the three bolts and fit the new shims, then recheck the endfloat. If it is correct then continue the rebuilding.

12 Refit the crownwheel to the half casings if it was removed. Make sure that it is the right way round.

13 Refit and tighten all the half casing mounting bolts to the specified torque.

14 Refit the differential bearings to the side gear shafts. Start by refitting the inner races. They will have to be heated in an oven to approximately 80°C (176°F) in order that they fit onto the shafts correctly. Be careful when handling them after they have been heated.

15 With the inner races correctly fitted, refit the outer races to them.

16 The differential unit is now reassembled and ready for fitting into the carrier. This is covered in Section 2.

4 Differential side gear oil seals – renewal

1 The oil seals which are housed in the cover plates on the sides of the differential unit carrier should be renewed as a matter of course when the differential unit is overhauled.

Fig. 7.7 Checking the side gear endfloat (Sec 3)

Fig. 7.8 Removing the left-hand half casing from the crownwheel (Sec 3)

1 Crownwheel

4.6 The differential unit drain plug (arrowed)

4.7 The seal fits into the carrier from inside

Chapter 7 Differential unit

2 However if they start to leak they can be renewed separately.
3 The right-hand seal can be quite easily renewed in situ, however the left-hand seal is harder to get at because of the engine/gearbox layout, and the left-hand front subframe member restricts the access from underneath the car. However access is possible, after removing the roadwheel, or alternatively if the battery, battery carrier, cooling system expansion tank, clutch cable and its mounting bracket are removed then it can be reached from above. In later models the windscreen washer reservoir will also have to be removed for this operation.
4 In order to renew the seals in situ the first task is to undo the driveshaft inner CV joints where they are mounted to the drive flanges. In the later models with a three-piece driveshaft layout the centre section will also have to be withdrawn. All these operations are covered in detail in Chapter 8.
5 With the driveshafts removed the next stage is to remove the drive flange(s). In the later models there is of course only one drive flange to remove. Uncrimp the locking ring nut, make up a bracket to stop the flange from turning and remove the nut. Use a puller to remove the drive flange.
6 Remove the drain plug in the rear end of the differential casing and drain the oil out into a container (photo).
7 Undo and remove the four bolts holding the seal cover plate in place and remove one cover plate and seal at a time and renew the seal. Tap the old seal out and fit a new seal. It fits from the inside of the carrier (photo).
8 Also renew the O-ring between the cover plate and the differential carrier casing.
9 When removing the right-hand cover plate and seal note that there are shims inside. These are for setting the differential bearings preload and must not be removed.
10 Lightly smear the inner lips of the seal with oil and refit the cover plate seal and O-ring to the casing. Then refit and torque tighten the four retaining bolts.
11 Refit the drive flange(s) and/or driveshaft centre section and torque tighten the retaining nuts to the specified torques.
12 Refit the driveshafts inner CV joints to the drive flanges as described in Chapter 8.
13 Finally refill the gearbox/differential unit with the correct grade and quantity of oil.

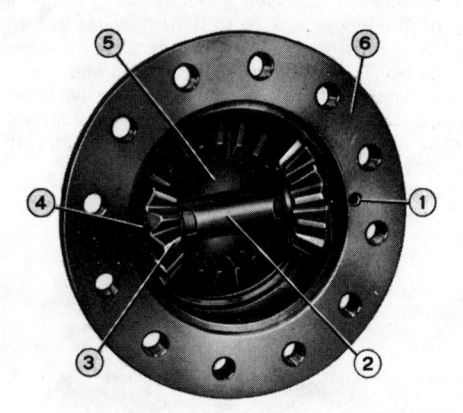

Fig. 7.9 Dismantling the pinion gears (Sec 3)

1 Retaining pin
2 Shaft
3 Pinion gears
4 Thrust washer
5 Side gear
6 Half casing

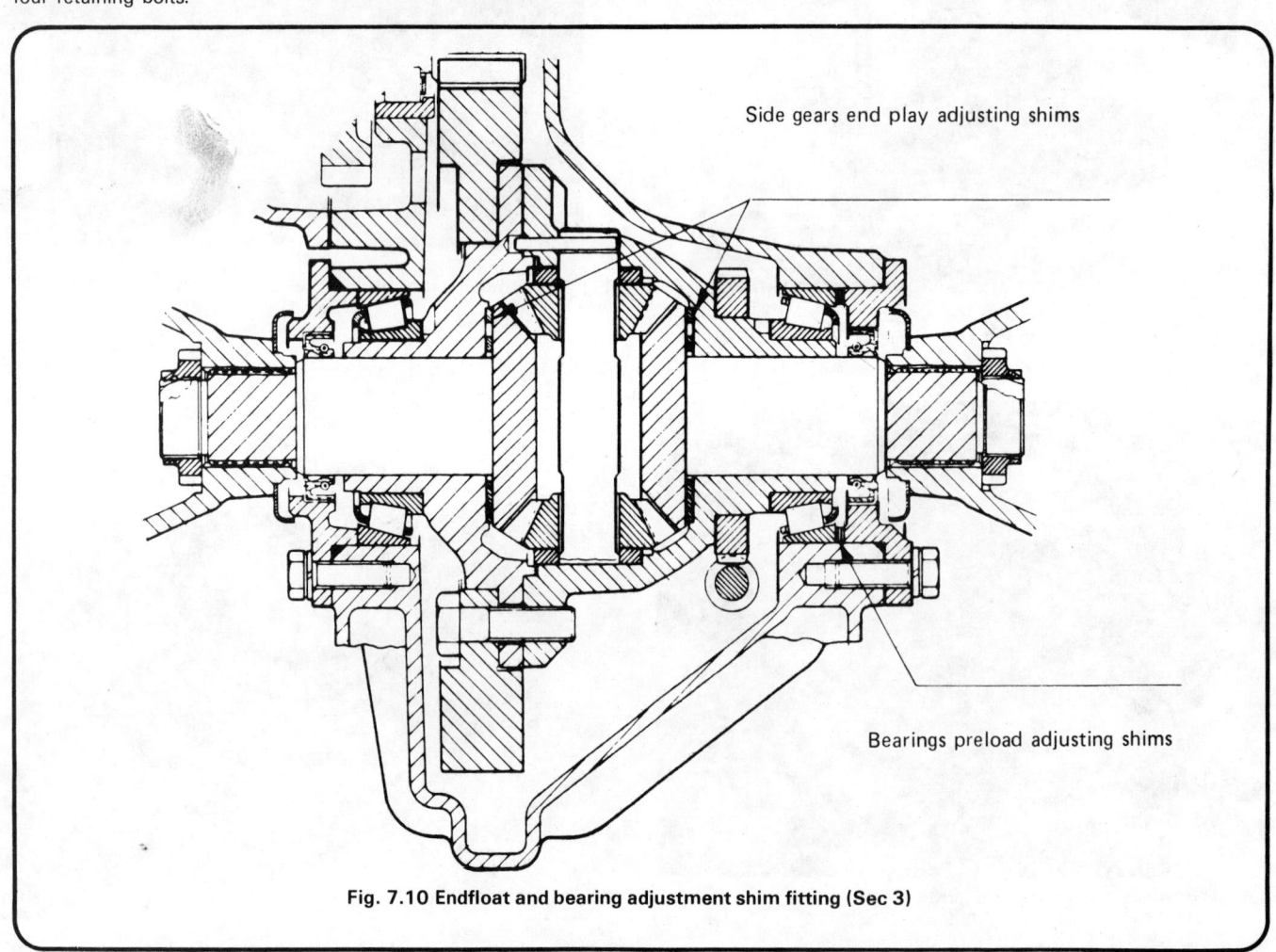

Fig. 7.10 Endfloat and bearing adjustment shim fitting (Sec 3)

5 Fault diagnosis – differential unit

The differential fault diagnosis Section should be used in conjunction with the gearbox fault diagnosis Section in Chapter 6 as the two units are so closely related. Sometimes it is very difficult to tell in which assembly the problem lies

Symptom	Reason(s)
Noisy differential unit with engine pulling at high speed	Worn crownwheel and pinion
Noisy differential unit, noise persistant in all gears	Badly worn differential bearings
Oil leaks from side gear shafts/cover plates	Oil seals defective or O-rings defective
Noisy differential unit	Oil level too low

Chapter 8 Driveshafts and CV joints

Contents

Driveshaft – dismantling, inspection and reassembly 5	Fault diagnosis – driveshafts and CV joints 7
Driveshafts (except right-hand driveshafts on early models) 2	General description 1
Driveshaft centre section – removal and refitting 4	Right-hand driveshaft (early models) – removal and refitting 3
Driveshaft centre section bearing – removal and refitting 6	

Specifications

Driveshaft type
Early models .... 2 driveshafts of unequal length with CV joints at inner and outer ends
Later models .... 2 driveshafts of equal length with CV joints at inner and outer ends. A separate centre section is used with the right-hand driveshaft

Driveshaft markings
Class A .... Blue marking
Class C .... Red marking

CV joint markings
Class A .... Blue marking
Class B .... White marking
Class C .... Red marking

CV joint greasing
Type of grease .... Lithium based
Quantity of grease .... 80 grams for each boot

Torque wrench settings
	lbf ft	Nm
Inner CV joint to drive flange:		
2-piece driveshaft assembly	22	30
3-piece driveshaft assembly	23	31
Front wheel hub retaining nut	231	314
Front disc to hub	9	12
Drive flange retaining nut	11	16

1 General description

Drive is transmitted from the differential unit to the front wheels by means of two driveshafts. On early models the shafts are of unequal length, the right-hand one being more than twice the length of the left. This makes articulation at the ends of the shafts a problem and because of the differing lengths of the shafts balance could also be a difficulty.

On later models this problem was overcome by making the right-hand shaft into a two section assembly. This means that the system now employed in all models has two driveshafts of equal length, thus solving the earlier problems. The centre section is a separate unit, with a splined end fitting into the differential unit and a flange on the outer end. It runs in a bearing at the outer end, which is mounted in a casting bolted to the engine block. This later driveshaft layout is known as the three-piece system (photo).

At either end of the driveshafts are constant velocity type joints. The driveshaft fits inside the circular outer CV joint which is also the driven shaft. Drive is transmitted from the driveshaft to the driven shaft by six steel balls which are located in curved grooves, machined in line with the axis of the shaft on the inside of the driven shaft. Both joints are covered with rubber boots, which are filled with grease. Provided that the boots are kept in good order the driveshafts and joints have a very long life expectancy.

The CV joints allow for the vertical movement of the front suspension units while the sliding splines on either end of the driveshaft itself allow for horizontal movement.

1.2 The later type right driveshaft joined to the flanged end of the centre section

Chapter 8 Driveshafts and CV joints

Fig. 8.1 Driveshaft layout – three piece assembly (Sec 1)

1 Rubber boots
2 Inner CV joints
3 Transmission/differential dipstick
4 Centre section of driveshaft assembly

2 Driveshafts (except right-hand driveshafts on early models) – removal and refitting

1 This procedure is for the left-hand driveshaft in early models or for both driveshafts in later models. The procedure for the early right-hand driveshaft is covered in the next Section.
2 Jack up the front end of the car and support it securely on stands or blocks, having loosened the wheel bolts.
3 Remove the front roadwheel(s).
4 Undo the rubber fasteners and pull the steering arm boot inwards to gain easier access to the six mounting bolts for the driveshaft inner CV joint to the flange. On the left, the flange is straight off the differential unit and on the right the flange lies at the outer end of the extension driveshaft (photo).
5 The bolts have Allen type heads. There is not a lot of space to work in and it is easier if a long extension and flexi-joint are used to reach the bolt heads. The nuts are on the inside (photo).
6 Remove the six nuts and bolts and place them in a container.
7 Move the CV joint away from the flange to which it was joined.
8 Remove the hub cap in the centre of the wheel hub next. This will almost certainly require a slide hammer with clamping attachment as shown in Fig. 8.2.
9 Open up the crimped sections of the wheel hub retaining nut and

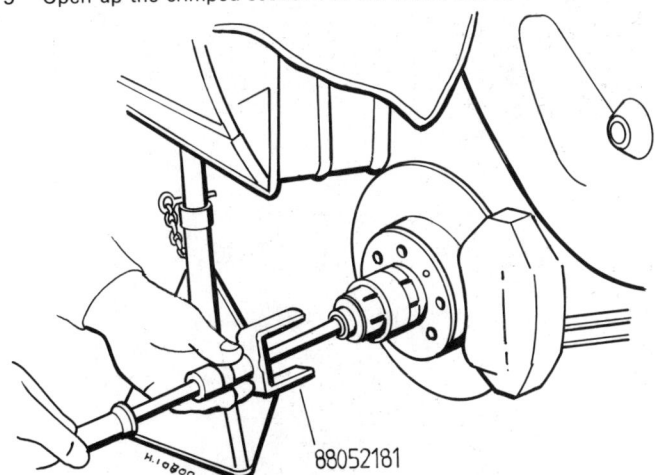

Fig. 8.2 Removing the wheel hub centre cap using a slide hammer/puller (Sec 2)

Special tool No – 88052181

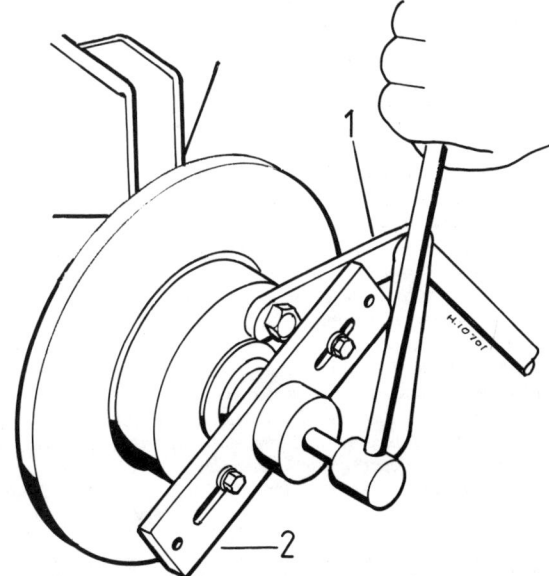

Fig. 8.3 Removing the front driveshaft splined end from the hub (Sec 2)

1 Arm for locking hub – Tool No – 88053151
2 Removing tool – Tool No 88052009

remove the nut. This will require some method of preventing the hub and shaft from turning, as the nut is done up extremely tight. Fig. 8.3 shows the use of an arm arrangement (1) which would stop the hub from rotating.
10 With the hub nut removed the driveshaft/joint outer end needs to be pressed out of the hub. Fig. 8.3 shows a special tool for doing this. The flat plate is screwed to the hub and disc, when the front disc mounting bolts have been removed. Use the same holes, but different bolts.
11 Screw the central bolt in a clockwise direction to push the splined end of the stub shaft out of the hub. Make sure, before commencing this part of the procedure, that the inner CV joint is well clear of the flange and has space inside it to move as the outer end of the driveshaft assembly is pushed out of the hub.
12 When the outer end of the driveshaft assembly is free the whole unit can be lifted away.
13 Remove the extractor tool and arm from the hub and disc. Refit the disc mounting bolts and tighten them to the specified torque.
14 Refit the assembly to the car in the reverse order to removal. Insert

Chapter 8 Driveshafts and CV joints

2.4 The right-hand inner CV joint viewed from inside the wheel arch

2.5 Left-hand inner CV joint with nuts removed (photo taken from above during engine refitting)

the outer splined end into the hub and push it in as far as it will go. Refit the nut and tighten it up thus pulling the splined shaft into the hub. Tighten the nut to the specified torque. Crimp the retaining sections of the nut.
15 Refit the wheel hub cap.
16 Refit the inner CV joint to the flange and secure it with the six bolts and nuts. Remember that each pair of bolts has a locking plate between them. Push all the bolts through first and lightly fit the nuts on the ends, then tighten them up to the correct torque as specified.

3 Right-hand driveshaft (early models) – removal and refitting

1 As has already been said, the early models had a short left-hand driveshaft and a long right-hand one. This procedure covers the latter.
2 Jack up the front of the car, having first loosened the right-hand front wheel bolts, and support it on stands. Remove the roadwheel.
3 Undo the six nuts and bolts which secure the inner CV joint to the right-hand drive flange of the differential unit (Fig. 8.4).
4 The nuts and bolts are the same as for the other CV joints with Allen type bolts, as described in Section 2.
5 Separate the CV joint from the drive flange so that it has free space to move when the outer end of the assembly is driven out of the hub.
6 The next task is to separate the outer CV joint splined end from the wheel hub. The procedure for this is exactly the same as for the short driveshafts. Refer back to Section 2 and follow the instructions from paragraph 8 onwards.
7 Refitting is a reversal of removal but note the instructions given in Section 2 paragraphs 13 to 16.

4 Driveshaft centre section – removal and refitting

1 Although it is called the centre section, it is of course the inner section of the right-hand driveshaft assembly.
2 The left-hand end is splined and fits into the differential unit. The right-hand end has a flanged fitting onto which is mounted the right-hand inner CV joint. The right-hand end of the centre section runs in a bearing housed in a casting, which is bolted to the engine block (photo).
3 Jack up the front of the car, having loosened the right-hand front roadwheel bolts and support the front of the car on stands.
4 Turn the steering onto full left lock.
5 Remove the roadwheel and the steering arm boot. Push it as far up the steering rod as it will go. Remove the splashguard where fitted.
6 Undo the inner CV joint to flange mounting nuts and bolts described in Section 2 (photo).
7 Separate the outer driveshaft from the flange.
8 Undo and remove the three bolts which secure the centre section

bearing retaining plate to the casting.
9 The centre section can now be withdrawn through the casting and right-hand wheel arch, complete with bearing.
10 Refitting is the reverse procedure to removal, but lubricate the rubber sleeve on the splined end of the extension shaft to make refitting easier. The splined end of the centre section has to locate in the splined section of the differential joint. This can sometimes be a fiddly job and if necessary it is easier to pull back the rubber sleeve to line up the splines.
11 Once the splined end is lined up and pushed home, the bearing retaining plate can be bolted back into position in the casting. Refit the inner CV joint to the flange as described in Section 2.

5 Driveshafts – dismantling and reassembly

1 Undo and remove the two clips which secure the inner CV joint rubber boot and then slide the boot along the shaft (Fig. 8.5).

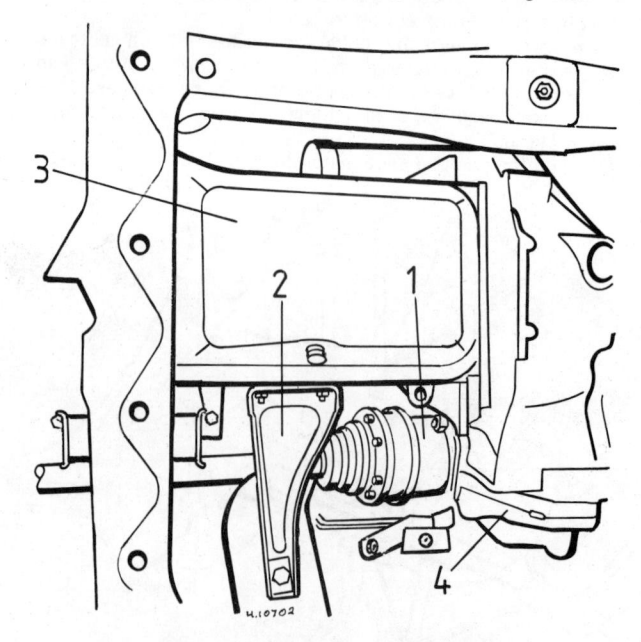

Fig. 8.4 Early type of long right-hand driveshaft (Sec 3)

1 Inner CV joint and differential flange
2 Exhaust mounting plate
3 Sump
4 Differential unit

Chapter 8 Driveshafts and CV joints

4.2 The driveshaft centre section viewed from underneath

4.6 Flanged end of centre section and bearing housing with right-hand inner CV joint nuts loosened

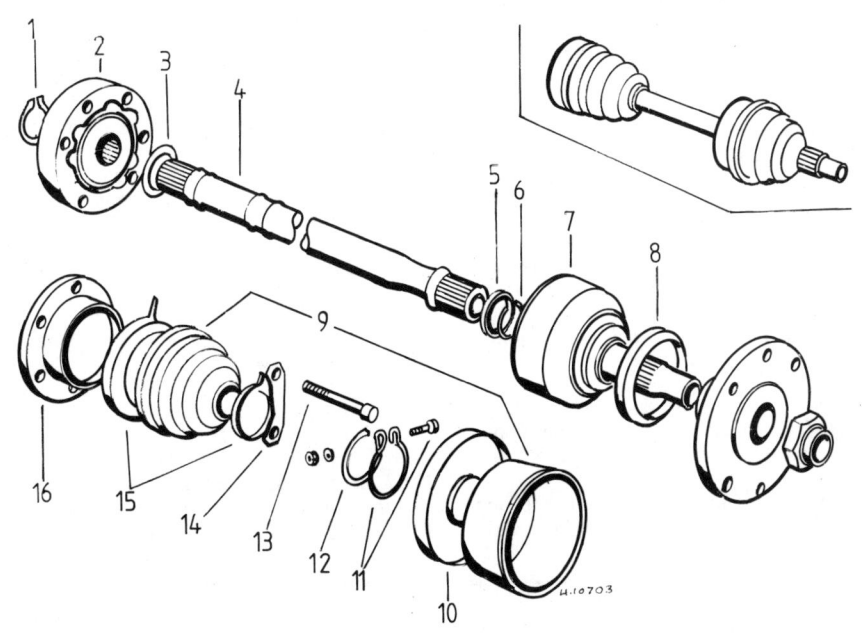

Fig. 8.5 Front driveshaft and CV joints – exploded view (Sec 5)

1	Circlip	5	Spacer	9	Rubber boots	13	Joint mounting bolt
2	Inner constant velocity joint	6	Circlip	10	Retainer	14	Lockplate
3	Washer	7	Outer constant velocity joint	11	Rubber boot clip	15	Rubber boot clips
4	Shaft	8	Bearing dust cap	12	Retainer	16	Rubber boot collar

2 Remove the circlip from the inner end of the driveshaft that holds the inner CV joint in place.
3 Clamp the driveshaft vertically in a vice with protective jaws with the inner CV joint at the bottom.
4 Using a soft faced hammer drive the joint off the shaft. If it is too tight then it will have to be pressed off. For this you will have to take it to a local engineering firm or someone who has a fly-press. Be careful when driving off the joint that the balls do not fall out.
5 With the inner CV joint removed from the shaft, undo the vice and slide off the rubber boot mounting collar, the boot and the clips.
6 Now remove the clips from the outer CV joint boot and slide them all down the driveshaft and off the inner end.
7 Refit the driveshaft in the vice this time with the outer CV joint downwards. Using a soft faced hammer, drive the outer CV joint assembly off the driveshaft splines. The circlip and spacer will come off as well. Check the splines for wear if they part too easily.
8 If the CV joints are known to be badly worn (ie they have been

knocking when driving, or on hard lock) then they should be changed anyway. The same applies to the rubber boots. If they are worn, perished, split or damaged in any way then renew them.
9 To check the CV joints, first thoroughly clean them in paraffin. Check that the balls and tracks have an unmarked quality about them. There should be no sign of scoring or wear. If there has been a loss of lubricant in the joint because of a split or cracked rubber boot then be very careful when examining the joint. Remember that a split boot not only lets the grease out, it also lets grit and abrasive matter in.
10 If the balls fall out of the inner CV joint when it is removed then they must be carefully refitted after cleaning and inspection have been carried out.
11 To refit them to the race tilt the race outwards and slip the ball into its slot in the outer edge of the race. When all the balls are back in place make sure that the outer casing is aligned with the inner casing as shown in Fig. 8.6. The larger sections on the outer casing must be opposite the smaller sections in the outer casing and vice versa as in

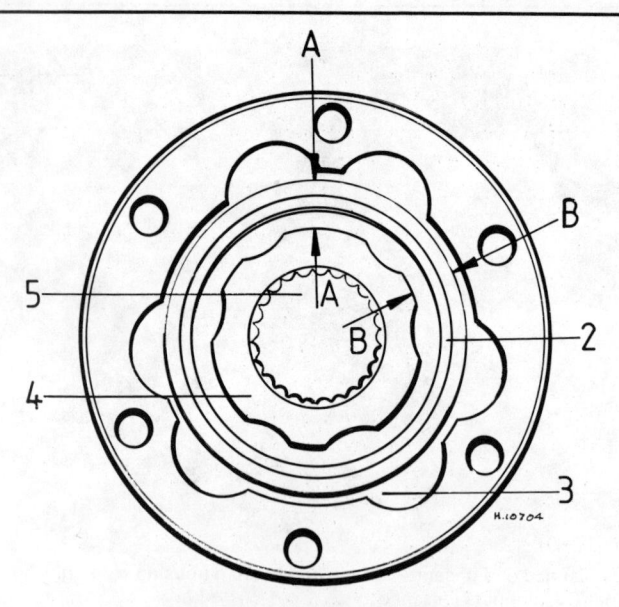

Fig. 8.6 Inner CV joint (Sec 5)

1 Outer casing
2 Ball cage
3 Ball
4 Inner casing
5 Driveshaft

A to A and B to B = correct alignment for inner to outer casing

A to A or B to B. If the outer and inner casings are not lined up correctly like this the joint will lock up.

12 If the driveshaft or CV joints are being renewed then it must be remembered that the shafts and joints can only be mated in certain ways. A class A driveshaft can be mated with a class A or B CV joint. A class C driveshaft can be mated with a class C or B CV joint. The markings are:- A – blue, B – white and C – red. The only type of joints to be supplied as spares are class B (white) which can be mated with either class A or C shafts (Fig. 8.7).
13 Having obtained the necessary new parts the assembly should be rebuilt. It is worth stating here that it is advisable to renew the rubber boots anyway, whatever their condition, unless they have been renewed very recently.
14 Start with the outer end of the driveshaft. Refit the circlip and clamp it in place in order to slide the spacer over it to hold it in position.
15 Clamp the driveshaft tightly in a vice in the vertical plane and press on the outer CV joint, so that it is fully engaged with the splines.
16 Pack the CV joint with grease of the recommended amount and type, then refit the rubber boot. This has to be slid on from the inner end of the shaft. Secure the rubber boot with the clip and retainers.
17 Use a punch to impress the large retainer in four places, thus securing it in position.
18 Refit the inner joint rubber boot and collar to the shaft and then slide them down it to the other end.
19 Clamp the driveshaft firmly in the vice once again with the inner end upwards. Place the locking washer on the splined end and slide it down.
20 Refit the inner CV joint to the driveshaft. Note that the groove, which runs round the circumference of the joint, must face the drive flange.
21 Having driven the CV joint onto the driveshaft refit the circlip.
22 Repack the joint with the correct grade and quantity of grease (see the Specifications).
23 Refit the collar to the CV joint and the rubber boot to the collar and secure it in place with the clips. Before tightening the clips check that the holes in the collar and the holes in the joint are correctly lined up.
24 The narrower end of the boot should be located between the two ridges on the driveshaft.

6 Driveshaft centre section bearing – removal and refitting

1 Remove the centre section assembly complete with the bearing and retaining plate.
2 Fit the driveshaft vertically in a vice with protective jaws. Uncrimp the locking ring nut which retains the drive flange to the shaft. Remove the ring nut. This will probably mean making up a special peg spanner, using a length of tube of a suitable size.
3 Drive the flange off the shaft by inverting the shaft in the vice and using a soft faced hammer to drive the flange off.
4 Be careful, because as the drive flange comes off so will the retaining plate and seal.
5 Use a puller to remove the bearing itself. Once the bearing is off the shaft it can be examined, after it has been cleaned in paraffin. Renew it if it is worn.
6 Examine the seal and renew it if necessary.
7 Refitting is the reverse procedure to removal. Place the bearing on the end of the driveshaft and drive it home using a length of tube slightly larger in internal diameter than the diameter of the driveshaft itself. Refit the seal and retaining plate, followed by the drive flange. Tighten the ring nut to the specified torque and then crimp it.
8 The centre section driveshaft assembly is now ready to be refitted to the car. This is covered in Section 4.

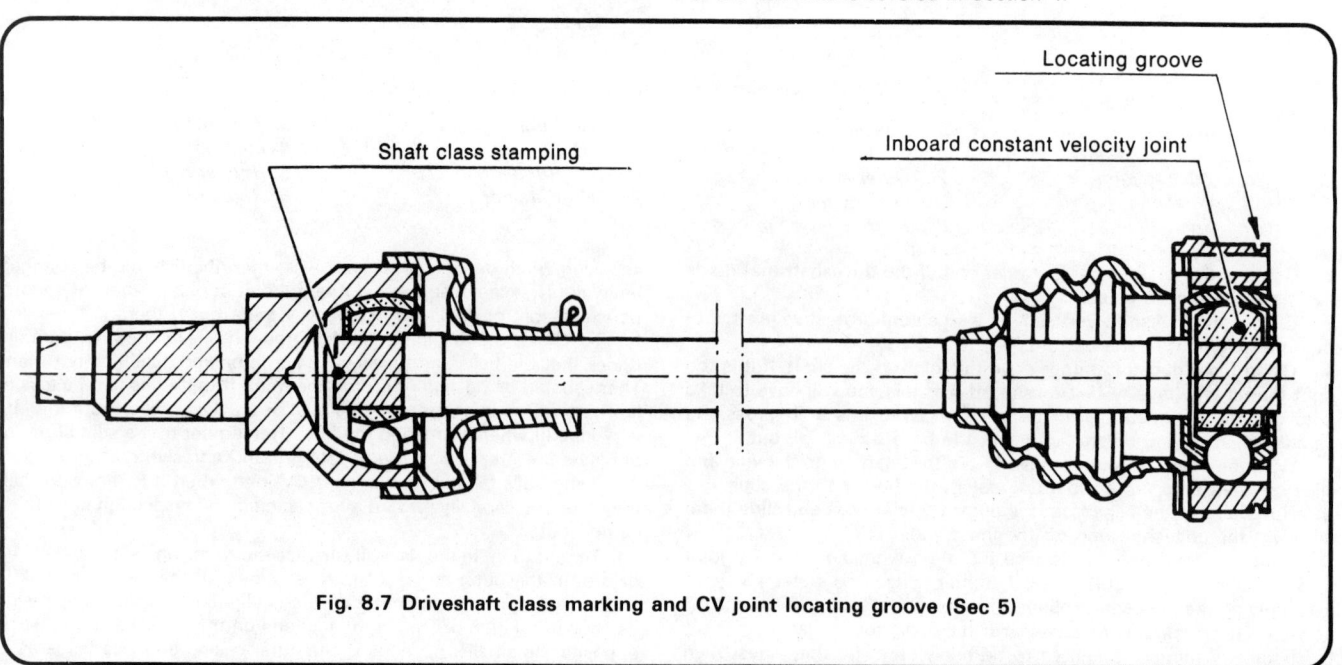

Fig. 8.7 Driveshaft class marking and CV joint locating groove (Sec 5)

7 Fault diagnosis – driveshafts and CV joints

Symptom	Reason(s)
Noisy driveshafts	Lack of lubricant due to split rubber boot Worn or seized-up bearing in driveshaft centre section
Knocking or rattling noise from driveshafts when cornering	Outer CV joint worn or rubber boot split and grease lost causing wear
Grease on inside of tyre or roadwheel	Split rubber boot on outer CV joint

Chapter 9 Braking system

Contents

Bleeding the hydraulic system	2
Brake and clutch pedal assembly – removal, renovation and refitting	15
Brake discs – inspection, removal and refitting	11
Brake servo unit – description	16
Brake servo unit – removal and refitting	17
Brake servo unit air filter – renewal	18
Fault diagnosis – braking system	22
Front brake caliper – overhaul	5
Front brake caliper block and yoke – removal and refitting	4
Front disc brake pads – removal, inspection and refitting	3
General description	1
Handbrake cable – adjustment	12
Handbrake cable – inspection, removal and refitting	13
Handbrake lever assembly and warning light switch – removal, refitting and adjustment	14
Master cylinder – removal and refitting	9
Rear brake caliper – overhaul	8
Rear brake caliper block – removal and refitting	7
Rear brake compensator device – description	19
Rear brake compensator device – removal, refitting and adjusting	20
Rear brake compensator device – overhaul	21
Rear disc brake pads – removal, inspection and refitting	6

Specifications

System type
Lancia Super duplex dual circuit system with servo assistance. Disc brakes on all four wheels with load compensator acting on rear brakes. Cable operated handbrake on the rear wheels only

Front brakes
Type .... Disc and floating caliper (DBA – Bendix)
Disc thickness (new) .... 12.6 to 12.8 mm (0.496 to 0.503 in)
Minimum disc thickness after refacing .... 11.5 mm (0.453 in)
Disc run-out (fitted) .... Must not exceed 0.05 mm (0.00196 in) measured 2 mm (0.079 in) from outside edge
Pad thickness (new) .... 11 mm (0.433 in) (actual friction material)
Minimum pad thickness .... 1 mm (0.039 in)

Rear brakes
Type .... Disc and floating caliper (DBA – Bendix)
Disc thickness (new) .... 9.865 to 10.135 mm (0.388 to 0.399 in)
Minimum disc thickness after refacing .... 9 mm (0.354 in)
Disc run-out (fitted) .... Must not exceed 0.05 mm (0.00196 in) measured at a point 2 mm (0.079 in) from outside edge
Pad thickness (new) .... 7 mm (0.275 in) actual friction material
Minimum pad thickness .... 1 mm (0.039 in)

Handbrake
Type .... Cable operated to rear discs from centrally mounted floor lever
Adjustment .... Self-adjusting, but can be taken up on yoke to offset cable stretch

Vacuum servo unit
Servo unit stroke .... 36 to 38 mm (1.42 to 1.50 in)
Servo unit control valve closing stroke .... 2 mm (0.79 in)

Master cylinder
Type .... Duplex – tandem
Bore diameter .... 22.22 mm (0.875 in)
Mixed circuit piston stroke .... 19 mm (0.748 in)
Front circuit piston stroke .... 16.5 mm (0.649 in)

Calipers
Front:
  Number of pistons 2
  Front circuit piston diameter .... 45 mm (1.77 in)
Mixed circuit piston diameter .... 34 mm (1.33 in)
Rear:

Chapter 9 Braking system

Torque wrench settings	lbf ft	Nm
Front caliper to hub carrier	48	65
Rear caliper to stub axle	32	44
Front and rear discs to hubs	9	12
Bleeder nipples	9	12

1 General description

All models have disc brakes of the single floating caliper type operating on all four wheels. These brakes are controlled by a tandem master cylinder, connected to a vacuum servo unit. The front circuit is connected to the large pistons in the front calipers, and the mixed circuit is connected to the smaller pistons in the front calipers and the single pistons in the rear calipers as well. Therefore if one circuit fails there is always a back up braking system in operation.

Front brake pad wear is monitored by built-in contacts in the pads, which are linked to a warning light in the instrument panel. This gives immediate warning when the front pads have reached the minimum permissible thickness. This warning indicator on the instrument panel also serves as a reminder when the brake fluid level is too low.

The handbrake is a conventional floor mounted lever which operates the rear disc calipers, through a special linkage. As far as brake pad wear is concerned the system is self-adjusting. However, there is provision for adjustment to take up the slack in the handbrake cable and linkage to the rear brakes.

The rear brakes are fitted with a load compensating device to stop skidding and prevent locking up. When the car is unladen the amount of pressure let through the valve is small, but as the load is increased so greater quantities of fluid are let through to apply greater pressure to the caliper pistons, in order to stop the vehicle.

USA models are equipped with a pressure drop warning switch mounted inside the car beneath the fusebox. This operates a warning light when there is a leak in either of the two hydraulic circuits.

2 Bleeding the hydraulic system

1 This without doubt will be one of the most frequent tasks to be performed on the brake system. You will require some small $\frac{1}{8}$ or $\frac{3}{16}$ inch bore rubber or clear plastic tubing and a clean dry glass jar. The tubing should be at least 15 inches long. You will also require a quantity of brake fluid probably between $\frac{1}{4}$ and $\frac{1}{2}$ a pint.

2 It will be necessary to bleed the brakes whenever any part or all of the brake system has been overhauled or when a brake pipe connection has been undone in the course of performing tasks on other assemblies on the car. Bleeding of the brakes will also be necessary if the level of fluid in the brake system reservoirs has fallen too low and air has been taken into the system. During the task of bleeding the brakes, the level of fluid should not be allowed to fall below half way, or air will be drawn into the system again.

3 Although not necessary for the point of view of a successfully completed task, it improves access to remove the road wheel adjacent to the brake to be bled. Beginning at the front of the car, place a jack underneath the lower arm of the suspension. Raise the roadwheel off the ground and remove the wheel. Chock the other wheels and release the handbrake.

4 Since the front wheels have two separate brake circuits there are two bleed screws on each front caliper. Bleed the front circuit line on both sides first, then the mixed circuit. The bleed screws are easily identified. The front circuit bleed screw is the outer one (nearest the disc) and the mixed circuit bleed screw is the inner one (at the end of the caliper housing) (photo).

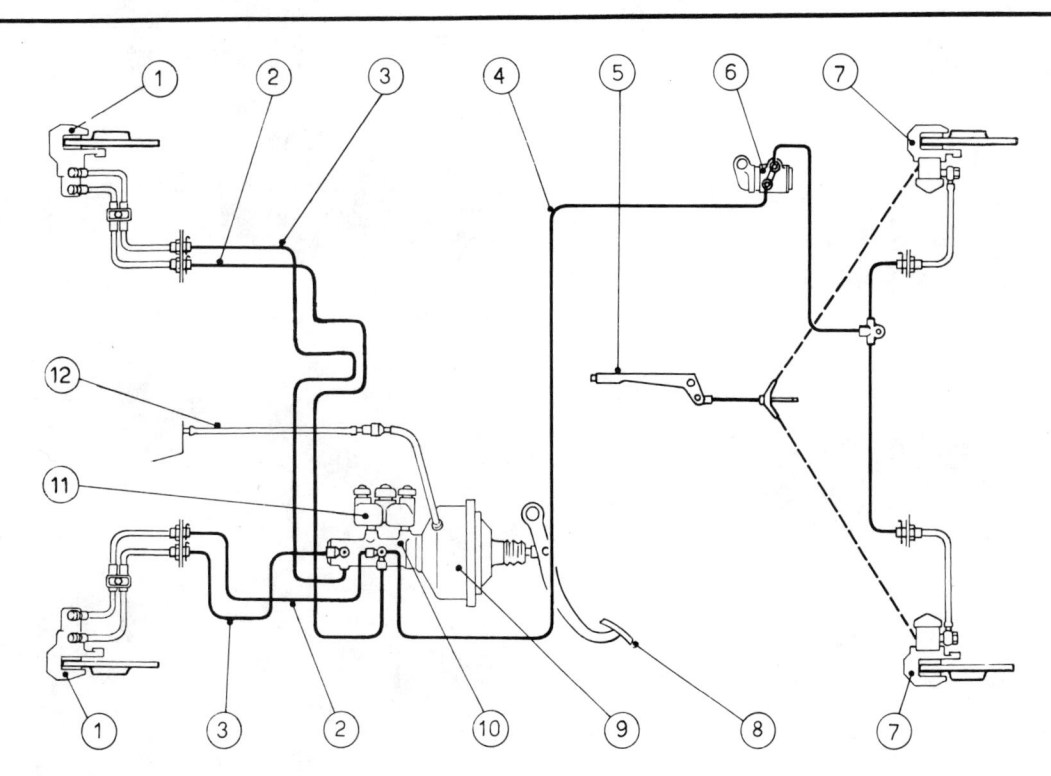

Fig. 9.1 Braking system layout (Sec 1)

1 Front caliper
2 Front caliper mixed circuit line
3 Front circuit line
4 Rear caliper mixed circuit line
5 Handbrake lever
6 Balance limiter
7 Rear caliper
8 Pedal
9 Servo unit
10 Master cylinder
11 Reservoir
12 Vacuum inlet line from intake manifold with non-return valve

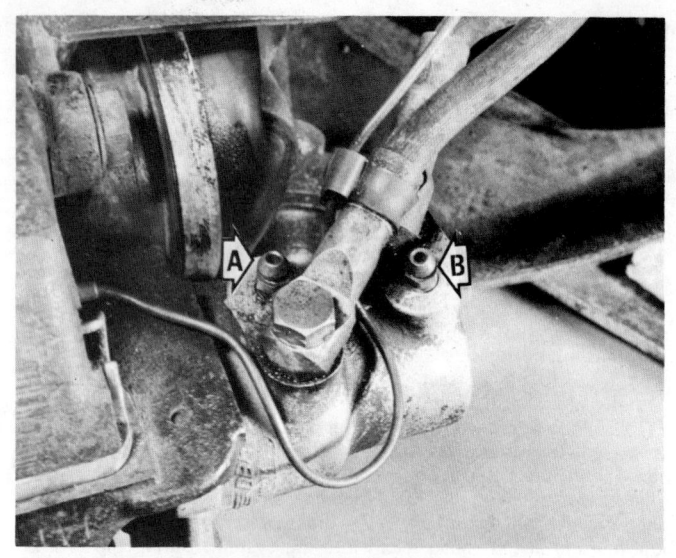

2.4 The bleed screws are easily identifiable
A   Front circuit bleed screw
B   Mixed circuit bleed screw

3.4 Removing the cotter pins which hold the wedges in place

3.5 With the wedges out the caliper block can be removed

3.6 Removing the pad with spring attached

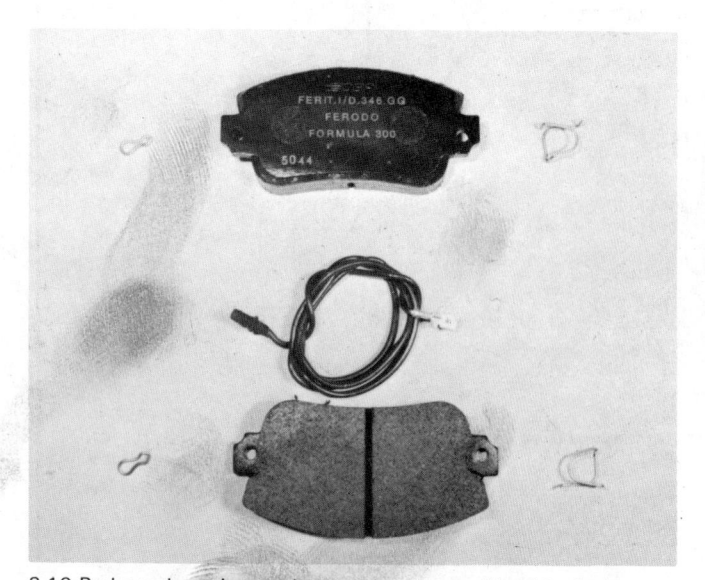

3.10 Brake pads, springs and cotter pins for one side of the front braking system with the warning lead

3.12 Tap the second wedge in if necessary

Chapter 9 Braking system

5 Remove the rubber dust cover from the bleed screw and wipe the screw head clean. Push the rubber/plastic tubing onto the screw head and drop the other end into the glass jar placed nearby on the floor. Remove the appropriate reservoir cap and ensure that the fluid level is near the top of the reservoir. Pour a little brake fluid into the glass jar, sufficient to cover the end of the tube lying in the jar. Take great care not to allow any brake fluid to come into contact with the paint on the bodywork; it is highly corrosive.

6 Raise the jar so that it is approximately 8 inches (200 mm) higher than the bleed screw.

7 Use a suitable open ended spanner and unscrew the front circuit bleed screw about one half to a full turn.

8 An assistant should now pump the brake pedal by depressing it in one full stroke, allowing the pedal to return of its own accord each time. Check the level of fluid in the reservoir and replenish if necessary with new fluid. *Never* re-use fluid.

9 Carefully watch the flow of fluid into the glass jar and when the air bubbles cease to emerge from the bleed screw and braking system through the plastic pipe, tighten the bleed screw during a down stroke on the pedal. It may be necessary to repeat pumping detailed in paragraph 8 if there was a particularly large accumulation of air in the brake system.

10 Repeat the operations detailed in paragraph 5 to 9 for the mixed circuit. When bleeding the rear brakes, place the jacks under the suspension and remove the wheels. **Do not use chassis stands acting on the bodyshell otherwise the rear brake regulator system will be brought into operation and prevent the flow of fluid to the rear brakes.** A last additional point with regard to the rear brakes, is to pump the pedal slowly and allow one or two seconds between each stroke.

11 Sometimes it may be found that the bleeding operation for one or more cylinders is taking a considerable time. The cause is probably due to air being drawn past the bleed screw threads, back into the system during the return stroke of the brake pedal and master cylinder, when the bleed screw is still loose. To counteract this occurrence, it is recommended that at the end of the downward stroke the bleed screw be temporarily tightened and loosened only when another downstroke is about to commence.

12 Once all the brakes have been bled, recheck the level of fluid in the reservoir(s) and replenish as necessary. Always use new brake fluid – *never* re-use fluid.

13 If after the bleed operation, the brake operation still feels spongy, this is an indication that there is still some air in the system, or that the master cylinder is faulty.

3 Front disc brake pads – removal, inspection and refitting

1 Inspection of the disc pads on the brakes fitted to this series of car cannot be readily undertaken with the brakes in place. It will be necessary to remove the caliper block in order to expose the pads and enable an inspection to be made.

2 The caliper block seats in a yoke fitting which is bolted to the hub carrier assembly and 'wraps around' part of the disc. The pads and their retaining springs, and the caliper block which acts on the pads, are held between the upper and lower parts of the enveloping yoke. The caliper is held so that it is free to move axially to centre itself on the pads and disc. It is retained radially by two wedges which are held axially to the caliper yoke by two cotter pins.

3 The procedure for removal of the disc brake pads is as follows: Jack up the appropriate suspension and remove the road wheel. Chock the other wheels.

4 Pull out the cotter pins which retain the caliper block wedges (photo). Wedges are fitted above and below the caliper block. Mark the wedges before removing them to be sure that you will be able to refit them in exactly the same position from which they were taken.

5 Once the wedges are out, the caliper block can be lifted away from the brake assembly. Be very careful not to strain the flexible brake pipe joining the caliper to the pipe system on the bodyshell. Rest the caliper on the lower suspension member, or tie it up so that the pipes are supported (photo).

6 Remove the brake pads. The pad anti-vibration springs may be removed if desired (photo). Make sure that you note which way they fit before doing so.

7 Mark the pads so that they may be refitted into the exact positions from which they were taken.

8 Measure the pad material thickness, if less than 0.039 in (1 mm) or quite close to that thickness, renew the pads. Disconnect the electrical leads from the connectors.

9 Attention should be paid to the colour coding on the back plate of the pad. These codings indicate the type of material the pad is made of and as usual there is only one combination of pads which is safe to use.

10 The colour code (and hence pad material) on all eight pads used in the four brakes on the car *MUST* be the same. Make sure that the colour coding of the pads being discarded and the new pads is the same. Note that the electrical leads come as part of the new pad assembly (photo).

11 While you have the brake assembly off the disc it is a good opportunity to inspect the disc for scores and excessive wear. You may only have the disc reground if the thickness after machining will not be less than 0.453 in (11.5 mm)

12 The refitting of the disc pads follows the reversal of the removal sequence. The second or last wedge may be a tight fit. Use a plastic hammer to tap it in (photo).

13 Before the caliper block is refitted over the new pads, the slave piston will need to be gently pushed back into its bore in the caliper, to provide a sufficient gap between the piston and the outer claw of the caliper to accommodate the new pads and the disc brake. It is as well to remember that as the piston is moved back into the block brake fluid will be displaced and returned to the reservoir. The reservoir will overflow unless fluid is taken out with a device such as a pipette as the pistons are moved.

14 Finally depress the brake pedal three or four times to seat the pads and pistons.

4 Front brake caliper block – removal and refitting

1 There is no need to remove the caliper block yoke, if it is only the caliper that is to be attended.

2 The yoke needs little attention and the only occasions when it is necessary to remove it, is when it is desired to remove the brake disc which is mounted on the wheel hub.

3 Remove the brake caliper block as follows: jack up the appropriate lower suspension arm and remove the road wheel.

4 Working inside the engine compartment, remove the cap on the appropriate brake fluid reservoir, and stretch a thin sheet of polythene over the top of the reservoir. Refit the cap. This measure will prevent excessive loss of brake fluid when brake pipe connections are subsequently undone.

5 Disconnect the two brake hoses from the caliper block by undoing the banjo unions. Do this one hose at a time and collect any fluid that runs out in a container. This must be disposed of and not re-used (photo). Remove the front circuit brake hose first.

6 Tape over the ends of both brake hoses and block the ports in the caliper to prevent the ingress of dirt.

7 Next remove the four cotter pins which hold the two caliper wedges in position and withdraw the wedges from between the caliper block and yoke. Mark them to ensure that you can refit them into the position from which they have been taken

8 The caliper block can now be lifted free and taken to a clean bench for overhaul.

9 Refitting the caliper block follows the exact reversal of the removal procedure except the following tasks are added: once the caliper block has been refitted the brake system must be bled, to expel all the air from the system, as described in Section 2.

10 To remove the caliper yoke proceed as directed in paragraphs 3 to 8 of this Section and remove the caliper block. Lift the brake pads away. The two bolts which secure the yoke to the hub carrier assembly can now be undone and the yoke lifted free (photo).

11 The block retaining springs should be removed and the yoke brush cleaned. Renew the yoke only if cracks or serious wear is found.

12 Refitting is the reversal of removal and remember to use new lock washers and to tighten the retaining bolts to the specified torques.

5 Front brake caliper – overhaul

1 Once the caliper block has been removed as described in Section 4 the next problem is to extract the slave piston so that the fluid seals may be removed and all components cleaned for inspection. Some

Chapter 9 Braking system

4.5 The front circuit brake hose to caliper union (arrowed) is removed to make the mixed circuit union easier to reach

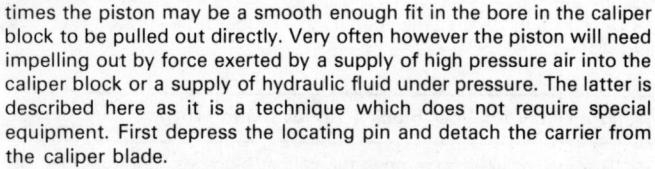

4.10 Front caliper yoke is retained to the hub carrier by two bolts

times the piston may be a smooth enough fit in the bore in the caliper block to be pulled out directly. Very often however the piston will need impelling out by force exerted by a supply of high pressure air into the caliper block or a supply of hydraulic fluid under pressure. The latter is described here as it is a technique which does not require special equipment. First depress the locating pin and detach the carrier from the caliper blade.

2 Temporarily reconnect the caliper block to the brake system and bleed the lines to the block. Having bled the lines, continue to pump the brake pedal to push the large piston in the caliper block out of its cylinder. It is as well to have an old tray beneath the area of work to catch any brake fluid split. Once the piston protrudes 0.5 in (13 mm) or so it may be pulled out with your fingers. Once the piston has been removed, remove the dust guard and disconnect the front circuit pipe and plug it off.

3 The next step is to remove the retaining ring which holds the parting piece in place (Fig. 9.2). This can be a tricky operation and great care must be taken not to damage or score the cylinder. Make up some form of protection, such as tin foil, that can be inserted into the cylinder so that any lever used to release the retaining ring will not damage the cylinder walls. It is also possible to insert a piece of wire into the cylinder through the front bleed screw hole, when the bleed screw has been removed, to help remove the ring.

4 When the ring has been removed, refit the main piston to act as a guide for the parting piece. Then refit the mixed circuit brake hose and pump the brake pedal gently. The hydraulic pressure will drive out the main piston, the parting piece and the secondary (rear) piston. Disconnect the brake hose from the cylinder and cover the end with tape.

5 Remove the seals from the cylinder and also from the parting piece and rear piston. Do not use a sharp instrument for doing this, or you may scratch the piston bore.

6 Wash all the components in brake cleaning fluid or methylated spirit. Do not use any other cleaner fluid because it may damage the seals and contaminate the hydraulic fluid when the block is re-assembled.

7 Inspect the caliper bore and piston for scoring and wear; renew the whole caliper block assembly if such wear is found.

8 To reassemble the caliper, wet the new 'O'-ring seals with clean brake fluid and carefully insert into its groove near the rim of the bore in the caliper, and the other into the groove by the parting piece position.

9 Insert new 'O'-rings to the inner bore of the parting piece and the outer edge of the mixed circuit or rear piston, coating them with clean brake fluid.

10 Refit the rear piston and parting piece to the cylinder and then refit the retainer ring. Take care that the cylinder is well protected. Also make sure that the front circuit bleed screw hole is not blocked by the ring. The gap in the ring must line up with the bleed screw hole.

11 Fit the new dust seal onto its seating at the rim of the bore in the

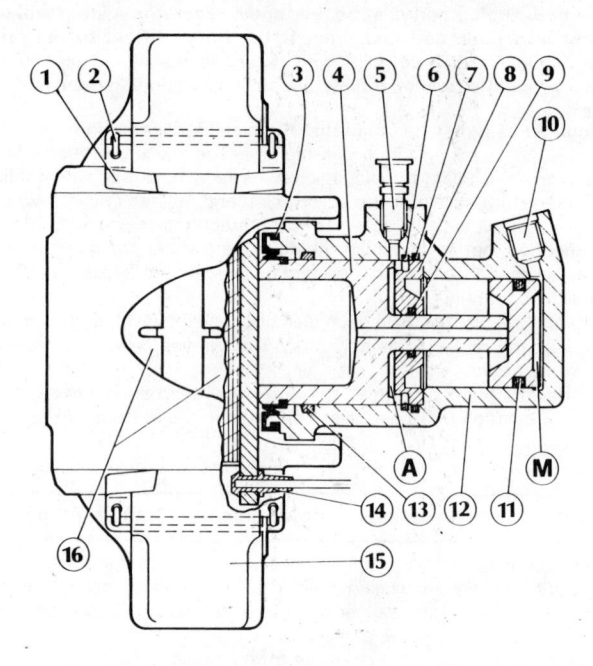

Fig. 9.2 Sectional view of front caliper assembly (Sec 5)

A   Front circuit chamber
M   Mixed circuit chamber

1   Slider
2   Cotter key
3   Dust guard
4   Front circuit piston
5   Bleed screw
6   Retainer
7   Seal
8   Parting piece
9   Stem seal
10 Mixed circuit feed
11 Rear piston seal
12 Cylinder block
13 Front piston seal
14 Friction pad wear indicator
15 Caliper
16 Friction pads

caliper. Coat the side of the piston with hydraulic fluid and carefully insert it into the bore in the caliper, until it protrudes by about 0.5 in (13 mm). Fit the dust seal onto the top of the piston and then push the piston into the bore as far as it will go.

12 Refit the carrier to the caliper block using a nut and bolt to expand the carrier as necessary. File or cut the bolthead so that it seats in the carrier groove.

13 The caliper block is ready to be fitted to the brake assembly. Reconnect the hoses and bleed the brake system, as described in Section 2.

6 Rear disc brake pads – removal, inspection and refitting

1 As with the front brake, the caliper block will need to be removed from the yoke to expose the pads for inspection and removal when necessary.
2 The rear pads must be the same type and material as the front brakes. Colour stripes on the rear of the pad indicate type and material and it follows therefore that the colours on all pads must be the same.
3 The removal and refitting procedure for the caliper block and brake pads is the same as with the brakes as detailed in Section 3, except that the rear pads have no wear thickness warning mechanism fitted to them. Pads are normally supplied in axle sets (ie 4 pads per box) and included in the accessories bag will be the cotter pins, anti vibration springs and wiring.
4 The slave piston in the caliper will need screwing clockwise back into the caliper so there is a sufficient gap between the piston and outer claw of the caliper to pass over new pads and the disc brake.
5 Remember that as the slave piston is moved back brake fluid will be pushed out of the caliper cylinder along the brake lines and into the reservoir. The reservoir might easily overflow if, as the slave pistons are moved back, fluid is not drawn out of the reservoir with a device like a pipette.
6 Once the new pads have been fitted, try the brakes and pay particular attention to the feel of the brake pedal action. If it is at all spongy bleed the brakes, because you might have been unlucky and air might have been drawn into the brake system while the slave pistons were being moved to accommodate the new pads.

7 Rear brake caliper block – removal and refitting

1 This job is almost identical to the operation described in Section 4. However, there are differences.
2 The rear braking system has only one brake hose each side that has to be disconnected.
3 The handbrake cable has to be disconnected from the rear caliper operating arm. To do this first release the handbrake so that the cable is slack.
4 The rear caliper block is retained in exactly the same manner to the yoke member, as the front brake caliper block is to its yoke member.

8 Rear brake caliper overhaul

1 Although the mode of operation of the rear brake caliper is similar to the front, the complication of the handbrake system incorporated in the block means that the overhaul procedure is rather different from that for the front brake.
2 Begin by removing the boot which covers the handbrake lever pivot assembly on the caliper block. Once the lever pivot is exposed, the pivot pin may be driven from the block to release the lever and cam.
3 Unscrew the plunger which follows the cam at the lever pivot to press the brake pads against the brake disc. Retrieve the plunger return spring and the thrust washer. There will be an 'O'-ring oil seal on the large shank of the plunger.
4 Depress the locating pin and detach the carrier from the caliper block. The slave piston can now be driven from the bore in the caliper block with a slim drift inserted through the hole which took the plunger right in to contact the face of the piston. Be careful not to damage the threads in the special nut in which the plunger is engaged. Remove the dust seal and 'O'-ring seal from the bore in the caliper block with a plastic knitting needle.
5 Do not attempt to remove the special nut and its associated parts from inside the piston. You will certainly damage the spun in disc which retains the nut in position and the manufacturers do not supply the individual parts.
6 Once all the components have been separated they may be cleaned with the brake cleaning fluid or methylated spirits. Do not use any other solvent. Dry the parts and inspect for wear.
7 If any wear or scores are found on the piston or in the bore in the caliper block, the whole assembly should be renewed. You should not attempt to polish out scratches since the seals rely on a close fit to be effective.
8 Reassembly of the caliper block begins with the fitting of the new

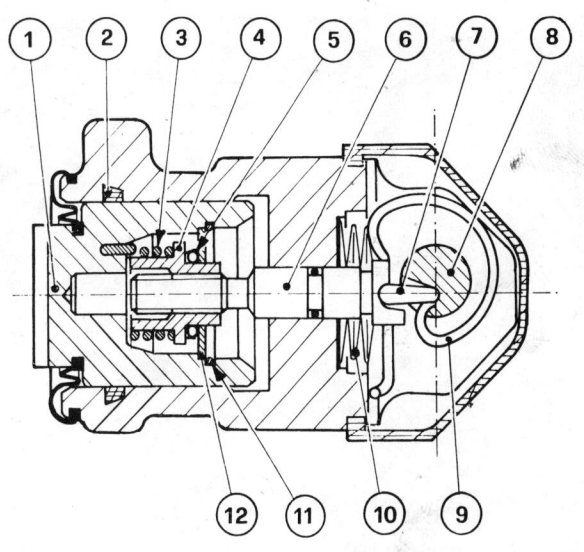

Fig. 9.3 Sectional view of the rear caliper cylinder and piston (Sec 8)

1 Piston
2 Piston seal
3 Adjuster spring
4 Clearance self-adjuster
5 Bearing
6 Clearance self-adjusting rod
7 Handbrake piston drive block
8 Handbrake control shaft
9 Handbrake control shaft return spring
10 Washer
11 Circlip
12 Thrust washer

'O'-ring oil seal into its groove in the bore in the caliper block. The seal should be wetted in hydraulic fluid before fitment. Following the fitting of the 'O'-ring with the fitting of the dust seal.
9 Wet the sides of the slave piston assembly and push gently into the bore. When only 0.5 in (13 mm) is left protruding fit the dust cover onto the top of the piston and then push the piston into the caliper block as far as it will go.
10 Wet the small new 'O'-ring that fits onto the plunger with brake fluid and slip over the plunger shank into its groove.
11 Place the plunger spring thrust washers and the springs themselves into position on the caliper block around the hole into which the plunger fits.
12 Insert the plunger into the caliper block and push and screw simultaneously to work it right into the special nut which is in the slave piston.
13 Place the lever pivot into position in the fork on the outside of the caliper block. You may need to slip a 'G'-clamp over the lever pivot point and caliper block bearing onto the top face of the slave piston, to hold the lever in the position where the pivot pin may be driven into place.
14 The protective boot is now slipped over the handbrake lever and into place on the caliper block.
15 Refit the carrier to the caliper block using a nut and bolt to expand the carrier as necessary.
16 The caliper assembly is ready to be fitted to the yoke and wheel hub.

9 Master cylinder – removal and refitting

1 All models are fitted with a tandem cylinder allowing dual circuit braking. The front section of the master cylinder controls the front braking circuit and the rear section controls the mixed circuit (front and rear).
2 Disconnect and remove the battery, as described in Chapter 10, before starting work.
3 Pull off the electric leads to the master reservoir. These control the low fluid warning light (photo).
4 Place a container under the master cylinder to collect the brake fluid. If removing the battery has not created enough working room then remove the expansion tank for the cooling system, and also on later models the large windscreen washer reservoir. On USA models

Chapter 9 Braking system

9.3 The master cylinder reservoir has a central communal filler cap (removed) and a cap for each braking section. The warning wires are connected to each section

9.7 The master cylinder is secured to the servo unit by two nuts (viewed from underneath with the engine removed for clarity)

11.5 The brake disc is secured to the hub by two bolts

remove the emission control diverter valve and solenoid valve (see Chapter 3).
5 Remove the cap(s) from the brake fluid reservoir(s) and stretch a thin sheet of polythene over the reservoir top(s) and refit the cap(s). This measure will prevent excessive loss of fluid when the brake fluid pipes are disconnected from the master cylinder.
6 Remove the two banjo bolts which secure the front braking circuit and the mixed braking circuit to the master cylinder. Cover the banjo unions to prevent the ingress of dirt.
7 Finally undo and remove the two nuts which secure the master cylinder to the servo unit and lift the master cylinder away to a clean bench for inspection and overhaul (photo).
8 Refitting the master cylinder follows the reversal of the removal procedure except that the whole brake system must be bled before the car is taken on the road as described in Section 2. Always renew the master cylinder mounting gasket.

10 Master cylinder – overhaul

1 With the master cylinder and reservoir removed from the car as described in Section 9 begin by emptying the brake fluid in the reservoir into the container. This fluid must *NOT* be re-used.
2 Then remove the reservoir from the master cylinder.
3 Clean any accumulated dust or grime from the outside of the master cylinder using petrol and a small stiff brush. Then dry it thoroughly. Remember that when dealing with the braking system absolute cleanliness must be observed.
4 Place the master cylinder in a vice with protective jaws and remove the circlip at the rear end, and the caps if fitted (Fig. 9.5).
5 Then remove the piston assembly for the rear (mixed circuit) section of the braking system. The last item to be withdrawn is the return spring.
6 Now undo and remove the piston stop screw (19) and the front circuit piston assembly can be withdrawn.
7 With the master cylinder and components stripped down throughly check the pistons and cylinder for scoring or wear. If any scoring is visible the component must be renewed. Assuming that the components are unmarked, check that the pistons slide freely in the cylinder. Clean all the components in methylated spirit and dry them thoroughly using air pressure if possible.
8 Check the reservoir. It must be free from any dirt or foreign matter. The floats must be in good condition, and the switches must operate correctly. They must close approximately 1 to 4 mm (0.039 to 0.15 in) before the fluid level reaches the top of the reservoir division (Fig. 9.4).
9 Before reassembling the components in the reverse order to removal, fit new seals to the pistons. Then lightly smear all the components with brake fluid to help the reassembly process.
10 When refitting the front circuit piston ensure that the groove lines up with the hole in the bottom of the master cylinder for the piston stop screw.
11 Make sure when the rear piston assembly has been fitted that the circlip is a good tight fit and located properly. Then refit the caps (if fitted) and then the reservoir. Make sure that it is a tight fit.
12 The unit is now ready for refitting to the car.

11 Brake discs – inspection, removal and refitting

1 The discs can easily be inspected by removing the roadwheels. Check that the surface is not badly scored or pitted. As discs wear from salt corrosion in the winter they tend to flake at the edges. This will impair the braking efficiency of the car and cause the brakes to judder and grate. Also the brake pads will wear very quickly on such a rough surface. The discs can be removed and refaced provided after doing this that they are not below the minimum thickness quoted in the specifications section. However, both front or both rear discs must be refaced or renewed at the same time.
2 To remove the disc, begin by removing the brake caliper, block and disc brake pads as detailed in Section 3 and 6 of this Chapter.
3 It should be appreciated that the caliper blocks must be carefully supported not to strain the flexible hose which connects it to the brake pipe system on the car.
4 The caliper yoke members may be unbolted from the hub carrier at the front or the stub axle at the rear.

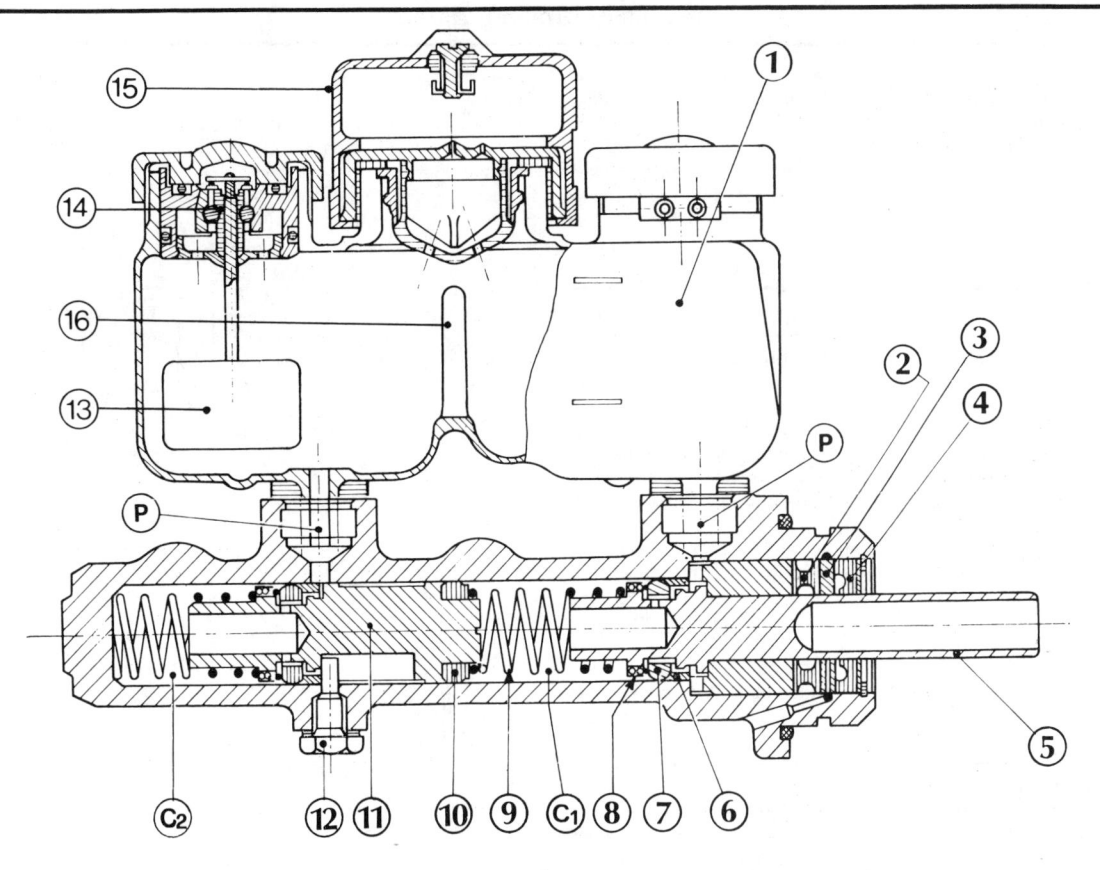

Fig. 9.4 Sectional view of master cylinder and reservoir (Secs 9 & 10)

1 Reservoir
2 Plunger guide bush seal
3 Intermediate washer
4 Seal
5 Mixed circuit piston
6 Floating spacer
7 Piston seal
8 Seal spring
9 Rear piston return spring
10 Main seal between front and mixed circuits
11 Front circuit piston
12 Piston stop screw
13 Float for front circuit
14 Warning light switch
15 Filler cap for both sections
16 Reservoir division
C1 Mixed circuit chamber
C2 Front circuit chamber
P Front and mixed circuit fluid supply passages

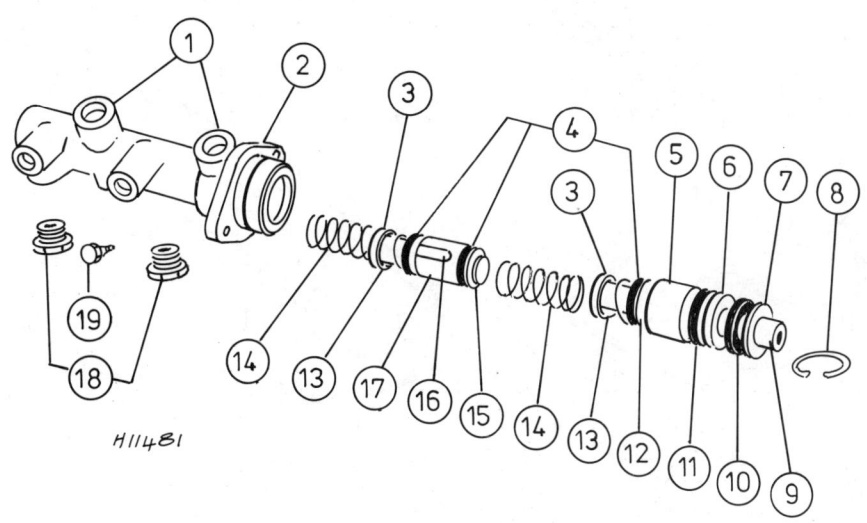

Fig. 9.5 Master cylinder – exploded view (Sec 10)

1 Reservoir mounting and supply ports
2 Mounting flange
3 Spring cap
4 Piston seals
5 Piston guide bush
6 Intermediate washer
7 Washer
8 Assembly retaining circlip
9 Mixed circuit (rear) piston
10 Seal
11 Piston guide bush seal
12 Floating spacer
13 Seal spring
14 Return springs
15 Spring thrust washer
16 Main seal between chambers
17 Front circuit piston
18 Caps
19 Piston stop screw

Chapter 9 Braking system

5 Loosen the brake disc mounting bolts and tap the disc free using the mounting bolts as guides. Remove the two bolts and the disc (photo). Note that the front discs are fitted with spacers.
6 Clean off the brake disc and hub before refitting commences.
7 Refit the discs in the reverse order to removal and tighten the mounting bolts to the correct torques. Before refitting the wheels check the discs for run-out using a dial gauge or feeler blades with a fixed block (see Specifications).

12 Handbrake cable – adjustment

1 The rear brakes are self-adjusting so the only adjustment necessary on the handbrake is to accommodate cable stretch and wear in the mechanical linkage.
2 Adjustment will be necessary when the handbrake lever passes more than 4 notches on the pivot ratchet. Jack the rear of the car up and support it. Block the front wheels.
3 The adjustment mechanism can be found underneath the car and forward of the rear anti-roll bar (Fig 9.6).
4 Loosen the locknut and turn the adjuster rod until the slack in the cable has been removed. This should be done with the handbrake lever pulled up three clicks on the ratchet from the rest or off position. Tighten the cables using the adjuster rod until the cables are tight and the roadwheels locked. Tighten the locknut.
5 Check that the wheels are free when the handbrake is released and that they are locked when the lever is pulled up in three clicks. The handbrake is now correctly adjusted.

13 Handbrake cable – inspection, removal and refitting

1 The cable will need renewal when it has been necessary to adjust it regularly to accommodate an increasing amount of stretch in the cable. Normally the handbrake linkage should only need adjustment once a year but if it becomes necessary to adjust at monthly intervals,

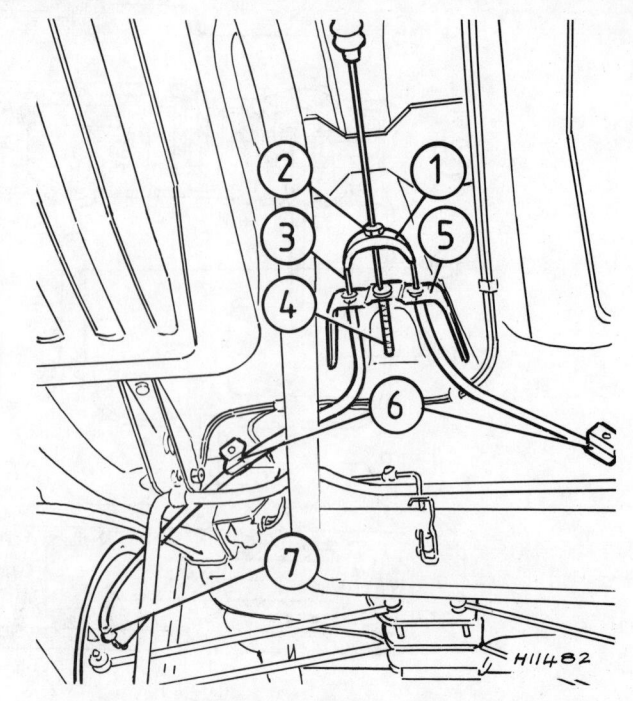

Fig. 9.6 Handbrake cable linkage (Secs 12 and 13)

1 Cable guide bracket
2 Locknut
3 Control cable
4 Adjuster rod
5 Cable sheath thrust bracket
6 Cable holders
7 Cable sheath retainer and cable-to-caliper mounting lever

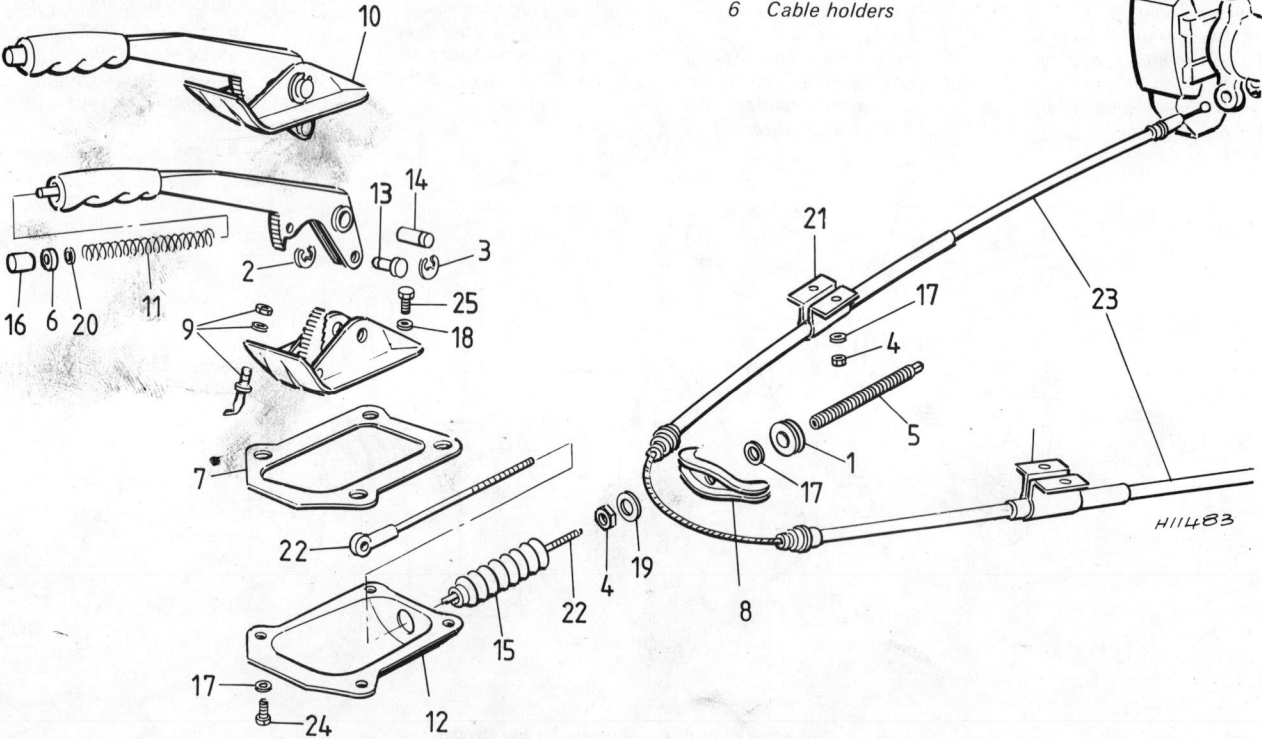

Fig. 9.7 Exploded view of the handbrake lever, cable and linkage (Secs 13 and 14)

1 Bush
2 Circlip
3 Circlip
4 Locknut
5 Adjuster rod
6 Bush
7 Seal
8 Cable guide
9 Handbrake warning light switch assembly
10 Handbrake lever assembly (inset)
11 Return spring
12 Cover plate
13 Linkage pin
14 Pivot pin
15 Rubber gaiter
16 Push button
17 Washer
18 Washer
19 Washer
20 Washer
21 Handbrake cable brackets
22 Link rod
23 Handbrake cable
24 Bolt
25 Bolt

Chapter 9 Braking system

then the cable merits renewal.
2 Begin the removal task by raising the rear of the car onto car ramps or chassis stands.
3 Slacken the locknut on the adjuster rod and unscrew the adjuster rod sufficiently so that the handbrake cable can be removed from the guide bracket.
4 Free the two outer cable ends from the thrust bracket, thus freeing the front loop of the handbrake inner cable.
5 Remove the two cable brackets from the floor pan.
6 Disconnect the cable ends from the brake caliper operating mechanism and lift the complete cable away.
7 Refitting is the reverse procedure, although it may be found easier to refit the cable brackets to the floor pan after the cable has been fitted to the thrust bracket and guide bracket.
8 Tighten up the cable adjuster when the whole assembly has been refitted and adjust the tension of the handbrake cable as described in Section 12.

14 Handbrake lever assembly and warning light switch – refitting and adjusting

1 Jack the car up and support it on stands. Then release the handbrake.
2 Undo the four bolts and washers which retain the cover plate and lower the cover plate and seal (1) so that it hangs freely on the link rod. (Fig. 9.8). On USA models remove the exhaust system main silencer heat shield.
3 Remove the circlip and push out the pin so that the link rod is disconnected from the lower end of the lever.
4 Remove the centre console inside the car, to reach the handbrake itself. With the seats forward the rear retaining bolts can be removed and vice versa.
5 Disconnect the handbrake warning light switch supply cable.
6 Undo and remove the handbrake lever and ratchet assembly retaining bolts and lift the unit out of the car.
7 Refit the assembly to the car in the reverse order to removal, and check that the handbrake operates correctly. Adjust the handbrake if necessary as described in Section 12.
8 Check that the handbrake warning light operates correctly. With the handbrake lever in the off or rest position the light should be out. The light on the dashboard should come on when the handbrake is pulled up two, or, at the most, three clicks. If necessary unscrew the locknut which holds the switch in position and adjust it as necessary.

15 Brake and clutch pedal assembly – removal, renovation and refitting

1 The brake and clutch pedals are mounted on a substantial channel member which at its lower end is bolted to the engine compartment bulkhead, and at the top end to the dashboard to support the upper section of the steering column.
2 The pedals themselves pivot on bushes running on a common long bolt which passes from one side of the channel to the other.
3 Begin removal of either pedal by removing the return spring acting on the clutch cable beside the gearbox underneath the car, and the return spring on the brake pedal situated immediately above the panel.
4 Undo and remove the nut on the end of the long bolt and then draw the bolt from the channel.
5 Remove the clutch pedal from the channel and disconnect the cable from the fork piece in the pedal lever.
6 The brake pedal can be removed once the clip that retains the servo pushrod linkage has been removed. The linkage can then be detached from the pedal assembly.
7 Once the pedals are free, an inspection can be made of the bolt, bushes and spacers. If the bushes are oval and worn, renew them.
8 The brake pushrod linkage and pivot are available as individual spares and therefore both should be inspected for wear.
9 The reassembly procedure is the reversal of the dismantling procedure, except that it is as well to check the clutch pedal free travel (Chapter 5) if the pedal assembly was repaired, also the operation of the brake switch which is mounted adjacent to the brake pedal pivot, as described in Chapter 10.

Fig. 9.8 The handbrake lever to linkrod assembly (Sec 14)

1 Cover plate and seal
2 Circlip
3 Pin

16 Brake servo unit – description

1 The vacuum servo unit is fitted into the brake system in series with the master cylinder and brake pedal to provide power assistance to the driver when the brake pedal is depressed.
2 The unit operates by vacuum odtained from the induction manifold, and comprises basically a booster diaphragm and a non return valve.
3 The servo unit and hydraulic master cylinder are connected together so that the servo unit pushrod acts as the master cylinder pushrod. The driver's braking effort is transmitted to the servo unit either directly, in the case of a left-hand drive car, or by a complicated linkage system in a right-hand drive car. Because of the cramped layout under the bonnet there is no room to mount the servo unit and master cylinder assembly on the right-hand side of the engine bay. Therefore it is mounted on the left hand side of the rear engine bay bulkhead and connected by linkage to the brake pedal (photo).
4 The servo unit piston does not fit tightly into the cylinder but has a strong diaphragm to keep its periphery in contact with the cylinder wall so assuring an air tight seal between the two parts. The forward chamber is held under vacuum conditions created in the inlet manifold of the manifold of the engine and during periods when the brake is not in use the controls open a passage to the rear chamber so placing it under vacuum. When the brake pedal is depressed, the vacuum passage to the rear chamber is cut off and the chamber is opened to atmospheric pressure. The consequent rush of air into the rear chamber pushes the servo piston forward into the vacuum chamber and operates the pushrod to the master cylinder. The controls are designed so that assistance is given under all conditions. When the brakes are not required, vacuum is re-established in the rear chamber when the brake pedal is released.
5 Air from the atmosphere passes through a small filter before entering the control valves and rear chamber and it is only this filter that will require periodic attention.

17 Brake servo unit – removal and refitting

1 Refer to Section 9 of this Chapter and remove the brake master cylinder.
2 Slacken the hose clip and remove the vacuum hose from the inlet manifold from the union on the forward face of the servo unit.

Chapter 9 Braking system

7 Pull back the dust cover over the unit body and refit the pedal assembly.

Right-hand drive models

8 The complete servo unit will have to be removed to carry out this operation, because of the way the unit is fitted to the bulkhead in front of the passenger's footwell.

19 Rear brake compensator device – description

1 The brake compensator device is fitted to a bracket on the right-hand side of the rear floor pan and is connected to the anti-roll bar by a torsion bar and linkage (Fig 9.10).
2 The device controls the pressure of the brake fluid to the rear wheel disc brakes. When the car is empty only very limited pressure is allowed through. When the car is laden full pressure is allowed through, as the rear wheel grip will be improved by the load and a greater effort can be applied to the brakes without locking them up and causing the car to skid.
3 The control linkage continuously monitors the variations in distance between the body and suspension. Therefore it can alter the compensator and the fluid pressure not only according to load but also according to the bumpiness of the road surface.

20 Rear brake compensator device – removal, refitting and adjusting

1 Raise the rear of the car on ramps and block the front wheels.
2 From the engine compartment remove the brake fluid reservoir filler cap, stretch a sheet of polythene over the top of the reservoir and refit the cap. This measure will prevent excessive loss of brake fluid when the pipes to the unit are subsequently disconnected.
3 Remove the top shackle bolt which connects the torsion bar to the control lever and loosen the bottom shackle bolt (Fig. 9.10).
4 Undo the bolts which secure the torsion bar guide bracket to the floor pan so that the bracket is free.
5 Undo the brake pipe inlet and outlet couplings to the compensator device.
6 Undo and remove the two bolts which secure the compensator device to the mounting bracket, and then free the brake pipes from the unit and lift it away complete with the torsion bar and bracket.
7 Remove the rubber boot on the front end of the unit and then withdraw the bolt which holds the torsion bar pivot bracket to the compensator. The compensator and torsion bar can then be separated.
8 Refitting is the reverse procedure to removal but the compensator must be correctly adjusted before the mounting bolts or brake pipe connectors are tightened.
9 To do this the car must be empty. This means that the spare wheel must be removed, along with the jack and tools, and the petrol tank must be drained.
10 Rotate the compensator in the opposite direction to that shown as 'E' in Fig 9.10. Then turn it back in the direction of arrow 'E' until the plunger contacts the torsion bar end. Lock the compensator mounting bolts. Do not place any load on the torsion bar, the compensator plunger must just rest against the end of the rod but must not move it at all. Tighten up the brake pipe inlet and outlet connections.
11 Refit the rubber boot to the front end of the compensator having smeared a little grease on the fork of the plunger. Then bleed the rear brakes as described in Section 2.
12 Refill the tank with petrol and place the jack, tools and spare wheel in the car.

21 Rear brake compensator unit – overhaul

1 The overhaul of the regulator is quite straightforward; with the regulator and torsion bar on a clean bench proceed as follows.
2 Slip the boot which protects the acting pivot end of the torsion bar and the regulator plunger off the regulator body and down the torsion bar. Undo and remove the bolt which secures the torsion bar retaining fitting and separate the torsion bar and regulator block.
3 Unscrew and remove the large end plug and retrieve the washer seal.

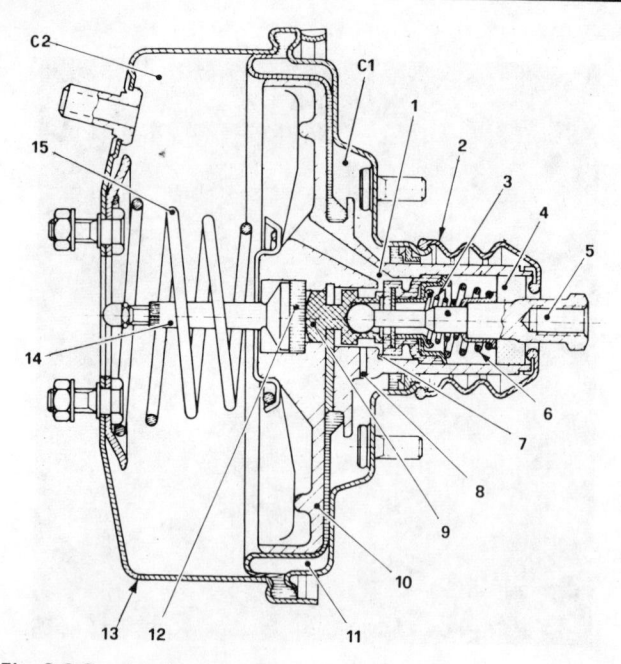

Fig. 9.9 Sectional view of the servo unit at rest (Secs 16 and 17)

1 Connecting gallery for chambers C1 and C2
2 Rubber boot
3 Valve return spring
4 Air inlet filter
5 Control rod
6 Seal return spring
7 C1 to C2 chamber connection at rest
8 C1 to C2 chamber connection gallery at rest and C1 chamber air inlet with unit working
9 Valve
10 Diaphragm carrier
11 Diaphragm
12 Reaction disc
13 Casing
14 Master cylinder control rod
15 Return spring

3 Next remove the clip that retains the small pin that joins the servo pushrod to the pedal link. Remove the pin and the pedal return spring and separate the pushrod and pedal linkage.
4 The servo unit is attached to the mounting plate by four nuts on studs in the servo unit. Remove the nuts and then withdraw the servo unit from the bulkhead in the engine bay.
5 Refitting the servo unit is the reverse procedure. Make sure that the mounting surface is clean before refitting it. Ensure that the feed pipe from the inlet manifold is refitted correctly. If it has been removed from the manifold end as well as the servo unit end, then it must be refitted with the arrow on the non-return valve pointing towards the engine.
6 Once the unit is refitted, refill and bleed the braking system completely as described in Section 2.

18 Brake servo unit – air filter renewal

1 Under normal operating conditions the servo unit is very reliable and does not require overhaul except possibly at very high mileages. In this case it is better to obtain a service exchange unit, rather than repair the original.
2 However, the air filter may need renewal and fitting details are given below.

Left-hand drive models

3 Remove the brake and clutch pedals as detailed in Section 15 of this Chapter.
4 Working inside the car in the driver's foot well, push back the dust cover from the pushrod and control valve housing to expose the endcap and air filter element.
5 Using a screwdriver ease out the end cap and then with a pair of scissors cut out the air filter.
6 Make a diagonal cut through the air filter element and push over the pushrod. Hold in position and refit the end cap.

Chapter 9 Braking system

4 The spring, washer, slotted ring, plunger, primary seal, seal spring and rest ring and finally the secondary seal can all be extracted in that order.
5 Clean all components with methylated spirit and wipe dry with a non-fluffy rag.
6 Inspect the seals, plunger shank and regulator bores for wear and surface deterioration. If only the seals are worn, renew the seals and reassemble the regulator. If the regular plunger or bores are worn the regulator must be renewed as a whole assembly.

7 Reassembly of the regulator follws the reversal of the dismantling procedure, except that the primary and secondary seals should be wetted with clean brake fluid before they are refitted. It would be as well to use a new plug seal washer on each occasion the regulator is reassembled.
8 Refit the torsion bar and regulator block, slip the protective boot over the block and torsion bar end and commence refitment of the regulator assembly to the vehicle (Section 20).

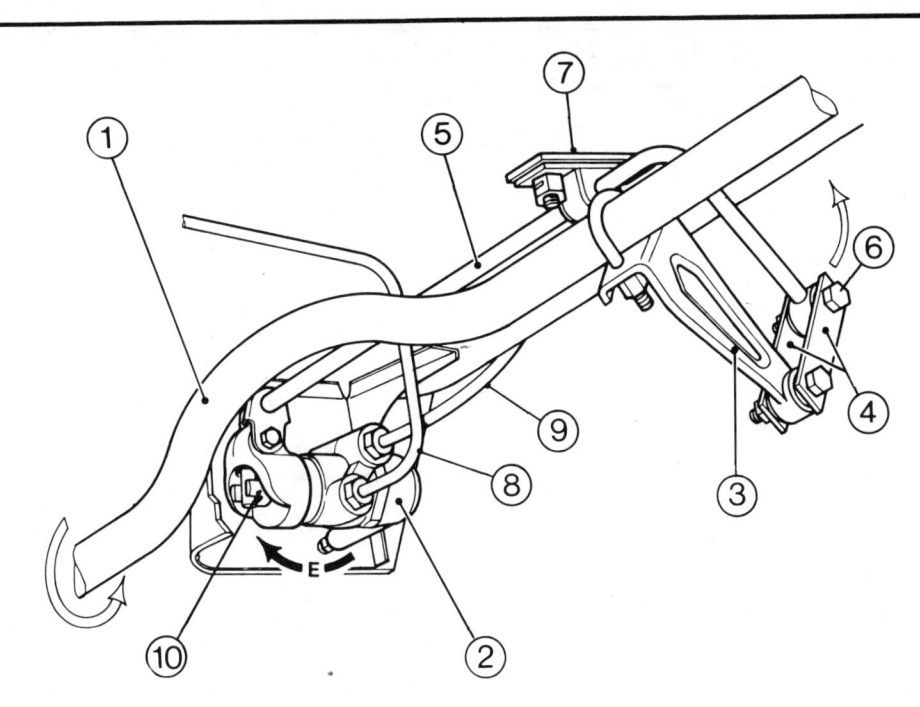

Fig. 9.10 Rear brakes compensator device (Secs 19 and 20)

E See Section 20

1 Rear anti-roll bar
2 Compensator device
3 Compensator device control lever
4 Shackle bars
5 Torsion bar
6 Shackle bolt
7 Torsion bar guide bracket
8 Inlet brake pipe
9 Outlet brake pipe
10 Plunger

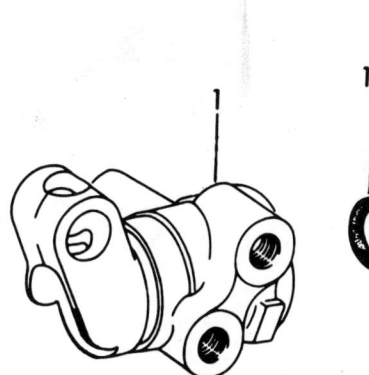

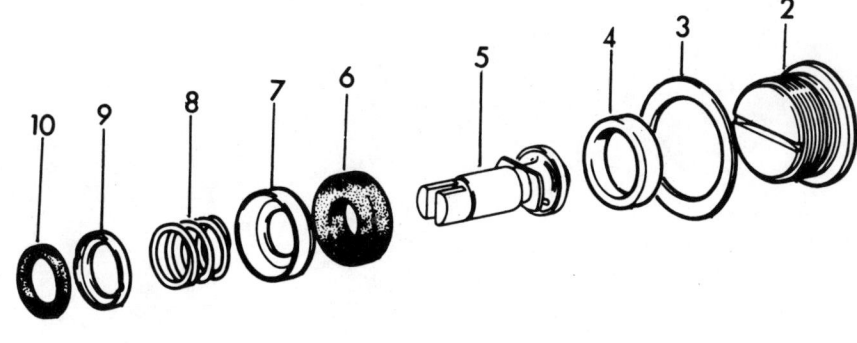

Fig. 9.11 Brake compensator unit – exploded view (Sec 21)

1 Regulator block
2 End plug
3 Seal washer
4 Spacer ring slotted
5 Plunger
6 Primary seal
7 Seal cup
8 Seal spring
9 Rest ring
10 Secondary seal

22 Fault diagnosis – braking system

Before diagnosing faults from the following chart, check that any braking irregularities are not caused by:-
1. *Uneven and incorrect tyre pressures*
2. *Incorrect 'mix' of radial and cross-ply tyres*
3. *Wear in the steering mechanism*
4. *Defects in the suspension*
5. *Misalignment of the chassis*

Symptoms	Reason(s)
Stopping ability poor, even though pedal pressure is firm	Pads and/or discs badly worn or scored One or more wheel hydraulic pistons seized, resulting in some brake pads not pressing against the discs Brake pads contaminated with oil Wrong types of pads fitted (too hard) Brake pads wrongly assembled Servo unit faulty
Car veers to one side when the brakes are applied	Brake pads on one side are contaminated with oil Hydraulic pistons on one side partially or fully siezed A mixture of pad materials fitted between sides Unequal wear between sides caused by badly worn or corroded discs
Pedal feels spongy when the brakes are applied	Air is present in the hydraulic system
Pedal feels springy when the brakes are applied	Brake pads not bedded into the discs (after fitting new ones) Master cylinder or servo mounting bolts loose Severe wear in the brake discs causing distortion when the brakes are applied
Pedal travels right down with little or no resistance and brakes are virtually non-operative	Leak in the hydraulic systems If no signs of leakage are apparent all the master cylinder internal seals are failing to sustain pressure
Binding, juddering, overheating	One or a combination of causes given in the foregoing sections

Chapter 10 Electrical system

Contents

Alternator – brush renewal	9
Alternator drivebelt – checking, adjustment and renewal	8
Alternator – general description	5
Alternator regulator – renewal	10
Alternator – removal and refitting	7
Alternator – testing and maintenance	6
Auxiliary switches, illumination and warning lamps (Coupe, Spider and HPE up to 1979 – removal and refitting	31
Auxiliary switches and indicator lamps (Coupe, Spider and HPE models, 1979 onwards) – removal and refitting	32
Auxiliary switches and indicator lamps (Saloon models) – removal and refitting	33
Auxiliary switch panel (Coupe, Spider and HPE) – removal and refitting	37
Battery – charging	4
Battery – maintenance and care	3
Battery – removal and refitting	2
Brake stop light switch – removal and refitting	42
Courtesy light switches – removal and refitting	41
Front parking and flasher lamps – removal and refitting	23
Fault diagnosis – electrical system	52
Flasher units – removal and refitting	44
Fuses – general	16
General description	1
Glovebox light – removal and refitting	30
Headlamps – alignment	21
Headlamp bulbs – renewal	22
Headlamp self levelling system – general	20
Headlamp unit (Coupe, Spider and HPE models) – removal and refitting	19
Headlamp unit (Saloon models) – removal and refitting	18
Horns and compressor (Coupe, Spider and HPE models) – removal and refitting	47
Horns (Saloon models) – removal and refitting	46
Horn switch – removal and refitting	48
Ignition switch – removal and refitting	39
Instrument panel (Coupe, Spider and HPE models) – removal and refitting	36
Instrument panel (Saloon models) – removal and refitting	35
Instrument panel warning/indicator lights – bulb renewal	34
Instruments – removal and refitting	38
Interior courtesy lamp – removal and refitting	27
Lighting, direction indicators, wiper/washer control switches assembly – removal and refitting	40
Luggage compartment light – removal and refitting	40
Luggage compartment light – removal and refitting	26
Number plate lamps – removal and refitting	29
Radios and tape players – fitting (general)	54
Radios and tape players – suppression of interference (general)	55
Rear lamp assembly – removal and refitting	25
Reversing light switch – removal and refitting	43
Sender units – removal and refitting	45
Side repeater lamps – removal and refitting	24
Solenoids (relays) – general	17
Starter motor – dismantling, repair and reassembly	14
Starter motor – circuit testing	12
Starter motor – general description	11
Starter motor – removal and refitting	13
Starter motor drive pinion – inspect and repair	15
Tailgate wiper assembly (HPE models) – removal and refitting	53
Under bonnet lamp – removal and refitting	28
Windscreen washer pump – removal and refitting	49
Windscreen wiper arms and blades – removal and refitting	50
Windscreen wiper motor and mechanism – removal and refitting	51
Windscreen wiper motor and mechanism – dismantling and reassembly	52

Specifications

System type .... 12 volt, negative earth

Battery type (standard) .... 12 volt 45 Ampere hour at 20 hour rate (special version 55 to 60 Ampere hour)

Alternator

	Bosch	Marelli	Ducellier
Models	K1-14V-55A 20	AA 124-14V-42A or AA 125-14V-55 A	514.001 A or 12V 43A
Cut in speed at 12V (25°C)	1000 rpm	1000 rpm	1150 rpm (14V)
Speed at $\frac{2}{3}$ max amp output	2000 rpm (37 amps)	–	2250 rpm (37 amps)
Speed at maximum amp output	6000 rpm	7000 rpm	8000 rpm
Maximum shaft speed	12000 rpm	13000 rpm	12500 rpm
Maximum amperage	55 amps	44 amps (AA 124)/ 55 amps (AA 125)	43 amps
Direction of rotation	Clockwise	Clockwise	Clockwise
Induction coil resistance – ohms	–	4.3 + 0.2 ohms at 20°C (68°F)	4 + 0.2 ohms at 20°C (68°F)

Chapter 10 Electrical system

Alternator drivebelt tension		10 to 15 mm (0.39 to 0.59 in) deflection under a load of 5 kg (11 lbs) on centre of horizontal section of belt	

Starter motor

	Bosch	Marelli
Model	12V 1.1CV	E 100 1.3/12
Rated output	1.1 kW	1.3 kW
Number of poles	4	4

Bulbs

	UK and Europe	USA
Headlights:		
Main beam	55W (halogen)	
Dipped beam	55W (halogen)	
Direction indicators:		
Front and rear	21W	5/21W
Side repeaters	4W	3W
Reversing lights	21W	21W
Stoplights	21W	21W
Side or parking lights	5W	5W
Engine bay light	5W	4W
Courtesy light	5W	
Boot light	4W	4W
Cigar lighter light	4W	4W
Number plate lights	4W	5W
Fibre optic light for switches	3W	
Clock illumination	3W	3W
Glovebox light	3W	4W
Instrument lights	3W or 1.2W	3W
Heating control lights	3W or 1.2W	3W
Power window lift switch lights	3W or 1.2W	
Warning lights	1.2W	3W
Door safety lights		4W
Fasten seat belts light	Not fitted	3W
EGR renewal light	Not fitted	3W
Windscreen wiper switch light		1.2W
Hazard flasher switch light		1.2W
Warning lights for heated rear window light, side lights, main beam light, indicator light, oil light, engine overheating light, no charge light, fuel reserve light, brakes and brake pad wear limit indicator light, EGR service (Coupe) light, fasten seat belt light (Coupe)		1.2W

Fuses – UK and European models (up to 1979)

Fuse No	Amps	Circuits protected
1	8	Front right side light, right number plate light and rear left side light
2	8	Engine compartment light, boot light, cigarette lighter light, front left side light, instruments lights, side light warning light, rear right side light and left number plate light
3	8	Right dipped beam
4	8	Left dipped beam
5	8	Right main beam
6	8	Left main beam and main beam warning light
7	8	Stop lights, windscreen washer motor and radio (if fitted)
8	16	No. 14 solenoid switch, heating and ventilation booster and windscreen wipers
9	16	No. 12 solenoid switch, turn indicators, ignition coil, handbrake warning light, instruments and warning lights and engine cooling fan solenoid switch
10	8	Courtesy lights, door safety lights, glove locker light and clock light
11	25	Glove locker light, door safety lights, courtesy light, horns, plug in socket, cigarette lighter, rear heated window and clock

Fuse identification   **Colour coding**
8 amp .... Black
16 amp .... Green
25 amp .... Hazel
Numbering .... 1 to 11 from left to right

Fuses – UK and European models (1979 on models)

Fuse No	Amps	Circuits protected
1	8	Left front side light, right rear tail light, number plate light, side light indicator, instruments light, engine bay light, boot light
2	8	Right side light, parking light, front and rear light, clock light, cigar lighter light, foglamps, fog lamp indicator light, fibre optics for switches light, heater controls light, glovebox light
3	8	Left dipped beam
4	8	Right dipped beam
5	8	Left main beam and indicator lamp
6	8	Right main beam
7	8	Rear fog lamp and indicator lamp

Chapter 10 Electrical system

8	16	Front fog lights
9	16	Left power window lift
10	16	Right power window lift
11	16 or 25	Reversing lights, solenoid switches for power window lifts, horns, radiator fan, cigar lighter, heated rear window and indicator lamp, clock, fan booster, idling fuel cut-out, instruments cluster
12	16	Wiper motor and relay, windscreen washers, stop lights, direction indicators, handbrake warning light
13	16	Radiator fan (on some models this is a 16 amp in-line fuse between battery + terminal and fan)
14	16	Horns
15	16	Cigar lighter, heated rear window and indicator lamp
16	16	Hazard flasher and warning light, plug-in socket, radio, powered aerial, digital clock (later models), courtesy lights

Fuse identification
8 amp ....
16 amp ....
25 amp ....
Numbering ....

Colour coding
Black
Green
Hazel
1 to 16 from left to right

Fuses – USA models

Fuse No	Amps	Circuits protected
1	8	Front right side light, front right clearance light, rear left side light, rear left clearance light, right number plate light (Saloon), left number plate light (Coupe), boot light (Saloon)
2	8	Lighting: engine compartment, cigarette lighter, instruments cluster, heating and ventilation controls, wipe speed and hazard switches; front left side light, front left clearance light, rear right side light, rear right clearance light, side lights warning light, left number plate light (Saloon), right number plate light (Coupe), boot light (Coupe), clock light (Coupe)
3	8	Right low beam
4	8	Left low beam
5	8	Right main beam, main beams warning light
6	8	Left main beam
7	8	Stop lights, radio (if fitted)
8	16	Reversing lights, engine ignition remote control switch, solenoid switches for electric window lifts (if fitted), rear heated window, cigarette lighter, air conditioner fan motor (if fitted), horns, engine coolant motor-driven fan, emission control system solenoid switches, air conditioning system solenoid switches (if fitted), air ventilation and heating fan, windscreen wiper, windscreen washer motor
9	16	Direction indicators and tell-tales, ignition coil, carburettor fuel idling cut-off device, emission control system solenoid switch, EGR warning light device. Feed for following warning lights: engine oil low pressure, engine overheating, EGR, brake, brake pad, fasten belts, fuel reserve and alternator. Feed for following instruments: oil pressure gauge, voltmeter (Coupe), fuel gauge, coolant temperature gauge and engine rev counter, EGR circuit and driver's seat belts circuit solenoid switches and electric delaying device
10	8	Plug-in socket, front and rear interior lights, open door safety lights, clock and glove locker
11	25	Solenoid switches for: horns, rear heated window, heating and ventilation fan and rear heated window warning light

Fuse identification
3 amp ....
8 amp ....
16 amp ....
25 amp ....

Colour coding
White
Black
Green
Hazel

In-line fuses – USA (Saloon models)

Amps	Circuits protected
16	Engine coolant fan
16	Air conditioner condenser motor-driven fan
25	Air conditioner
8	Fuel lift pump
3	Driver's seat belt unfastened and open door buzzer with ignition key on sited below dashboard right-hand side
16	Hazard signalling system located below dashboard right-hand side
3	EGR warning light system and relative solenoid switch located below dashboard right-hand side
25	Electric window lifts (if fitted)

Fuse identification
3 amp ....
8 amp ....
16 amp ....
25 amp ....

Colour coding
White
Black
Green
Hazel

In-line fuses – USA (Coupe models)

Amps	Circuits protected
16	Engine coolant fan
16	Air conditioner condenser motor-driven fan
25	Air conditioner
16	Electric window lifts (if fitted)
8	Fuel lift pump
3	Driver's seat belt unfastened and open-door buzzer with ignition key on located below dashboard right-hand side

Chapter 10 Electrical system

16	Hazard signalling sited below dashboard right-hand side
3	EGR warning light and relevant solenoid switch located below dashboard right-hand side

Fuse identification
3 amp ....
8 amp ....
16 amp ....
25 amp ....

Colour coding
White
Black
Green
Hazel

Torque wrench settings

	lbf ft	Nm
Alternator main lower pivot bolt	51	69
Alternator adjuster nut and bolt:	32	43
Alternator bracket to engine		
Series A engines	38	52
Series B engine 1300/1600	32	43
USA 1800 and 2000 models	16	22
Temperature transmitter to cylinder head	36	49
Oil pressure switch to support	24	32

1 General description

1 The electrical systems on the Lancia Beta models reflect the trend towards ever more profusion as is apparent on so many new cars. Following this greater complexity comes the likelihood of failure of those systems, and therefore whereas the mechanics of your car will be more reliable than in years past, the electrical system will most probably demand more attention than any others on the vehicle. Fortunately, more often than not, it will be the electrical lead connections and electrical device mountings that will give trouble, rather than the electrical devices themselves. It is from this factor that the need comes for cleanliness and neatness when dealing with electrical systems.

2 In this Chapter details are given of the maintenance and repair of each system, including any special precautions that may be taken to guard against premature failure of those systems.

3 As mentioned earlier, because you will probably have to deal with electrical faults more than anything else on this car it may be worth while investing in a few items of electrical test equipment. One very useful item of equipment is a voltmeter (0 to 20 volts) or at very often not much more expense, a multi-meter. This last device is really useful, not only for the car but also around the house. A multimeter can measure voltage, current or resistance on a variety of scales.

4 So much for the equipment and rationale. The electrical systems on the Lancias are conventional 12 volt systems with a negative earth. They are powered by a battery and an alternator driven off the engine. A pre-engaged starter is fitted to the front of the engine.

5 The battery provides power for starting, and reserve power should the demand from the systems exceed the output of the alternator. When the alternator is not fully loaded, the voltage regulator ensures that the battery is kept charged.

6 The potential power from the alternator or battery is sufficient to severely damage the electrical wiring, if faults occur which allow that power through the circuits. Therefore most are protected with fuses. Those not protected with fuses are associated with engine systems and include the ignition system and starter motor circuit.

7 When fitting accessories to cars with negative earth systems it is important, if they contain silicon diodes or transistors, that they are connected correctly, otherwise they may be irreparably damaged. Before purchasing any accessory check that the polarity is suitable or can be changed.

8 Always disconnect the battery leads, if the battery is to be boost charged in situ, or if any welding work is to be carried out using arc-welding equipment.

2 Battery – removal and refitting

1 The battery is situated on the left-hand side of the engine compartment, and is held in place by a clamp plate and bolt (photo).

2 To remove the battery begin by disconnecting the negative earth lead from the battery and bodyshell. Then disconnect the positive lead from the battery.

3 Once the leads have been removed, the bolt, which secures the clamp onto the battery base ledge, may be unscrewed so that the battery is free to come out.

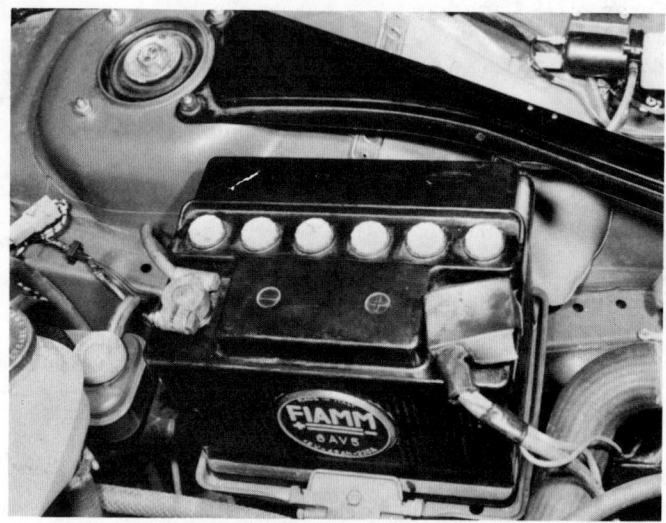

2.1 Battery locations and fitting (HPE model shown)

4 Lift the battery from its seating in the bodyshell, taking great care not to spill any of the highly corrosive electrolyte.

5 With the battery out of the car, clean the battery tray and the clamps if they are dirty or corroded. Also clean the battery itself and the posts.

6 Refitting is the reversal of this procedure. Refit the positive lead first and smear the clean terminal posts and lead clamp assembly beforehand with petroleum jelly (Vaseline) in order to prevent corrosion. Do not use ordinary grease.

3 Battery – maintenance and care

1 At weekly intervals, the electrolyte level should be checked and topped up if required.

2 One of three types of battery may be encountered. On a battery with a non-removable cover, raise the cover as far as it will go. If the electrolyte level is below the bottom of the filling tubes, top up using distilled water only.

3 On a battery with a removable cover or screw plugs, lift off the cover and inspect the electrolyte level. This should be just above the perforated splash guard. If it is below this, top it up with distilled water.

4 It should be noted that with the non-removable battery cover, distilled water is poured into the trough on top of the cells and will not actually enter them until the cover is partially refitted.

5 On modern cars, the addition of distilled water is required very infrequently. If the need for it becomes excessive, suspect overcharging caused by a fault in the alternator control section.

6 Periodically inspect the battery tray and mounting bolts for corrosion. This will appear as a white deposit and must be neutralised before permanent damage to the metalwork occurs.

Chapter 10 Electrical system

7 Disconnect the battery leads, release the battery mountings and remove the battery from the car.
8 Sponge any deposits with ammonia, dry the surface and protect with anti-corrosive paint such as underseal.
9 Refit the battery, clean the terminals and clamps if necessary by careful scraping. Connect the leads, making quite sure that they are fitted to the correct polarity terminals by having the battery the right way round.
10 Smear the battery terminals and clamps with petroleum jelly (never grease) and wipe away any moisture or dirt from the top of the battery.
11 The foregoing operations are all that should be required to keep a battery in first class condition.

4 Battery – charging

1 As a result of the efficiency of the alternator, the battery will be kept in a good state of charge even if only short journeys are undertaken.
2 Only in a case where electrical equipment has been left on by mistake, should mains charging be necessary.
3 On cars with a battery charge indicator, the battery charge condition will be visually apparent whenever the ignition is on or the engine running.
4 On cars without this instrument, the state of charge can be ascertained using a hydrometer. The specific gravity of the electrolyte for fully charged conditions at the electrolyte temperature indicated, is listed in **Table A**. The specific gravity of a fully discharged battery at different temperatures of the electrolyte is given in **Table B**.

Type A
Specific gravity – battery fully charged
1.268 at 100°F or 38°C electrolyte temperature
1.272 at 90°F or 32°C electrolyte temperature
1.276 at 80°F or 27°C electrolyte temperature
1.280 at 70°F or 21°C electrolyte temperature
1.284 at 60°F or 16°C electrolyte temperature
1.288 at 50°F or 10°C electrolyte temperature
1.292 at 40°F or 4°C electrolyte temperature
1.296 at 30°F or -1.5°C electrolyte temperature

Table B
Specific gravity – battery fully discharged
1.098 at 100°F or 30°C electrolyte temperature
1.102 at 90°F or 32°C electrolyte temperature
1.106 at 80°F or 27°C electrolyte temperature
1.110 at 70°F or 21°C electrolyte temperature
1.114 at 60°F or 16°C electrolyte temperature
1.118 at 50°F or 10°C electrolyte temperature
1.122 at 40°F or 4°C electrolyte temperature
1.126 at 30°F or -1.5°C electrolyte temperature

5 When the time comes that regular battery charging from a mains charger is required to give the battery enough life to start the car in the mornings and it is known that the alternator is working correctly and reasonable mileages are being run, then the battery should be checked for failure by your dealer and if found to be at the end of its useful life, renewed.

5 Alternator – general description

The main advance of the alternator lies in its ability to provide a relatively high power output at low revolutions. Driving slowly in traffic with a dynamo fitted invariably means a very small or even no charge at all reaching the battery. In similar conditions even with the wipers, heater, lights and perhaps radio switches on the alternator will still ensure a charge reaches the battery.

The alternator is of the rotating field ventilated design and comprises principally a laminated stator on which is wound a 3 phase output winding and a twelve pole rotor carrying the field windings. Each end of the rotor shaft runs in ball race bearings which are lubricated for life. Aluminium end brackets hold the bearings and incorporate the alternator mounting lugs. The rear bracket supports the silicone diode rectifier pack which converts the AC output to DC for battery charging and output to the voltage regulator.

The rotor is belt driven from the engine through a pulley keyed to the rotor shaft. A special fan adjacent to the pulley draws air through the alternator. This fan forms an integral part of the alternator specification. It has been designed to provide adequate flow of air with the minimum of noise and to withstand the stresses associated with the high rotational speeds of the rotor. Rotation is clockwise when viewed from the drive end.

The rectifier pack of silicone diodes is mounted on the inside of the rear end casing, the same mounting is used by the brushes which contact the slip rings on the rotor to supply the field current. The slip rings are carried on a small diameter moulded drum attached to the rotor. By keeping the circumference of the slip rings to a minimum, the contact speed and therefore the brush wear is minimised.

Maintenance consists of occasionally wiping away any oil or dirt which may have accumulated on the outside of the unit.

Take extreme care when connecting the battery to ensure that the polarity is correct, and never run the engine with a battery charger connected. Do not stop the engine by disconnecting the battery leads, as this will almost certainly damage the alternator. When jump starting from another battery ensure that the jump leads are connected positive to positive and negative to negative.

6 Alternator – testing and maintenance

1 The alternator has been designed for the minimum amount of attention during service. The only items subject to wear are the brushes and bearings.
2 If the ignition warning light on the instrument panel lights up indicating that the battery is no longer being charged, check the continuity of the leads to and from the alternator voltage regulator and battery.
3 Ensure all the connections of those leads are clean, and check for breaks in each cable by disconnecting both ends of a cable and reconnecting in series with a small battery and bulb. If the cable is complete the bulb will light up. If once the continuity checks have been completed and nothing found at fault, the alternator may be checked in situ as follows.
4 Connect one end of a jumper lead to the large terminal on the alternator, and the other to a 0 to 20 v voltmeter. Connect the other (-ve) terminal on the voltmeter to the bodyshell. Start the engine and monitor the voltmeter reading as the engine speed is increased and decreased.
5 If only a few volts registered on the voltmeter, remove the alternator and inspect more cloely. If the reading obtained was between 12 and 14½ volts the alternator would seem to be all right. It must be stated however that the diode bridge network which rectifies the output of the stator makes it impossible to be certain that the alternator is serviceable, even if the output volts appear to be between 12 and 14½ volts. Therefore, if the checks detailed in paragraphs 3 and 4 do not reveal a clear fault, take the car to the nearest auto-electrician.

7 Alternator – removal and refitting

1 Disconnect the negative and positive cables from the battery to isolate the electrical system.
2 With Saloon models remove the radiator grille, as described in Chapter 12 to get better access to the alternator (Fig. 10.1).
3 With Coupe, Spider and HPE models and all USA models the grille cannot easily be removed and therefore the operation has to be carried out from above. However, it may be found to be advantageous if the right-hand body bracer strut is removed first (photo). On all USA models remove the air cleaner (see Chapter 3).
4 Either working through the grille opening or from above, disconnect the leads to the alternator. Note which wires fit where (photo).
5 Slacken off the adjuster/tensioner nut and bolt, and the lower pivot bolt and nut (photo). Pivot the alternator in towards the engine and slip the drivebelt off the pulley.
6 Undo and remove the adjuster/tensioner nut and bolt and the main pivot bolt and nut and lift the alternator away.
7 Refitting is basically the reverse procedure to removal, but do not tighten the pivot bolt and nut or the adjuster bolt and nut until the tension of the drive belt has been checked, as described in Section 8.
8 Finally tighten the mounting bolts and nuts to their specified torques.

Chapter 10 Electrical system

7.3 The alternator is easier to get at if this body bracer is first removed (HPE model shown)

7.4 This is how the wires are connected to the back of the Marelli type alternator (note the radio capacitor fitted to the alternator)

7.5 Alternator adjustment and pivot bolt locations

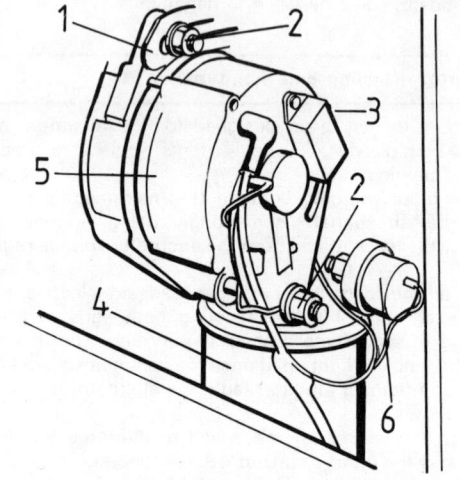

Fig. 10.1 View of alternator with the grille removed (Saloon models) (Sec 7)

1 Alternator belt adjusting bracket
2 Alternator adjuster nut, bolt and pivot nut
3 Electronic voltage regulator
4 Oil filter
5 Alternator
6 Oil pressure gauge transmitter

8 Alternator drivebelt – checking, adjustment and renewal

1 If the drivebelt is slack then it will slip and the alternator will not produce its correct output. If it is too tight then the alternator and water pump bearings may be ruined.
2 When correctly tensioned, the drivebelt should be able to be depressed 10 to 15 mm (0.39 to 0.59 in) in the centre of its longest run when a force of 5 kg (11 lb) is placed on it. This is equivalent to firm pressure of the forefinger or thumb (Fig. 10.2).
3 If the belt is too slack or needs renewing proceed as follows. Loosen the adjuster/tensioner nut and bolt on the top of the alternator. Push the alternator in so that the old belt can be lifted off, then fit a new one. Pull the alternator away from the engine to tighten the drivebelt. If necessary the lower pivot bolt may have to be slackened slightly as well.
4 When the belt is tensioned correctly lock the adjuster nut and bolt. Recheck the tension (photo).
5 Tighten the lower pivot bolt if it was slackened off.
6 If a new belt has been fitted, the tension should be checked after approximately 150 miles (250 km) of motoring.

9 Alternator – brush renewal

Bosch type
1 With the alternator removed from the car, undo the two screws and remove the voltage regulator from the rear of the alternator. The voltage regulator will come away complete with brushes and brush holder (Fig. 10.3).
2 Using a piece of pointed nose pliers, relieve the tension of the lead clamps and then using a soldering iron separate the leads from the brushes (Fig. 10.4).
3 Withdraw the brushes and springs.
4 Fit the new brushes and springs and solder the leads to the brushes.
5 Tighten the lead clamps and then refit the regulator and brush assembly to the alternator.

Marelli type
6 With the alternator removed from the car as described in Section 7, proceed as follows. Disconnect the single spade connector from the brush holder at the rear of the alternator (photo). This wire runs to the rectifier. Disconnect the double connector from the brush holder. This runs to the regulator (photo).
7 Remove the crosshead screw, which holds the brush holder in position and prise the brush holder out of the casing (photo).
8 Lever out the spring clips at the rear end of the brushes and remove the brushes and springs.
9 Fit new brushes and springs, secure them in position and refit the brush holder in the reverse order to removal.
10 Finally refit the alternator to the car as described in Section 7.

10 Alternator regulator – renewal

Bosch type
1 With the alternator removed from the car, withdraw the regulator assembly, as described in the previous Section.
2 The regulator and brush holder are a one-piece assembly and must be renewed as such.
3 Refit the regulator as described in the previous Section.

Marelli type
4 The electronic regulator is mounted to the rear alternator casing by two screws. Remove the screws, unplug the connector at the brush holder and lift it away (photos).
5 Refit a new regulator in the reverse order.
6 Refit the alternator to the car as described in Section 7 and test it in position.

11 Starter motor – general description

1 The starter motors are of the pre-engaged type, in which the solenoid switch is also employed to physically move the drive pinion along the starter motor shaft, into contact with the ring gear on the flywheel before power is supplied to the motor for turning the engine.
2 There is a spring between the pinion and the actuating lever from the solenoid, so that in the event of an exact abutment of gearteeth as the pinion is impelled to engage with the flywheel ring gear; the solenoid switch will still continue and make power contact. The pinion will fall into engagement as soon as the motor shaft turns.
3 The location of the starter motor is as usual on the left-hand side of the engine, and it is bolted to the clutch bellhousing. The

8.4 Check the tension and lock the adjuster nut and bolt

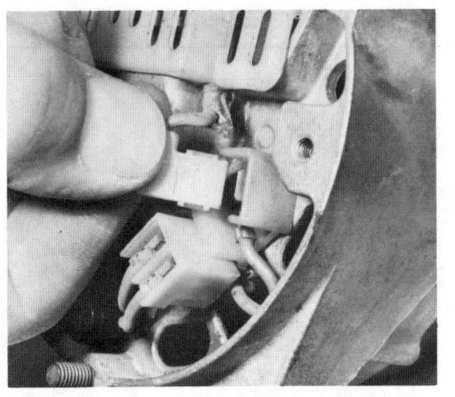

9.6a Disconnect the lead from the rectifier

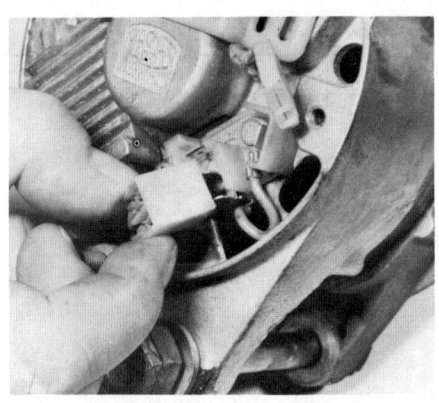

9.6b Disconnect the lead from the regulator

9.7a Removing the brush holder retaining screw

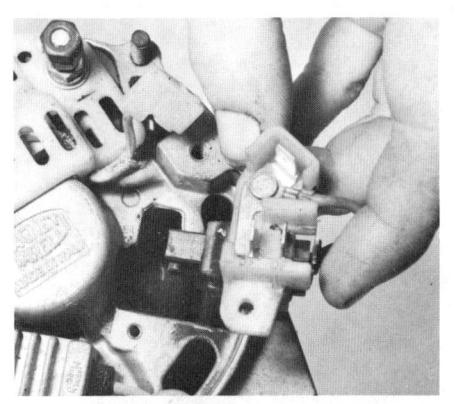

9.7b The brush holder removed from the casing

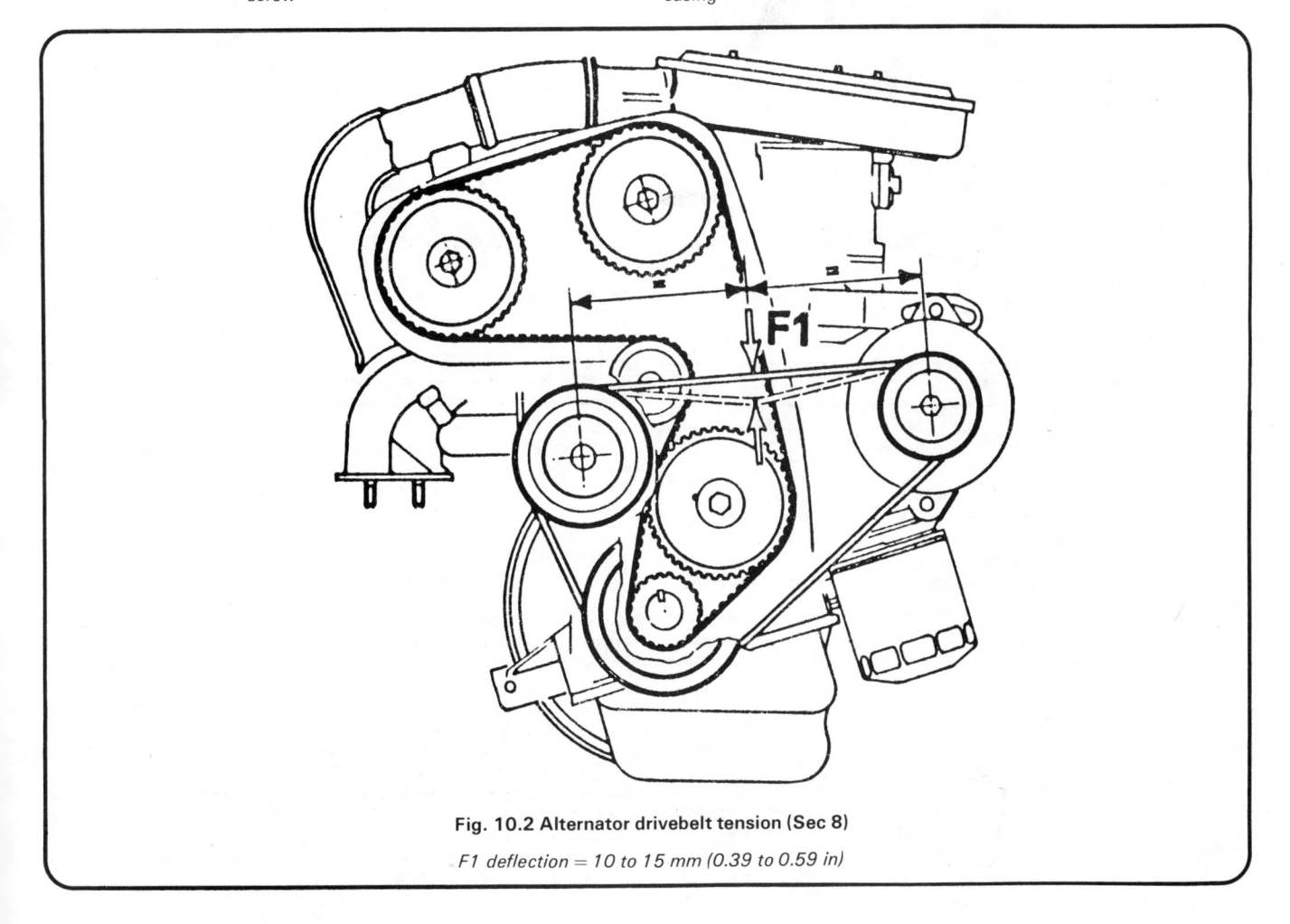

Fig. 10.2 Alternator drivebelt tension (Sec 8)

F1 deflection = 10 to 15 mm (0.39 to 0.59 in)

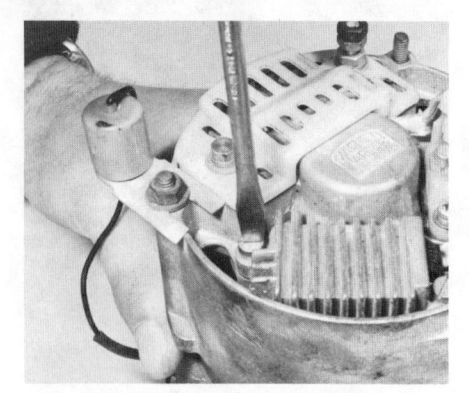

10.4a Removing one of the regulator retaining screws

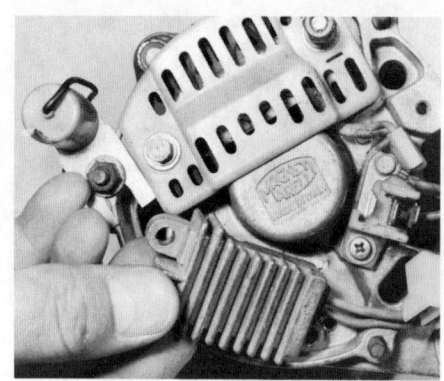

10.4b Removing the regulator

13.7 Withdrawing the starter motor (engine removed for clarity)

Fig. 10.3 Removing the Bosch voltage regulator (Secs 9 and 10)

1 Voltage regulator   2 Brushes

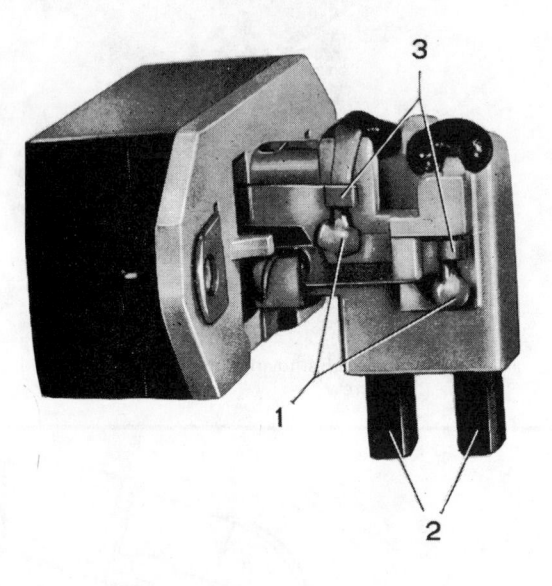

Fig. 10.4 Bosch voltage regulator and brush holder (Secs 9 and 10)

1 Brush lead soldered joints
2 Brushes
3 Lead clamps

motor/pinion shaft projects into the bellhousing near the periphery of the flywheel.

12 Starter motor circuit – testing

1 If the starter motor fails to turn the engine when the switch is operated, there are four possible reasons why:
   (a) The battery is no good
   (b) The electrical connections between switch, solenoid, battery and starter motor are somewhere failing to pass the necessary current from the battery through the starter to earth
   (c) The solenoid switch is no good
   (d) The starter motor is either jammed or electrically defective

2 To check the battery, switch on the headlights. If they go dim after a few seconds the battery is definitely suspect. If the lamps glow brightly, next operate the starter switch and see what happens to the lights. If they go dim then you know that power is reaching the starter motor but failing to turn it. Therefore, check that it is not jammed by placing the car in gear and rocking it to and fro. If it is not jammed the starter will have to come out for examination. If the starter should turn very slowly go on to the next check.

3 If, when the starter switch is operated, the lights stay bright, then the power is not reaching the starter. Check all connections from battery to solenoid switch and starter for perfect cleanliness and tightness. With a good battery installed this is the most usual cause of starter motor problems. Check that the earth link cable between the engine and frame is also intact and cleanly connected. This can sometimes be overlooked when the engine has been taken out.

13 Starter motor – removal and refitting

1 The starter motor is fitted to the front of the engine beneath the inlet manifold and is mounted onto the bellhousing. It is secured in place by three bolts through the rear of the bellhousing. Like most of the components in this tightly packed engine bay it is not that easy to reach.
2 Start by disconnecting the battery negative lead and by removing the radiator and fan assembly as described in Chapter 2. It may be found easier if the air cleaner is also removed as described in Chapter 3.
3 With the Saloon model the radiatr grille can be removed for easier access to the starter motor, as described in Chapter 12.
4 Disconnect the wiring from the starter motor solenoid and note which wires fit onto which terminals.
5 Unhook the clutch control arm return spring from its bracket, which is attached to the lower starter motor mounting bolt.
6 Undo and remove the three mounting bolts for the starter motor.
7 Withdraw the starter motor from the bellhousing (photo). Pull it out and slightly downwards to clear the other components and then manoeuvre it out of the engine bay through the space created by removing the radiator and fan.
8 If any difficulty is experienced in actually withdrawing the starter motor from the bellhousing on earlier models, because of the lack of

Chapter 10 Electrical system

space, it may be found necessary to remove the oil pressure warning switch and oil temperature gauge transmitter assemblies which are located differently on the oil filter mounting block. This will give you sufficient room to withdraw the starter motor.

9 Refit the starter motor in the reverse order to removal. Remember to tighten the various mounting bolts to the specified torques, and don't forget to reconnect the clutch control arm return spring, and refill the cooling system as described in Chapter 2.

14 Starter motor – dismantling, repair and reassembly

1 The starter motor assembly comprises three sub-assemblies; the motor itself, the solenoid switch and the pinion actuator housing. The actuator housing forms the mechanical link between the motor and the solenoid switch.

2 Such is the inherent reliability and strength of the starter motors fitted that it is very unlikely that a motor will ever need dismantling until it is totally worn out and in need of renewal as a whole.

3 The solenoid which is usually available individually as a spare is attached to the actuator housing by three nuts on three long bolts passing the length of the solenoid. Undo and remove these three end nuts and lift the solenoid from the starter motor assembly.

4 There is no possibility of repairing the solenoid and therefore if after reconnecting across the battery with two stout leads the unit remains lifeless or the switch part fails to work, the whole solenoid must be renewed.

5 Starter motor brushes: On the forward end of the motor there is a

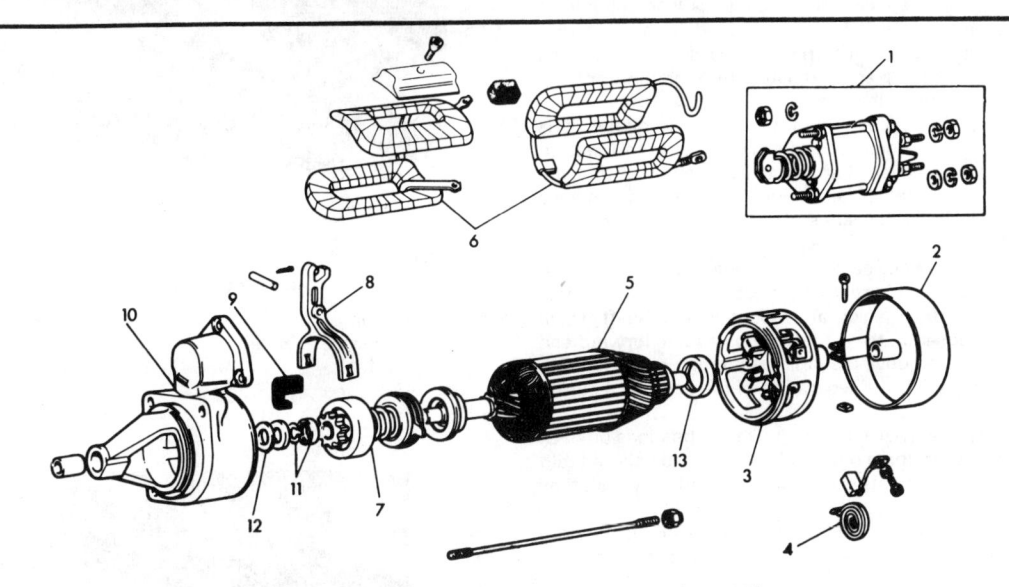

Fig. 10.5 Starter motor – exploded view (Sec 13)

1 Solenoid
2 Brush aperture cover strap
3 Brush and terminal cage
4 Spiral spring and brushes
5 Motor armature
6 Field windings
7 Pinion and pinion carriage
8 Pinion carriage actuating lever
9 Lever buffer, fitted to motor casing
10 Pinion actuator housing
11 Pinion carriage retaining ring and sleeve
12 Thrust washer
13 Thrust washer front end

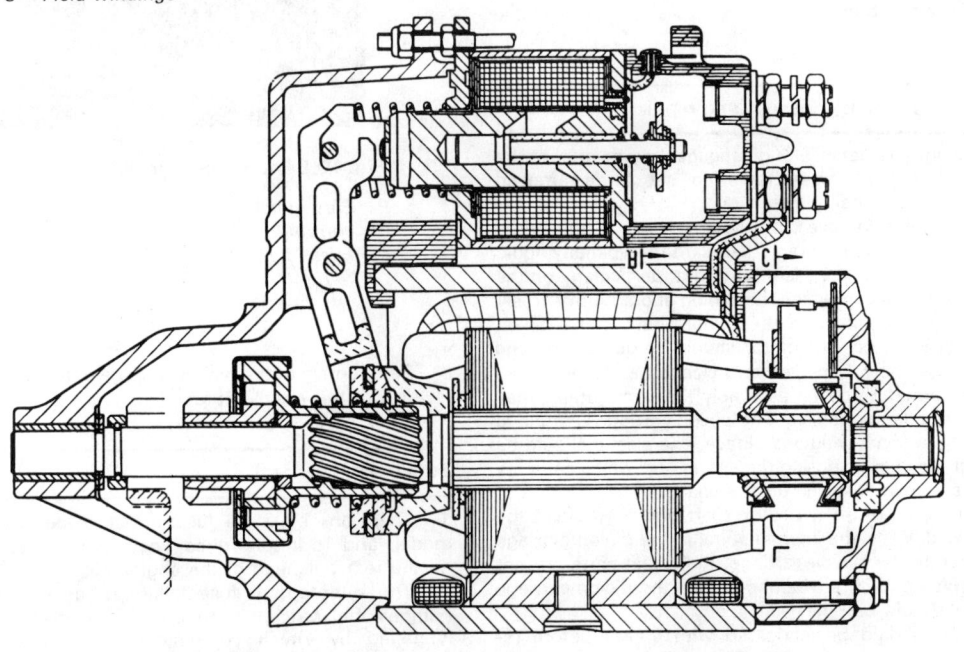

Fig. 10.6 Longitudinal section of the starter motor (Sec 13)

wide strap with a single screw to tighten it in position and it covers the aperture which allows access to the motor brushes.

6 The procedure for inspection and renewal of the brushes is straightforward. With the starter motor removed from the car and on a bench, slacken the single screw which clamps the end strap in postion and slip the strap along the motor casing to uncover the brush access apertures.

7 The brushes are retained in their mountings by spiral springs. Move the ends of the spiral springs to allow the carbon brushes to be extracted from their mountings. Undo the small screw which secures the small lead from the brush to its terminal on the forward end fitting and remove the brush from the motor assembly. If the brush was worn to the extent that the spiral spring applies little force, then the brush should be renewed.

8 Motor dismantling: having already removed the solenoid, the front end of the motor is the next unit to be separated from the motor assembly. The motor is held together by tie rods screwed into the actuator housing at the rear end and projecting through the front end fitting to accommodate nuts at the other.

9 Before proceeding to separate the motor sub-assemblies it is necessary to disconnect the electrical connections between them. In particular the forward end cover strap should be removed and the brushes removed. The electrical leads running from the field windings in the motor, to terminals in the forward end cover, should be detached from those terminals.

10 Once the nuts on the tie-rods have been removed, the forward end fitting, motor casing and the pinion actuator housing can be separated. The motor armature and drive pinion are mounted on a shaft which runs in bush bearings housed in the actuator housing and forward end fitting. It will be necessary to drive the pinion actuating pivot pin from the actuator housing to permit the separation of the actuator housing and motor armature assembly.

11 Be careful to retrieve the spacer shims and thrust bearings on each end of the motor shaft when the motor shaft is freed. You should refit the spacers and thrust components exactly in the positions which they occupied before dismantling.

12 The field windings are held to the inside of the motor main casing by special blocks which are in turn secured by screws passing through the casing.

13 The armature and pinion assembly will usually be separated from the motor components in order to gain access to the pinion assembly. Inspection and repair of the pinion assembly is described in Section 15.

14 Reassembly of the starter motor follows the reversal of the dismantling procedure. Fortunately there is little in the way of adjustments to make on the motor, the assemblies usually fit together to their correct relative positions.

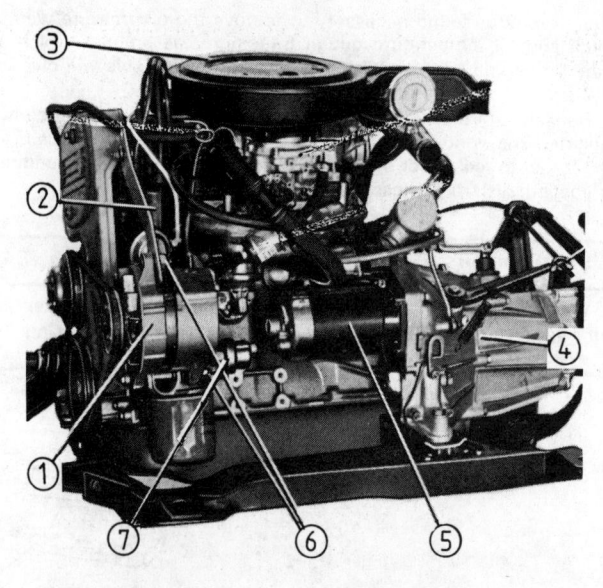

Fig. 10.7 View of starter motor and alternator locations with engine out of the car (Secs 13 and 7)

1 Alternator
2 Alternator adjuster bracket
3 Air cleaner
4 Clutch control arm return spring
5 Starter motor
6 Alternator mounting bolts
7 Oil transmitter

15 Starter motor drive pinion – inspect and repair

1 Persistent jamming or reluctance to disengage may mean that the starter pinion needs attention. The starter motor should be removed from the car first of all for general inspection.

2 With the starter motor removed, thoroughly clean all the grime and grease off with a petrol soaked rag. Take care to avoid any liquid running into the motor itself. If there is a lot of dirt, particularly on the pinion itself, this could be the trouble. The pinion should move freely along a spiral which is machined on the motor shaft. If the pinion motion is not smooth and easy, the motor should be dismantled and the armature/pinion assembly inspected and cleaned as follows.

3 Having removed armature/pinion assembly the commutator may be cleaned with a petrol dampened rag. The pinion carriage is retained on the motor shaft by a spring ring and sleeve. The sleeve should be driven off the end of the shaft exposing the spring ring which can now be slipped out of its groove seat and off the shaft.

4 Slide the pinion carriage off the rotor shaft and then clean the spiral which is exposed. Wipe the internal spiral in the pinion carriage clean. Do not dismantle the pinion carriage. Individual parts are not available and if the pinion teeth are damaged then the pinion carriage or the whole starter should be renewed.

5 The spiral splines should be lubricated with grease before reassembly of the carriage onto the shaft. The intermediate disc that forms the thrust bearing between the actuating lever ring and the pinion carriage sleeve, should also be lubricated with grease.

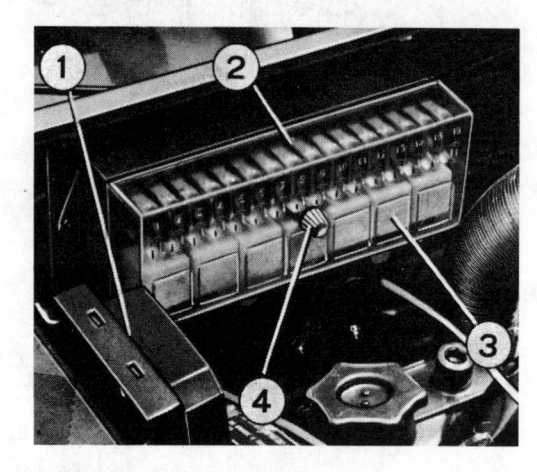

Fig. 10.8 Later type fuse box (with 16 fuses) (Sec 16)

1 Electronic ignition control unit
2 Fuses 1 to 16 from left to right
3 Solenoid switches
4 Cover retaining knob

6 Reassembly of the pinion carriage onto the motor shaft follows the reversal of the removal procedure.

16 Fuses – general

1 There are 11 or 14 fuses fitted in the electrical system in early models and 16 in later ones. They are mounted on a holder mounted on the rear bulkhead of the engine bay (Fig. 10.8 and photo).

2 The symptom of fuse failure is the simultaneous 'failure' of a number of electrical systems. The fuse which has broken can then be identified by which combination of electrical systems does not operate.

3 Some of the fuses are 8 amp, some 16 amp and some 25 amp. All are colour coded.

Chapter 10 Electrical system

16.1 Removing the fuse box cover on an earlier model (fuses are numbered from right to left)

17.1 Solenoid (relay) locations on early models

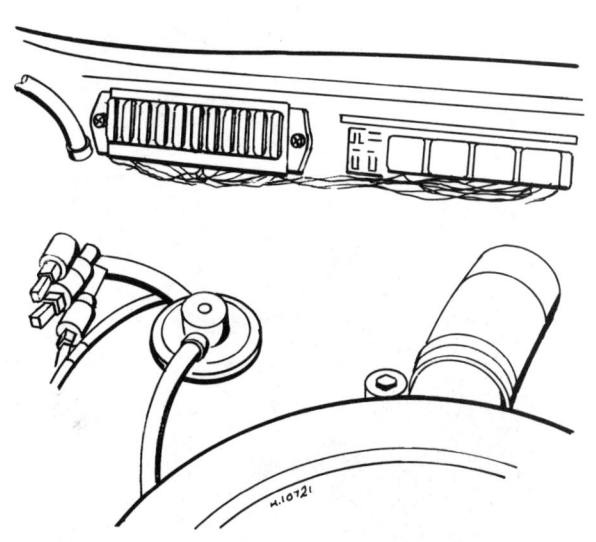

Fig. 10.9 Layout of fuses and solenoids on the USA models (Secs 16 and 17)

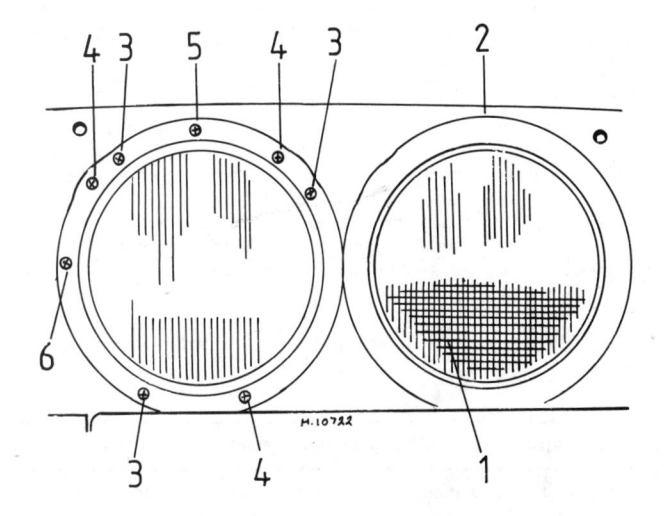

Fig. 10.10 Headlamp mounting layout (1400 Saloon models) (Sec 18)

1 Headlamp
2 Water gaurd
3 Lamp unit mounting screws
4 Headlamp-to-bodywork mounting bolts
5 Vertical aiming screw
6 Horizontal aiming screw

4 The Specifications section at the beginning of this Chapter details the identification of each fuse and the circuits protected by them. Ensure that you check the relevant table for your model, or refer to the car's handbook.
5 Never think you can leave fuses out or by-pass them, or substitute a fuse with a piece of tin foil or similar. A fuse blows for a reason and if the fault is not righted immediately, you will do serious damage to the wiring on the circuit involved and even adjacent wiring.
6 The plastic electric insulation material is also a heat insulation material and if excessive currents flow through the wires, they will soon heat up and melt the insulation.

17 Solenoids (relays) – general

1 Early models have three solenoids mounted on the rear bulkhead of the engine bay next to the fuse holder (photo). Later models, 1979 onwards, have 7 solenoids, as shown in Fig. 10.8. Models produced for the USA market have 4 solenoids mounted on the bulkhead as shown in Fig. 10.9.
2 They are easily removed by undoing the mounting nuts appropriate to the unit which needs renewing. Disconnect the wires running to the solenoid and it is free. Refitting is a reversal of the removal procedure.

18 Headlamp unit (Saloon model) – removal and refitting

1 Disconnect the battery negative cable.
2 Remove the knurled screws, from inside the engine bay, which hold the headlamps twim panel or rectangular lens cover in place.
3 The headlamp units are mounted individually in the front panel of the bodywork, and inner or outer headlamp can be removed separately (Fig. 10.10).
4 Disconnect the earth lead from the headlamp to the body.
5 Disconnect the snap connector in the power supply cable to the headlamp unit, leaving the bulb in place.
6 Remove the bolts which retain the headlamp to the body panel and lift the headlamp away. This is simple for the inner, or main beam lamp, but for the outer, or dipped beam units in models fitted with self-levelling headlamp systems, the self-levelling control tube has to be disconnected before the headlamp can be removed.
7 Refit the headlamp units in the reverse order, and check that they operate correctly.

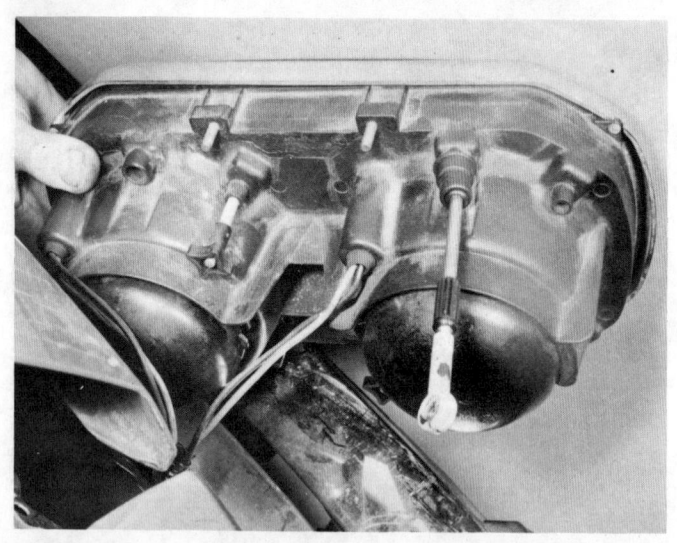

19.2 Headlamp unit removed (Coupe, Spider and HPE models)

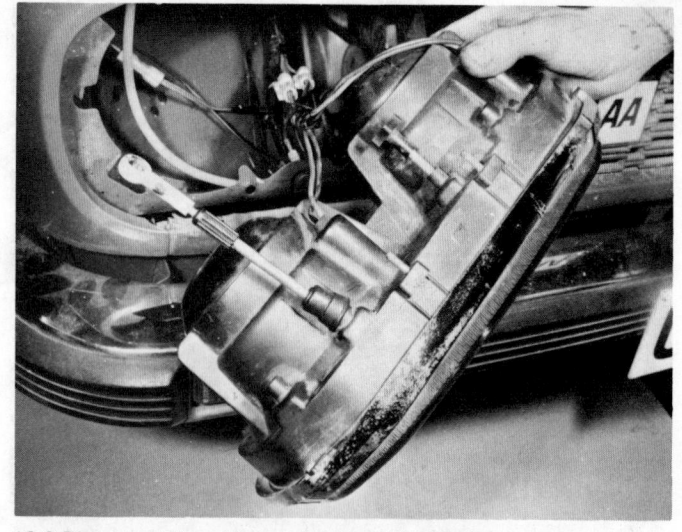

19.3 Disconnect the wiring at the rear of the headlamp unit and its free

19 Headlamp unit (Coupe, Spider and HPE) – removal and refitting

1　Disconnect the battery negative lead.
2　The headlamp unit comprises both inner and outer headlamps in one moulded assembly. This assembly has four bolts in it, two at the top and two at the bottom. These are located through holes in the front body panel and secured in place by nuts (photo).
3　Remove the four headlamp assembly retaining nuts. Disconnect the self-levelling system linkage and remove the connector covers. Disconnect the power supply cables to the headlamps. Disconnect also the earth wire spade connector. The headlamp assembly is now free to be removed (photo).
4　Refit the headlamp assembly in the reverse order, and check that they operate correctly.

20 Headlamp self-levelling system – general

1　The outer, or dipped, units of the headlamp system are automatically set to the correct level according to load by a simple hydraulic system. The control units are connected to sensors linked to the left-hand side front transverse link of the rear suspension, as described in Chapter 11.
2　As the suspension moves up and down under load the sensors alter the fluid pressure and this is transmitted to the two control units mounted one on either wing (photo). These units move back and forth in a horizontal plane thus altering the beam height by pushing the adjustable link rod forwards or, if the load is removed, pulling it back.
3　The pipes which connect the sensors to the headlamp units and each other must never be removed, or the fluid will leak out and the pressure will be altered, thus ruining the self-levelling system alignment.
4　The adjustable links which connect the headlamp units to the control units are easily removed from the control unit ball arms by carefully levering them off.
5　If any major problems occur with the self-levelling headlamp system then it must be repaired by your local Lancia dealer.

21 Headlamp – alignment

1　It is strongly recommended that the adjustment of the headlamps is only carried out by a Lancia dealer or garage that has special beam setting equipment. This should always be carried out if a new headlamp unit or bulb has been fitted.
2　However, if for some reason the headlamp beam has been altered dramatically then it can be roughly aligned using the large plastic knobs on the rear of the headlamp unit (Figs. 10.11 and 10.12).

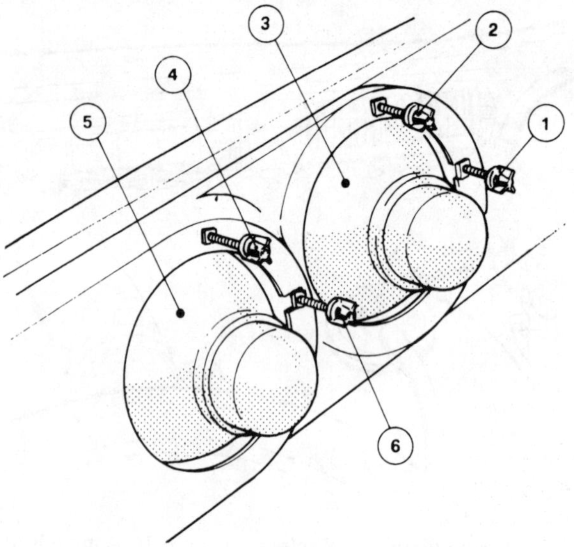

Fig. 10.11 Headlamp aiming knobs (models without self-levelling system) (Sec 21)

1　Main beam horizontal aiming knob
2　Main beam vertical aiming knob
3　Left-hand main beam headlamp
4　Dipped beam vertical aiming knob
5　Left-hand dipped beam headlamp
6　Dipped beam horizontal aiming knob

3　On models fitted with the self-levelling system, if the beam appears too high or too low then it can be altered using the plastic screw adjuster in the centre of the control link.
4　Any rough adjustment that is made shoud only be considered as a temporary measure. The car must be taken to a garage and have the headlights set properly as soon as possible.

22 Headlamp bulbs – renewal

1　Disconnect the battery negative cables.
2　Remove the rear cap of the appropriate headlamp unit (photo).
3　Disconnect the spade connectors from the rear terminals of the bulb.
4　Squeeze the retaining clip and withdraw the bulb (photo). Do not touch the glass part if the bulb is to be re-used.

20.2 Headlamp self-levelling unit (HPE model shown)

22.2 Removing the right-hand main beam headlamp rear cap (HPE model shown)

22.4 Removing the bulb

23.1 Removing the front parking and flasher lens

23.2 The lamp holder is held into the front bumper from the rear

24.6 The wiring for the repeater lamp comes through the inner wing

Chapter 10 Electrical system

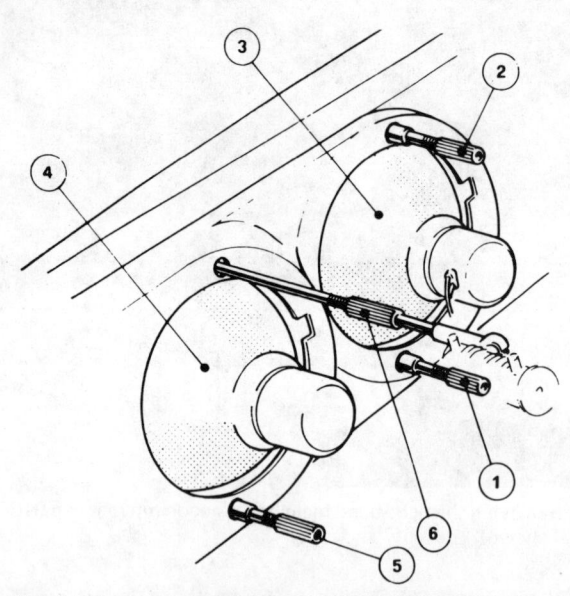

Fig. 10.12 Headlamp aiming knobs (1600, 1800 and 2000 Saloon models) (Sec 21)

1 Main beam horizontal aiming knob
2 Main beam vertical aiming knob
3 Left-hand main beam headlamp
4 Left-hand dipped beam headlamp
5 Dipped beam horizontal aiming knob
6 Dipped beam vertical aiming knob on automatic adjuster control link

5 Fit the bulb ensuring that you do not touch the glass part. If it is accidentally touched then it must be cleaned using methylated spirit and dried with a lint-free cloth. The bulb can only fit one way.
6 Refit the retaining spring clip and reconnect the cables, then refit the rear cap.
7 Standard British sealed beam headlamp units may be available as a spare for your car.

23 Front parking and flasher lamps – removal, refitting and bulb renewal

Bulb renewal
1 Undo the two retaining screws and lift away the dual lens cover. Either bulb can then be removed. Fit a bulb of the appropriate wattage with a bayonet fitting (photo).

Removing and refitting the lamp holder
2 The lamp holder or body is secured to the bumper from behind by a nut and bolt and a bolt to a bracket. Remove these and the lamp holder can be withdrawn rearwards (photo).
3 Disconnect the electrical connectors and the lamp holder can be removed from the car.
4 Refit the unit in the reverse order. Check before refitting the lens that the rubber seal is in good order and correctly fitted.

24 Side repeater lamps – removal, refitting and bulb renewal

Bulb renewal
1 Turn the steering wheel to full lock in order to reach up inside the wing to pull back the rubber boot over the repeater lamp bulb holder (Fig. 10.13).
2 Pull the bulb holder out of the socket and then remove the old bulb.
3 Fit a bulb of the correct type and wattage and refit the rubber boot, making sure that the boot fits securely.

Removing and refitting the lamp assembly
4 Disconnect the battery negative cable.

5 Lever the flasher repeater light assembly out of the wing using a flat bladed screwdriver. Take great care not to damage the paintwork.
6 Remove the plastic connector cover from the power supply wire and disconnect the cables. Also disconnect the earth spade connector from the earth lead point below the headlamps (photo).
7 Withdraw the grommet from the inner wing and remove the whole assembly.
8 Refit in the reverse order to removal. To secure the repeater lamp in the wing push it firmly home. Ensure before doing so that the lamp base is clean and that the fitting recess in the wing has been cleaned off.

25 Rear lamp assembly – removal, refitting and bulb renewal

Bulb removal (Saloon and HPE models)
1 Unscrew and remove the rear lamp lens retaining screws. The Saloon has four and the HPE three (photo). It is easier if the boot lid or tailgate is opened first.
2 Lift away the lens and identify and remove the bulb to be renewed (photo).
3 Fit a new bulb of the same wattage (photo).
4 Check that the lamp holder rubber seal is in good condition and refit the lens and the retaining screws.

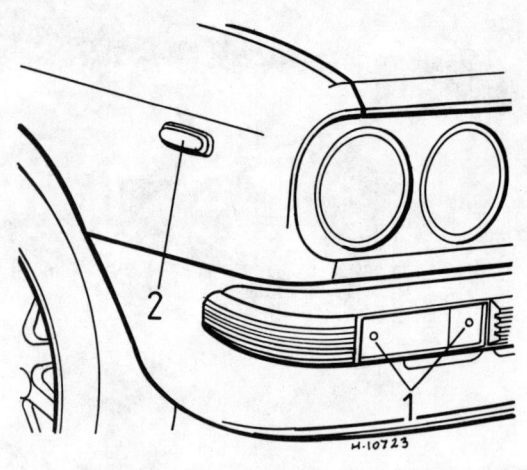

Fig. 10.13 Side flasher and repeater lamp layout (Coupe, Spider and HPE models) (Secs 23 and 24)

1 Side light and flasher lens retaining screws
2 Repeater lamp

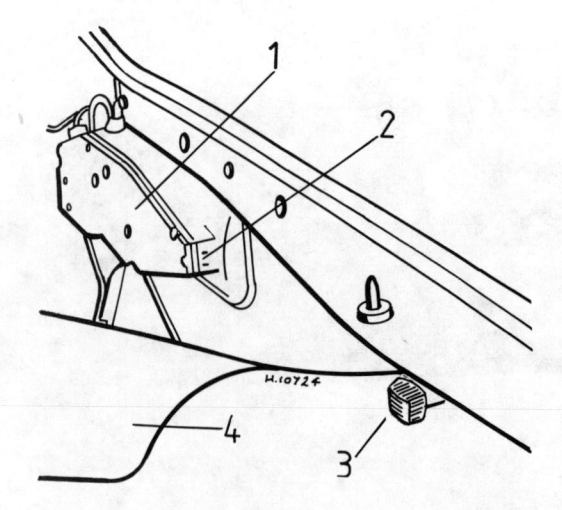

Fig. 10.14 Rear lamp assembly (Coupe and Spider) (Sec 25)

1 Lamp cover
2 Tab
3 Boot lamp
4 Boot carpet

Chapter 10 Electrical system

Bulb removal (Coupe and Spider models)
5 Open the boot and release the studs to remove the boot lining on the appropriate side to get at the lamp holder (Fig. 10.14). The bulbs are changed from inside the boot.
6 Lift up the catch on the outside of the lamp assembly and lift off the cover.
7 Identify and change the appropriate bulb. Ensure that the new bulb is of the same fitting and wattage as the original.
8 Refit the cover; the inner end slots in first and the outer end is held by the catch, which has to be lifted up so that the outer end can seat properly.

Removing and refitting the lamp assembly
9 Disconnect the battery negative lead.
10 Remove the luggage compartment lining on the left or right side, or the spare wheel on the right (Saloon and HPE models) to reach the rear lamp assembly (photo).
11 Disconnect the electrical snap connector or plug-in type, depending on model.
12 Remove the lamp to body mounting nuts. There are four on each lamp, and then remove the lamp assembly from the car. Take care that the seal is not damaged.
13 With the Saloon, HPE and early Spider models the lamp assembly is withdrawn from the outside of the car (ie the lamp holder is held into the rear body panel). With the Coupe and later Spider models the lamp holder is mounted onto the rear panel from inside the boot.

26 Luggage compartment light – removal, refitting and bulb renewal

Bulb renewal (Saloon, Coupe and Spider models)
1 Slide the lamp to the right to release it from the guides and the bulb can then be removed.
2 Fit a new bulb of the same fitting and wattage and fit the lamp in the guides.

Bulb renewal (HPE models)
3 The HPE model is fitted with a lamp, similar to the interior courtesy lamp, over the luggage area (photo). Refer to Section 27 for the removal, refitting and bulb renewal procedure.

Removing and refitting the lamp assembly
4 Slide the lamp out of its guides, disconnect the wiring plugs and remove it.
5 Refit it in the reverse order to removal.

27 Interior courtesy lamp – removal, refitting and bulb renewal

Bulb renewal
1 Squeeze the two long sides of the lamp lens together if the lamp is of the type with a protruding lens cover, or if it is a flush fitting lens cover then prise it out using a small electrical screwdriver (photo).
2 With the lens removed the festoon bulb can be removed.
3 Refitting is the reverse procedure. The flush fitting type of lens cover needs only to be pressed home.

Removing and refitting the lamp assembly
4 Disconnect the battery negative lead.
5 Remove the lens cover and bulb.
6 Disconnect the two wires to the lamp holder. On some models the wires to the lamp holder are connected from the rear, and cannot be disconnected until the lamp holder has been released from the roof panel. The HPE rear lamp over the luggage area has only one wire.
7 Unscrew and remove the two crosshead screws and remove the lamp holder assembly.
8 Refitting is a reversal of the removal procedure.

28 Under bonnet lamp – removal, refitting and bulb renewal

Bulb renewal
1 Open the bonnet and support it.

25.1 Removing the rear lamp lens retaining screws

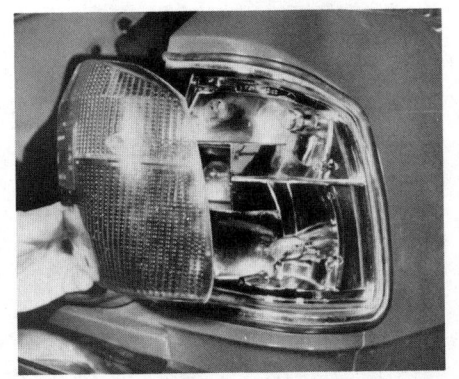

25.2 Removing the lens

25.3 Fitting a new tail lamp bulb

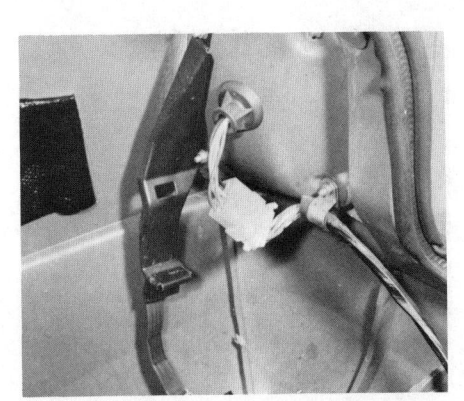

25.10 The rear lamp assembly is fitted like this on the right-hand side

26.3 The HPE has a rear courtesy lamp over the luggage area

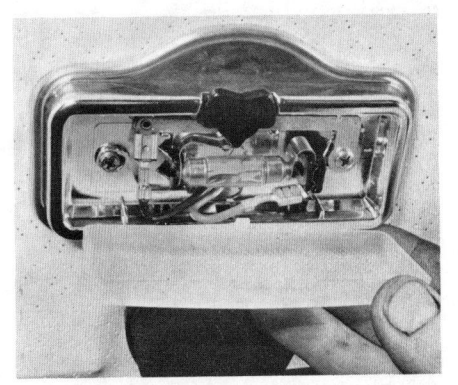

27.1 Removing the courtesy lamp lens to change a bulb

2 Push in and twist to remove the bulb. Fit a new one of the same type and wattage.

Removing and refitting the lamp assembly
3 The under bonnet lamp is mounted to the rear bulkhead of the engine compartment by a single mounting. Undo the nut and disconnect the electrical plug and remove the unit (photo).
4 Refitting is the reverse of the removal procedure.

29 Number plate lamps – removal, fitting and bulb renewal

Bulb renewal
1 All models have two rear number plate lamps. On the Saloon models they are fitted in the boot lid itself, and on the Coupe, Spider and HPE they are fitted into the upper surface of the bumpers.
2 To remove a bulb and socket simple pull it out of the lamp holder (photo).
3 Remove the bulb from the socket and fit a new one of the same wattage and fitting.
4 Push the new bulb and socket firmly home into the holder.

Removing and refitting the lamp assemblies
5 Remove the bulb and socket.
6 Squeeze the retaining lugs inwards to release the lamp holder and lens unit from the boot lid or bumper.
7 Refitting is simply a matter of pressing the lamp holder into the top edge of the bumper or the boot lid opening. Make sure that the seal is in good condition first.

30 Glovebox light – removal, refitting and bulb renewal

Bulb renewal
1 Pull the bulb and socket rearwards out of its fitting; the bulb can now be renewed. Use a bulb of the same wattage and fitting.

2 Push the bulb and socket back into the fitting (photo).

Removing and refitting the lamp assembly
3 Undo the retaining screw next to the switch button.
4 Disconnect the connector from the rear of the bulb socket.
5 Pull the assembly forward through the mounting bracket and disconnect the black wire from the switch.
6 Refitting is a reversal of the removal procedure.

31 Auxiliary switches, illumination and warning lights (Coupe, Spider and HPE models up to 1979)

1 In the Coupe, HPE and Spider models the auxiliary switches are all mounted in a wood veneer panel in the centre of the dashboard. This is held in place by four knurled screws in the corners. Remove these and the panel can be carefully eased forward but disconnect the battery negative lead first of all (photo).
2 To renew a bulb in the cigar lighter or hazard flasher, heated rear window switch or rear fog warning lamp (as fitted) is a simple operation. Pull the bulb and socket out of the rear of the switch. Refit the bulb and socket in the reverse procedure to removal.
3 To remove a switch first remove the centre panel. Remove the switch knob by unscrewing it. Remove the knurled ring on the outside of the panel and clear plastic or metal collar behind it. Withdraw the switch complete and disconnect the wiring, noting which wire fits where.
4 Refit the switch to the panel and secure it with the knurled collar. Refit the wiring and the knob. Refit the panel.

32 Auxiliary switches and indicator lamps (Coupe, Spider and HPE models, 1979 onwards) – removal and refitting

1 In later models the old type of centre panel was discontinued, but the new auxiliary switches, of the rocker type, are still fitted in the centre of the dashboard, although the indicator or warning lamps are

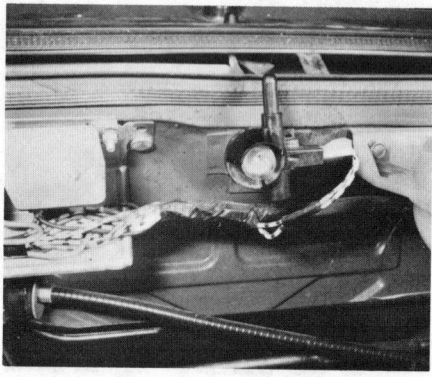

28.3 The under-bonnet light has an integral switch

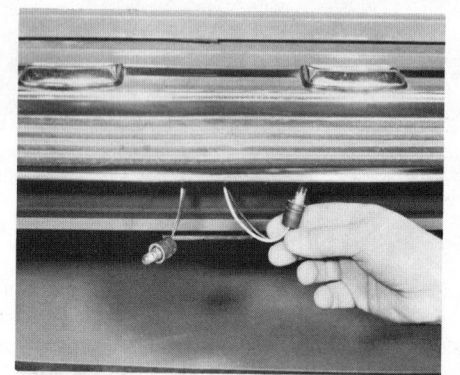

29.2 Removing the second rear number plate lamp

30.2 The integral light and switch for the glovebox

31.1 Easing the auxiliary switch panel forward to reach the switches and bulbs

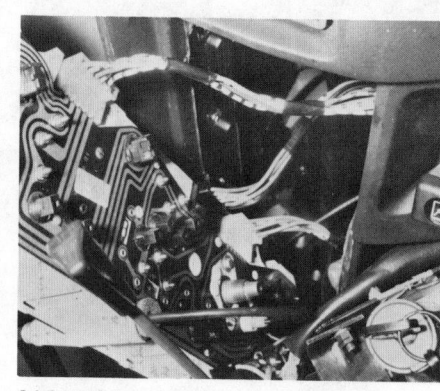

34.2 With the panel pulled out the bulbs can be reached

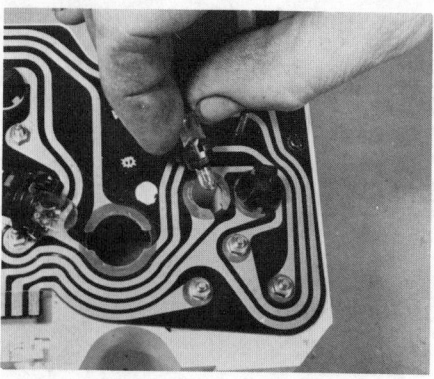

34.3 Twist and pull out the bulb and holder

Chapter 10 Electrical system

refer to Section 34, which covers the instrument illumination and panel warning lights.
4 To change any of the rocker switches, prise the appropriate switch out of the dashboard panel, disconnect the wiring from the rear and fit it to the new switch. Insert the switch in the panel and push it home.
5 To renew the cigar lighter, remove the knob, undo the securing collar and withdraw the unit through the panel. Disconnect the wiring and fit it to a new unit. Fit the new cigar lighter in the reverse sequence to removal.

33 Auxiliary switches and indicator lamps (Saloon models) – removal and refitting

1 On the Saloon models the auxiliary switches for the heated rear window, rear fog warning lights and hazard flasher are mounted on an auxiliary panel under the dashboard to the right of the steering wheel.
2 The indicator lamps for these switches are incorporated in the instrument panel layout. To change a bulb in one of these indicator lamps, refer to Section 34, which deals with instrument illumination and panel warning lights.

Switch removal and refitting
3 Identify the switch that needs replacing and remove the knob. Undo the knurled ring which secures it to the auxiliary switch panel and withdraw it from the rear.
4 Disconnect the wiring to the switch and note which wire fits where. Alternatively take the new switch and refit each wire in turn as it is removed so that there can be no mistake.
5 Fit the new switch to the panel and secure it in place with the knurled collar. Then refit the knob.

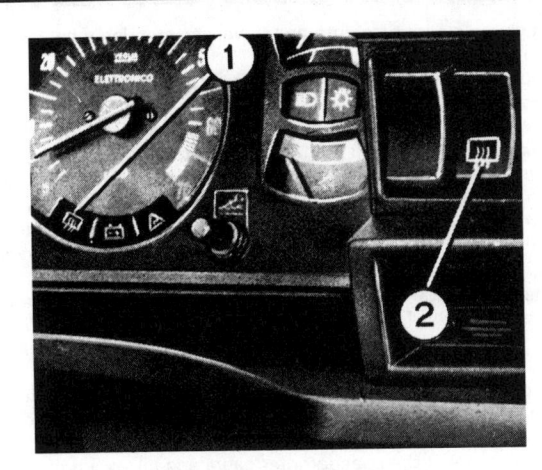

Fig. 10.15 The heated rear window switch and warning lamp (later models) Sec 32)

1 Warning or indicator lamp
2 Heated rear window switch

now incorporated in the instrument panel itself (Fig. 10.15).
2 The heated rear window and hazard flasher switches are mounted in the upper centre section of the dashboard on either side of the heater controls, while the fog lamps switches and cigar lighter are mounted in the lower panel in the centre console forward of the gear lever (Fig. 10.16).
3 To change an indicator or warning bulb for either of the switches

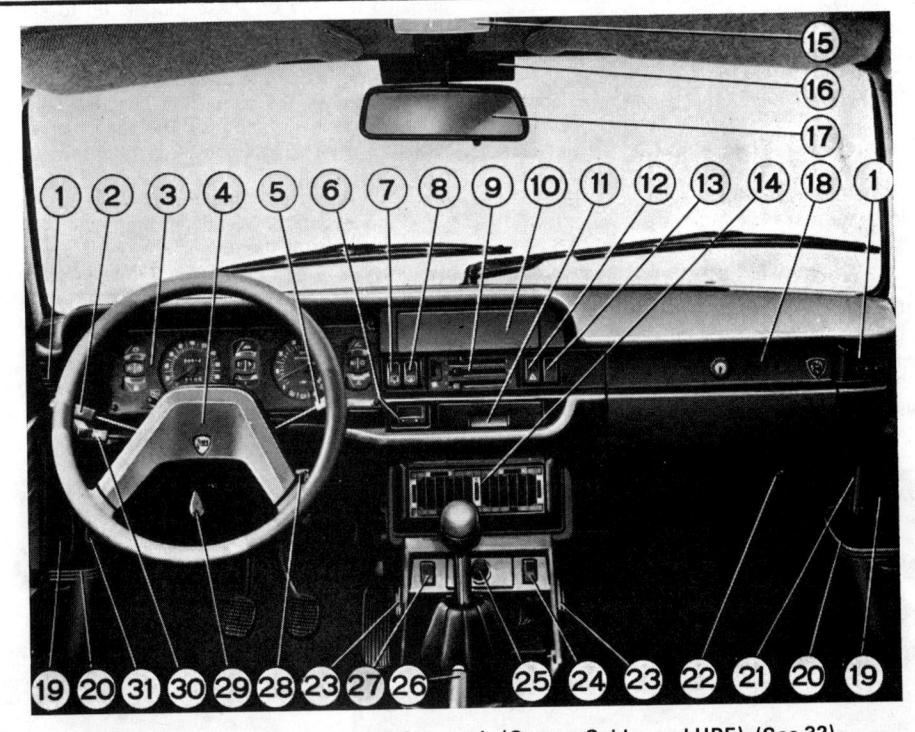

Fig. 10.16 Layout of later model controls (Coupe, Spider and HPE) (Sec 32)

1 Heating and ventilation directional outlets
2 Outside lights switch
3 Instruments panel
4 Horn switch
5 Windscreen wiper and control
6 Digital clock
7 Extra switch
8 Rear heated window switch
9 Ventilation and heating controls
10 Radio blank
11 Ash-tray
12 Hazard signalling switch
13 Spare switch
14 Heating and ventilation outlets and controls
15 Inside light
16 Licence and insurance tag holder (Italian market only)
17 Rear view mirror
18 Glove box
19 Speakers
20 Bins
21 Bonnet emergency release lever
22 Plug-in socket
23 Front seat footwell outlets
24 Rear fog lamp (if fitted) switch
25 Cigarette lighter
26 Handbrake lever
27 Fog lamps (if fitted) switch
28 Ignition key switch
29 Steering wheel position setting lever
30 Turn indicators control
31 Bonnet release lever

34 Instrument panel warning indicator lights – bulb renewal

1 Remove the instrument panel as described in Section 35 or 36 but it is not absolutely necessary to withdraw it completely or disconnect the multi-plug connectors. Most of the bulbs can be reached quite easily (photo).
3 Having identified the bulb concerned, twist and remove the bulb holder and bulb (photo). Fit a new bulb of the correct type and wattage. As can be seen from the photo, there are different sizes.
4 All these small panel light bulbs are of the pull-out and push-in variety. The bigger bulbs are for the instrument panel illumination, and the smaller ones are for the warning and indicator lamps.
5 Refitting is a reversal of removal.

35 Instrument panel (Saloon models) – removal and refitting

1 Disconnect the battery negative cable.
2 Disconnect the speedometer drive cable from the gearbox/differential housing.
3 On early models unscrew the four screws which retain the instrument panel in position; these project from the panel itself. On later models 2 screws in the bottom corners hold the cowling in position and another two behind hold the panel in place.
4 Pull the panel away from the dashboard so that the multi-plug connectors can be detached from the instrument panel. There are 3 plugs and it is very important to note which fits where. Fully lower the steering wheel where necessary.
5 Disconnect the speedometer cable from the back of the speedometer and the oil level pressure pipe from the rear of the panel (if it is fitted).
6 The instrument panel, which is of the printed circuit type, can then be removed carefully.
7 Refitting the panel is the reverse procedure to removal. Make sure that the multi-plug connectors are fitted to the correct terminal positions. They can be interchanged without realising it as the terminals and plugs are in some cases similar.

36 Instrument panel (Coupe, Spider and HPE) – removal and refitting

1 Disconnect the battery negative cable.
2 Disconnect the speedometer cable at the gearbox/differential housing. Remove the dipstick.
3 Unlock the steering column and lower the steering wheel as far as it will go.
4 Undo the 4 retaining screws and lift off the steering column upper and lower casings (photo).
5 Undo the knurled rings from the three knobs in the lower section of the instrument panel and then unscrew and remove the 2 screws in the top front of the instrument panel. These retain the cowling, which can then be removed (photo). In later models (from 1979 onwards) there are four screws (one in each corner), which have to be removed first, as the instrument panel is a single assembly. There is no separate cowling.
6 Beneath the cowling (where fitted) there are two screws in the bottom edge of the instrument panel which hold the panel itself in position. Remove these and ease the panel away from the dashboard (photo).
7 Disconnect the multi-plug connectors from the instrument panel and note which fits onto which terminal. If in doubt refer to the wiring diagram in your handbook.
8 Also disconnect the speedometer cable and oil level pressure pipe from the rear of the instrument panel.
9 The instrument panel, which is of the printed circuit type, can now be manoeuvred out of its housing and can be placed to one side.
10 Refitting the instrument panel is the reverse process to removal. Remember to ensure that the multi-plug connectors are fitted to the correct terminals. Do not forget to reconnect the speedometer cable to the gearbox housing.

37 Auxiliary switch panel (Coupe, Spider and HPE) – removal and refitting

1 The centrally mounted auxiliary switch panel with a wood veneer finished is fitted on these models up to early 1979 when the facia design was altered.
2 Depending on the model and year there may be two, four or five switches fitted to the panel.
3 To remove the panel first unscrew and remove the four knurled screws (one in each corner). The panel can then be eased forward. This may take a bit of fiddling as it is a tight fit when five units/switches are fitted.
4 With the panel pulled forward it is easy to change any one switch by disconnecting the leads and removing the knob and knurled collar from the front of the panel.
5 When the panel is removed the clock is left behind as can be seen in the photo. To remove the clock first remove the ashtray and the speaker and speaker grille in the top of the dashboard. Undo the 2 nuts which hold the clock onto the bracket which runs behind it. With the nuts removed the clock can be pushed out into the car from behind. The switch panel has to be removed to carry out this operation.
6 Refitting both the clock and switch panel is the reverse procedure to removal.

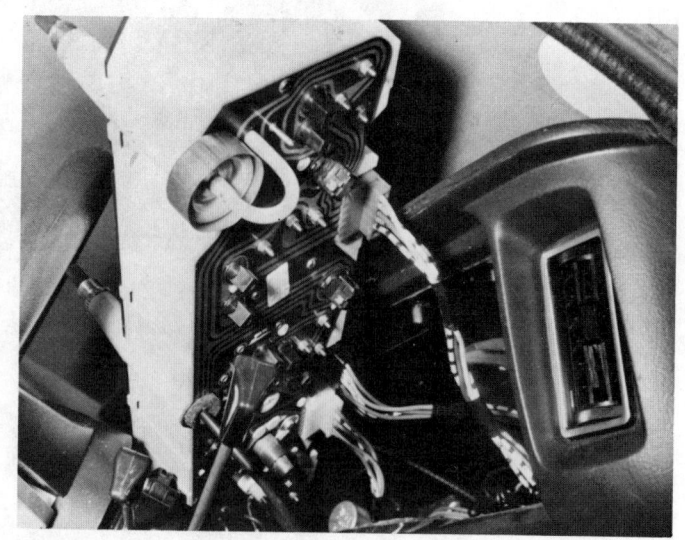

36.4 Removing the steering column lower casing

36.6 Removing the instrument panel

38.2 Remove the front panel from the main assembly

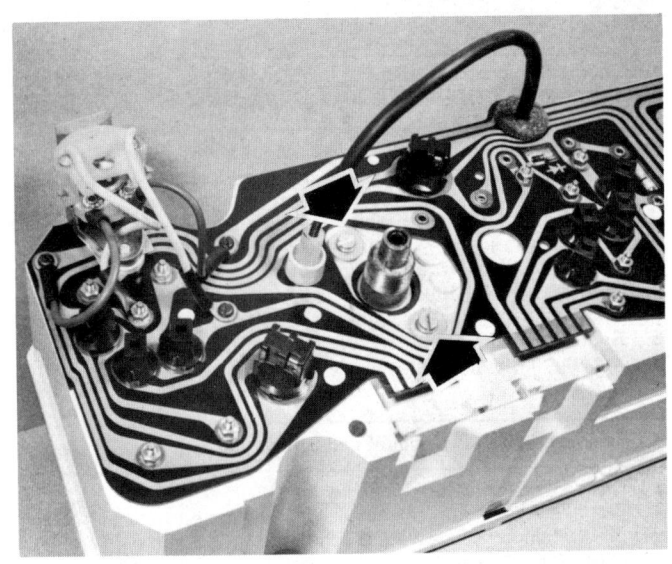

38.3 The instrument mounting nuts and screws can be easily identified (speedometer screws arrowed)

39.3 The ignition switch with casings removed

40.1 The lighting, direction indicators and wiper/washer control switch assembly is a one piece unit

41.2 The front left-hand door courtesy light switch

42.1 The brake stop light switch is fitted to the pedal box

38 Instruments – removal and refitting

1 Remove the instrument panel as described in Sections 35 and 36.
2 Working on the bench remove the panel glass sheet and front black panel (photo).
3 Identify the instrument which needs to be removed. The small ones are held in place with nuts and the large instruments (speedometer and tachometer) are retained by screws (photo).
4 With the screws removed or nuts undone the instrument can then be withdrawn from the panel, for further stripping or renewal.
5 Refit the instrument in the reverse order to removal and refit the panel to the dashboard as described in Sections 35 to 36.

39 Ignition switch – removal and refitting

1 Disconnect the battery negative lead.
2 Remove the steering column casings; there are two, split horizontally and retained by screws through the lower section.
3 It is easier to remove the upper casing with the steering column fully lowered (photo).
4 Undo the cable ties and trace the wires from the ignition switch back as far as the multi-plug connectors. Disconnect them.
5 Undo and remove the bolt which secures the ignition switch to the anti-theft locking device. Insert the key into the ignition lock, turn it to the GAR position and remove it. The complete ignition switch can be removed by depressing the locking spring pin with a pointed instrument and sliding it out.
6 Refitting is the reverse procedure to removal.

40 Lighting, direction indicators and wiper/washer control switch assembly – removal and refitting

1 The lights, direction indicators and wiper/washer controls are all mounted on the steering column. The assembly is a one piece unit and if one section goes wrong, then the whole assembly has to be renewed (photo).
2 Disconnect the battery negative lead.
3 Remove the steering column upper and lower casings. They are retained by screws underneath.
4 Remove the steering wheel as described in Chapter 11.
5 Undo the cable ties which keep the harness in place.
6 Trace the various harnesses back to their multi-plug connectors and separate them.
7 Undo the control switch assembly retaining screws and lift the whole switch combination off the steering column.
8 Fit a new switch assembly to the steering column in the reverse order to removal. Before refitting the steering wheel check that all the controls function correctly. The indicator cancellation mechanism cannot of course be checked until the steering wheel has been refitted, and provided that the 'straight ahead' position has been maintained then there should be no problem.

41 Courtesy light switches – removal and refitting

1 The courtesy lights are operated when any door is opened, or in the case of the HPE when the tailgate is opened.
2 The courtesy light switches are mounted in the front door pillars of each door (photo).
3 To remove one of these switches first remove the trim panel covering the inside of the pillar. Unscrew the retaining collar, remove the washer and retrieve the switch from inside the pillar.
4 These switches tend to suffer a great deal as they grow old. The commonest problem is that they get clogged up and won't release, which means that the light will not come on. By a little judicious cleaning and gentle lubrication using a product such as WD 40 spray the switch can be resuscitated.
5 Fit the switch in the reverse order and then check it for operation.

42 Brake stoplight switch – removal and refitting

1 The switch is located on the pedal carrier to the rear of the brake pedal (photo).
2 Disconnect the battery negative cable before commencing work.
3 Undo the retaining nut on the front of the mounting plate, remove the washer and withdraw the switch.
4 Disconnect the two spade connectors.
5 Refit the switch in the reverse sequence to removal. Reconnect the battery.
6 Set the switch, using the nut at the rear of the bracket to adjust it, so that the lights come on when the pedal pad has moved 12 mm (0.47 in) forward from its rest position, or if measured at the switch, the pedal arm should move only 2 to 3 mm (0.079 to 0.118 in) before the lights come on. When it is correctly set tighten the nuts.

43 Reversing light switch – removal and refitting

1 The reversing light switch is mounted on top of the gearbox beneath the clutch control cable bracket (photo).
2 To remove it first disconnect and remove the battery and battery tray and the large windscreen washer reservoir in later models. Remove the bolts which secure the clutch cable bracket to the gearbox. Release the clutch control arm. Lift the assembly to one side.

43.1 The reversing light switch is really well hidden

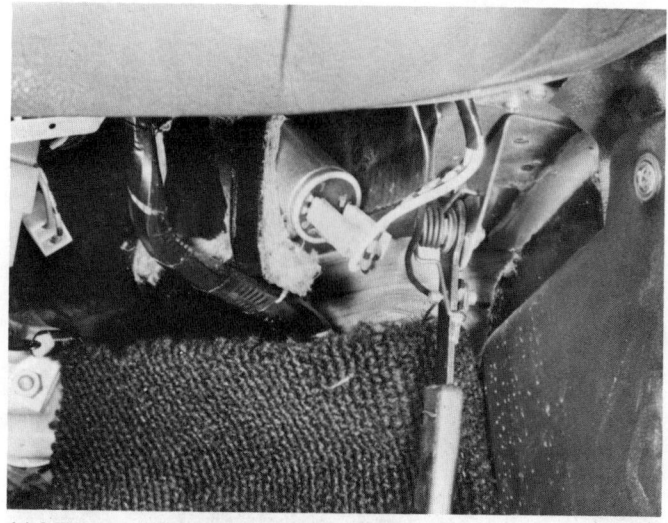

44.2 The round flasher unit is for the direction indicators

Chapter 10 Electrical system

3 Disconnect the cables from the switch and then remove the switch from the casing.
4 Refit the switch in the reverse order to removal. Do not forget to check the clutch cable adjustment when the cable and bracket have been refitted. This is covered in Chapter 5.

44 Flasher units – removal and refitting

1 There are two flasher units fitted to the Lancia Beta Series. Both are located under the dashboard, on the drivers side. Disconnect the battery negative cable.
2 The small round one is the flasher unit for the indicators (photo). Pull it out of its position and remove the connectors.
3 The hazard flasher unit is a square metal cased unit mounted on the outside of the pedal box. This can clearly be seen in photo 42.1.
4 Undo the two mounting nuts and remove the unit. Disconnect the five leads to it.
5 Refitting is the reverse procedure to removal.

45 Sender units – removal and refitting

Temperature gauge sender unit
1 Mounted in the top of the cylinder between the sparking plugs are two sender units. The one nearest the front of the engine is the temperature gauge transmitter (photo).
2 Disconnect the lead to it having first pulled back the protective rubber boot.
3 Using a socket remove the transmitter.
4 Immediately fit the new unit and screw it in. Tighten it to the specified torque, refit the lead and slide the boot over it. Check and if necessary top up the cooling system (see Chapter 2).

Engine overheating warning light switch
5 The other sender unit in the cylinder head, between Nos 3 and 2 cylinders, is the switch for the engine overheating warning light.
6 The procedure for removal and refitting is exactly the same as for the temperature gauge transmitter.

Oil pressure gauge transmitter and low oil level warning light switch
7 The oil pressure gauge transmitter is mounted onto the rear of the oil filter mounting block. On early models it is mounted horizontally and on later models it is mounted vertically on a right angled banjo union with the oil level warning light switch (photo).
8 Remove the wiring connector and apply a spanner to the mounting nut on the base of the transmitter. To reach the unit will involve the removal of the air cleaner assembly. On Saloon models, where the grille can be removed, the task is much simpler.

45.1 There are two sender units mounted in the cylinder head

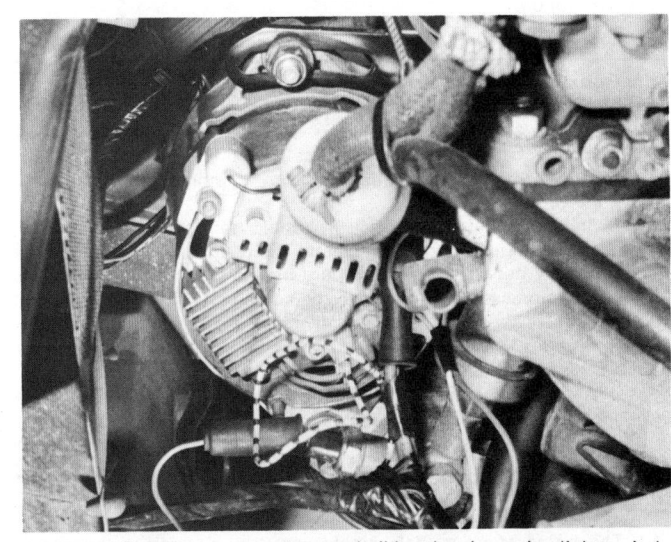

45.7 The oil pressure transmitter and oil low level warning light switch are mounted behind the alternator on the oil filter block

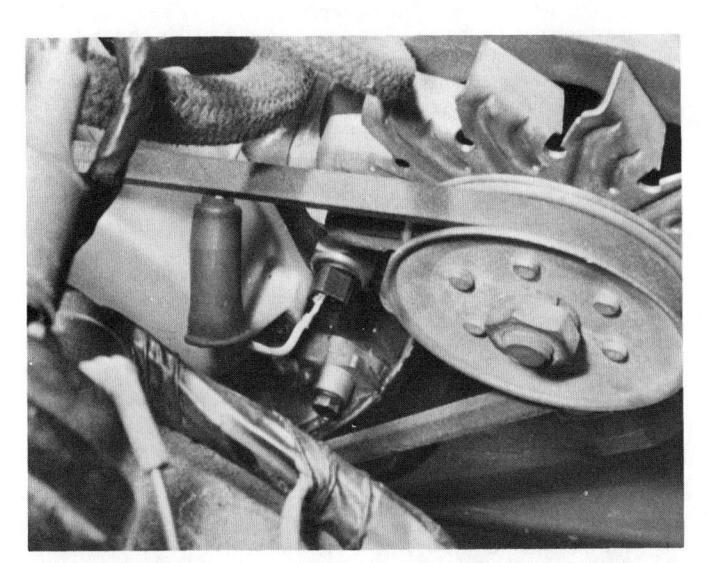

45.11 The oil temperature switch is mounted in the front of the oil filter mounting block

47.2 The horns and air compressor are mounted on a bracket behind the left-hand headlamps

9 To remove the low oil level switch, where it is fitted, pull back the rubber boot and disconnect the lead to the switch. Apply a spanner and remove the switch.
10 Refitting is the reverse procedure to removal but ensure all the units are tightened to the correct torques.

Oil temperature switch
11 The oil temperature switch is mounted in the front end of the oil filter mounting block (photo).
12 Disconnect the lead to it, having pulled back the rubber boot and remove the transmitter.
13 Fit a new switch in the reverse order to removal.

Fuel tank gauge transmitter
14 The removal and refitting procedure for the fuel tank transmitter unit is fully covered in Chapter 3.

46 Horns (Saloon models) – removal and refitting

1 Disconnect the battery negative cable.
2 Disconnect the electric leads from the horns.
3 Undo the mounting nuts and bolts and remove first the upper horn and then the lower horn.
4 Refit the horns in the reverse order to removal and then check their operation.

47 Horns and compressor (Coupe, Spider and HPE models) – removal and refitting

1 Disconnect the battery negative cable.
2 Disconnect the clear plastic tube from the horns and compressor unit (photo).
3 Disconnect the wiring from the compressor unit.
4 Undo the mounting nut and bolt for the upper air horn and remove it. Then carry out the same procedure for the lower horn.
5 The compressor is mounted to the same bracket by two nuts and bolts. Remove these and lift the unit away.
6 Refitting is the reverse procedure; remember to check the correct operation of the horns on reassembly.
7 Lubricate the compressor regularly as detailed in the maintenance Section, at the beginning of the manual.

48 Horn operating switch – removal and refitting

1 The horns of all models are actuated by pushing in the centre section of the steering wheel. However, the type of horn push varies considerably as does its fixing method.
2 With all Saloon models the horn push is retained by screws through the steering wheel spokes from behind. These screws hold the horn push against the spring.
3 To remove the horn push, first disconnect the battery negative lead then remove the screws and remove the horn push and spring.
4 With the Coupe, Spider and HPE models that have a small inverted T-shape horn push this is held in position by lugs. Release the lugs from the rear of the steering wheel and the horn push and spring can then be withdrawn.
5 Later (1979 onwards) versions of the Coupe and HPE have inverted V-shape horn pushes that are retained by screws, as in the Saloon models.
6 To refit the horn push, place the spring in position and refit the horn push and secure it. Then check the horns for correct operation.

49 Windscreen washer pump – removal and refitting

1 Various types of windscreen washer reservoir and sump system have been fitted. In very early models the water is housed in the traditional water bag and the pump is an integral part of the bag assembly.
2 Later models have the bag separate from the pump which is mounted on the left inner wing.
3 New Coupe, Spider and HPE models from 1979 onwards have a

Fig. 10.17 Latest type of washer reservoir and pump (Sec 49)

1 Windscreen washer motor
2 Brake fluid reservoir
3 Windscreen washer reservoir

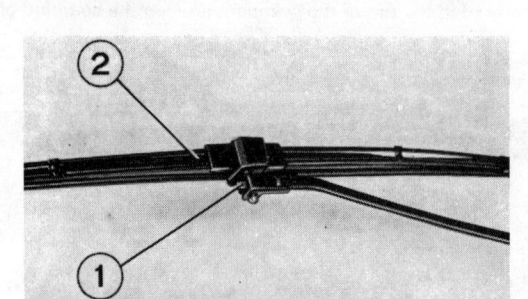

Fig. 10.18 Removing the wiper blade from the arm (Sec 50)

1 Blade retainer
2 Blade

modern type system with a large rectangular plastic water container with a built-in pump. This is mounted inboard of the battery and has a much greater capacity than the old type of bag (Fig. 10.17).
4 To remove the pump disconnect the wiring having first removed the battery negative cable.
5 Then remove the inlet and outlet plastic pipes from the pump, and note which is which for refitting purposes.
6 Next remove the pump. How this is done depends on the type of system fitted. In early models the bag and pump are simple lifted out when the wiring and pipes have been disconnected.
7 In later models the mounting screws, which retain the pump to the inner wing, have to be removed; then the pump can be withdrawn.
8 In the newest models the complete reservoir assembly has to be removed from the vehicle. Then the reservoir has to be emptied in order to reach the washer pump mounting nuts, which are inside the reservoir. With the nuts removed, the pump unit can be removed complete with its seal.
7 All models are fitted with a sealed pump and motor unit. If the motor malfunctions, it has to be renewed, as it cannot be repaired, unless of course it is suffering from a simple blockage in the pump. This can be cleaned out either by air pressure or a thin piece of wire.
6 Refit the unit in the reverse order to that in which it was removed. Finally check the washers for correct operation.

50 Windscreen wiper arms and blades – removal and refitting

Wiper blade
1 Lift the wiper arm away from the windscreen.
2 Release the blade connector from the arm. This is either by pulling the connector up to release it from the stud on the arm and then

Chapter 10 Electrical system

withdrawing it (photo), or if it is the alternative fitting which is to be found on later models, by lifting the retainer and pulling the blade upwards (Fig. 10.18).

3 Refitting is the reverse procedure to removal.

Wiper arm

4 The wiper arm is retained on the splined shaft of the wiper spindle by a chip which locates under the lower edge of the splined boss (photo) or a nut.

5 Lift the arm away from the windscreen and use a flat bladed screwdriver to release the clip or unscrew the nut. Pull the wiper arm off the splined boss.

6 To refit the wiper arm position the arm in relation to the windscreen so that it is in the 'parked' position. Then slide the arm onto the splined boss until the clip locks it in place or until the nut can be refitted.

7 Finally check the operation of the wipers to ensure that the splines are correctly located and that the wiper returns to its correct park position. If it doesn't remove the arm again and reposition it as necessary on the splines.

51 Windscreen wiper motor and mechanism – removal and refitting

1 There are minor differences between the Saloon models and the others in the range in the fitting of the windscreen wiper mechanism, although the basic procedure is the same.

2 Start by disconnecting the battery negative cable.

3 In the Saloon version begin by dismounting the washer jet supporting bracket, which is secured to the front of the section which houses the wiper motor under the bonnet, then remove the motor protective cover. If necessary disconnect the speedometer drive cable from the gearbox for access.

4 Disconnect the windscreen wiper motor multi-plug connector. On the Coupe, Spider and HPE models this is located beside the motor. The cable clamp to the right will also need undoing (photo).

5 On the Saloon models the multi-plug is located behind the instrument panel. This can be reached without removing the panel and the end can be pushed through the hole in the bulkhead, once the grommet has been removed. If difficulty is experienced in reaching or identifying the wiper motor multi-plug connector then remove the instrument panel.

6 Remove the wiper arms complete with blades, as described in the previous Section, and then undo and remove the spindle retaining nuts, washers and seals.

7 Remove the two wiper mounting plate retaining nuts, which support the mounting plate on rubber pads.

8 The whole assembly is now free and can be manoeuvred out of the car (Fig. 10.19).

9 With the assembly on the bench the motor can be removed from its mounting plate. Undo the drive arm retaining nut and lift off the arm. Unscrew the two motor retaining screws and the motor is free.

10 Refitting is the reverse procedure to removal, but make sure that the motor is in the automatic 'park' position before fitting the linkage to it. The linkage should be fitted with the drive arm and link bar in a straight line (Fig. 10.19).

11 Finally check, after refitting the whole assembly to the car, that the wipers operate correctly.

52 Windscreen wiper motor and mechanism – dismantling and reassembly

1 Remove the wiper motor and mechanism as described in Section 51.

2 If the motor is known to be faulty, there is no point in trying to effect a repair, a new motor/gearbox assembly should be purchased.

3 If the problem was one of sloppy operation with a great deal of slack in the operating mechanism, the bearing bushes at each end of both link rods should be closely inspected. The bearings are not renewable individually and it will be a matter of renewing the appropriate link rods. It should be remembered that if the bearings are worn, then the pins on which the bearings run could also be worn.

4 Still with the investigation into sloppy operation, it is worth removing the top of the gearbox so that the condition of the worm driven gear wheel can be inspected. This main gear wheel is listed as available as spare and may be renewed, if found to be worn.

5 The wiper spindles are not renewable individually and if the spindles are a shaky fit in their housings, it will be necessary to renew the whole main framework. This framework comprises both spindles, the housings and spacing member and finally the motor mounting.

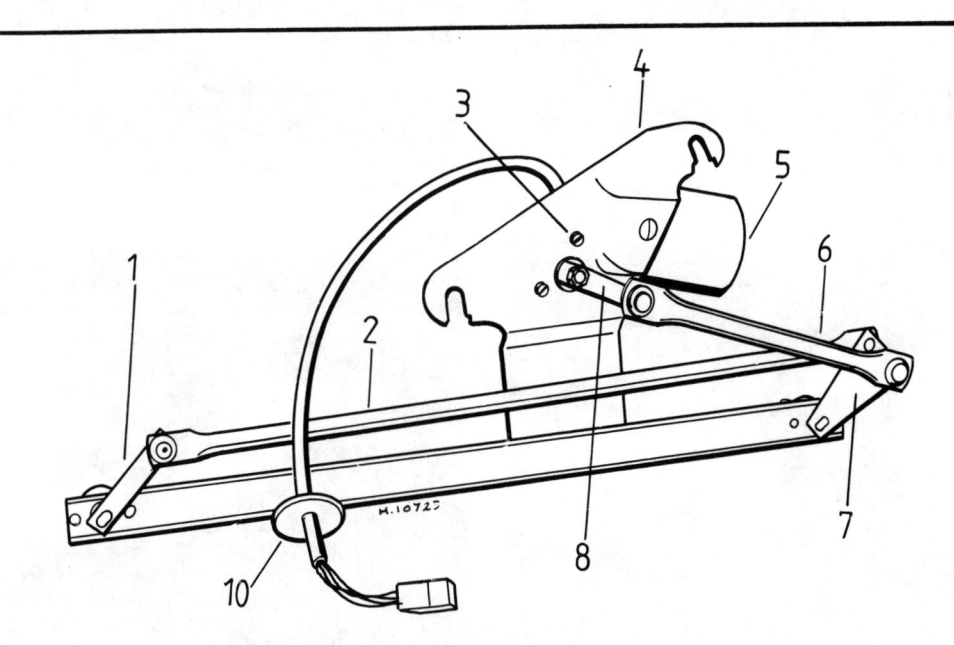

Fig. 10.19 Windscreen wiper motor and mechanism (Sec 51)

1 Lever	3 Motor-to-support bracket mounting bolts	5 Motor
2 Link	4 Motor support bracket	6 Motor-to-lever link
		7 Dual lever
		8 Drive arm
		9 Multiplug
		10 Grommet

53 Tailgate wiper assembly (HPE models) – removal and refitting

1 The tailgate wiper motor may be either of Ducellier or Lucas manufacture. The fitting is slightly different.
2 Disconnect the battery negative lead, and remove the wiper blade and arm as described in Section 50.
3 Undo the wiper spindle retaining nut and remove the washer, any spacers and seal.
4 Open the tailgate and remove the interior trim panel. Also remove the slotted grille and blind assembly.
5 With the Ducellier type motor, undo the cable clips and unplug the electric leads to the wiper motor (Fig. 10.20).
6 Remove the motor mounting bracket retaining screw and then the motor mounting screws. The bracket can now be withdrawn.
7 Remove the motor from the tailgate.
8 With the Lucas type motor, remove the two screws at either end of the long mounting plate (photo).
9 Disconnect the mechanism by moving it back.
10 Disconnect the electric leads to the motor and lift it away.
11 Refitting is the reverse procedure to removal.

54 Radios and tape players – fitting (general)

A radio or tape player is an expensive item to buy and will only give its best performance if fitted properly. If you do not wish to do the fitting yourself there are many in-line vehicle entertainment specialists who can do the fitting for you.

Make sure the unit purchased is of the same polarity as the car and ensure that units with adjustable polarity are correctly set before commencing installation.

It is difficult to give specific information with regard to fitting, as final positioning of the radio/tape player, speakers and aerial is entirely a matter of personal preference. However, the following paragraphs give guidelines, which are relevant to all installations.

Radios

Most radios are a standardised 7 in wide by 2 inches deep – this ensures that they will fit into the radio aperture provided in most cars. If your car does not have such an aperture, the radio must be fitted in a suitable position either in, or beneath, the dashpanel. Alternatively, a special console can be purchased which will fit between the dashpanel and the floor, or on the transmission tunnel. These consoles can also be used for additional switches and instrumentation if required. Where no radio aperture is provided, the following points should be borne in mind before deciding where to fit the unit:

(a) The unit must be within easy reach of the driver wearing a seat belt
(b) The unit must not be mounted in close proximity to an electric tachometer, the ignition switch and its wiring, or the flasher

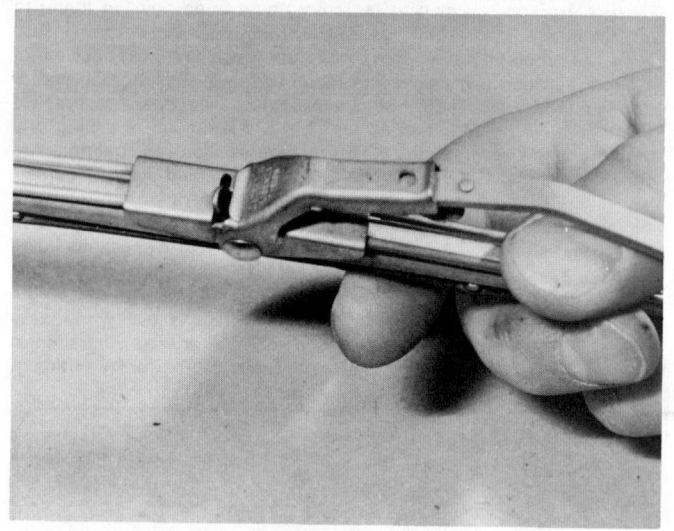

50.2 Removing the wiper blade from the arm

50.4 The wiper arm is mounted on the splined boss

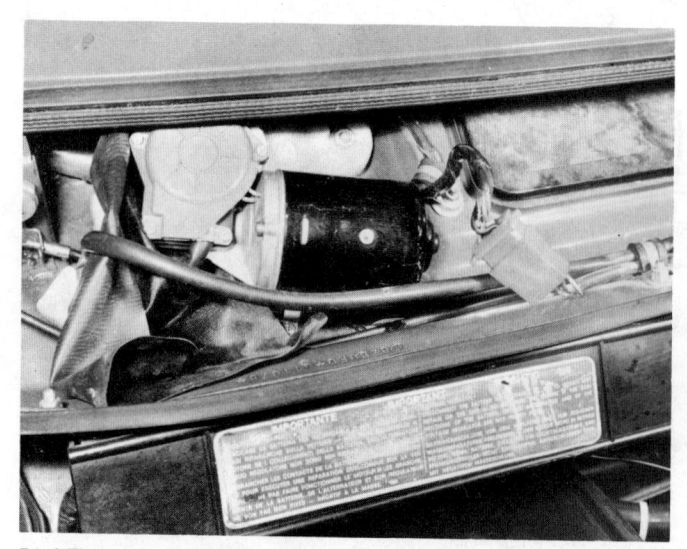

51.4 The wiper motor is mounted in the channel section between the scuttle and engine bay

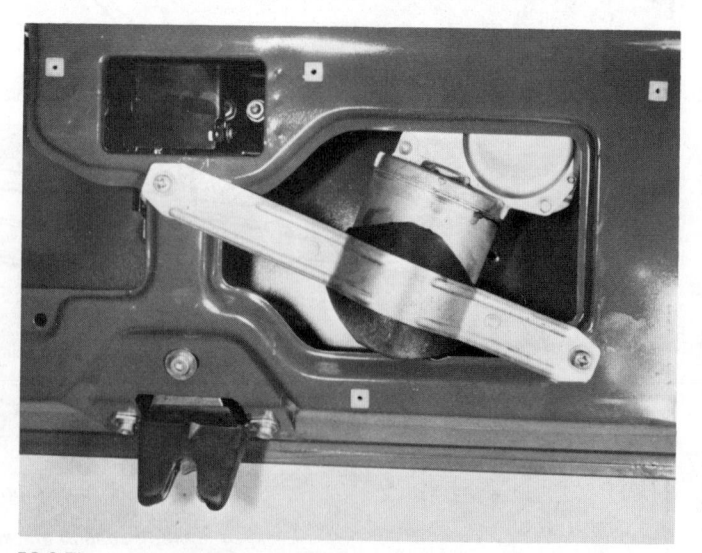

53.8 The Lucas type motor is held in place by the plate and two screws

Chapter 10 Electrical system

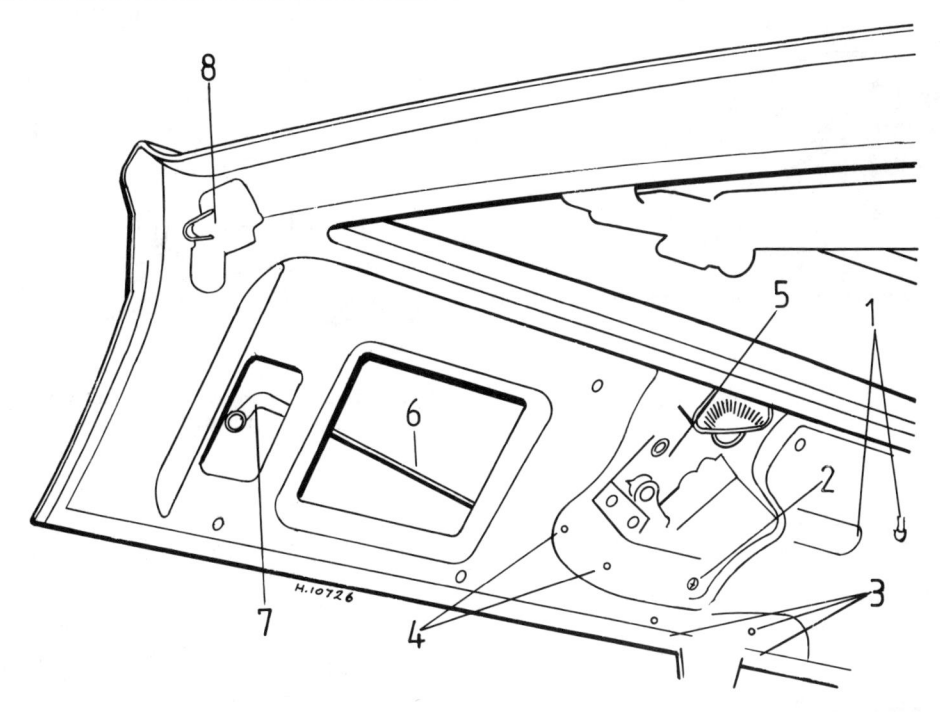

Fig. 10.20 Layout of the tailgate with the Ducellier wiper motor (Sec 53)

1. Electric cables clips
2. Motor bracket mounting screw
3. Latch mounting screws
4. Motor mounting screws
5. Wiper motor
6. Tailgate control
7. Tailgate control lever
8. Side pad

unit and associated wiring
(c) The unit must be mounted within reach of the aerial lead and in such a place that the aerial lead will not have to be routed near the component detailed in preceding paragraph 'b'
(d) The unit should not be positioned where it might cause injury to the car occupants in an accident; for instance, under the dashpanel above the driver's or passenger's legs
(e) The unit must be fitted really securely

Some radios will have mounting brackets provided, together with instructions: others will need to be fitted using drilled and slotted metal strips, bent to form mounting brackets. These strips are available from most accessory shops. The unit must be properly earthed by fitting a separate earthing lead between the casing of the radio and the vehicle frame.

Use the radio manufacturer's instructions when wiring into the vehicle's electrical system. If no instructions are available, refer to the relevant wiring diagram to find the location of the radio 'feed' connection in the vehicle's wiring circuit. A 1–2 amp 'in-line' fuse must be fitted in the radio's 'feed' wire; a choke may also be necessary (see next Section).

The type of aerial used and its fitted position, is a matter of personal preference. In general, the taller the aerial, the better the reception. It is best to fit a fully retractable aerial – especially, if a mechanical car-wash is used or if you live where cars tend to be vandalised. In this respect, electrical aerials which are raised and lowered automatically when switching the radio on or off are convenient, but are more likely to give trouble than the manual type.

When choosing a site for the aerial, the following points should be considered:

(a) The aerial lead should be as short as possible – this means that the aerial should be mounted at the front of the car
(b) The aerial must be mounted as far away from the distributor and HT leads as possible
(c) The part of the aerial which protrudes beneath the mounting point must not foul the roadwheels, or anything else
(d) If possible, the aerial should be positioned so that the coaxial lead does not have to be routed through the engine compartment
(e) The plane to the panel on which the aerial is mounted should not be so steeply angled that the aerial cannot be mounted vertically (in relation to the 'end-on' aspect of the car). Most aerials have a small amount of adjustment available

Having decided on a mounting position, a relatively large hole will have to be made in the panel. The exact size of the hole will depend upon the aerial being fitted, although, generally, the hole required is of $\frac{3}{4}$ inch (19 mm) diameter. On metal bodies cars, a 'tank-cutter' of the relevant diameter is the best tool to use for making the hole. This tool needs a small diameter pilot hole drilled through the panel, through which the tool clamping bolt is inserted. On GRP bodies cars a 'hole-saw' is the best tool to use. Again, this tool will require the drilling of a small pilot hole. When the hole has been made the raw edges should be de-burred with a file and then painted, to prevent corrosion.

Fit the aerial according to the manufacturer's instructions. If the aerial is very tall, or if it protrudes beneath the mounting panel for a considerable distance, it is a good idea to fit a stay between the aerial and the vehicle frame. This can be manufactured from the slotted and drilled metal strips previously mentioned. The stay should be securely screwed or bolted in place. For best reception it is advisable to fit an earth lead between the aerial body and the vehicle frame – this is essential for GRP bodied cars.

It will probably be necessary to drill one or two holes through bodywork panels in order to feed the aerial lead into the interior of the car. Where this is the case ensure that the holes are fitted with rubber grommets to protect the cable, and to stop possible entry of water.

Positioning and fitting of the speaker depends mainly on the type. Generally, the speaker is designed to fit directly into the aperture already provided in the car (usually in the shelf behind the rear seats, or in the top of the dashpanel). Where this is the case, fitting the speaker is just a matter of removing the protective grille from the aperture and screwing or bolting the speaker in place. Take care not to damage the speaker diaphragm whilst doing this. It is a good idea to fit a 'gasket' between the speaker frame and the mounting panel, in order to prevent vibration – some speakers will already have such a gasket fitted.

If a 'pod' type speaker was supplied with the radio, the best acoustic results will normally be obtained by mounting it on the shelf behind the rear seat. The pod can be secured to the panel with self-tapping screws.

When connecting a rear mounted speaker to the radio, the wires should be routed through the vehicle beneath the carpets, or floor

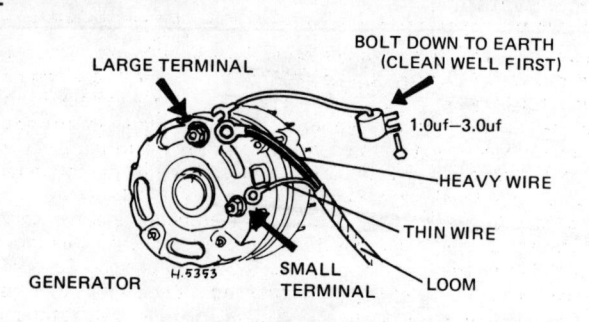

Fig. 10.21 The correct way to connect a capacitor to the alternator (Sec 55)

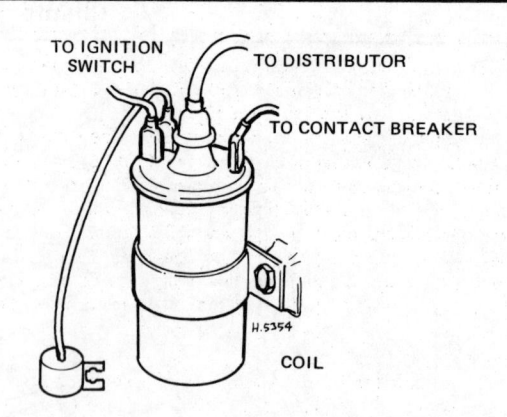

Fig. 10.22 The capacitor must be connected to the ignition side of the coil (Sec 55)

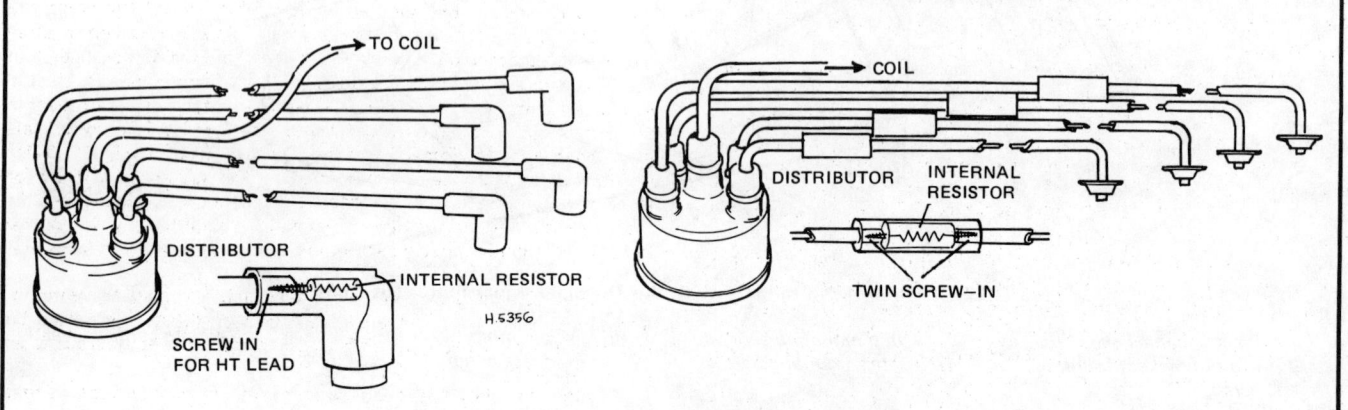

Fig. 10.23 Ignition HT lead suppressors (Sec 55)

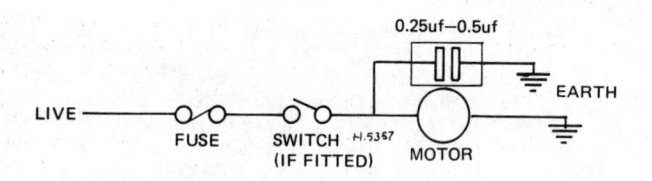

Fig. 10.24 Correct method of suppressing electric motors (Sec 55)

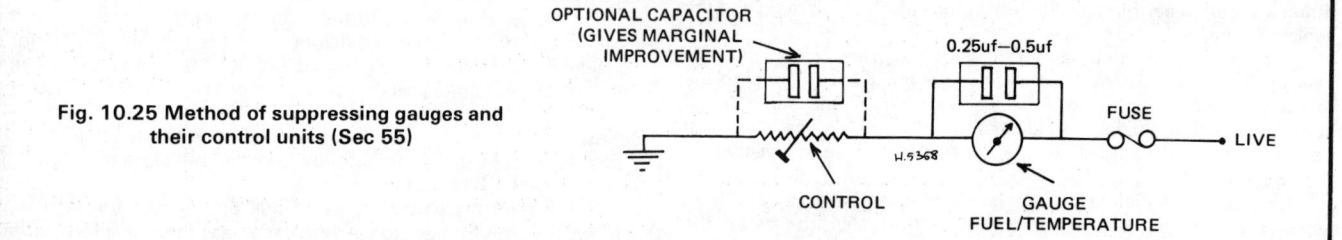

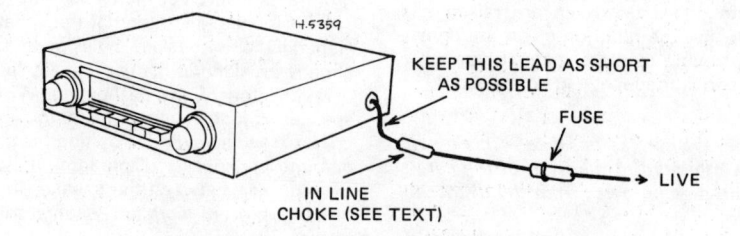

Fig. 10.25 Method of suppressing gauges and their control units (Sec 55)

Fig. 10.26 An 'in-line' choke should be fitted with the line supply lead as close to the unit as possible (Sec 55)

mats – preferably the middle, or along the side of the floorpan, where they will not be trodden on by passengers. Make the relevant connections as directed by the radio manufacturer.

Ensure that all the electrical connections have been made properly, so that there is a good electrical contact and that all the wiring is installed neatly and secured to the car with wiring clips, or PVC tape.

After completing the installation of the radio, it will be necessary to trim the radio to suit the aerial. If specific instructions on this are not given by the manufacturer of the radio, proceed as follows. Find a medium waveband station with a low signal strength and turn the trim screw on the radio in or out until the signal is received at maximum strength.

Tape players

Fitting instructions for both cartridge and cassette stereo tape players are the same and in general the same rules apply as when fitting a radio. Tape players are not usually prone to electrical interference like radios – although it can occur – so positioning is not so critical. If possible the player should be mounted on an 'even keel'. Also, it must be possible for a driver wearing a seat belt to reach the unit to change or turn over tapes.

For the best results from speakers designed to be recessed into a panel, mount them so that the back of the speaker protrudes into an enclosed chamber within the car (eg door interiors or the boot cavity).

To fit recessed type speakers in the front doors, first check that there is sufficient room to mount the speakers in each door without fouling the latch or window winding mechanism. Hold the speaker against the skin of the door, and draw a line around the periphery of the speaker. With the speaker removed draw a second 'cutting' line, within the first, to allow enough room for the entry of the speaker back, but at the same time, providing a broad seat for the speaker flange. When you are sure that the 'cutting-line' is correct, drill a series of holes around its periphery. Pass a hacksaw blade through one of the holes and cut through the metal between the holes until the centre section of the panel falls out.

De-burr the edges of the hole and paint the raw metal to prevent corrosion. Cut a corresponding hole in the door trim panel – ensuring that it will be completely covered by the speaker grille. Now drill a hole in the door edge and a corresponding hole in the door surround. These holes are to feed the speaker leads through – so fit grommets. Pass the speaker leads through the door trim, door skin and out through the holes in the side of the door and door surround. Refit the door trim panel and then secure the speaker to the door using self-tapping screws.

Note: *If the speaker is fitted with a shield to prevent water dripping on it, ensure that this shield is at the top.*

Pod type speakers can be fastened to the shelf behind the rear seat, or anywhere else offering a corresponding mounting point. If the pod speakers are mounted on each side of the shelf behind the rear seat, it is a good idea to drill several large diameter holes through to the boot cavity beneath each speaker; this will improve the sound reproduction. Pod speakers sometimes offer a better reproduction quality if they face the rear window – which then acts as a reflector – so it is worthwhile to do a little experimenting before finally fixing the speaker.

55 Radios and tape players – suppression of interference (general)

To eliminate buzzes and other unwanted noises costs very little and is not as difficult as sometimes thought. With a modicum of common sense and patience, and following the instructions in the following paragraphs, interference can be virtually eliminated.

The first cause for concern is the generator. The noise this makes over the radio is like an electric mixer and the noise speeds up when the engine is revved. (To prove the point, remove the drivebelt and try it). The remedy for this is to simple, connect a 1.0 to 3.0 mfd capacitor between earth (probably the bolt that holds down the generator base) and the positive (+) terminal on the alternator. This is most important for if it is connected to the small terminal, the generator will probably be damaged permanently.

A second common cause of electrical interference is the ignition system. Here a 1.0 mfd capacitor must be connected between earth and the SW or + terminal on the coil. This may stop the tick-tick sound that comes over the speaker. Next comes the spark itself.

There are several ways of curing interference from the ignition HT system. One is the use of carbon-cored HT leads. Where copper cable is used then resistive spark plug caps must be used. These should be of about 10 000 to 15 000 ohm resistance. If due to lack of room these cannot be used an alternative is to use 'in-line' suppressors. If the interference is not too bad, it may be possible to get away with only one suppressor in the coil to distributor line. If the interference does continue (a 'clacking' noise), then modify all HT leads.

At this stage it is advisable to check that the radio is well earthed also the aerial and to see that the aerial plug is pushed well into the set and that the radio is properly trimmed (see preceding Section). In addition, check that the wire which supplied the power to the set is as short as possible. At this stage it is a good idea to check that the fuse is of the correct rating. For most sets this will be about 1 to 2 amps.

At this point, the more usual causes of interference have been suppressed. If the problem still exists, a look at the cause of interference may help to pinpoint the component generating the stray electrical discharges.

The radio picks up electromagnetic waves in the air. Some are made by regular broadcasters and some, which we do not want, are made by the car itself. The home made signals are produced by stray electrical discharges floating around in the car. Common producers of these signals are electrical motors, ie the windscreen wipers, electric screen washers, electric window winders, heater fan or an electric aerial if fitted. Other sources of interference are flashing turn signals and instruments. The remedy for these cases is shown in Fig. 10.24 for an electric motor whose interference is not too bad and Fig. 10.25 for instrument suppression. Turn signals are not normally suppressed. In recent years, radio manufacturers have included in the live line of the radio, in addition to the fuse, an 'in-line' choke. If your circuit lacks one of these, put one in as shown in Fig. 10.26.

All the foregoing components are available from radio stores or accessory stores. If you have an electric clock fitted, this should be suppressed by connecting a 0.5 mfd capacitor directly across it as shown for a motor in Fig. 10.24.

If after all this you are still experiencing radio interference, first assess how bad it is, for the human ear can filter out unobtrusive unwanted noises quite easily. But if you are still adamant about eradicating the noise, then continue.

As a first step, a few 'experts' seem to favour a screen between the radio and the engine. This is OK as far as it goes, literally! The whole set is screened anyway and if interference can get past that then a small piece of aluminium is not going to stop it.

A more sensible way of screening is to discover if interference is coming down the wires. First, take the live lead; interference can get between the set and the choke (hence the reason for keeping the wires short). One remedy here is to screen the wire and this is done by buying screened wire and fitting that. The loudspeaker lead could be screened also to prevent 'pick-up' getting back to the radio, although this is unlikely.

Without doubt, the worst source of radio interference comes from the ignition HT leads, even if they have been suppressed. The ideal way of suppressing these is to slide screening tubes over the leads themselves. As this is impractical, we can place an aluminium shield over the majority of the lead areas. In a vee form engine this is relatively easy but for a straight engine results are not particularly good.

Now for the really impossible cases, here are ideas to try out. Where metal comes into contact with metal, an electrical disturbance is caused which is why good clean connections are essential. To remove interference due to overlapping or butting panels you must bridge the join with a wide braided earth strap (like that from the frame to the engine/transmission). The most common moving parts that could create noise and should be strapped are, in order of importance:

(a) Silencer to frame
(b) Exhaust pipe to engine block and frame
(c) Air cleaner to frame
(d) Front and rear bumpers to frame
(e) Steering column to frame
(f) Bonnet and boot lids to frame
(g) Hood frame to bodyframe on soft tops

These faults are most pronounced when the engine is idling or labouring under load. Although the moving parts are already connected with nuts, bolts etc, these do tend to rust and corrode, thus creating a high resistance interference source.

If you have a ragged sounding pulse when mobile, this could be wheel or tyre static. This can be cured by buying some anti-static powder and sprinkling it liberally inside the tyre.

If the interference takes the shape of a high pitched screeching noise that changes its note when the vehicle is in motion and only comes now and then, this could be related to the aerial, especially if it is of the telescopic or whip type. This source can be cured quite simply by pushing a small rubber ball on top of the aerial as this breaks the electric field before it can form; but it would be much better to buy yourself a new aerial of a reputable brand. If, on the other hand, you are getting a loud rushing sound every time you brake, this is brake static. This effect is most prominent on hot dry days and is cured only by fitting a special kit, which is quite expensive.

In conclusion, it is pointed out that it is relatively easy, therefore cheap, to eliminate 95 per cent of all noise, but to eliminate the final 5 per cent is time and money consuming. It is up to the individual to decide if it is worth it. Please remember also, that you cannot get a concert hall performance out of a cheap radio.

Finally, tape players and eight track players are not usually affected by vehicle noise, but in a very bad case, the best remedy is the first suggestion, plus using a 3 to 5 amp choke in the live line, and in incurable cases, screening the live and speaker wires.

Note: *If your car is fitted with electronic ignition, then it is not recommended that either the spark plug resistors or the ignition coil capacitor be fitted as these may damage the system. Most electronic ignition units have built in suppression and should, therefore, not cause interference.*

56 Fault diagnosis – electrical system

Symptom	Reason(s)
Starter motor fails to turn engine	Battery discharged Battery defective internally Battery terminal leads loose or earth lead not securely attached to body Loose or broken connections in starter motor circuit Starter motor switch or solenoid faulty Starter brushes badly worn, sticking, or brush wires loose Commutator dirty, worn or burnt Starter motor armature faulty Field coils earthed
Starter motor turns engine very slowly	Battery in discharged condition Starter brushes badly worn, sticking, or brush wires loose Loose wires in starter motor circuit
Starter motor turns without turning engine	Pinion or flywheel gear teeth broken or worn
Starter motor noisy or excessively rough engagement	Pinion or flywheel gear teeth broken or worn Starter motor retaining bolts loose
Battery will not hold charge for more than a few days	Battery defective internally Electrolyte level too low or electrolyte too weak due to leakage Plate separators no longer fully effective Battery plates severely sulphated Alternator belt slipping Battery terminal connections loose or corroded Alternator not charging properly* Short in ligthing crcuit circuit causing continual battery drain Regulator unit not working correctly
Ignition light fails to go out, battery runs flat in a few days	Drivebelt loose and slipping or broken Brushes worn, sticking, broken or dirty Brush springs weak or broken Alternator faulty*

* If all appears to be well but the alternator is still not charging, take the car to an automobile electrician to check the alternator

Failure of individual electrical equipment to function correctly is dealt with alphabetically below. In cases of electrical failure it is always worth checking the obvious, such as blown fuses (particularly if associated equipment has also failed) and loose or broken wires.

Symptom	Reason(s)
Fuel gauge gives no reading	Fuel tank empty! Electric cable between tank sender unit and gauge earthed or loose Fuel gauge case not earthed Fuel gauge supply cable interrupted Fuel gauge unit broken
Fuel gauge registers full all the time	Electric cable between tank unit and gauge broken or disconnected
Horn operates all the time	Horn push either earthed or stuck down Horn cable to horn push earthed
Horn fails to operate	Blown fuse Cable or cable connection loose, broken or disconnected Horn has an internal fault
Horn emits intermittent or unsatisfactory noise	Cable connections loose Horn incorrectly adjusted

Chapter 10 Electrical system

Lights do not come on	Blown fuse If engine not running, battery discharged Light bulb filament burnt out or bulbs broken Wire connections loose, disconnected or broken Light switch shorting or otherwise faulty
Lights come on but fade out	If engine not running battery discharged
Lights give very poor illumination	Lamp glasses dirty Reflector tarnished or dirty Lamps badly out of adjustment Incorrect bulb with too low wattage fitted Existing bulbs old and badly discoloured Electrical wiring too thin not allowing full current to pass
Lights work erratically — flashing on and off, especially over bumps	Battery terminals or earth connections loose Lights not earthing properly Contacts in light switch faulty
Wiper motor fails to work	Blown fuse Wire connections loose, disconnected or broken Brushes badly worn Armature worn or faulty Field coils faulty
Wiper motor works very slowly and takes excessive current	Commutator dirty, greasy or burnt Drive to spindles bent or unlubricated Drive spindle binding or damaged Armature bearings dry or unaligned Armature badly worn or faulty
Wiper motor works slowly and takes little cirrent	Brushes badly worn Commutator dirty, greasy or burnt Armature badly worn or faulty
Wiper motor works but wiper blades remain static	Linkage disengaged or faulty Drive spindles damaged or worn Wiper motor gearbox parts badly worn

See overleaf for Wiring Diagrams

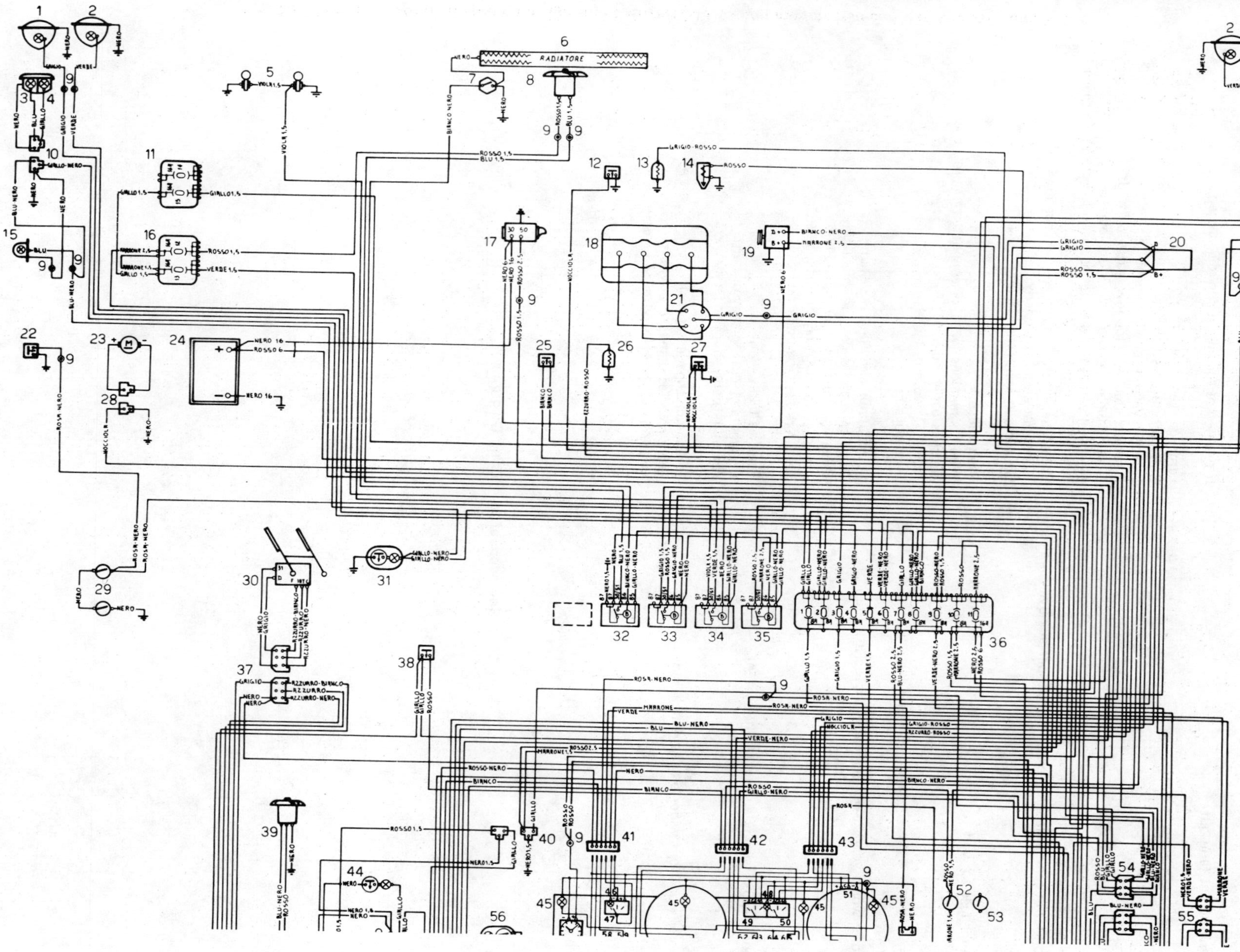

Fig. 10.27 Wiring diagram for right-hand drive 1300 Saloon models (typical). For key see page 188

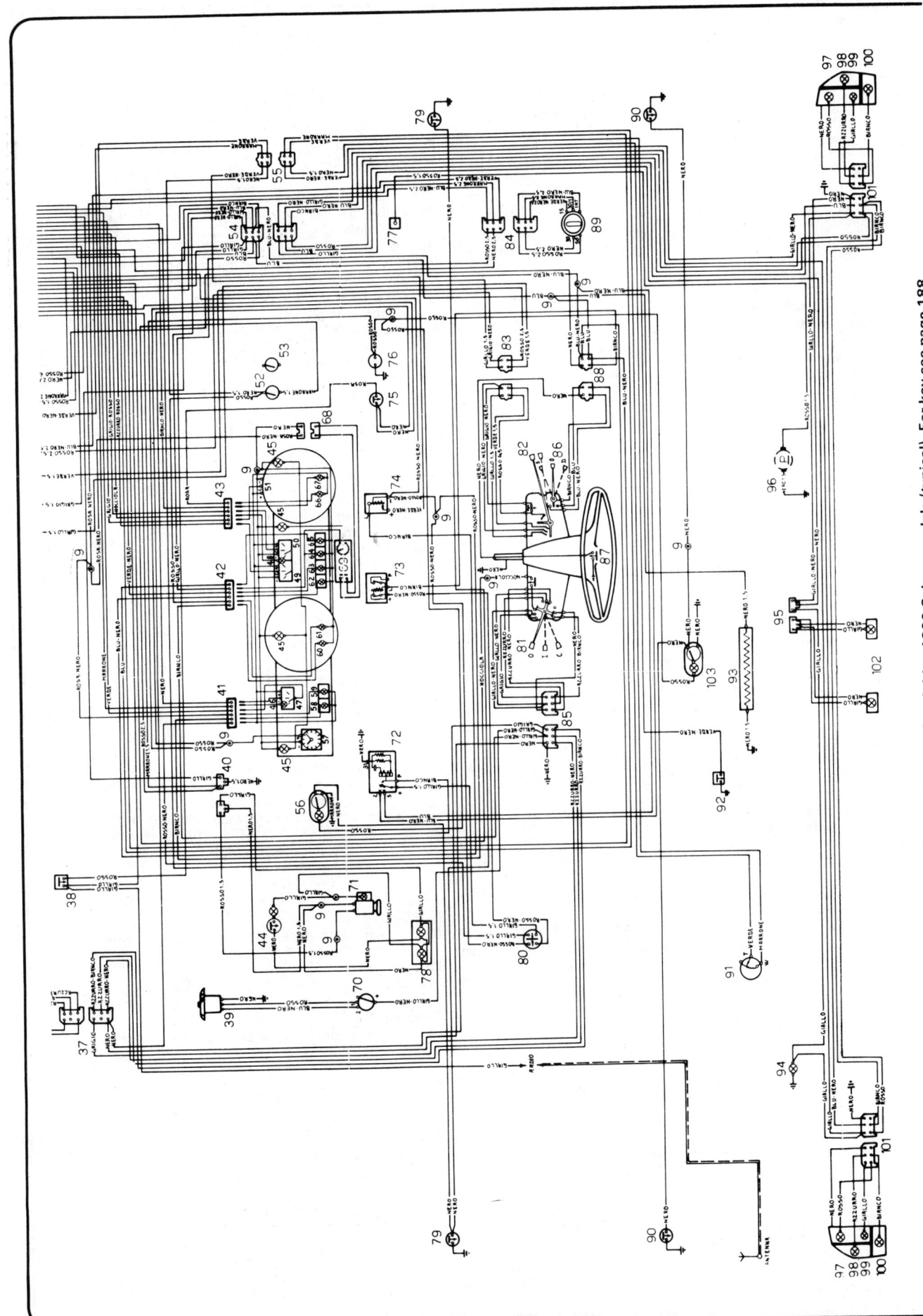

Fig. 10.27 (cont'd) Wiring diagram for right-hand drive 1300 Saloon models (typical). For key see page 188

Fig. 10.27 Key to Wiring Diagram for right-hand drive 1300 Saloon models (typical)

1 Dipped beam
2 Main beam
3 Front turn indicator
4 Front side light
5 Horn
6 Radiator
7 Radiator fan thermoswitch
8 Radiator fan
9 Plug-in junction
10 Front turn indicator and front side light three-outlet junction block (white)
11 Twin-fuse box for item 72
12 Engine oil low pressure warning light switch
13 Oil pressure gauge transmitter
14 Carburettor slow-running fuel cut-off device
15 Turn indicator repeater
16 Twin-fuse box for items 5 and 7
17 Starter motor
18 Spark plugs
19 Alternator
20 Ignition coil
21 Ignition distributor
22 Front brake pads wear limit warning light
23 Windscreen washer motr
24 Battery
25 Reversing light switch
26 Coolant temperature gauge transmitter
27 Engine overheating warning light switch
28 Washer motor two-outlet junction block (white)
29 Brake fluid low level warning light switch
30 Windscreen wiper two speed motor
31 Engine compartment light with switch
32 Radiator fan solenoid switch
33 Dipped beam solenoid switch
34 Horn solenoid switch
35 Heated rear window and cigarette lighter solenoid switch
36 Fuse box
37 Windscreen wiper motor six-outlet junction block (red)
38 Stop light switch
39 Heating and ventilation system fan two-speed motor
40 Three-outlet junction block (white) for items 44-71-78
42 Instruments cluster six-outlet junction block (red)
43 Instruments cluster six-outlet junction block (black)
44 Glove locker light with switch
45 Instruments cluster light
46 Fuel reserve warning light
47 Fuel gauge
48 Engine oil low pressure and engine overheating warning light
49 Coolant temperature gauge
50 Oil pressure gauge
51 Electronic rev counter
52 Heated rear window switch
53 Spare switch
54 Rear harness six-outlet junction block (white)
55 Rear harness four-outlet junction block (white)
56 Front interior light with switch
57 Clock
58 Handbrake warning light
59 Brake fluid low level and front friction pad wear limit warning light
60 Hazard signalling system warning light
61 Heated rear window warning light
62 Left-hand turn indicator warning light
63 Main beam warning light
64 Side light warning light
65 Right-hand turn indicator warning light
66 Choke warning light (if fitted)
67 Alternator warning light
68 Two-outlet junction block (white) for item 69
69 Instrument cluster light switch with intensity adjuster and push-button to check item 59
70 Heating and ventilation system fan motor change-over switch
71 Cigarette lighter with light
72 Hazard signalling system flasher
73 Turn indicator flasher
74 Handbrake warning light flasher
75 Choke warning light switch
76 Plug-in socket
77 Fuel lift pump one-outlet junction block (white)
78 Heater controls twin-lights
79 Interior front light press-switch on front door pillar
80 Hazard signalling system switch
81 Windscreen wiper two-speed motor and washer motor control
82 Side light, dipped beam, main beam and dipped beam signalling control
83 Light control four-outlet junction block (white)
84 Ignition key switch six-outlet junction block (black)
85 Windscreen wiper control six-outlet junction block (white)
86 Turn indicator control
87 Horn
88 Light control four-outlet junction block (red)
89 Ignition and accessories key switch with anti-theft device
90 Interior rear light press-switch on rear door pillar
91 Fuel gauge transmitter
92 Handbrake warning light switch
93 Heated rear window filament
94 Boot light
95 Number plate light three-outlet junction block (white)
96 Electric fuel lift pump (if fitted)
97 Stop light
98 Rear turn indicator
99 Rear side light
100 Reversing light
101 Tail lamp six-outlet junction block (white)
102 Number plate light

Colour code

Bianco	White	Nocciola	Hazel
Blu	Blue	Nero	Black
Giallo	Yellow	Rosa	Pink
Grigio	Grey	Rossa	Red
Marrone	Brown	Verde	Green

Fig. 10.28 Key to wiring diagram for right-hand drive 1600, 1800 and 2000 Saloon models (typical)

1. Dipped beam
2. Main beam
3. Dipped beam and main beam three-outlet junction block (white)
4. Front turn indicator
5. Front side light
6. Horn
7. Radiator
8. Radiator fan thermoswitch
9. Plug-in junction
10. Radiator fan
11. Front turn indicator and fromt side light three-outlet junction block (white)
12. Twin-fuse box for item 73
13. Engine oil low pressure warning light switch
14. Oil pressure gauge transmitter
15. Carburettor slow-running fuel cut-off device
16. Turn indicator repeater
17. Twin-fuse box for items 6 and 8
18. Starter motor
19. Spark plugs
20. Alternator
21. Ignition coil
22. Front brake pads wear limit warning light
23. Windscreen washer motor
24. Washer motor two-outlet junction block (white)
25. Battery
26. Reversing light switch
27. Coolant temperature gauge transmitter
28. Ignition distributor
29. Engine overheating warning light switch
30. Brake fluid low level warning light switch
31. Windscreen wiper two-speed motor
32. Engine compartment light with switch
33. Windscreen wiper motor six-outlet junction block (red)
34. Stop light switch
35. Radiator fan solenoid switch
36. Dipped beam solenoid switch
37. Horn solenoid switch
38. Heated rear window and cigarette lighter solenoid switch
39. Fuse box
40. Heating and ventilation system fan two-speed motor
41. Three-outlet junction block (white) for items 45, 72, 82
42. Instrument cluster six-outlet junction block (white)
43. Instrument cluster six-outlet junction block (red)
44. Instrument clutster six-outlet junction block (black)
45. Glove locker light with switch
46. Instrument cluster light
47. Fuel reserve warning light
48. Fuel gauge
49. Coolant temperature gauge
50. Engine oil low pressure and engine overheating warning light
51. Oil pressure gauge
52. Electronic rev counter
53. Heated rear window switch
54. Extra switch
55. Windscreen wiper motor high and low speed changeover switch
56. Rear harness six-outlet junction block (white)
57. Rear harness four-outlet junction block (white)
58. Front interior light with switch
59. Clock
60. Handbrake warning light
61. Brake fluid low level and front friction pad wear limit warning light
62. Hazard signalling system warning light
63. Heated rear window warning light
64. Left-hand turn indicator warning light
65. Main beam warning light
66. Side light warning light
67. Right-hand indicator warning light
68. Choke warning light (if fitted)
69. Alternator warning light
70. Two-outlet junction block (white) for item 76
71. Heating and ventilation system fan motor change-over switch
72. Cigarette lighter with light
73. Hazard signalling system flasher
74. Windscreen wiper solenoid switch with intermittence device
75. Turn indicator flasher
76. Instrument cluster light switch with intensity adjuster and push-button to check item 61
77. Handbrake warning light flasher
78. Choke warning light switch (if fitted)
79. Plug-in socket
80. Fuel lift pump one-outlet junction block (white)
81. Intyerior front light press-switch on front door pillar
82. Heater controls twin-lights
83. Windscreen wiper solenoid switch four-outlet junction block (white)
84. Hazard signalling system switch
85. Windscreen wiper control six-outlet junction block (white)
86. Windscreen wiper two-speed motor and washer motor control
87. Side light, dipped beam, main beam and dipped beam signalling control
88. Turn indicator control
89. Lights control four-outlet junction block (white)
90. Ignition key switch six-outlet junction block (black)
91. Horn control
92. Lights control four-outlet junction block (red)
93. Ignition and accessories key switch with anti-theft device
94. Interior rear light press-switch on rear door pillar
95. Fuel gauge transmitter
96. Handbrake warning light switch
97. Interior rear light with switch
98. Heated rear window filament
99. Boot light
100. Number plate light three-outlet junction block (white)
101. Electric fuel lift pump (if fitted)
102. Stop light
103. Rear turn indicator
104. Rear side light
105. Reversing light
106. Tail lamp six-outlet junction block (white)
107. Number plate light

Colour code

Bianco	White	Nocciola	Hazel
Blu	Blue	Nero	Black
Giallo	Yellow	Rosa	Pink
Grigio	Grey	Rosso	Red
Marrone	Brown	Verde	Green

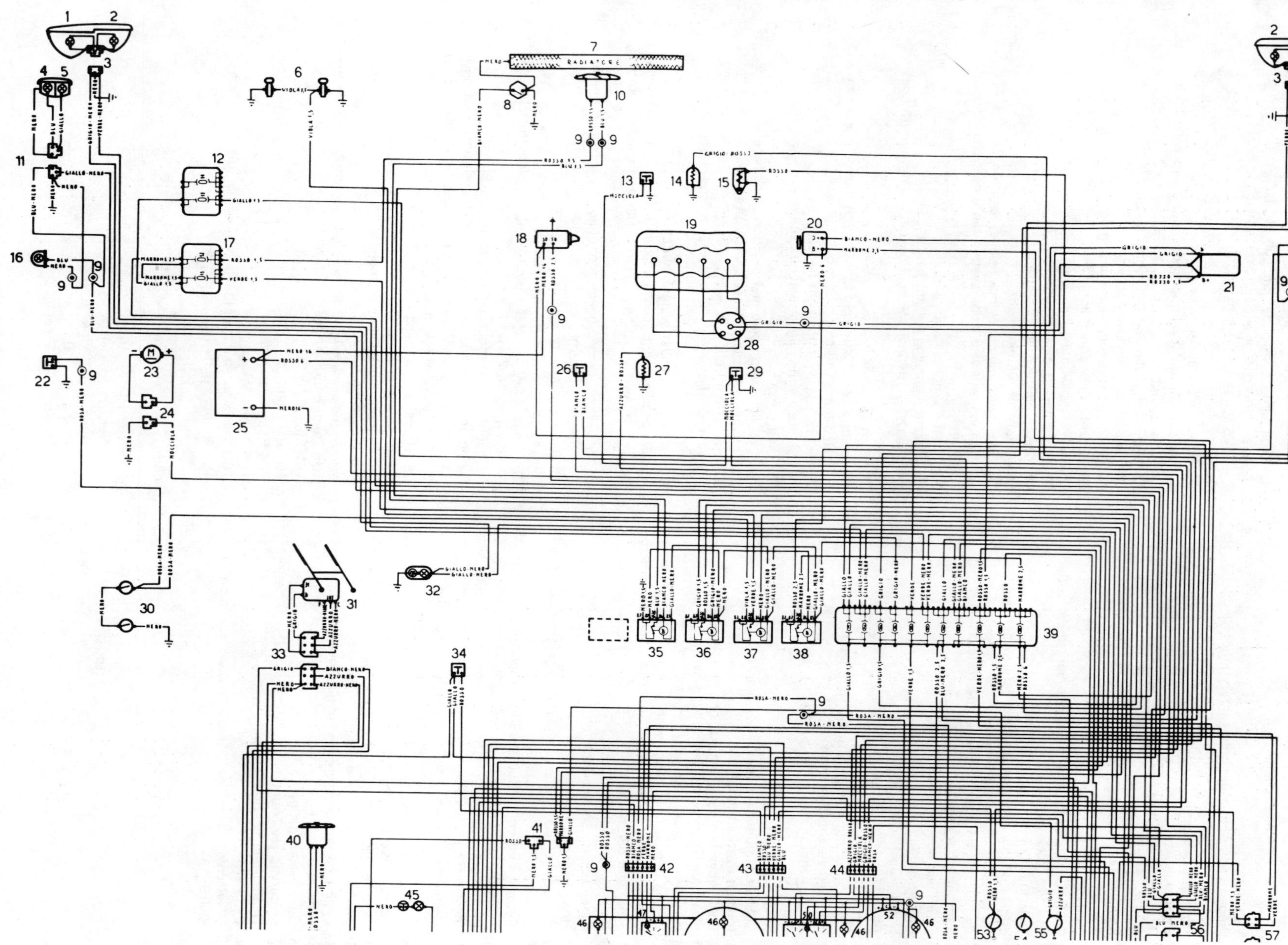

Fig. 10.28 Wiring diagram for right-hand drive 1600, 1800 and 2000 Saloon models (typical). For key see page 189

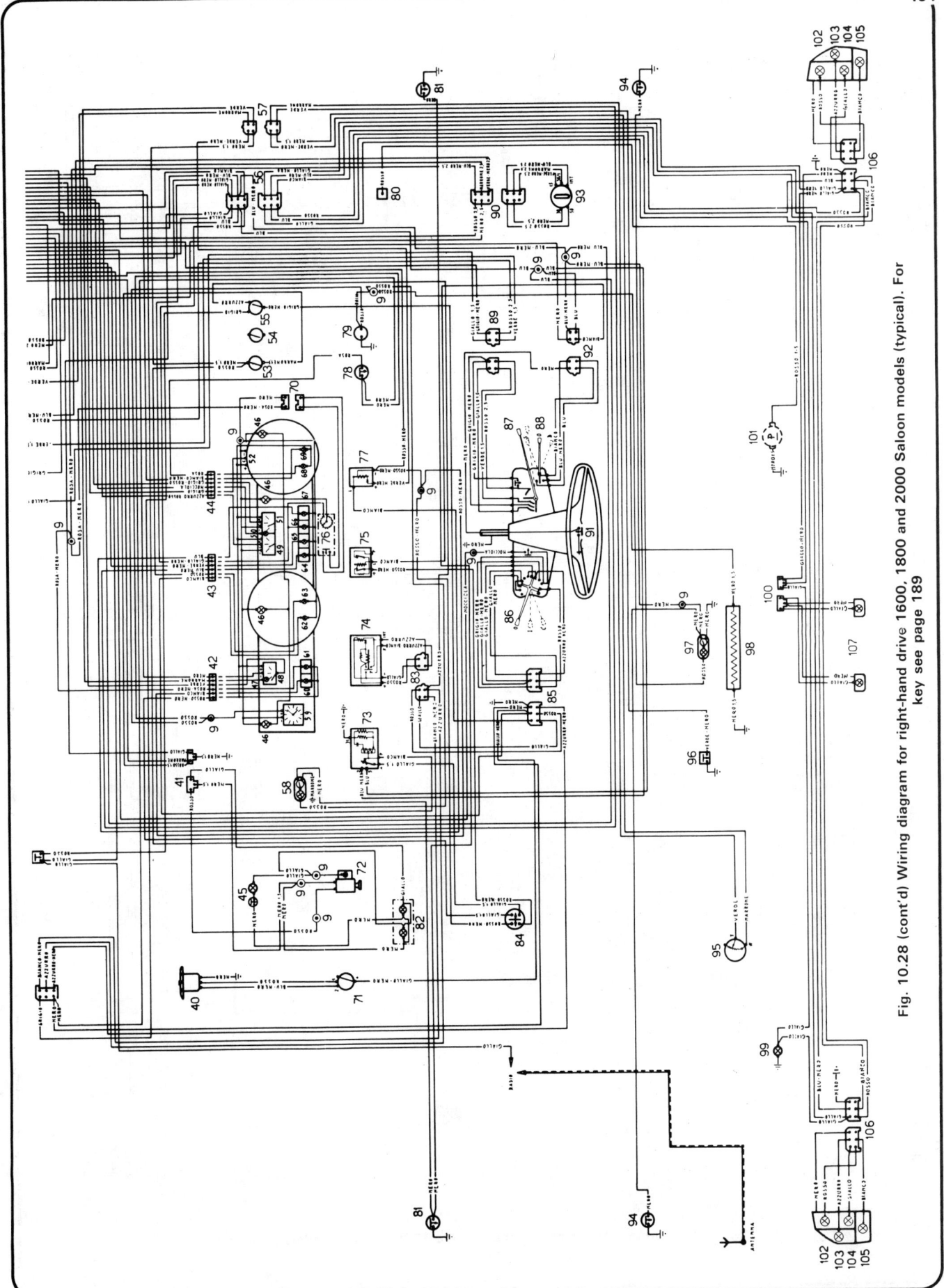

Fig. 10.28 (cont'd) Wiring diagram for right-hand drive 1600, 1800 and 2000 Saloon models (typical). For key see page 189

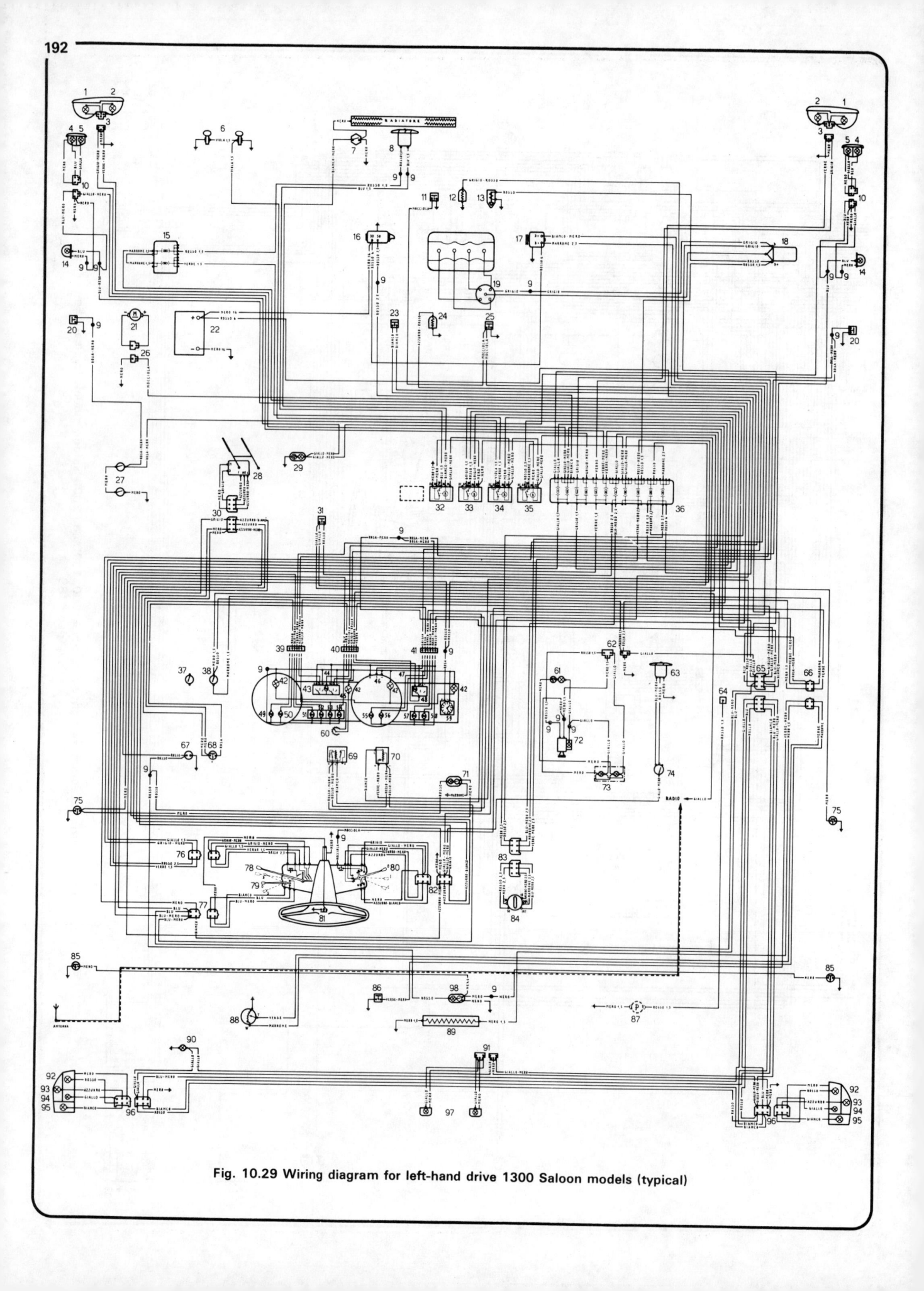

Fig. 10.29 Wiring diagram for left-hand drive 1300 Saloon models (typical)

Fig. 10.29 Key to wiring diagram for left-hand drive 1300 Saloon models (typical)

1 Dipped beam
2 Main beam
3 Three-outlet junction block (white)
4 Front turn indicator
5 Front side light
6 Horn
7 Radiator fan thermoswitch
8 Radiator fan
9 Plug-in junction
10 Front lamp three-outlet junction block (white)
11 Engine oil low pressure warning light switch
12 Oil pressure gauge transmitter
13 Carburettor slow-running fuel cut-off device
14 Turn indicator repeater
15 Twin-fuse box
16 Starter motor
17 Alternator
18 Ignition coil
19 Ignition distributor
20 Front brake pads wear limit warning light switch
21 Windscreen washer motor
22 Battery
23 Reversing light switch
24 Coolant temperature gauge transmitter
25 Engine overheating warning light switch
26 Windscreen washer motor two-outlet junction block (white)
27 Brake fluid low level warning light switch
28 Windscreen wiper motor
29 Engine compartment light with switch
30 Windscreen wiper motor six-outlet junction block (red)
31 Stop light switch
32 Radiator fan solenoid switch
33 Dipped beam solenoid switch
34 Horn solenoid switch
35 Heated rear window and cigarette lighter solenoid switch
36 Fuse box
37 Spare switch
38 Heated rear window switch
39 Instrument cluster six-outlet junction block (white)
40 Instrument cluster six-outlet junction block (red)
41 Instrument cluster six-outlet junction block (black)
42 Instrument cluster light
43 Coolant temperature gauge
44 Engine oil low pressure and engine overheating warning light
45 Oil pressure gauge
46 Electronic rev counter
47 Fuel reserve warning light
48 Fuel gauge
49 Extra warning light (available for hazard signalling system)
50 Heated rear window warning light
51 Left-hand turn indicator warning light
52 Main beam warning light
53 Side light warning light
54 Right-hand turn indicator warning light
55 Choke warning light
56 Alternator warning light
57 Handbrake warning light
58 Brake fluid low level and front friction pad wear limit warning light
59 Clock
60 Instruments cluster light switch with intensity adjuster
61 Glove locker light with switch
62 Three-outlet junction block (white) for items 61-72-73
63 Heating and ventilation system fan two-speed motor
64 Fuel lift pump one-outlet junction block (white)
65 Six-outlet junction block (white)
66 Four-outlet junction block (white)
67 Plug-in socket
68 Choke warning light switch (if fitted)
69 Turn indicator flasher
70 Handbrake warning light flasher
71 Interior front light with switch
72 Cigarette lighter with light
73 Two heater controls lights
74 Heating and centilation system fan motor changeover switch
75 Interior front light press-switch in front door pillar
76 Lights control four-outlet junction block (white)
77 Lights control four-outlet junction block (red)
78 Side light main beam, dipped beam and dipped beam signalling control
79 Turn indicator control
80 Windscreen wiper two-speed motor and washer motor control
81 Horn control
82 Windscreen wiper six-outlet junction block (white)
83 Ignition key switch six-outlet junction block (black)
84 Ignition and accessory key switch with anti-theft device
85 Interior rear light press-switch on rear door pillar
86 Handbrake warning light switch and to check warning light item 58
87 Fuel lift pump (on special versions only)
88 Fuel gauge transmitter
89 Heated rear window filament
90 Boot light
91 Number plate light three-outlet junction block (white)
92 Stop light
93 Rear turn indicator
94 Rear side light
95 Reversing light
96 Tail lamp six-outlet junction block (white)
97 Number plate light
98 Interior rear light with switch

Colour code

Bianco	White	Nocciola	Hazel
Blu	Blue	Nero	Black
Giallo	Yellow	Rosa	Pink
Grigio	Grey	Rosso	Red
Marrone	Brown	Verde	Green

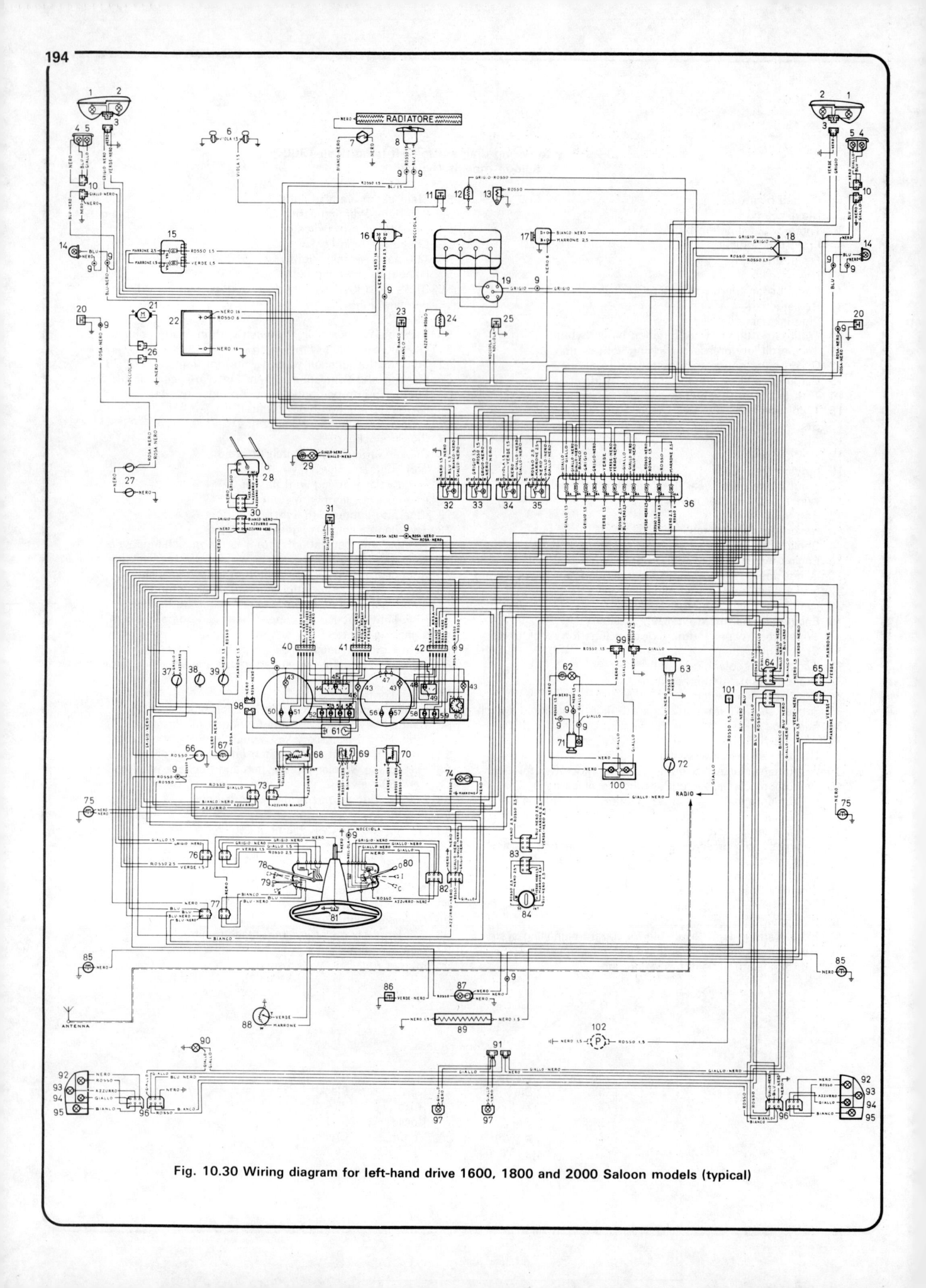

Fig. 10.30 Wiring diagram for left-hand drive 1600, 1800 and 2000 Saloon models (typical)

Fig. 10.30 Key to wiring diagram for left-hand drive 1600, 1800 and 2000 Saloon models (typical)

1. Dipped beam
2. Main beam
3. Three-outlet junction block (white)
4. Front turn indicator
5. Front side light
6. Horn
7. Radiator fan thermoswitch
8. Radiator fan
9. Plug-in junction
10. Front lamp three-outlet junction block (white)
11. Engine oil low pressure warning light switch
12. Oil pressure gauge transmitter
13. Carburettor slow-running fuel cut-off device
14. Turn indicator repeater
15. Twin-fuse box
16. Starter motor
17. Alternator
18. Ignition coil
19. Ignition distributor
20. Front brake pads wear limit warning light switch
21. Windscreen washer motor
22. Battery
23. Reversing light switch
24. Coolant temperature gauge transmitter
25. Engine overheating warning light switch
26. Windscreen washer motor two-outlet junction block (white)
27. Brake fluid low level warning light switch
28. Windscreen wiper two-speed motor
29. Engine compartment light with switch
30. Windscreen wiper motor six-outlet junction block (red)
31. Stop light switch
32. Radiator fan solenoid switch
33. Dipped beam solenoid switch
34. Horn solenoid switch
35. Heated rear window and cigarette lighter solenoid switch
36. Fuse box
37. Windscreen wiper motor high and low speed changeover switch
38. Spare switch
39. Heated rear window switch
40. Instrument cluster six-outlet junction block (white)
41. Instrument cluster six-outlet junction block (red)
42. Instrument cluster six-outlet junction block (black)
43. Instrument cluster light
44. Coolant temperature gauge
45. Engine oil low pressure and engine overheating warning light
46. Oil pressure gauge
47. Electronic rev counter
48. Fuel reserve warning light
49. Fuel gauge
50. Extra warning light (available for hazard signalling system)
51. Heated rear window warning light
52. Left-hand turn indicator warning light
53. Main beam warning light
54. Side light warning light
55. Right-hand turn indicator warning light
56. Choke warning light (if fitted)
57. Alternator warning light
58. Handbrake warning light
59. Brake fluid low level and front friction pad wear limit warning light
60. Clock
61. Instruments cluster light switch with intensity adjuster and push-button to check warning light (item 59)
62. Glove locker light with switch
63. Heating and ventilation system fan two-speed motor
64. Four-outlet junction block (white)
65. Four-outlet junction block (white)
66. Plug-in socket
67. Choke warning light switch (if fitted)
68. Windscreen wiper solenoid switch with intermittent device
69. Turn indicator flasher
70. Handbrake warning light flasher
71. Cigarette lighter with light
72. Heating and ventilation system fan motor change-over switch
73. Windscreen wiper solenoid switch four-outlet junction block (white)
74. Interior front light with switch
75. Interior front light press-switch on front door pillar
76. Lights control four-outlet junction block (white)
77. Lights control four-outlet junction block (red)
78. Side light, main beam, dipped beam and dipped beam signalling control
79. Turn indicator control
80. Windscreen wiper two-speed motor and washer motor control
81. Horn control
82. Windscreen wiper six-outlet junction block (white)
83. Ignition key switch six-outlet junction block (black)
84. Ignition and accessories key switch with anti-theft device
85. Interior rear light press-switch on rear door pillar
86. Handbrake warning light switch
87. Interior rear light with switch
88. Fuel gauge transmitter
89. Heated rear window filament
90. Boot lid
91. Number plate light three-outlet junction block (white)
92. Stop light
93. Rear turn indicator
94. Rear side light
95. Reversing light
96. Tail lamp six-outlet junction block (white)
97. Number plate light
98. Two-outlet junction block (white) for item 61
99. Three-outlet junction block (white) for items 62-71-100
100. Two heater controls lights
101. Fuel lift pump one-outlet junction block (white)
102. Fuel lift pump (on special versions only)

Colour code

Bianco	White	Nocciola	Hazel
Blu	Blue	Nero	Black
Giallo	Yellow	Rosa	Pink
Grigio	Grey	Rosso	Red
Marrone	Brown	Verde	Green

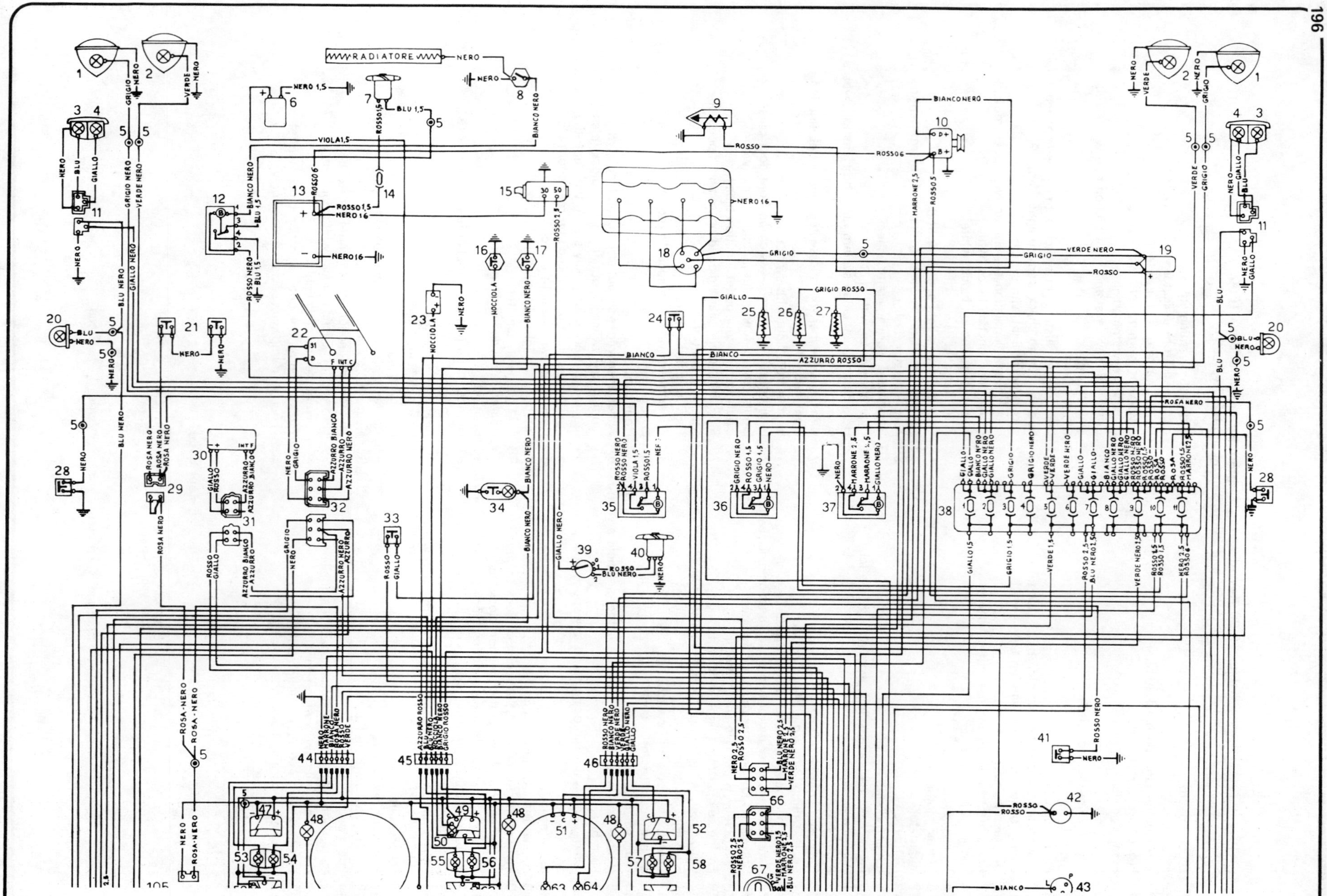

Fig. 10.31 Wiring diagram for right-hand drive 1600 and 2000 Coupe models (typical). For key see page 198

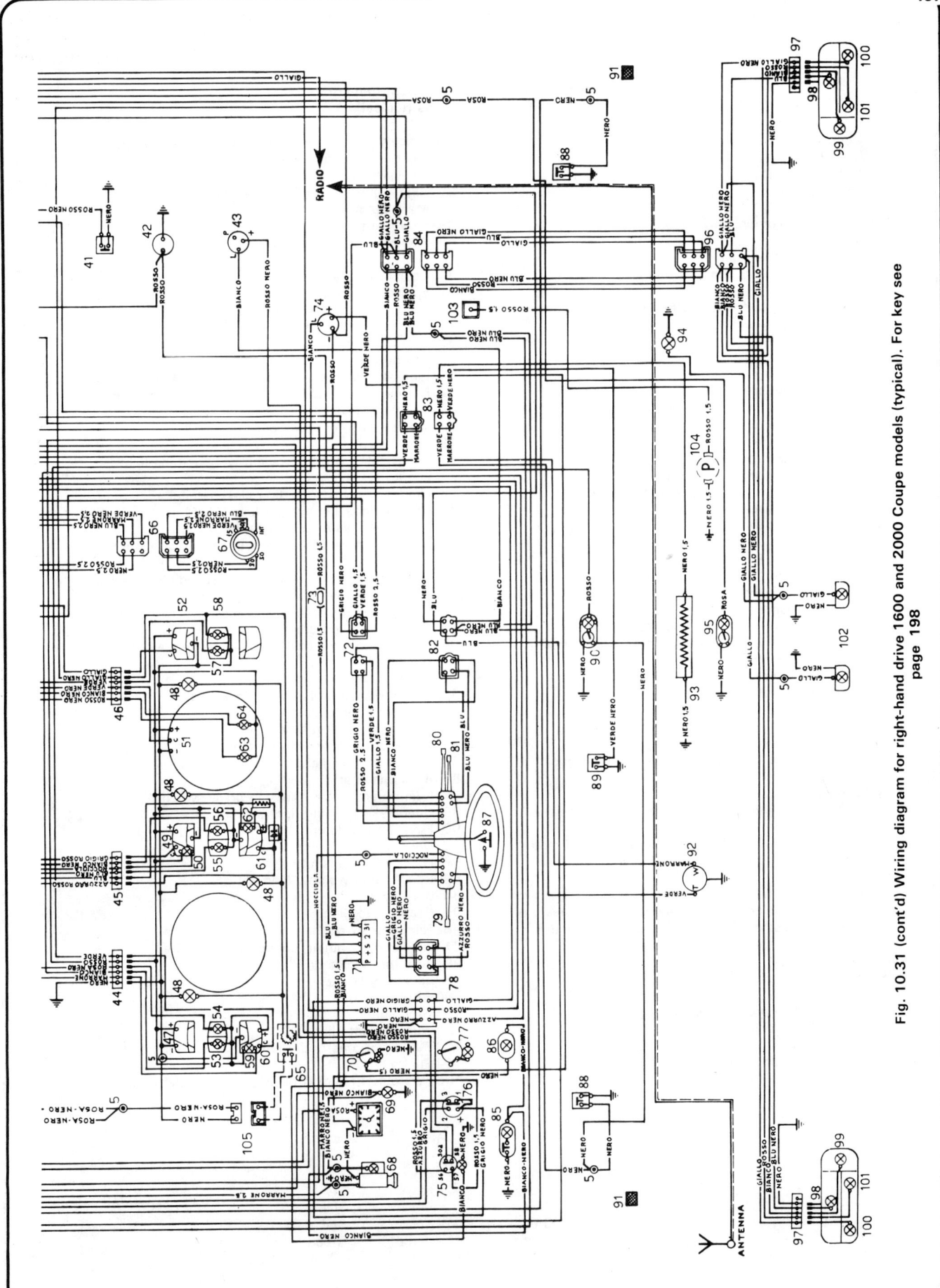

Fig. 10.31 (cont'd) Wiring diagram for right-hand drive 1600 and 2000 Coupe models (typical). For key see page 198

Fig. 10.31 Key to wiring diagrams for right-hand drive 1600 and 2000 Coupe models (typical)

1. Dipped beam
2. Main beam
3. Front turn indicator
4. Front side light
5. Plug-in junction
6. Horn electrocompressor
7. Radiator fan
8. Radiator fan thermoswitch
9. Carburettor slow-running fuel cut-off device
10. Alternator
11. Front lamp three-outlet junction block (white)
12. Radiator fan solenoid switch
13. Battery
14. Radiator fan fuse
15. Starter motor
16. Engine oil low pressure warning light switch
17. Engine overheating warning light switch
18. Ignition distributor
19. Ignition coil
20. Turn indicator repeater
21. Brake fluid low level warning light switch
22. Two-speed windscreen wiper motor
23. Windscreen washer motor
24. Reversing light switch
25. Oil temperature gauge transmitter
26. Oil pressure gauge transmitter
27. Coolant temperature gauge transmitter
28. Front brake friction pad wear limit warning light switch
29. Two-outlet junction block (white)
30. Windscreen wiper intermittence device
31. Windscreen wiper intermittence device four-outlet junction block (white)
32. Windscreen wiper motor six-outlet junction block (red)
33. Stop light switch
34. Engine compartment lamp with switch
35. Horn solenoid switch
36. Dipped beam solenoid switch
37. Heated rear window and cigarette lighter solenoid switch
38. Fuse box
39. Heating and ventilation system fan motor change-over switch
40. Heating and ventilation system fan two-speed motor
41. Tell-tale switch on special version only
42. Plug-in socket
43. Turn indicator flasher
44. Instrument cluster six-outlet junction block (black)
45. Instrument cluster six-outlet junction block (white)
46. Instrument cluster six-outlet junction block (red)
47. Voltmeter
48. Instrument cluster light
49. Oil pressure gauge
50. Engine oil low pressure warning light
51. Electronic rev counter
52. Oil temperature gauge
53. Handbrake warning light
54. Brake fluid low level and front brake friction pad wear limit warning light
55. Left-hand turn indicator warning light
56. Right-hand turn indicator warning light
57. Main beam warning light
58. Side lights warning light
59. Fuel reserve warning light
60. Fuel gauge
61. Coolant temperature gauge
62. Engine overheating warning light
63. Tell-tale on special version only
64. Alternator warning light
65. Instrument cluster light switch with intensity adjuster and push-button to check the warning light item 54
66. Ignition switch six-outlet junction block (black)
67. Engine ignition and accessories key switch with anti-theft device
68. Cigarette lighter with light
69. Clock
70. Rear heated window switch with warning light
71. Hazard signalling flasher
72. Lights four-outlet junction block (white)
73. Hazard signalling system fuse (16 A)
74. Handbrake warning light flasher
75. Hazard signalling change-over switch with warning light
76. Windscreen wiper high and low speed change-over switch
77. Spare switch with warning light
78. Windscreen wiper six-outlet junction block (white)
79. Two-speed windscreen wipers and washer motors control
80. Side lights, dipped beams, main beams and dipped beam signalling control
81. Turn indicator control
82. Lights four-outlet junction block (red)
83. Four-outlet junction block (white)
84. Six-outlet junction block (white)
85. Glove locker light with switch
86. Heater controls light
87. Air horns control
88. Interior lamp door-operated press-switch
89. Handbrake warning light switch
90. Front interior lamp
91. Door safety reflector
92. Fuel gauge transmitter
93. Rear heated window filament
94. Boot light
95. Rear interior lamp
96. Tail lamp six-outlet junction block (white)
97. Tail lamp six-outlet junction block (white)
98. Rear turn indicator
99. Reversing light
100. Rear side light
101. Stop light
102. Number plate light
103. Fuel pump one-outlet junction block (white)
104. Fuel pump
105. Two-outlet junction block (white) for item 65

Colour code

Bianco	White	Nocciola	Hazel
Blu	Blue	Nero	Black
Giallo	Yellow	Rosa	Pink
Grigio	Grey	Rosso	Red
Marrone	Brown	Verde	Green

Fig. 10.32 Key to wiring diagram for left-hand drive 1600, 1800 and 2000 Coupe models (typical)

1. Dipped beam
2. Main beam
3. Front turn indicator
4. Front side light
5. Plug-in junction
6. Horn electrocompressor
7. Radiator fan
8. Radiator fan thermoswitch
9. Carburettor slow-running fuel cut-off device
10. Alternator
11. Front lamp three-outlet junction block (white)
12. Radiator fan solenoid switch
13. Battery
14. Radiator fan fuse
15. Starter motor
16. Engine oil low pressure warning light switch
17. Engine overheating warning light switch
18. Ignition distributor
19. Ignition coil
20. Turn indicator repeater
21. Brake fluid low level warning light switch
22. Two-speed windscreen wiper motor
23. Windscreen washer motor
24. Reversing light switch
25. Oil temperature gauge transmitter
26. Oil pressure gauge transmitter
27. Coolant temperature gauge transmitter
28. Front brake friction pad wear limit warning light switch
29. Two-outlet junction block (white)
30. Windscreen wiper intermittence device
31. Windscreen wiper intermittence device four-outlet junction block (white)
32. Windscreen wiper motor six outlet junction block (red)
33. Stop light switch
34. Engine compartment light with switch
35. Horn solenoid switch
36. Dipped beam solenoid switch
37. Rear heated window and cigarette lighter solenoid switch
38. Fuse box
39. Heating and ventilation system fan motor change-over switch
40. Heating and ventilation system fan two-speed motor
41. Tell-tale switch on special versions only
42. Plug-in socket
43. Turn indicator flasher
44. Instrument cluster six-outlet junction block (black)
45. Instrument cluster six-outlet junction block (white)
46. Instrument cluster six-outlet junction block (red)
47. Voltmeter
48. Instrument cluster light
49. Oil pressure gauge
50. Engine oil low pressure warning light
51. Electronic rev counter
52. Oil temperature gauge
53. Handbrake warning light
54. Brake fluid low level and front friction pad wear limit warning light
55. Left-hand turn indicator warning light
56. Right-hand turn indicator warning light
57. Main beam warning light
58. Side light warning light
59. Fuel reserve warning light
60. Fuel gauge
61. Coolant temperature gauge
62. Engine overheating warning light
63. Tell-tale on special versions only
64. Alternator warning light
65. Instrument cluster light switch with intensity adjuster and push-button to check warning light (item 54)
66. Key switch six-outlet junction block (black)
67. Cigarette lighter with light
68. Clock
69. Rear heated window switch with warning light
70. Ignition and accessories key switch with anti-theft device
71. Spare switch with warning light
72. Windscreen wiper motor change-over switch
73. Heater control light
74. Glove locker light with switch
75. Handbrake warning light flasher
76. Light control four-outlet junction block (white)
77. Light control four-outlet junction block (red)
78. Main beam, dipped beam, light signalling and side light control
79. Turn indicator control
80. Horn control
81. Windscreen wiper two-speed motor and washer motor control
82. Windscreen wiper six-outlet junction block (white)
83. Four-outlet junction block (white)
84. Six-outlet junction block (white)
85. Open door safety reflector
86. Inside lamp press-switch on door pillar
87. Handbrake warning light switch
88. Front inside lamp
89. Fuel gauge transmitter
90. Rear heated window filament
91. Rear inside lamp
92. Boot light
93. Tail lamp six-outlet junction block (white)
94. Tail lamp six-outlet junction block (white)
95. Rear turn indicator
96. Reversing light
97. Rear side light
98. Stop light
99. Number plate light
100. Fuel pump one-outlet junction block (white)
101. Fuel lift pump
102. Two-outlet junction block (white) for item 65

Colour code

Bianco	White	Nocciola	Hazel
Blu	Blue	Nero	Black
Giallo	Yellow	Rosa	Pink
Grigio	Grey	Rosso	Red
Marrone	Brown	Verde	Green

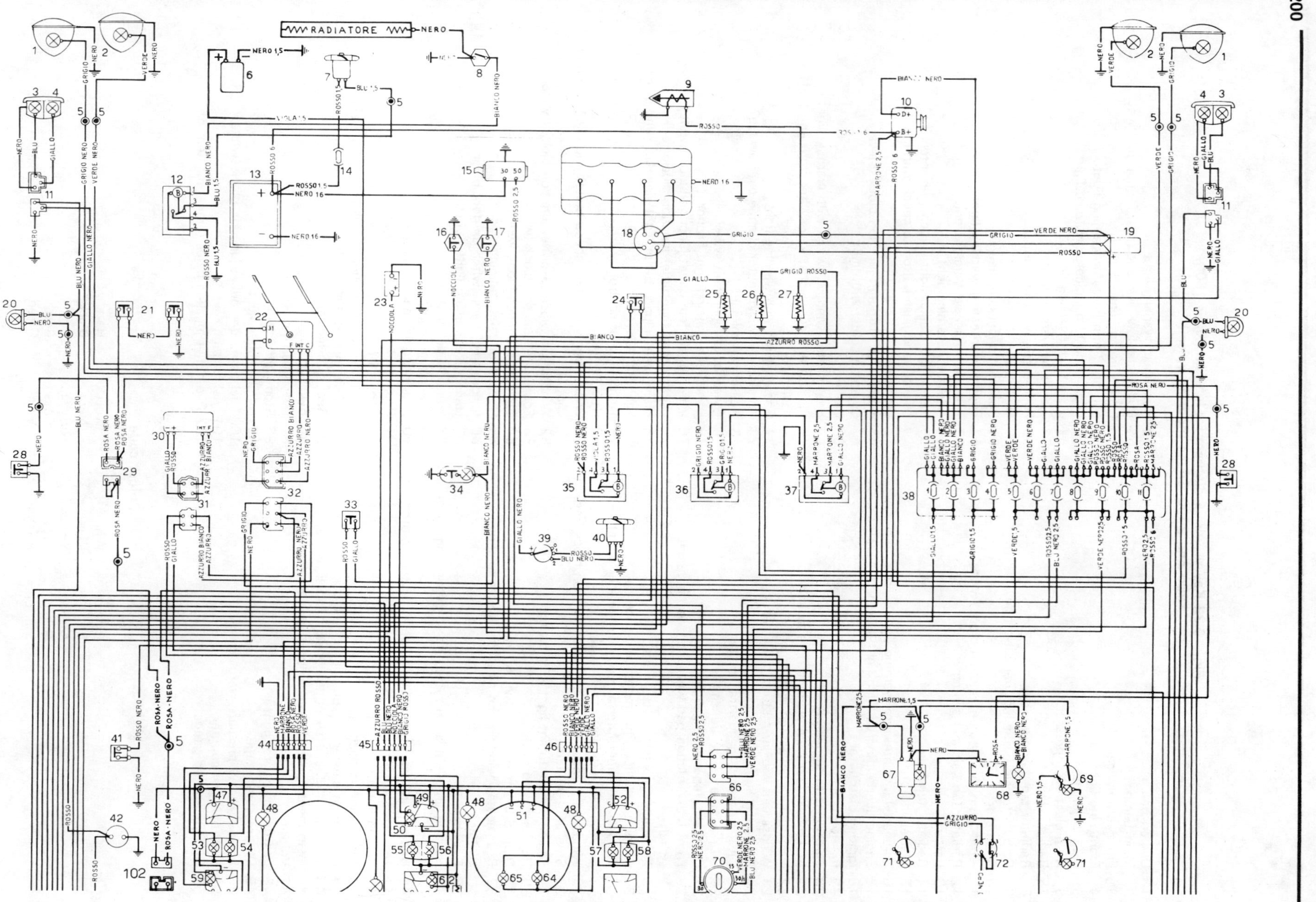

Fig. 10.32 Wiring diagram for left-hand drive 1600, 1800 and 2000 Coupe models (typical). For key see page 199

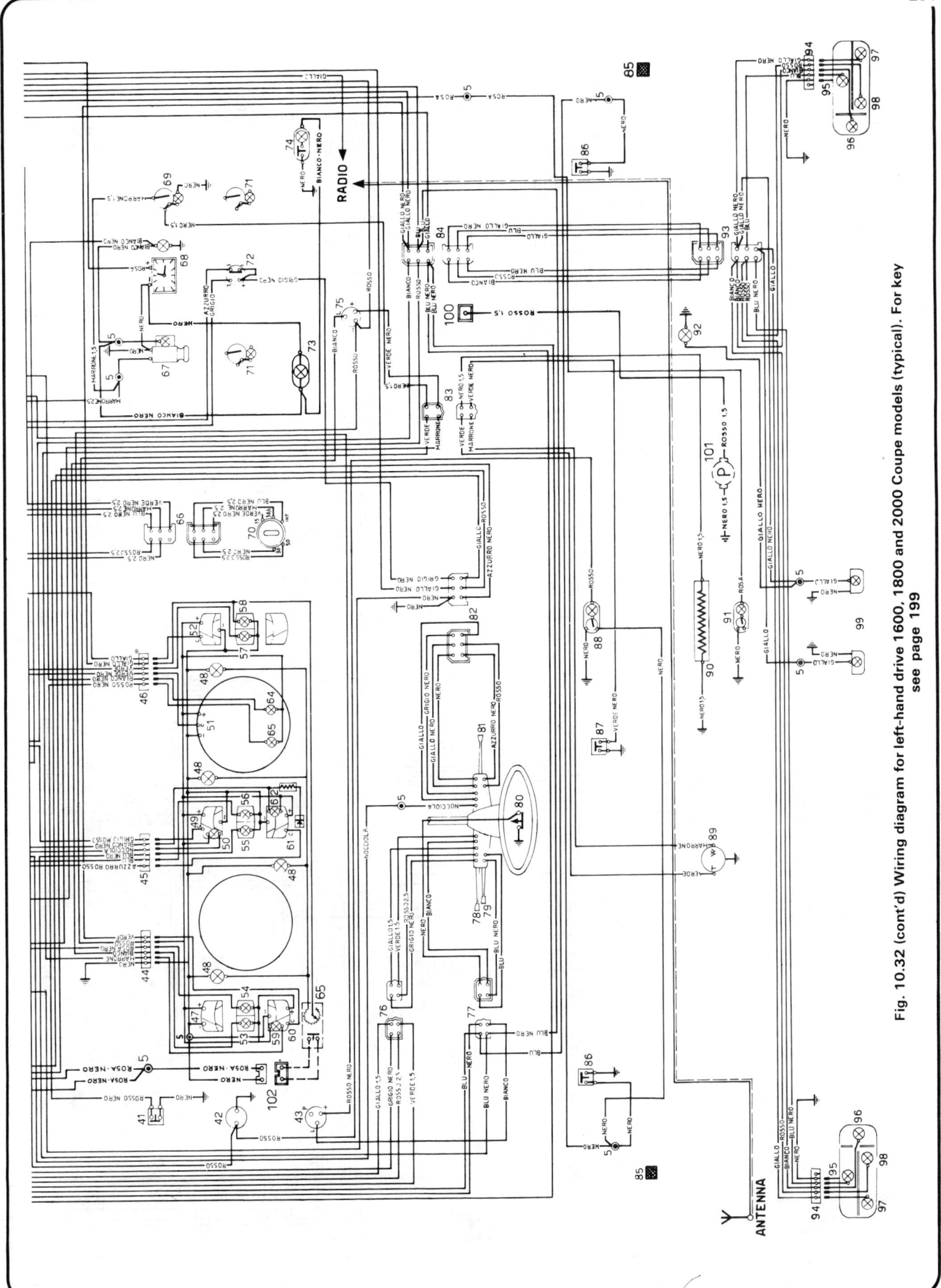

Fig. 10.32 (cont'd) Wiring diagram for left-hand drive 1600, 1800 and 2000 Coupé models (typical). For key see page 199

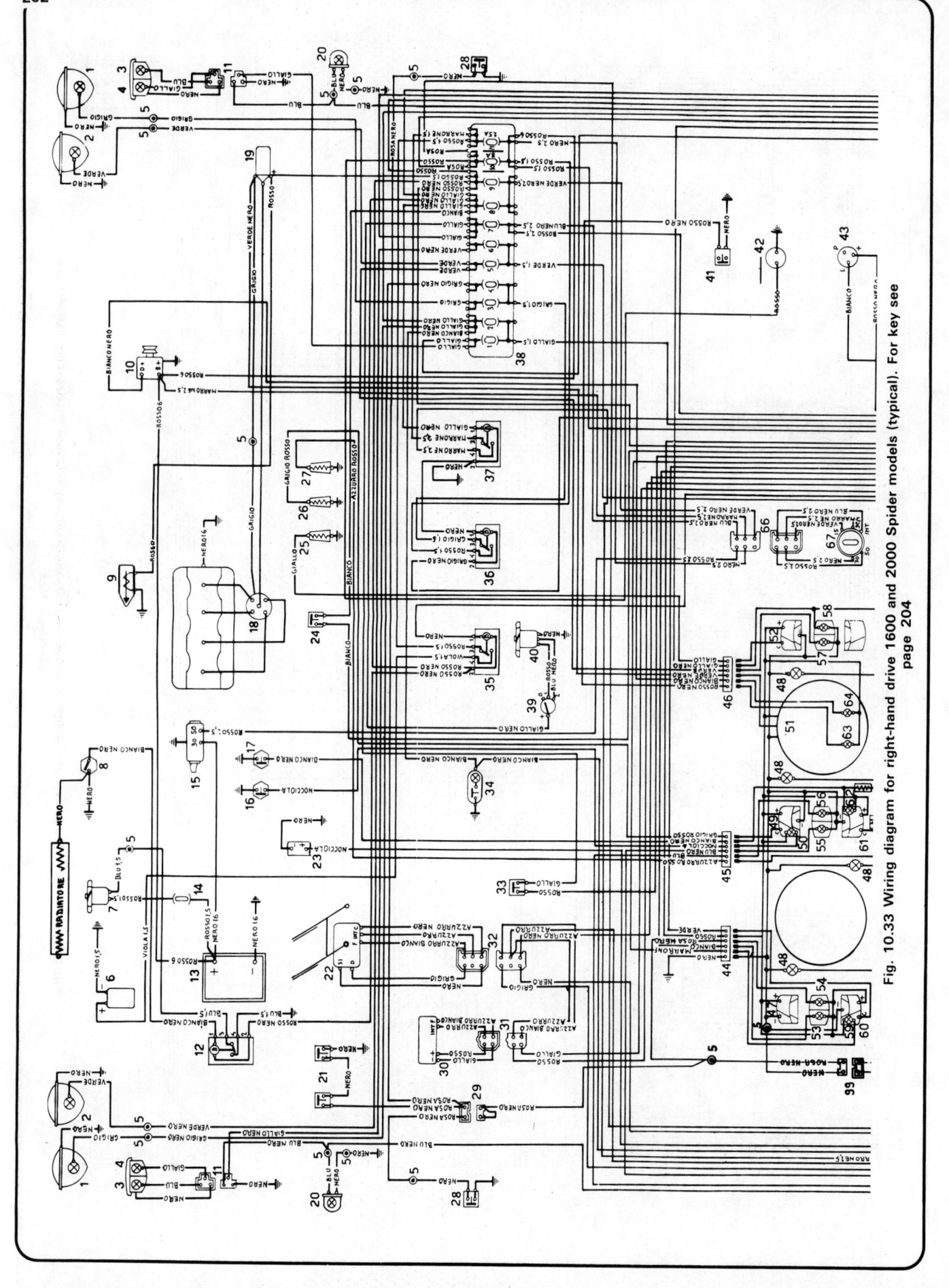

Fig. 10.33 Wiring diagram for right-hand drive 1600 and 2000 Spider models (typical). For key see page 204

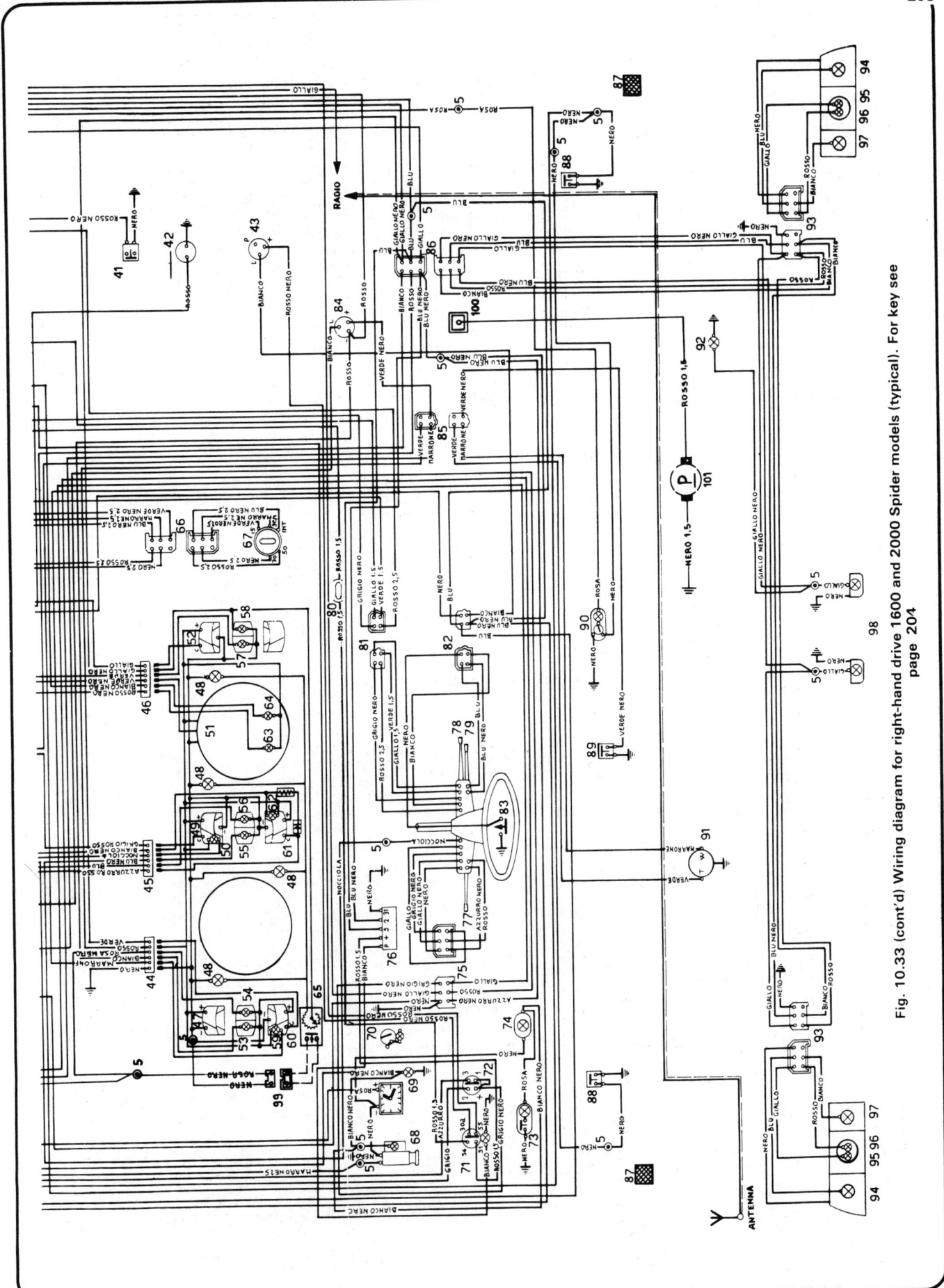

Fig. 10.33 (cont'd) Wiring diagram for right-hand drive 1600 and 2000 Spider models (typical). For key see page 204

Fig. 10.33 Key to wiring diagram for right-hand drive 1600 and 2000 Spider models (typical)

1. Dipped beam
2. Main beam
3. Front turn indicator
4. Front side light
5. Plug-in junction
6. Horn electrocompressor
7. Radiator fan
8. Radiator fan thermoswitch
9. Carburettor slow-running fuel cut-off device
10. Alternator
11. Front lamp three-outlet junction block (white)
12. Radiator fan solenoid switch
13. Battery
14. Radiator fan fuse
15. Starter motor
16. Engine oil low pressure warning light switch
17. Engine overheating warning light switch
18. Ignition distributor
19. Ignition coil
20. Turn indicator repeater
21. Brake fluid low level warning light switch
22. Two-speed windscreen wiper motor
23. Windscreen washer motor
24. Reversing light switch
25. Oil temperature gauge transmitter
26. Oil pressure gauge transmitter
27. Coolant temperature gauge transmitter
28. Front brake friction pad wear limit warning light switch
29. Two-outlet junction block (white)
30. Windscreen wiper intermittence device
31. Windscreen wiper intermittence device four-outlet junction block (white)
32. Windscreen wiper motor six outlet junction block (red)
33. Stop light switch
34. Engine compartment lamp with switch
35. Horn solenoid switch
36. Dipped beam solenoid switch
37. Cigarette lighter solenoid switch switch
38. Fuse box
39. Heating and ventilation system fan motor change-over switch
40. Heating and ventilation system fan two-speed motor
41. Tell-tale switch on special version only
42. Plug-in socket
43. Turn indicator flasher
44. Instrument cluster six-outlet junction block (black)
45. Instrument cluster six-outlet junction block (white)
46. Instrument cluster six-outlet junction block (red)
47. Voltmeter
48. Instrument cluster light
49. Oil pressure gauge
50. Engine oil low pressure warning light
51. Electronic rev counter
52. Oil temperature gauge
53. Handbrake warning light
54. Brake fluid low level and front friction pad wear limit warning light
55. Left-hand turn indicator warning light
56. Right-hand turn indicator warning light
57. Main beam warning light
58. Side light warning light
59. Fuel reserve warning light
60. Fuel gauge
61. Coolant temperature gauge
62. Engine overheating warning light
63. Tell-tale on special versions only
64. Alternator warning light
65. Instrument cluster light switch with intensity adjuster and push-button to check warning light (item 54)
66. Ignition switch six-outlet junction block (black)
67. Ignition and accessories key switch and anti-theft device
68. Cigarette lighter with light
69. Clock
70. Switch with warning light
71. Hazard signalling system change-over switch with warning light
72. Windscreen wiper motor change-over switch
73. Glove locker light with switch
74. Heating and ventilation control light
75. Windscreen wiper six-outlet junction block (white)
76. Hazard signalling system flasher
77. Windscreen wiper two-speed motor and washer motor control
78. Side light, dipped beam, main beam and headlamp flashing control
79. Turn indicator control
80. Hazard signalling system fuse (16 A)
81. Lights four-outlet junction block (white)
82. Lights four-outlet junction block (red)
83. Horn control
84. Handbrake warning light flasher
85. Four-outlet junction block (white)
86. Six-outlet junction block (white)
87. Open door safety reflector
88. Inside lamp press-switch on door pillar
89. Handbrake warning light switch
90. Inside lamp
91. Fuel gauge transmitter
92. Boot light
93. Tail lamp six-outlet junction block (white)
94. Rear turn indicator
95. Rear side light
96. Stop light
97. Reversing light
98. Number plate light
99. Two outlet junction block (white) for item 65
100. Fuel pump one-outlet junction block (white)
101. Fuel lift pump

Colour code

Bianco	White	Nocciola	Hazel
Blu	Blue	Nero	Black
Giallo	Yellow	Rosa	Pink
Grigio	Grey	Rosso	Red
Marrone	Brown	Verde	Green

Fig. 10.34 Key to wiring diagram for left-hand drive 1600 and 2000 spider models (typical)

1. Dipped beam
2. Main beam
3. Front turn indicator
4. Front side light
5. Plug-in junction
6. Horn electrocompressor
7. Radiator fan
8. Radiator fan thermoswitch
9. Carburettor slow-running fuel cut-off device
10. Alternator
11. Front lamp three-outlet junction block (white)
12. Radiator fan solenoid switch
13. Battery
14. Radiator fan fuse
15. Starter motor
16. Engine oil low pressure warning light switch
17. Engine overheating warning light switch
18. Ignition distributor
19. Ignition coil
20. Turn indicator repeater
21. Brake fluid low level warning light switch
22. Two-speed windscreen wiper motor
23. Windscreen washer motor
24. Reversing light switch
25. Oil temperature gauge transmitter
26. Oil pressure gauge transmitter
27. Coolant temperature gauge transmitter
28. Front brake friction pad wear limit warning light switch
29. Two-outlet junction block (white)
30. Windscreen wiper intermittence device
31. Windscreen wiper intermittence device four-outlet junction block (white)
32. Windscreen wiper motor six outlet junction block (red)
33. Stop light switch
34. Engine compartment lamp with switch
35. Horn solenoid switch
36. Dipped beam solenoid switch
37. Cigarette lighter solenoid switch
38. Fuse box
39. Heating and ventilation system fan motor change-over switch
40. Heating and ventilation system fan two-speed motor
41. Tell-tale switch on special version only
42. Plug-in socket
43. Turn indicator flasher
44. Instrument cluster six-outlet junction block (black)
45. Instrument cluster six-outlet junction block (white)
46. Instrument cluster six-outlet junction block (red)
47. Voltmeter
48. Instrument cluster light
49. Oil pressure gauge
50. Engine oil low pressure warning light
51. Electronic rev counter
52. Oil temperature gauge
53. Handbrake warning light
54. Brake fluid low level and front friction pad wear limit warning light
55. Left-hand turn indicator warning light
56. Right-hand turn indicator warning light
57. Main beam warning light
58. Side light warning light
59. Fuel reserve warning light
60. Fuel gauge
61. Coolant temperature gauge
62. Engine overheating warning light
63. Tell-tale on special version only
64. Alternator warning light
65. Instrument cluster light switch with intensity adjuster and push-button to check warning light (item 54)
66. Key switch six-outlet junction block (black)
67. Cigarette lighter
68. Cigarette lighter light
69. Clock
70. Spare switch with warning light
71. Ignition and accessories key switch with anti-theft device
72. Heater control light
73. Windscreen wiper motor change-over switch
74. Glove locker light with switch
75. Handbrake warning light flasher
76. Light control four-outlet junction block (white)
77. Light control four-outlet junction block (red)
78. Main beam, dipped beam, light signalling and side light control
79. Turn indicator control
80. Horn control
81. Windscreen wiper two-speed motor and washer motor control
82. Windscreen wiper six-outlet junction block (white)
83. Four-outlet junction block (white)
84. Six-outlet junction block (white)
85. Open door safety reflector
86. Inside lamp press-switch on door pillar
87. Handbrake warning light switch
88. Interior light
89. Fuel gauge transmitter
90. Boot lid
91. Tail lamp six-outlet junction block (white)
92. Rear turn indicator
93. Rear side light
94. Stop light
95. Reversing light
96. Number plate light
97. Fuel pump one-outlet junction block (white)
98. Fuel lift pump
99. Two-outlet junction block (white) for item 65

Colour code

Bianco	White	Nocciola	Hazel
Blu	Blue	Nero	Black
Giallo	Yellow	Rosa	Pink
Grigio	Grey	Rosso	Red
Marrone	Brown	Verde	Green

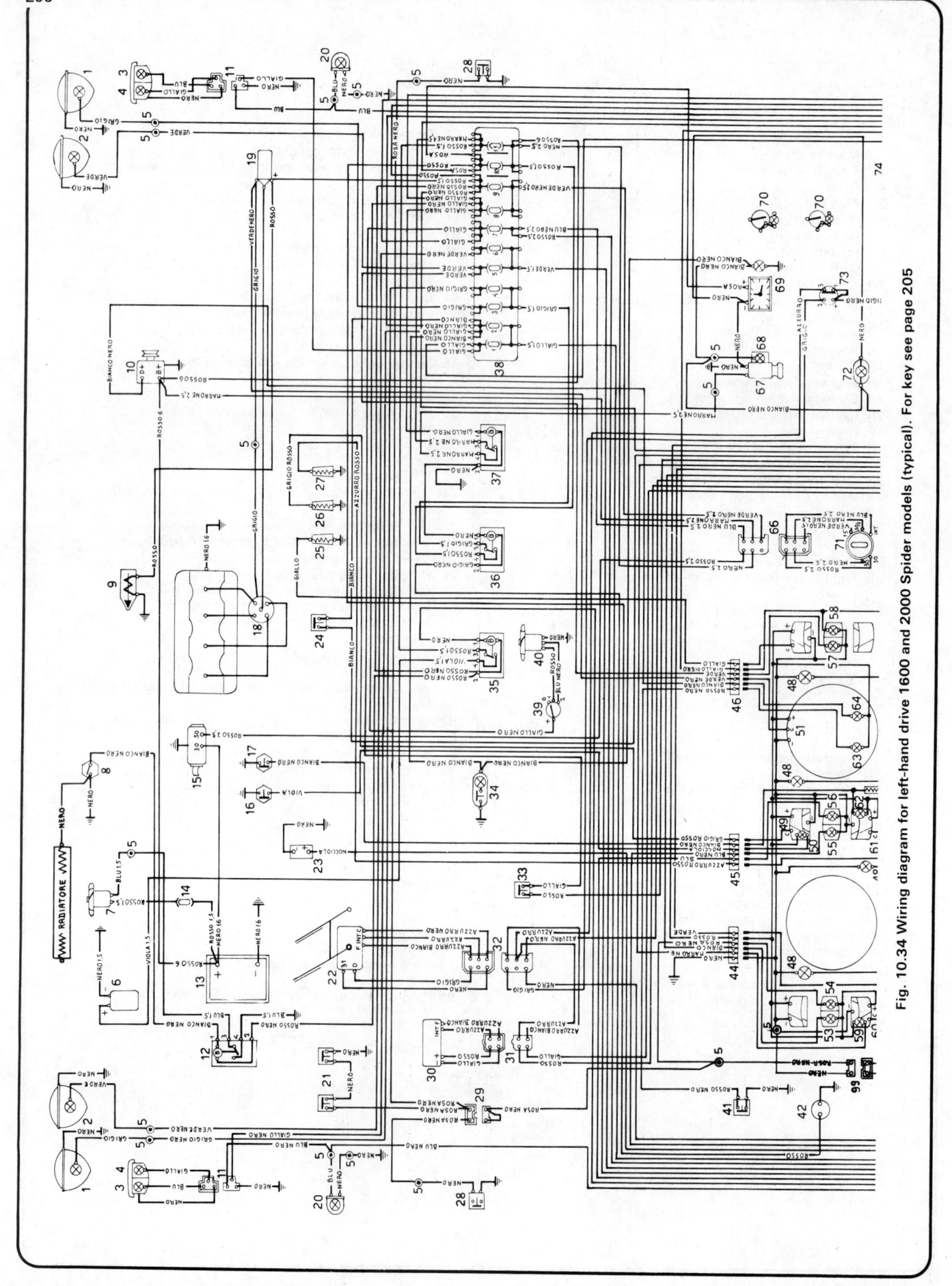

Fig. 10.34 Wiring diagram for left-hand drive 1600 and 2000 Spider models (typical). For key see page 205

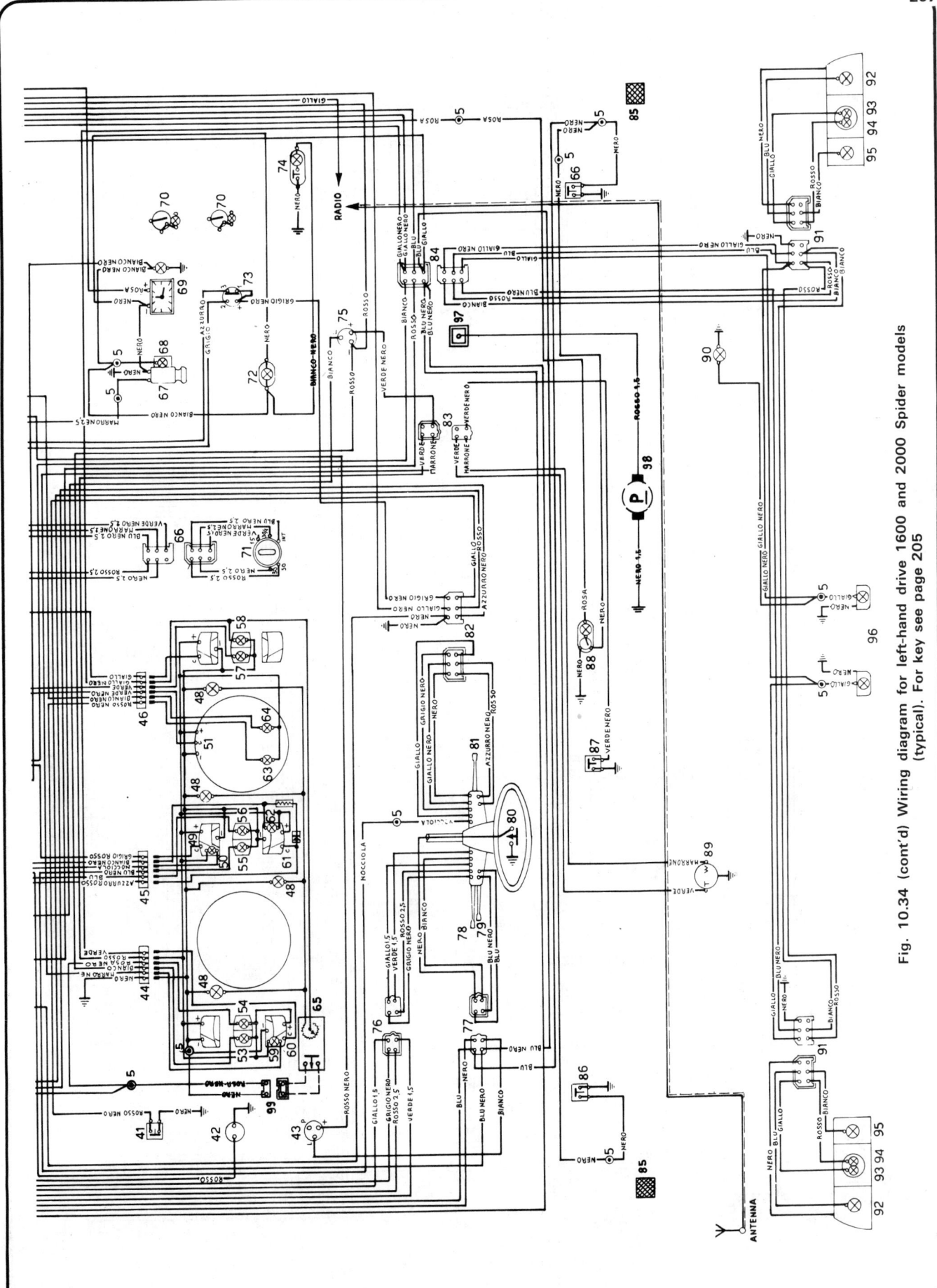

Fig. 10.34 (cont'd) Wiring diagram for left-hand drive 1600 and 2000 Spider models (typical). For key see page 205

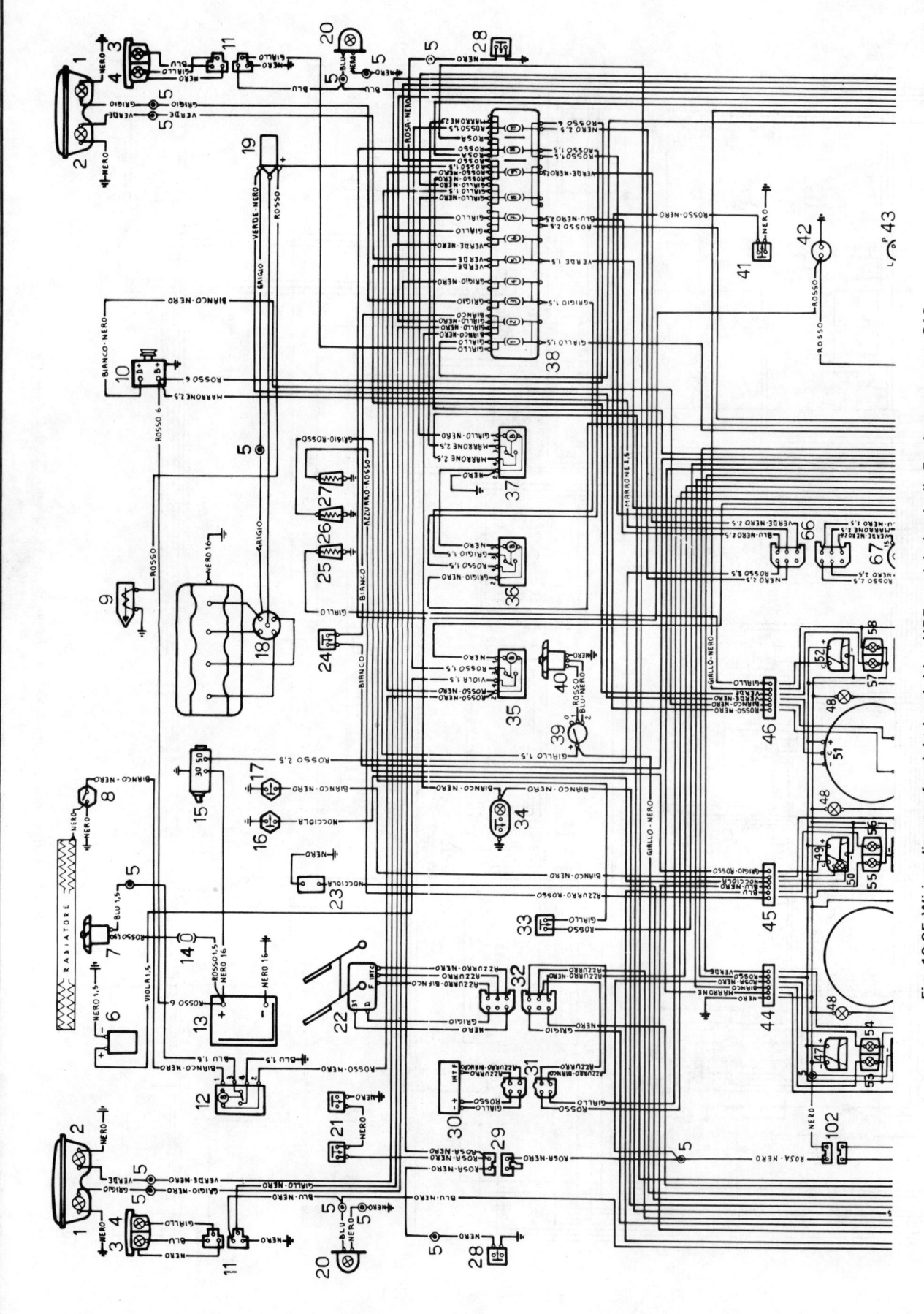

Fig. 10.35 Wiring diagram for right-hand drive HPE models (typical). For key see page 210

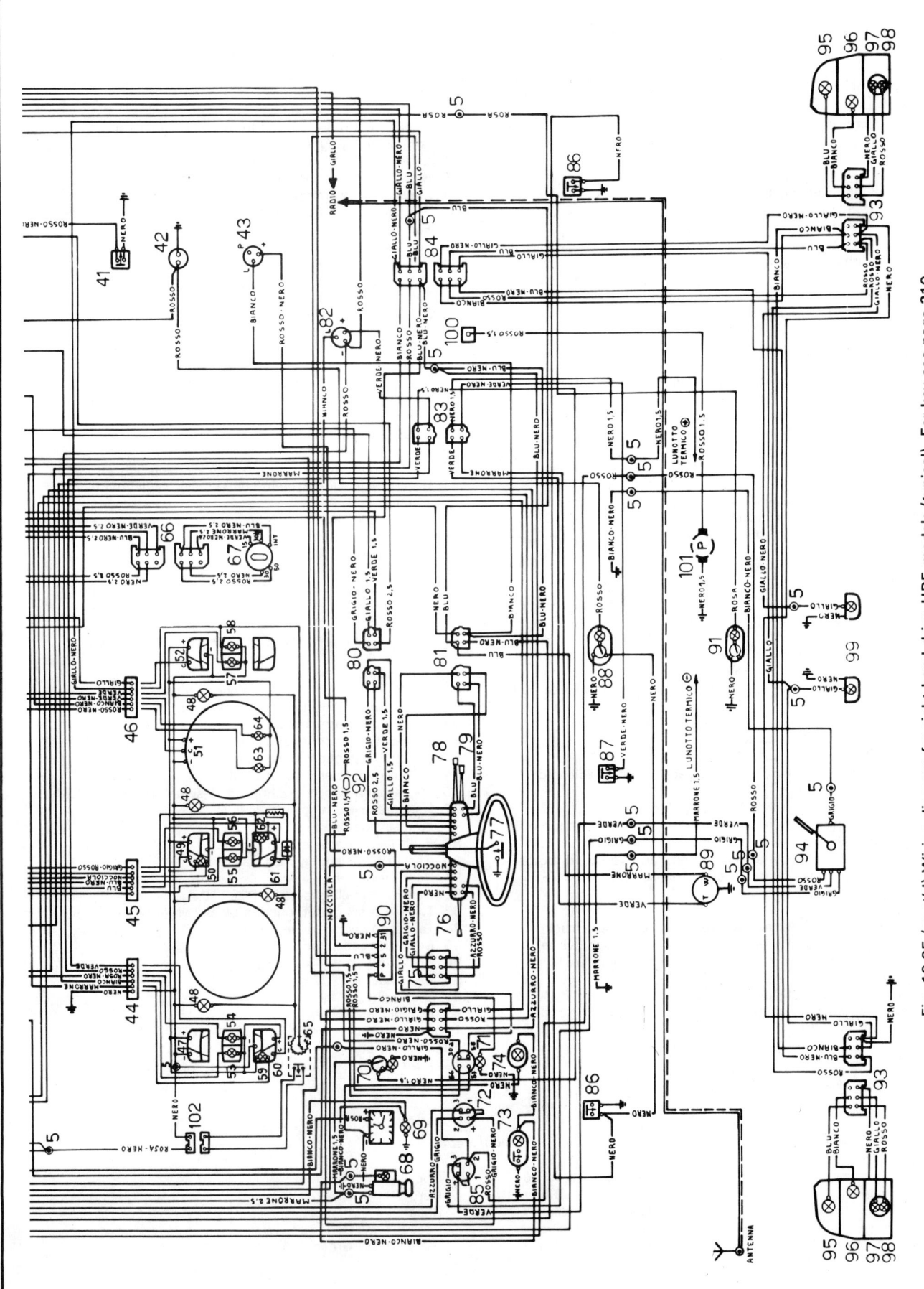

Fig. 10.35 (cont'd) Wiring diagram for right-hand drive HPE models (typical). For key see page 210

Fig. 10.35 Key to wiring diagram for right-hand drive HPE models (typical)

1. Dipped beam
2. Main beam
3. Front turn indicator
4. Front side light
5. Plug-in junction
6. Horn electrocompressor
7. Radiator fan
8. Radiator fan thermoswitch
9. Carburettor slow-running fuel cut-off device
10. Alternator
11. Front lamp three-outlet junction block (white)
12. Radiator fan solenoid switch
13. Battery
14. Radiator fan fuse
15. Starter motor
16. Engine oil low pressure warning light switch
17. Engine overheating warning light switch
18. Ignition distributor
19. Ignition coil
20. Turn indicator repeater
21. Brake fluid low level warning light switch
22. Two-speed windscreen wiper motor
23. Windscreen washer motor
24. Reversing light switch
25. Oil temperature gauge transmitter
26. Oil pressure gauge transmitter
27. Coolant temperature gauge transmitter
28. Front brake pad wear limit warning light switch
29. Two-outlet junction block (white)
30. Windscreen wiper intermittence device
31. Windscreen wiper intermittence device four-outlet junction block (white)
32. Windscreen wiper motor six outlet junction block (red)
33. Stop light switch
34. Engine compartment lamp with switch
35. Horn solenoid switch
36. Dipped beam solenoid switch
37. Heated rear window and cigarette lighter solenoid switch
38. Fuse box
39. Heating and ventilation system fan motor change-over switch
40. Heating and ventilation system fan two-speed motor
41. Tell-tale switch on special versions only
42. Plug-in socket
43. Turn indicator flasher
44. Instrument cluster six-outlet junction block (black)
45. Instrument cluster six-outlet junction block (white)
46. Instrument cluster six-outlet junction block (red)
47. Voltmeter
48. Instrument cluster light
49. Oil pressure gauge
50. Engine oil low pressure warning light
51. Electronic rev counter
52. Oil temperature gauge
53. Handbrake warning light
54. Brake fluid low level and front brake pad wear limit warning light
55. Left-hand turn indicator warning light
56. Right-hand turn indicator warning light
57. Main beam warning light
58. Side light warning light
59. Fuel reserve warning light
60. Fuel gauge
61. Coolant temperature gauge
62. Engine overheating warning light
63. Tell-tale on special versions only
64. Alternator tell-tale
65. Instrument cluster light switch with intensity adjuster and push-button to check warning light (item 54)
66. Ignition switch six-outlet junction block (black)
67. Ignition and accessories key switch with anti-theft device
68. Cigarette lighter with light
69. Clock
70. Heated rear window switch with warning light
71. Hazard signalling system change-over switch with warning light
72. Windscreen wiper two-speed motor change-over switch
73. Glove locker light with switch
74. Heater controls light
75. Windscreen wiper six-outlet junction block (white)
76. Windscreen wiper two-speed motor and washer motor motor control
77. Horn control
78. Side light, dipped beam, main beam and headlamp flashing control
79. Turn indicator control
80. Lights four-outlet junction block (white)
81. Lights four-outlet junction block (red)
82. Handbrake warning light flasher
83. Four-outlet junction block (white)
84. Six-outlet junction block (white)
85. Tailgate window wiper switch
86. Inside lamp and door lamp press-switch on door pillar
87. Handbrake warning light switch
88. Front inside lamp
89. Fuel gauge transmitter
90. Hazard signalling system flasher
91. Rear inside lamp
92. Hazard signalling system fuse (16 A)
93. Tail lamp six-outlet junction block (white)
94. Tailgate window wiper motor
95. Rear turn indicator
96. Reversing light
97. Rear side light
98. Stop light
99. Number plate light
100. Fuel lift pump single junction block (white)
101. Fuel lift pump
102. Two-outlet junction block (white) for item 65

Colour code

Bianco	White	Nocciola	Hazel
Blu	Blue	Nero	Black
Giallo	Yellow	Rosa	Pink
Grigio	Grey	Rosso	Red
Marrone	Brown	Verde	Green

Fig. 10.36 Key to wiring diagram for left-hand drive 1600, 1800 and 2000 HPE models (typical)

1. Dipped beam
2. Main beam
3. Front turn indicator
4. Front side light
5. Plug-in junction
6. Horn electrocompressor
7. Radiator fan
8. Radiator fan thermoswitch
9. Carburettor slow-running fuel cut-off device
10. Alternator
11. Front lamp three-outlet junction block (white)
12. Radiator fan solenoid switch
13. Battery
14. Radiator fan fuse
15. Starter motor
16. Engine oil low pressure warning light switch
17. Engine overheating warning light switch
18. Ignition distributor
19. Ignition coil
20. Turn indicator repeater
21. Brake fluid low level warning light switch
22. Two-speed windscreen wiper motor
23. Windscreen washer motor
24. Reversing light switch
25. Oil temperature gauge transmitter
26. Oil pressure gauge transmitter
27. Coolant temperature gauge transmitter
28. Front brake friction pad wear limit warning light switch
29. Two-outlet junction block (white)
30. Windscreen wiper intermittence device
31. Windscreen wiper intermittence device four-outlet junction block (white)
32. Windscreen wiper motor six outlet junction block (red)
33. Stop light switch
34. Engine compartment lamp with switch
35. Horn solenoid switch
36. Dipped beam solenoid switch
37. Heated rear window and cigarette lighter solenoid switch
38. Fuse box
39. Heating and ventilation system fan motor change-over switch
40. Heating and ventilation system fan two-speed motor
41. Tell-tale switch on special versions only
42. Plug-in socket
43. Turn indicator flasher
44. Instrument cluster six-outlet junction block (black)
45. Instrument cluster six-outlet junction block (white)
46. Instrument cluster six-outlet junction block (red)
47. Voltmeter
48. Instrument cluster light
49. Oil pressure gauge
50. Engine oil low pressure warning light
51. Electronic rev counter
52. Oil temperature gauge
53. Handbrake warning light
54. Brake fluid low level and front brake pad wear limit warning light
55. Left-hand turn indicator warning light
56. Right-hand turn indicator warning light
57. Main beam warning light
58. Side light warning light
59. Fuel reserve warning light
60. Fuel gauge
61. Coolant temperature gauge
62. Engine overheating warning light
63. Tell-tale on special versions only
64. Alternator warning light
65. Instrument cluster light switch with intensity adjuster and push-button to check warning light (item 54)
66. Key switch six-outlet junction block (black)
67. Cigarette lighter with light
68. Electronic clock with light
69. Rear heated window switch with warning light
70. Ignition and accessories key switch with anti-theft device
71. Tailgate window wiper pull-switch
72. Windscreen wiper change-over switch
73. Spare switch with warning light
74. Heater controls light
75. Glove locker light with switch
76. Handbrake warning flasher
77. Light control four-outlet junction block (white)
78. Light control four-outlet junction block (red)
79. Side light, dipped beam, main beam and light signalling control
80. Turn indicator control
81. Horn control
82. Windscreen wiper two-speed motor and washer motor control
83. Windscreen wiper six-outlet junction block (white)
84. Four-outlet junction block (white)
85. Six-outlet junction block (white)
86. Front inside lamp press-switch on door pillar
87. Handbrake warning light switch
88. Front inside lamp
89. Fuel gauge transmitter
90. Rear inside lamp
91. Tailgate window wiper motor and heated rear window plug-in junction
92. Rear turn indicator
93. Reversing light
94. Rear side light
95. Stop light
96. Tail lamp six-outlet junction block
97. Tailgate window wiper motor
98. Number plate light
99. Fuel pump one-outlet junction block (white)
100. Fuel lift pump
101. Two-outlet junction block (white) for item 65

Colour code

Bianco	White	Nocciola	Hazel
Blu	Blue	Nero	Black
Giallo	Yellow	Rosa	Pink
Grigio	Grey	Rosso	Red
Marrone	Brown	Verde	Green

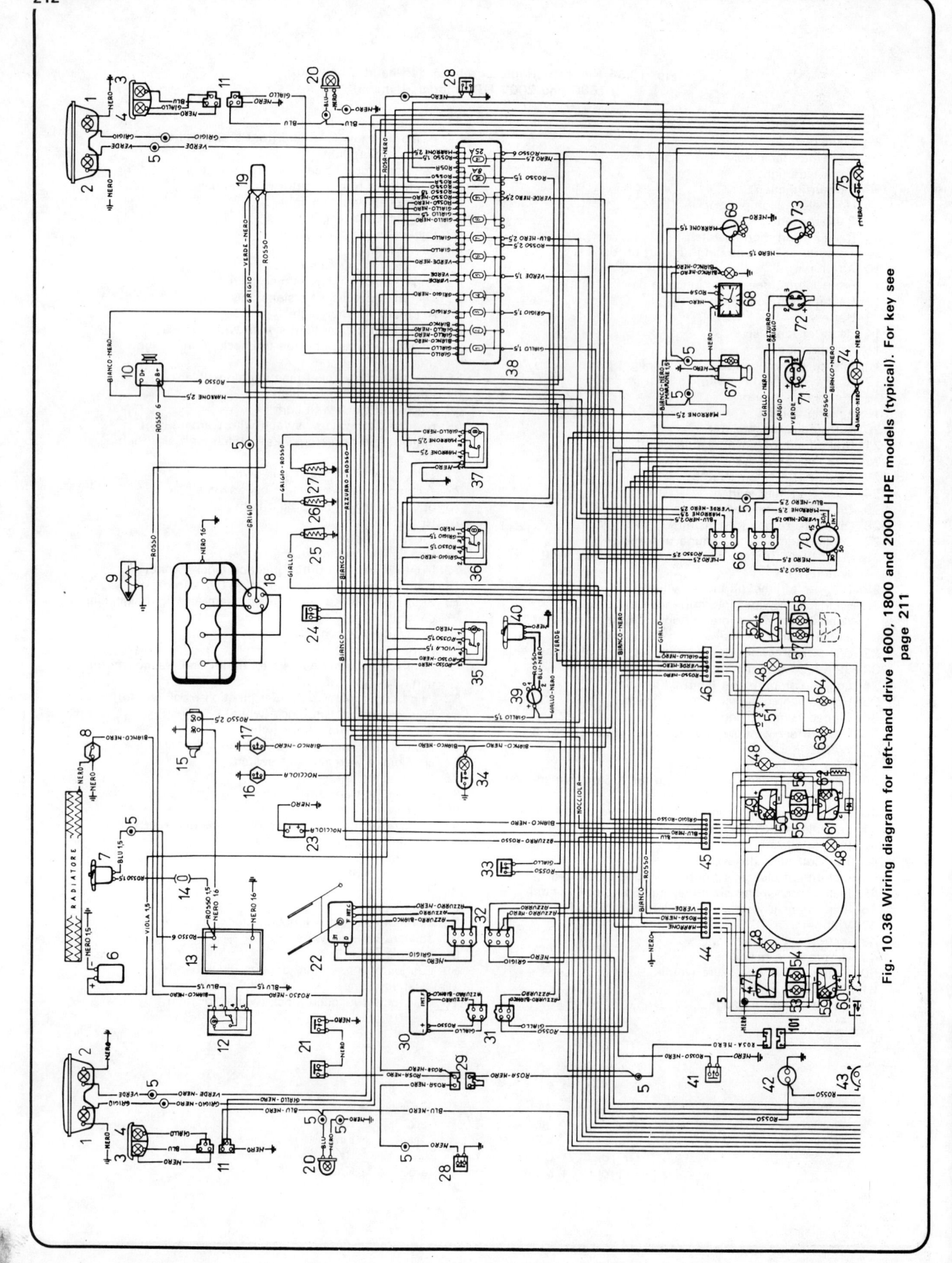

Fig. 10.36 Wiring diagram for left-hand drive 1600, 1800 and 2000 HPE models (typical). For key see page 211

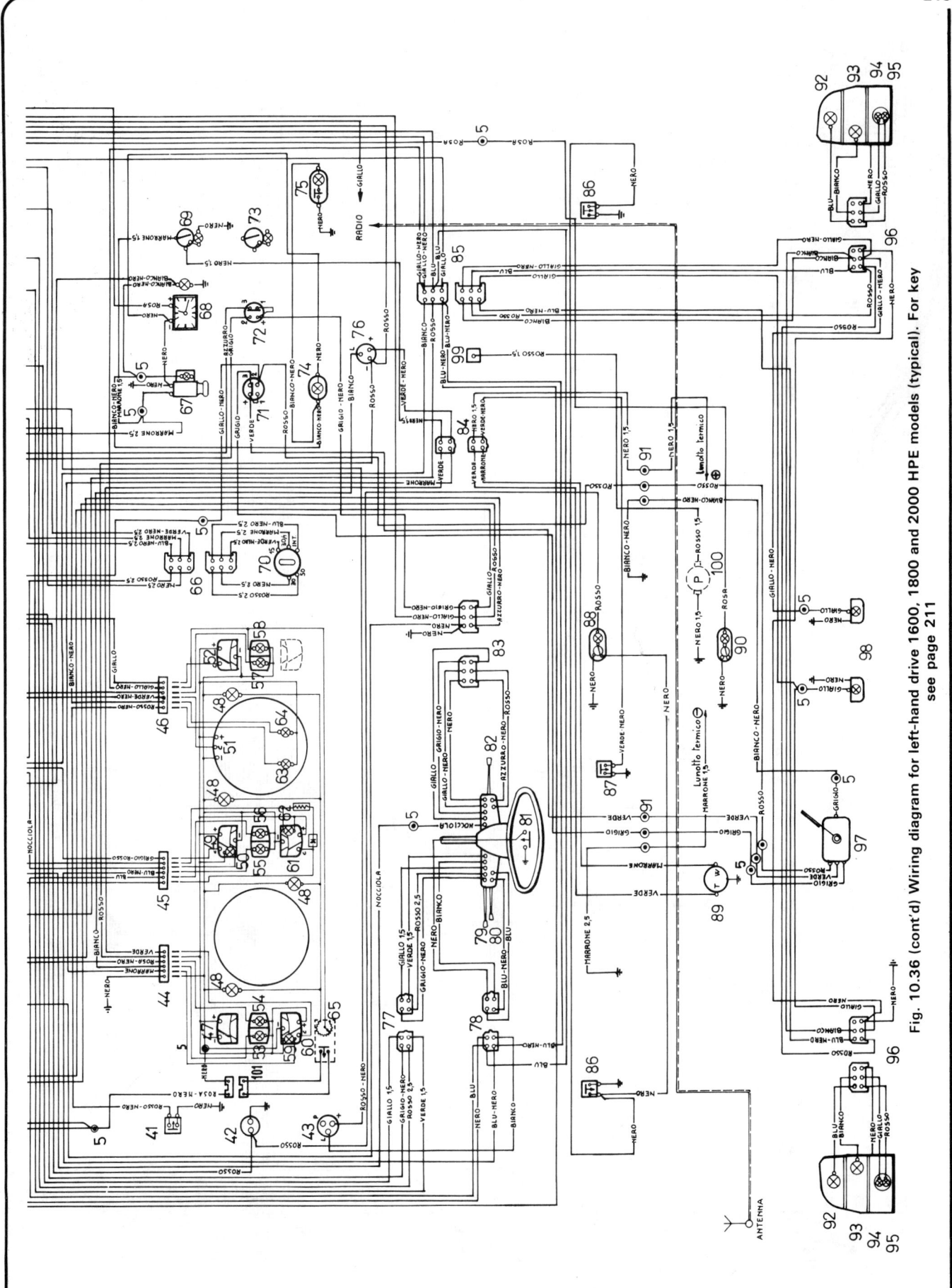

Fig. 10.36 (cont'd) Wiring diagram for left-hand drive 1600, 1800 and 2000 HPE models (typical). For key see page 211

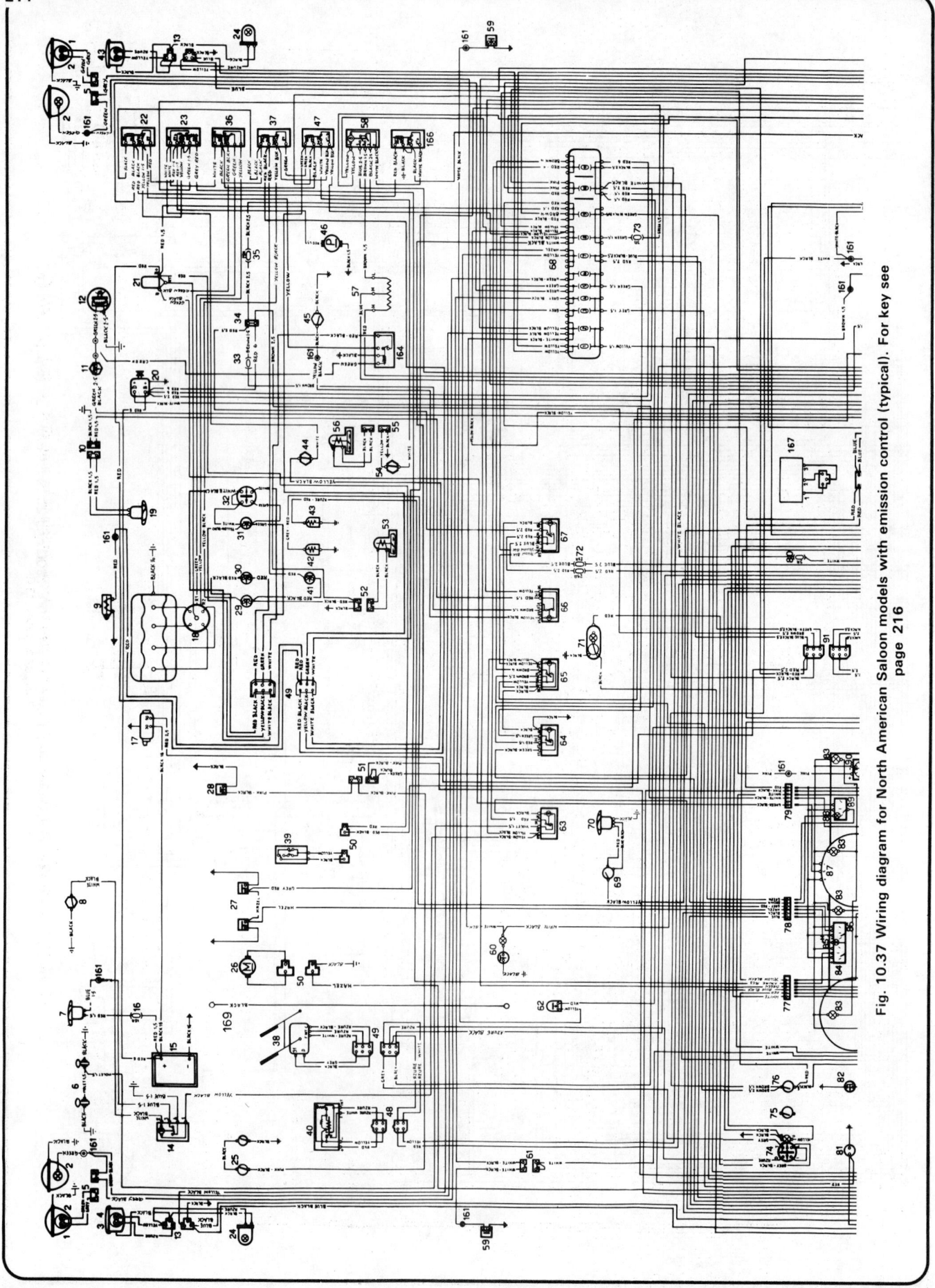

Fig. 10.37 Wiring diagram for North American Saloon models with emission control (typical). For key see page 216

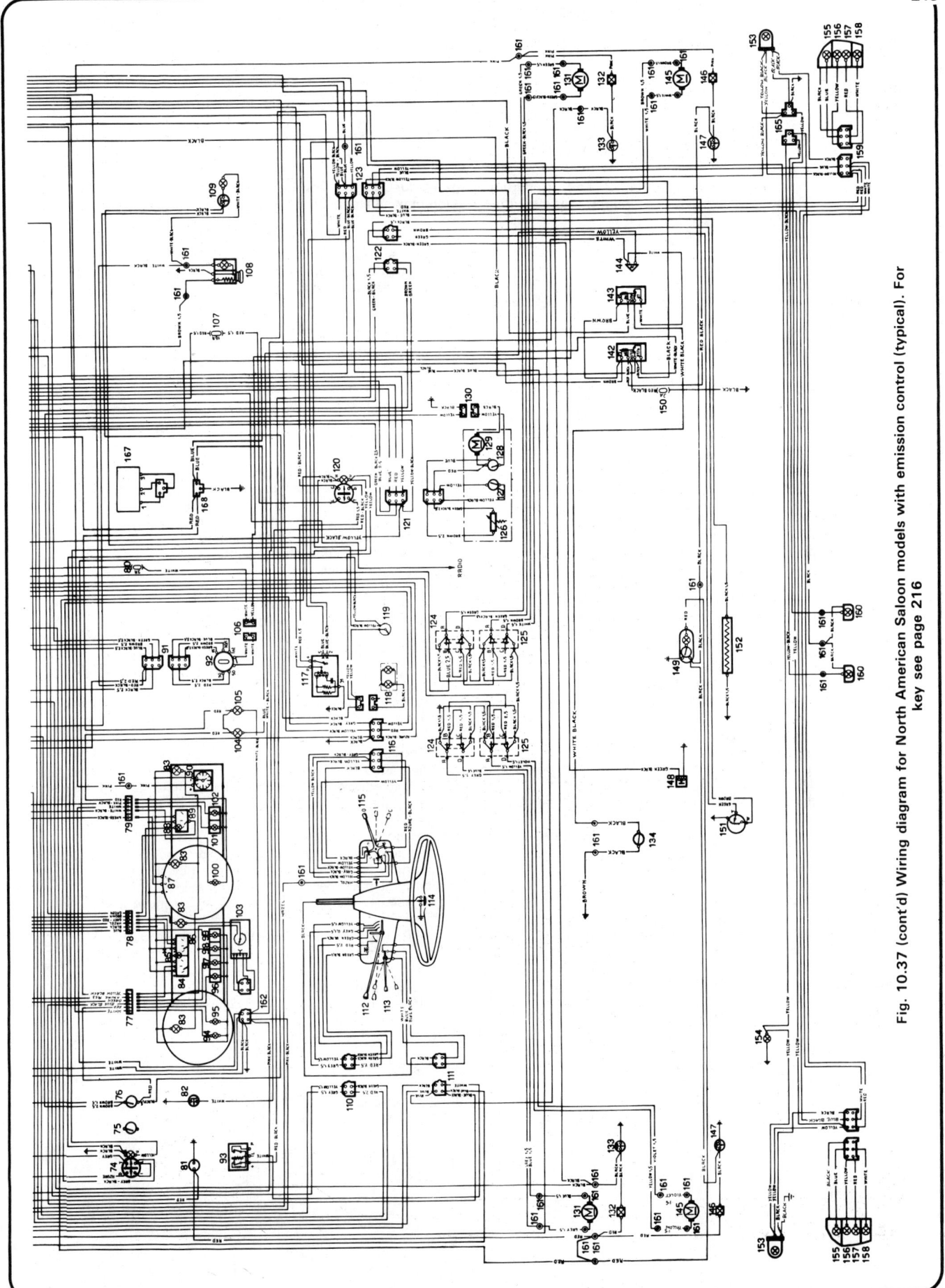

Fig. 10.37 (cont'd) Wiring diagram for North American Saloon models with emission control (typical). For key see page 216

Fig. 10.37 Key to wiring diagram for North American Saloon models with emission control (typical)

1. Dipped beam
2. Main beam
3. Front turn indicator
4. Front side light
5. Dipped and main beams two-outlet junction block (white)
6. Horn
7. Radiator fan
8. Radiator fan thermoswitch
9. Idling fuel cut-off device
10. Air conditioner condenser fan two-outlet junction block (white)*
11. Air conditioner minimum pressure switch*
12. Air conditioner compressor electro-magnetic coupling *
13. Front light three-outlet junction block (white)
14. Radiator fan solenoid switch
15. Battery
16. Radiator fan fuse (16 A)
17. Starter motor
18. Ignition distributor
19. Air conditioner condenser fan *
20. Alternator
21. Ignition coil
22. Fuel pump solenoid switch 12 V – 20 A (NO)
23. Ignition and fuel pump change-over switches 12 V – 20 A
24. Front clearance light
25. Brake fluid low level warning light
26. Windscreen washer motor
27. Engine oil low pressure and engine overheating warning light
28. Brake system pressure drop tell-tale
29. Emission control system press-switch (5 A)
30. Emission control system press-switch (5 A) on gearbox (3rd and 4th speed)
31. Emission control system press-switch (5 A) on gearbox (1st and 2nd speed)
32. Emission control system change-over switch (5 A) on gearbox (reverse and 5th speed)
33. Air conditioner motor fuse (16 A) *
34. Junction block *
35. Air conditioner circuit fuse (25 A) *
36. Ignition switch 12 V – 20 A
37. Emission control system solenoid switch (N.O.)
38. Windscreen wiper two-speed motor
39. EGR tell-tale device
40. Windscreen wiper intermittent device with remote control switch
41. Emission control system press-switch (5 A) on clutch pedal
42. Oil pressure gauge transmitter
43. Coolant temperature gauge transmitter
44. Ignition switch control thermo-switch
45. Air conditioner fan control temperature sensor *
46. Fuel pump 12 V
47. Emission control system solenoid switch (NC)
48. Windscreen wiper remote control switch four-outlet junction block (white)
49. Six-outlet junction block (red)
50. Two-outlet junction block
51. Brake system pressure drop tell-tale two-outlet junction block (white)
52. Emission control system electro-valve two-outlet junction block (white)
53. Fast idling control electro-valve (12 V)
54. Emission control system electro-valve thermo-switch
55. Emission control system electro-valve two-outlet junction block (white)
56. Diverter control electro-valve (12 V)
57. Air conditioner system resistor *
58. Air conditioner unit solenoid switch *
59. Front brake pad wear limit warning light contact
60. Engine compartment light with switch
61. Front brake pad wear limit warning light contact two-outlet junction block (white)
62. Stop light switch
63. Horn solenoid switch
64. Main beam solenoid switch
65. Heated rear window and cigarette lighter solenoid switch
66. Air conditioner fan solenoid switch *
67. Window lift control solenoid switch *
68. Fuse box
69. Heating and ventilation fan motor change-over switch
70. Heating and ventilation fan two-speed motor
71. Car front interior lamp with switch
72. Window lift fuses (25 A) *
73. Fuel pump fuse (3 A)
74. Windscreen wiper high and low speed change-over switch with light
75. Extra switch
76. Heated rear window switch
77. Instrument cluster six-outlet junction block (white)
78. Instrument cluster six-outlet junction block (red)
79. Instrument cluster six-outlet junction block (black)
80. Unfastened driver's belt warning buzzer loose fuse (3 A)
81. Socket
82. Ignition key inserted, front left door pillar fitted press-switch
83. No. 4 light for instrument cluster
84. Coolant temperature gauge
85. Engine oil low pressure and engine overheating warning light with diode
86. Oil pressure gauge
87. Electronic rev. counter
88. Fuel reserve warning light
89. Fuel gauge
90. Electronic clock
91. Ignition key switch six-outlet junction block (black)

92 Ignition and accessory key switch with anti-theft device
93 Turn indicator flasher
94 Hazard signalling system warning light
95 Heated rear window warning light
96 Left turn indicator warning light
97 Main beams warning light
98 Side lights warning light
99 Right turn indicator warning light
100 Alternator warning light
101 BRAKE tell-tale (brake system pressure drop, brake fluid level and handbrake)
102 BRAKE-PAD tell-tale (front brake pads wear)
103 Instruments light intensity adjuster with push-button to check BRAKE PAD - BRAKE - EGR tell-tale.
104 Driver's belt FASTEN-BELTS warning light
105 EGR tell-tale (scheduled maintenance)
106 Buzzer circuit two-inlet junction block (white)
107 Hazrd signalling system fuse (16 A)
108 Cigarette lighter with light
109 Glove locker light with switch
110 Lights four-outlet junction block (white)
111 Lights four-outlet junction block (red)
112 Side light, dipped beam, main beam and headlamp flashing control
113 Turn indicator control
114 Horn control
115 Windscreen wiper two-speed motor and washer motor control
116 Windscreen wiper six-outlet junction block (white)
117 Hazard signalling system flasher
118 Heating and ventilation controls light with two-outlet junction block (white)
119 Heating and ventilation controls light switch with intensity adjuster
120 Hazard signalling system lights switch with tell-tale
121 Air conditioner six-outlet junction block (white) *
122 Four-outlet junction block (white) for items 148 - 151 - 152
123 Six outlet junction block (white) for exterior lights
124 Front door window lifts switches fitted to centre console rear section *
125 Rear door window lifts switches fitted to centre console rear section *
126 Air conditioner thermostat *
127 Air conditioner limit switch (fitted to the recirculation control lever) *
128 Air conditioner booster fan motor change-over switch *
129 Air conditioner two-outlet junction block (white) *
131 Front door window lift motor *
132 Front open door safety lamp
133 Front door pillar press-switch
134 Driver's seat belt switch
142 EGR system solenoid switch (NC)
143 Driver's belt circuit solenoid switch (NO)
144 Unfastened driver's belt and ignition key fitted buzzer (12 V)
145 Rear door window lift motor *
146 Rear open door safety lamp
147 Rear door pillar press-switch
148 Handbrake warning light contact
149 Car rear interior lamp
150 EGR tell-tale solenoid switch fuse (3 A)
151 Fuel gauge transmitter
152 Heated rear window filament
153 Rear clearance light
154 Boot light
155 Rear turn indicator
156 Rear side light
157 Stop light
158 Reversing light
159 Rear light six-outlet junction blocl (white)
160 Number plate light
161 Plug-in junction
162 Four outlet junction block (white) for item 103
164 Safety system relay six-outlet junction block (red)
165 Number plate light three-outlet junction block (white)
166 Air conditioner cut-out solenoid switch (NC) when starting the engine
167 Delaying device
168 Delaying device three-outlet junction block (white)
169 Steering shaft ground

* Fitted as an optional extra

Colour code

Bianco	White	Nocciola	Hazel
Blu	Blue	Nero	Black
Giallo	Yellow	Rosa	Pink
Grigio	Grey	Rosso	Red
Marrone	Brown	Verde	Green

Please note that, at the time of writing, wiring diagrams for the HPE and Zagato models with emission control were not available

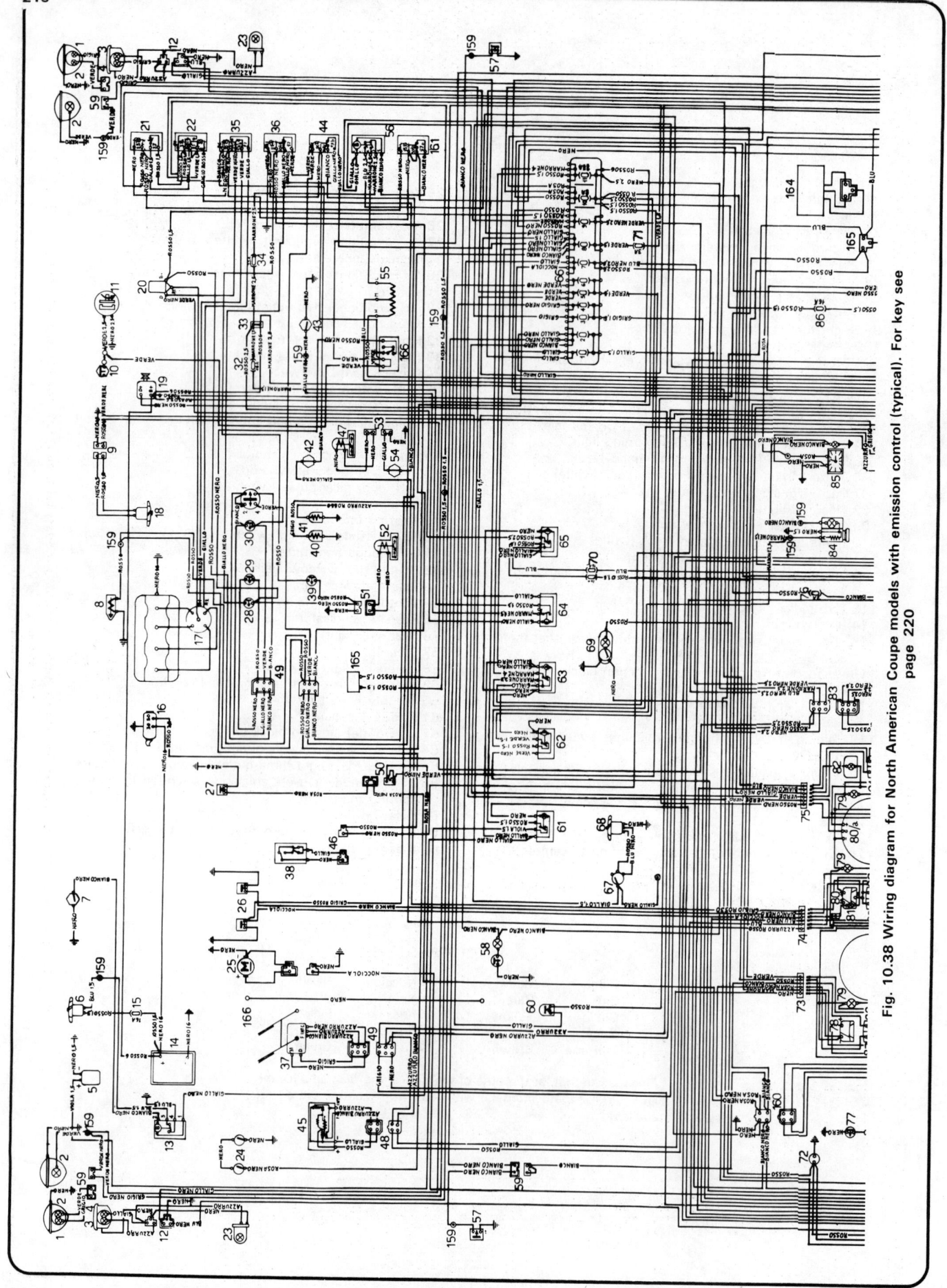

Fig. 10.38 Wiring diagram for North American Coupe models with emission control (typical). For key see page 220

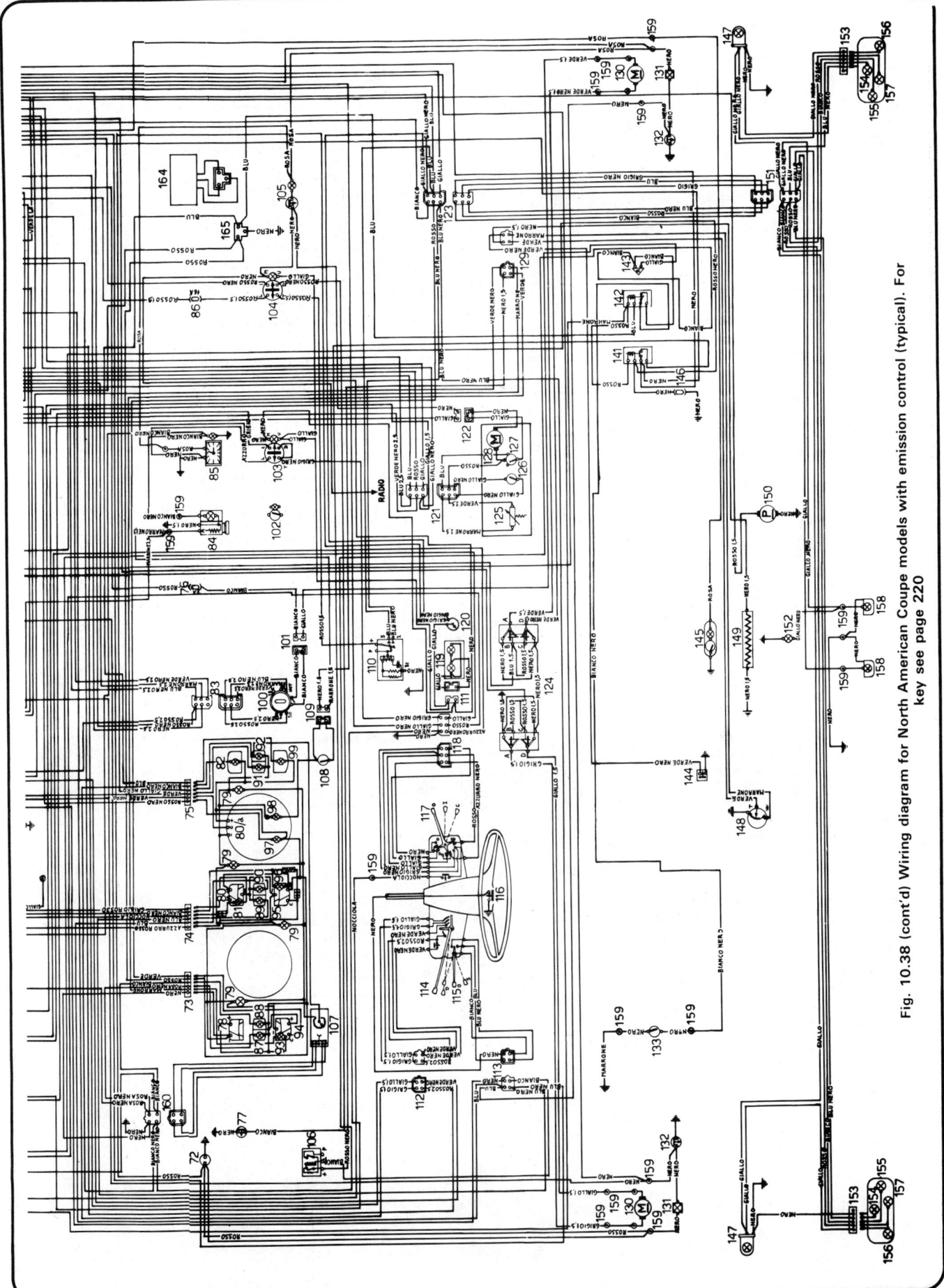

Fig. 10.38 (cont'd) Wiring diagram for North American Coupe models with emission control (typical). For key see page 220

Fig. 10.38 Key to wiring diagram for North American Coupe models with emission control (typical)

1. Dipped beam
2. Main beam
3. Front turn indicator
4. Front side light
5. Horn electro-compressor
6. Radiator fan
7. Radiator fan thermoswitch
8. Idling fuel cut-off device
9. Air condenser fan two-outlet junction block (white)
10. Air conditioned minimum pressure switch
11. Air conditioner compressor electro-magnetic coupling *
12. Front light three-outlet junction block (white) *
13. Radiator fan solenoid switch
14. Battery
15. Radiator fan fuse (16 A)
16. Starter motor
17. Ignition distributor
18. Air conditioner condenser fan *
19. Alternator
20. Ignition coil
21. Fuel pump solenoid switch 12 V - 20 A (NO)
22. Ignition and fuel pump change-over switches 12 V - 20 A
23. Front clearance light
24. Brake fluid low level warning light
25. Windscreen washer motor
26. Engine oil low pressure and engine overheating warning light
27. Brake system pressure drop tell-tale
28. Emission control system press-switch (5 A)
29. Emission control system press-switch (5 A) on gearbox (3rd and 4th speed)
30. Emission control system press-switch (5 A) on gearbox (1st and 2nd speed)
31. Emission control system change-over switch (5 A) on gearbox (Reverse and 5th speed)
32. Air conditioner motor fuse (16 A) *
33. Junction block *
34. Air conditioner circuit fuse (25 A)
35. Ignition switch 12 V - 20 A
36. Emission control system solenoid switch (N.O.)
37. Windscreen wiper two-speed motor
38. EGR tell-tale device
39. Emission control system press-switch (5 A) on clutch pedal
40. Oil pressure gauge transmitter
41. Coolant temperature gauge transmitter
42. Ignition switch control therm-switch
43. Air conditioner fan control temperature sensor *
44. Emission control system solenoid switch (PC)
45. Windscreen wiper intermittence device with remote control switch
46. Scheduled maintenance device two-outlet junction block (red)
47. Diverter control electro-valve (12 V)
48. Windscreen wiper remote control switch four-outlet junction block (white)
49. Six-outlet junction block (red)
50. Brake system pressure drop tell-tale two-outlet junction block (white)
51. Emission control system electro-valve item 52 two-outlet junction block (white)
52. Fast idling control electro-valve (12 V)
53. Two-outlet junction block (white)
54. Emission control system electro-valve thermo-swiitch
55. Air conditioner system resistor *
56. Air conditioner unit solenoid switch *
57. Front brake pad wear limit warning light contact
58. Engine compartment light with switch
59. Dipped and main beams two-outlet junction block (white)
60. Stop light switch
61. Air-horns solenoid switch
62. Main beams solenoid switch
63. Heated rear window and cigarette lighter solenoid switch
64. Air conditioner fan solenoid switch *
65. Window lift control solenoid switch *
66. Fuse box
67. Heating and ventilation fan motor change-over switch
68. Heating and ventilation fan two-speed motor
69. Car front interior lamp with switch
70. Window lift fuses (16 A)
71. Fuel pump fuse (3 A)
72. Plug-in socket
73. Instrument cluster six-outlet junction block (black)
74. Instrument cluster six-outlet junction block (white)
75. Instrument cluster six-outlet junction block (red)
76. Seat belts and start-block 12 V electronic control unit fuse (3 A)
77. Ignition key inserted, front left door pillar fitted press-switch
78. Voltmeter
79. No 4 lights for instrument cluster
80. Oil pressure gauge
80a. Electronic rev. counter
81. Engine oil low pressure warning light
82. EGR tell-tale (scheduled maintenance)
83. Ignition key switch six-outlet junction block (black)
84. Cigarette lighter with light
85. Electronic clock with light
86. Hazard signalling system fuse (16 A)
87. BRAKE tell-tale (handbrake and brake system failure)
88. BRAKE-PAD tell-tale (front brake pads wear)
89. Left turn indicator warning light

90 Right turn indicator warning light
91 Main beams warning light
92 Side lights warning light
93 Fuel reserve warning light
94 Fuel gauge
95 Coolant temperature gauge
96 Engine overheating warning light
97 Alternator warning light
98 Heated rear window warning light
99 Driver's seat belts FASTEN-BELTS tell-tale
100 Ignition and accessory key switch with anti-theft device
101 Seat belts circuit two-outlet junction block (white)
102 Extra switch with warning light
103 Windscreen wiper high and low speed change-over switch
104 Hazard signalling system switch with warning light
105 Glove locker light with switch
106 Turn indicator flasher
107 Instrument light intensity adjuster with push-button to check BRAKE PAD – BRAKE – EGR tell-tale
108 Heated rear window switch
109 Heated rear window two-outlet junction block (white)
110 Hazard signalling system flasher
111 Heating and ventilation controls light two-outlet junction block
112 Lights four-outlet junction block (white)
113 Lights four-outlet junction block (red)
114 Side light, dipped beam, main beam and headlamp flashing control
115 Turn indicator control
116 Air-horns control
117 Windscreen wiper two-speed motor and washer motor control
118 Windscreen wiper six-outlet junction block (white)
119 Heating and ventilation controls light
120 Control switches light intensity adjuster switch
121 Air conditioner six-outlet junction block (white) *
122 Seat belt two-outlet junction block (white)
123 Exterior lights six-outlet junction block (white)
124 Door window lifts switches fitted to centre console rear section *
125 Air conditioner thermostat *
126 Air conditioner limit switch (fitted to the recirculation control lever) *
127 Air conditioner booster fan motor change-over switch *
128 Air conditioner booster fan motor *
129 Four-outlet junction block (white) for items 144-148-149
130 Door window lift motor *
131 Open door safety lamp
132 Interior light door pillar press-switch
133 Driver's seat belts switch
141 EGR system solenoid switch (NC)
142 Driver's seat belts circuit, solenoid switch (NO)
143 Ignition on and seat belt unfastened 12 V buzzer
144 Handbrake warning light contact
145 Car rear interior lamp
146 EGR tell-tale solenoid switch fuse
147 Rear clearance light
148 Fuel gauge transmitter
149 Heated rear window filament
150 Fuel pump 12 V
151 Rear light six-outlet junction block (white)
152 Boot light
153 Rear light six-outlet junction block (white)
154 Rear turn indicator
155 Reversing light
156 Rear side light
157 Stop light
158 Number plate light
159 Plug-in junction
160 Four-outlet junction block (white) for item 107
161 Air conditioner cut-out solenoid switch (NC) when starting engine
162 Fuel electric pump feed system mercury switch
163 Fast idling control solenoid switch when engaging the electro-magnetic coupling item 11
164 Electric delaying device
165 Electric delaying device three-outlet junction block (white)
166 Steering column earth

* Fitted as an optional extra

Colour code

Bianco	White	Nocciola	Hazel
Blu	Blue	Nero	Black
Giallo	Yellow	Rosa	Pink
Grigio	Grey	Rosso	Red
Marrone	Brown	Verde	Green

Please note that, at the time of writing, wiring diagrams for the HPE and Zagato models with emission control were not available.

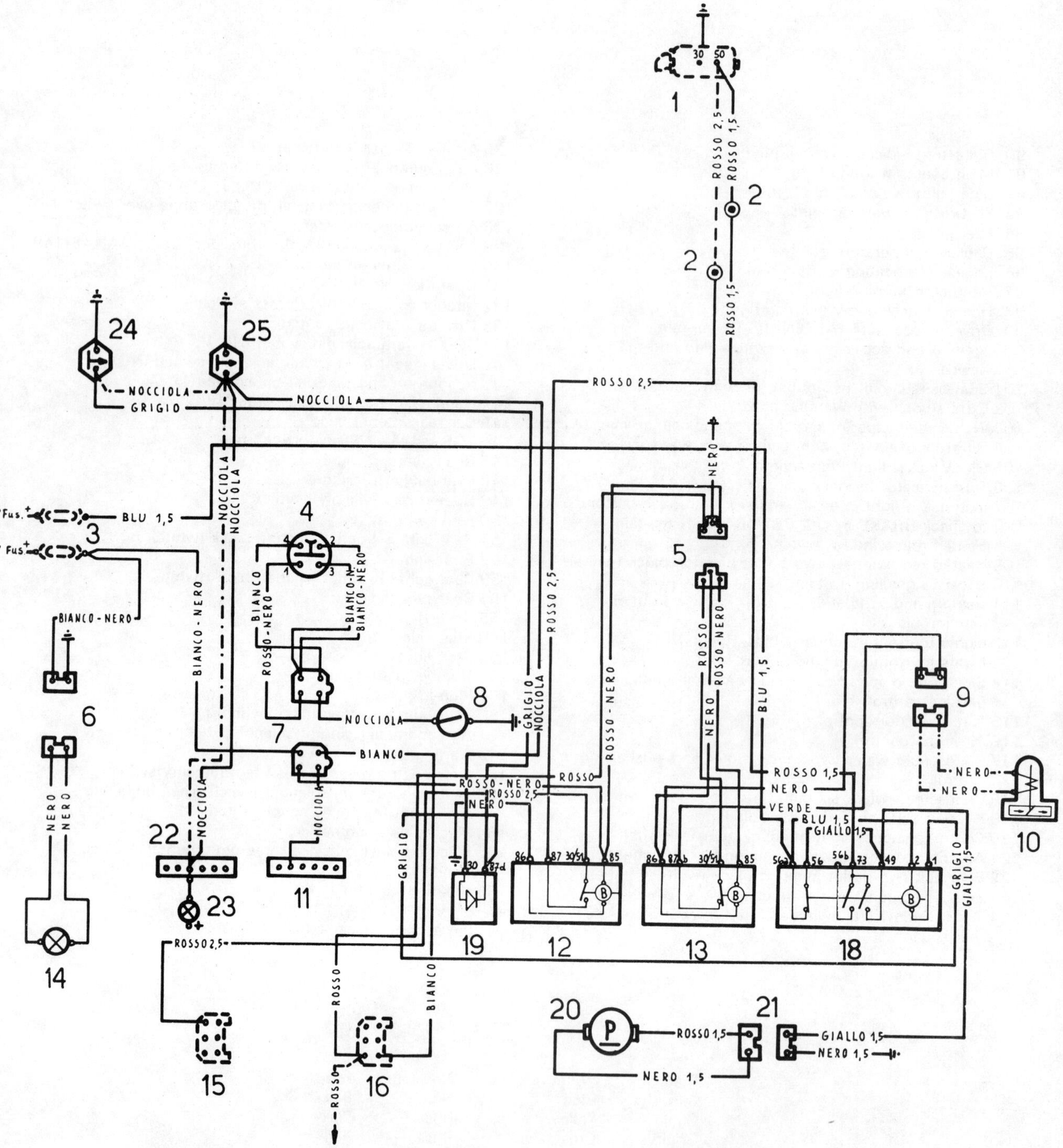

Fig. 10.39 Wiring diagram variations for cars fitted with automatic transmission

Key to Fig. 10.39

1 Starter motor
2 Plug-in junction
3 Fuse
4 Neutral and reverse tell-tale switch
5 Three-outlet junction block (white) for item 13
6 Two outlet junction block (white) for item 14
7 Four-outlet junction block (white) for items 4 and 8
8 Transmission fluid overheating thermoswitch
9 Two-outlet junction block (white) for item 10
10 Carburettor fast-idling electrovalve
11 Instrument cluster six-outlet junction block (black)
12 Starter motor block solenoid switch (N.O.) operating if gear selected
13 Solenoid switch (N.C.) operating the carburettor fast-idling electrovalve when braking with gear engaged
14 Gear selection light
15 Ignition switch six-outlet junction block (black)
16 Tail light cluster six-outlet junction block (white)
17 To stop light switch (connection already fitted)
18 Fuel pump solenoid switch
19 Diode to disconnect items 24 and 25
20 Fuel pump
21 Item 20 two-outlet junction block (white)
22 Instruments six-outlet junction block (white)
23 Engine oil low pressure and overheating warning light
24 Engine oil low pressure warning light switch
25 Engine overheating warning light switch

Colour code

Bianco	White	Nocciola	Hazel
Blu	Blue	Nero	Black
Giallo	Yellow	Rosa	Pink
Grigio	Grey	Rosso	Red
Marrone	Brown	Verde	Green

Chapter 11 Suspension and steering

Contents

Anti-roll bar (front) – removal and refitting	5
Anti-roll bar (rear) – removal and refitting	13
Anti-roll bar bushes (rear) – removal and refitting	14
Fault diagnosis – suspension and steering	36
Front hub and bearings – renewal	9
Front strut assembly – dismantling and reassembly	6
Front suspension spring and strut – removal and refitting	2
Front suspension wishbones – removal and refitting	3
Front wheel hub and carrier – removal and refitting	8
Front wishbones – renewing the bushes	7
General description	1
Power steering pump – removal and refitting	33
Power steering pump drivebelt – removal, refitting and tensioning	34
Power steering rack assembly – removal and refitting	31
Power steering rack overhaul – general	32
Power steering system – filling and bleeding	35
Rear hub and bearings – renewal	18
Rear suspension crossmember – removal and refitting	12
Rear suspension spring and strut – removal and refitting	10
Rear suspension strut – dismantling and reassembly	16
Rear suspension transverse links – removal and refitting	11
Rear wheel hub and stub axle – removal and refitting	17
Rear wheel toe-in – adjustment	19
Roadwheels and tyres – general	20
Steering column – removal and refitting	22
Steering column – dismantling and reassembly	24
Steering column adjustable bracket – removal and refitting	23
Steering column anti-theft lock – removal and refitting	25
Steering geometry – checking the front wheel alignment	29
Steering rack damper – removal and refitting	27
Steering rack (manual) – removal and refitting	26
Steering rack (manual) – overhaul	30
Steering tie-rod balljoint – removal and refitting	28
Steering wheel – removal and refitting	21
Transverse link bushes – renewal	15
Wishbone outer balljoint – removal and refitting	4

Specifications

Front suspension type .... Independent with MacPherson struts, offset coil springs lower wishbones and anti-roll bar

Rear suspension type .... Independent, with telescopic struts, coil springs, twin transverse links and anti-roll bar

Steering – manual
Type .... Rack and pinion
Manufacture .... TRW or ZF
Steering column .... 3 sections with universal joints. Top section adjustable for height
TRW steering rack:
  Oil type .... SAE 85W/90 gear oil
  Oil quantity .... 200 cc
  Pinion pre-load setting .... 0.025 to 0.130 mm (0.00098 to 0.0051 in)
  Rack to pinion backlash clearance – new rack .... 0.02 to 0.06 mm (0.00078 to 0.0025 in)
  Rack to pinion backlash clearance – worn rack .... 0.09 mm (0.0035 in)
ZF steering rack
  Lubrication .... High melting point grease
  Rack locating support clearance .... 0.08 mm (0.00315 in)

Steering – power assisted
Type .... Rack and pinion
Manufacture .... ZF

Number of turns lock to lock

	TRW	ZF
Manual steering:		
Early Saloon models	4.04	3.90
Late Saloons and HPE	3.82	3.90
Early Coupe models	3.82	–
Coupe and Spider	3.50	–
Saloon and Coupe (USA)	3.82	3.90
Power steering	–	3.13

Steering geometry

	Laden	Unladen
Camber	0°10' to 0°50'	0° 40' to 1° 20'
Castor	1°10' to 1°50'	1°05' to 1°45'
Toe-out	0 to 3.5 mm (0 to 0.137 in)	1 to 2.5 mm (0.0039 to 0.098 in)

Chapter 11 Suspension and steering

King-pin inclination
   Saloon .... 9°26' to 10°06'
   Coupe, Spider and HPE .... 9°42' to 10° 22'
Car laden weight:
   Saloon and HPE .... 4 occupants or 280 Kg (616 lb)
   Coupe and Spider .... 3 occupants or 210 Kg (462 lb)

Rear suspension geometry   **Laden**   **Unladen**
Camber (non-adjustable) .... −1°±20'   −0° 30' ± 20'
Toe-in:
   Early models .... 0.5 to 4 mm (0.0196 to 0.157 in)
   Later models .... 3 to 6 mm (0.118 to 0.236 in)   2 to 5 mm (0.079 to 0.197 in)
Car laden weight:
   Saloon and HPE .... 4 occupants or 280 Kg (616 lb)
   Coupe and Spider .... 3 occupants or 210 Kg (462 lb)

Roadwheels and tyres
Road wheel size – standard:
   Saloon models
      1300 and 1400 .... 5J x 14 – pressed steel
      1600, 1800 and 2000 .... 5½J x 14 – pressed steel
      2000 ES .... 5½J x 14 – light alloy
   Coupe models
      1300 .... 5J x 14 – pressed steel
      1600, 1800 and 2000 .... 5½J x 14 – light alloy
   Spider models
      1600 .... 5½J x 14 – pressed steel
      2000 .... 5½J x 14 – light alloy
   HPE
      1600 and 2000 .... 5½J x 14 – light alloy
Tyre sizes – standard:
   1300 Saloon and Coupe .... 155 SR 14
   1600, 1800 and 2000 Saloon .... 175/70 SR 14
   1600 Coupe, 1600 and 2000 HPE 175/70 SR14
   1800, 2000 Coupe and Spider 175/70 HR14
   Snow tyres – all models .... 155 SR 14 (snow type)
Tyre pressures – front and rear:
   Lightly laden .... 1.7 kgf/cm^2 (24 lbf/in^2)
   Fully laden or high speed .... 1.9 kgf/cm^2 (27 lbf/in^2)
   Consistent high speed driving .... 2.2 kgf/cm^2 (31 lbf/in^2)
   With snow tyres fitted increase pressures by ... 0.2 kgf/cm^2 (3 lbf/in^2)

Torque wrench settings   **lbf ft**   **Nm**

Front suspension

	lbf ft	Nm
Wishbone to subframe	43.5	59
Anti-roll bar to wishbone	16.5	22.5
Balljoint in wishbone to wishbone	18	24.2
Hub carrier to strut	72	98
Hub carrier-lower swivel to balljoint	70	95
Front hub bearing locking ring nut	231	314
Shock absorber upper mounting nut	65	88
Strut to wing-mounting nut	13	18

Rear suspension

	lbf ft	Nm
Shock absorber upper mounting nut	65	88
Strut to body mounting nuts	13	18
Stub axle to strut – upper bolt	42	57
Bearing to stub axle locking ring nut	231	314
Transverse links – inner and outer mounting bolts and nuts:		
Early models	42	57
Late models	39	54
Crossmember to body	18	24.5
Anti-roll bar to stub axle	67	91
Anti-roll bar mounting bracket to body	40	55
Anti-roll bar front mounting to bracket	40	55
Anti-roll bar to front mounting bush (early models with adjustable transverse link)	16.5	22.5
Transverse link/tie rod clamp bolts (early models)	11	15

Steering

	lbf ft	Nm
Roadwheel nuts or bolts	65	88
Steering wheel nut	21	29
Steering column lower section to pinion shaft and lower section to centre section	19	26
Track rod end balljoints to rod	14	19
Balljoint to steering arm – manual rack	36	49

Chapter 11 Suspension and steering

Balljoint to steering arm – power steering	57	78
Steering rack to bulkhead	21	29
Power steering rack to bulkhead	39	54
Track rods to steering rack	72	98
Banjo bolts to power steering box and rack	18	24
Banjo bolt to power steering pump	25	34
Power steering pump pulley retaining bolt	61	83
Pinion cover plate to rack housing bolts	14	19
Pinion adjuster – ZF type	16 to 19	21 to 25
Rack locating support adjuster – ZF type	4.5	6

1 General description

The front suspension is of the MacPherson strut type with telescopic shock absorbers, offset coil spring rubber bump stops and an anti-roll bar. The single lower wishbones are pivoted at their inner ends in rubber bushes set in the subframe assembly. The anti-roll bar, which is attached at either end to the lower wishbones, is also carried by the subframe. At the upper ends, the springs and shock absorbers are retained in housings in the front wing assemblies.

The rear suspension is of the telescopic strut type with independent units incorporating shock absorbers, coil springs and rubber bump stops. There is also a trailing anti-roll bar and each strut is located separately by twin transverse link bars. The upper ends of the struts are retained in housings in the rear wings.

The steering is rack and pinion. The rack is mounted on the front bulkhead. Power steering is only available in certain versions of left-hand drive cars due to the under bonnet layout. Twin steering rods are connected to the steering arms by balljoints. The steering wheel, which is adjustable for height, is connected to the steering rack by a three section universally jointed and energy absorbing steering column.

The steering and suspension is relatively maintenance free. On cars fitted with power steering the belt tension on the steering pump and the fluid level need to be regularly checked. The strut upper mounting nuts should also be occasionally checked for tightness.

2 Front suspension spring and strut – removal and refitting

1 Remove the front hub cap (if fitted) and loosen the front roadwheel nuts or bolts. Jack up the front of the car.
2 Support the front of the car on stands under the subframe and make sure the handbrake is firmly applied.
3 Remove the roadwheel.
4 Undo and remove the bracket which secures the brake pipe hoses to the base of the strut (Fig. 11.1).
5 Undo the nuts and then withdraw the two bolts which secure the upper end of the hub carrier to the bottom of the strut. The two parts can then be separated (photo).
6 Open the bonnet and undo and remove the four nuts which secure the top end of the strut to the wheelhousing (photo). Be careful not to lose the washer.
7 With the nuts removed, the front suspension assembly can be withdrawn downwards through the wheel arch.
8 Refitting the front strut is the reverse procedure to removal, but tighten the various nuts and bolts to the correct torques as given in the Specifications.

3 Front suspension wishbones – removal and refitting

1 Apply the handbrake, loosen the wheel nuts or bolts, jack the car up at the front end and support it under the subframe using axle stands.
2 Remove the roadwheel.
3 Undo the two bolts which retain the end clamp for the anti-roll bar and bush. Remove the clamp, bolts, washers and bush (Fig. 11.2).
4 If the wishbone being removed is the left-hand side unit and the car is fitted with automatic headlamp levelling system the next task is to disconnect the link between the levelling unit and the wishbone (photo).
5 Undo the nut which secures the bottom end of the hub carrier to the balljoint swivel in the outer end of the wishbone. Separate the carrier from the balljoint using a balljoint separator or split wedges. If you use the simple split wedge this task can be performed without having to remove the driveshaft, which lies directly above the balljoint swivel.
6 Undo the nuts and withdraw the bolts from the inner wishbone pivots, and lift the wishbone away.
7 Refitting the lower wishbone is basically the reverse procedure to removal. However the inner pivot bolts and nuts should only be tightened after all other components have been reassembled when doing this the wishbones must be horizontally in line with the subframe; that is to say with the car horizontal and laden as in the Specifications for checking the steering geometry. This can be achieved by jacking up the outer end of the wishbone once it has been assembled to achieve the correct relationship between the wishbone and the subframe (Fig. 11.3).
8 Tighten all the mounting bolts involved to this specified torque.

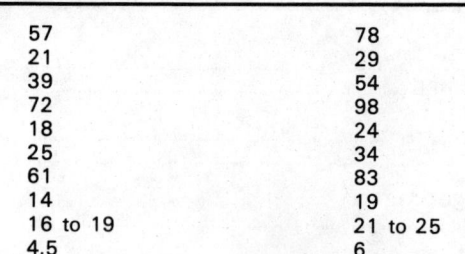

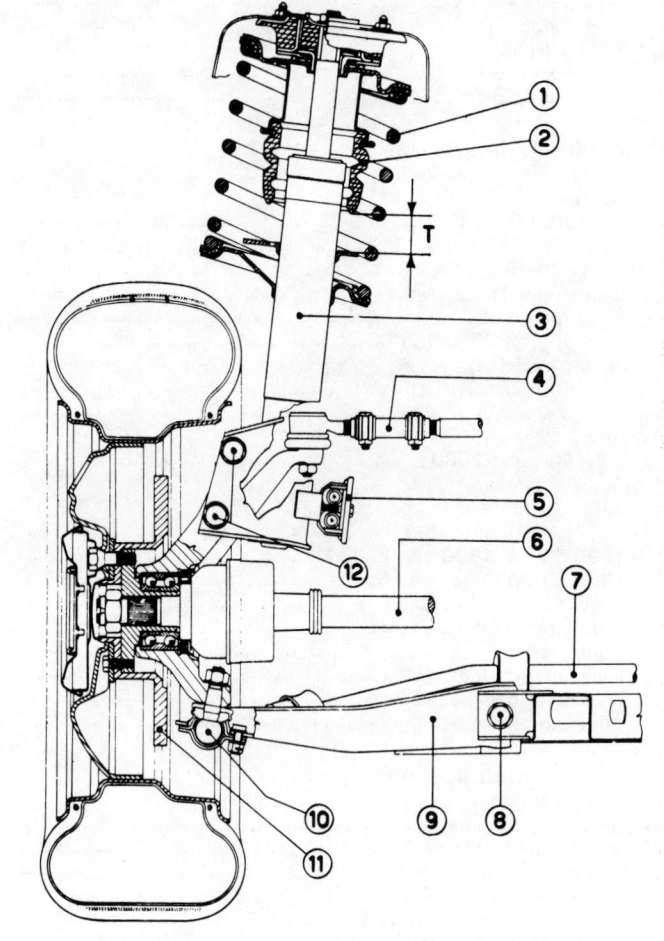

Fig. 11.1 Cross-section of the front suspension system (Sec 2)

1 Front coil spring
2 Bumper
3 Strut
4 Steering rod
5 Brake hoses bracket
6 Wheel driveshaft
7 Anti-roll bar
8 Wishbone mounting bolt
9 Wishbone
10 Wishbone balljoint
11 Brake disc
12 Strut-to-hub carrier bolts
T Bumper-to-cup static load distance

Chapter 11 Suspension and steering

2.5 These two bolts retain the hub carrier to the bottom end of the strut

2.6 The front strut is secured at the upper end by four nuts

3.4 Remove the link (arrowed) between the wishbone and levelling unit (picture taken with engine out of car)

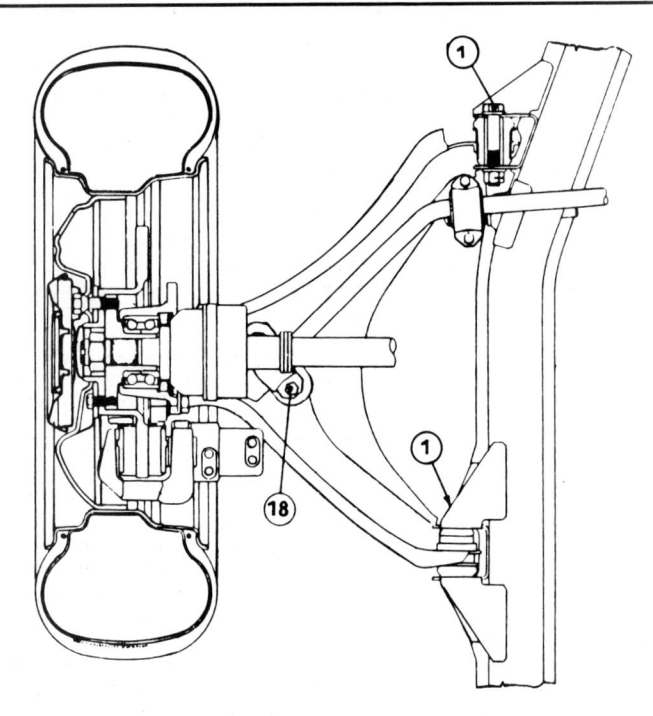

Fig. 11.2 Plan view of the front wishbone (Sec 3)

1 Inner pivot mounting bolts   18 Anti-roll bar end clamp bolts

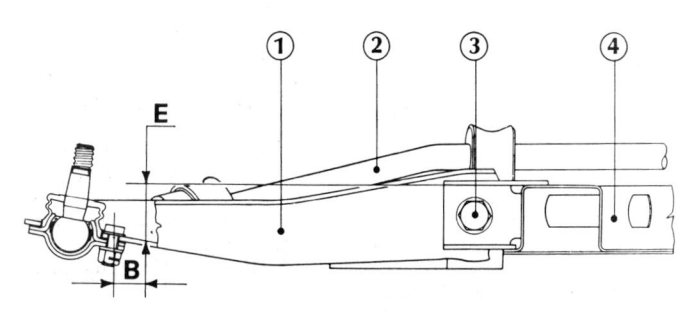

Fig. 11.3 Correct alignment of wishbone for refitting (Sec 3)

1 Wishbone   3 Wishbone mounting bolt
2 Anti-roll bar   4 Subframe

E – B distance should measure 42 mm (1.654 in) for the Saloon model or 37 mm (1.457 in) for the Coupe, Spider and HPE. Point E is level with the top of the subframe and point B is 22 mm (0.8662 in) from the centre line of the balljoint mounting bolt shown in the figure

4 Wishbone outer balljoint – removal and refitting

1 The balljoint fitted in the outer end of the wishbone acts as the lower pivot or swivel for the front hub carrier and the steering.
2 To remove it follow the instructions given in paragraphs 1, 2 and 5 in the previous Section.
3 Undo the three lock nuts and bolts which hold the balljoint assembly to the wishbone (photo) and lower the balljoint out of the wishbone. It should come out as one unit with its dustguard and felt piece.
4 Before refitting the balljoint clean the wishbone and lower pivot for the hub carrier.
5 Refit the balljoint to the wishbone, and ensure that the felt piece and dust guard are in position. Refit the nuts and bolts to secure it in position.
6 Offer up the balljoint to the lower end of the hub carrier and refit the washer and nut.

Chapter 11 Suspension and steering

7 When reassembly is complete check that all nuts are tightened to the correct torque as given in the Specifications.
8 Refit the driveshaft if it has been removed.
9 Refit the roadwheel and remove the axle stands.

5 Anti-roll bar (front) – removal and refitting

1 This section also covers the fitting of new bushes for the anti-roll bar, and needs to be carried out on a firm level surface.
2 Apply the handbrake and chock the rear wheels. Loosen the front wheel nuts or bolts, jack the car up and support it with stands under the side members of the subframe.
3 Remove the front roadwheels.
4 Using two jacks, one under the outer end of each wishbone, raise the wishbones slightly on each side until they are the same height on each side but not horizontal with the subframe, as shown in Fig. 11.3.
5 Undo the bolts which secure the clamps at the end of the anti-roll bar, and their bushes, to the wishbones.
6 Remove the clamps from the bushes, and the bushes from the ends of the anti-roll bar.
7 Undo the bolts and nuts which secure the clamps for the inner bushes. Note that the inner bushes are fitted in two halves with a plate in between. One half of the bush assembly is fitted to the anti-roll bar itself. The other half is mounted to a stub protruding from the subframe. The whole assembly is held together by two semi-circular clamps, secured by two bolts and nuts (Fig. 11.4).
8 With the clamps on both sides removed the complete anti-roll bar can be withdrawn from one side of the car. It is not necessary however to remove the bar unless it is to be renewed.
9 In order to renew the bushes the bar may be left in position and the bushes can be pulled off on either side. Clean the ends of the anti-roll bar first and then, to make it easier, use a concentrated soap and water solution which will enable the bushes to slide off the anti-roll bar quite easily. The other halves of the inner bushes can be easily removed from the subframe mounting points. Do not use oil or grease of any sort.

4.3 The balljoint is secured in place by three locknuts

10 The fitting procedure is basically the reverse of the removal procedure. Use only a concentrated soapy water solution to help slide the new bushes onto the anti-roll bar. Wipe off any excess.
11 Refit the clamps to the inner and outer bushes but do not tighten the inner ones yet. Don't forget the plate that fits between the two inner bushes either.
12 With the outer end clamps tightened to the specified torque, jack up the wishbones so that they now come level with the subframe as shown in Fig. 11.3.
13 Now the inner bush clamps can be tightened to their specified torque.

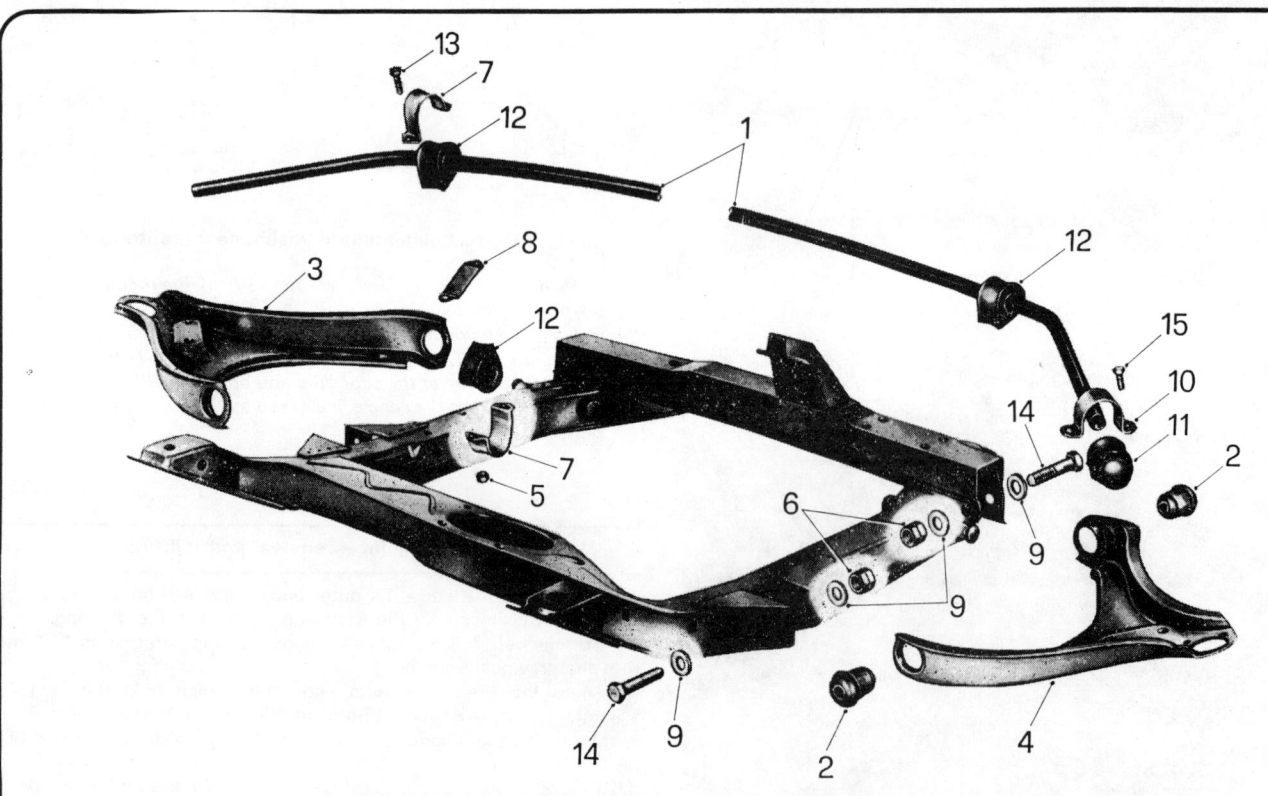

Fig. 11.4 Anti-roll bar and wishbone to subframe mountings (Sec 5)

1 Anti-roll bar	5 Nut	9 Washers	13 Bolt for inner bush clamp
2 Bush	6 Locknuts	10 Clamp	14 Bolts for wishbone pivots
3 Wishbone	7 Half clamp	11 End bush	15 Bolt for end bush clamp
4 Wishbone	8 Centre plate	12 Inner bush sections	

Chapter 11 Suspension and steering

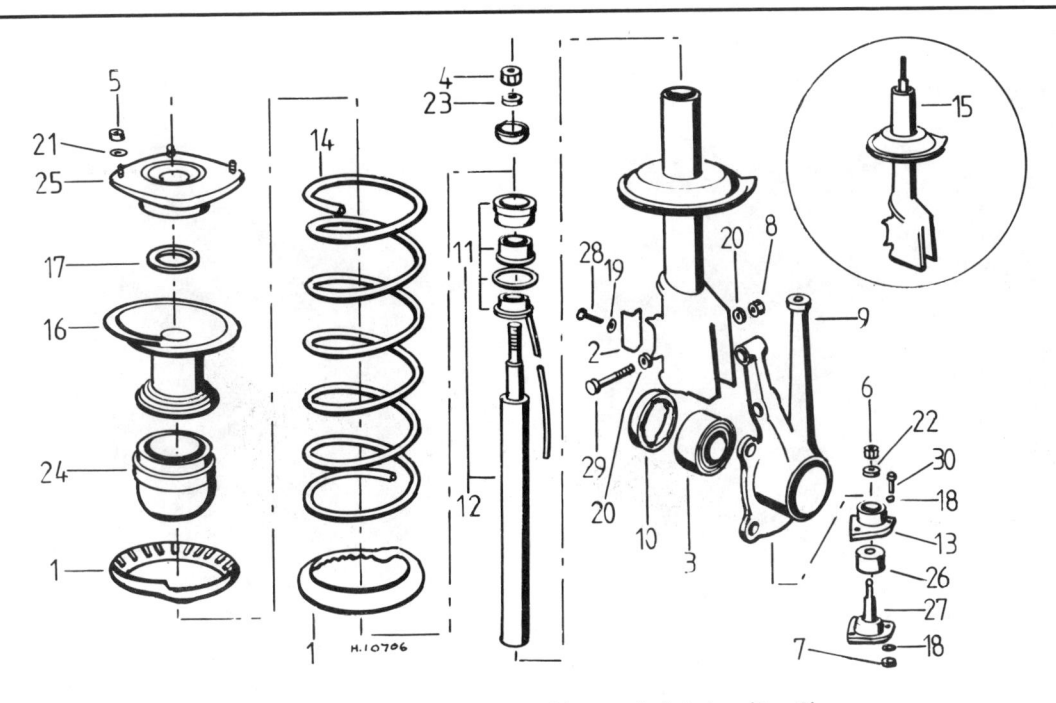

Fig. 11.5 Front strut assembly – exploded view (Sec 6)

1 Gaskets
2 Brake pipe bracket
3 Bearing
4 Damper upper mounting nut
5 Strut upper mounting nut
6 Hub carrier to lower balljoint retaining nut
7 Balljoint retaining nut
8 Hub carrier upper mounting nut
9 Hub carrier
10 Locking ring nut
11 Damper seals and cap
12 Stem
13 Balljoint mounting plate
14 Coil spring
15 Inset view of complete replacement strut
16 Upper seat
17 Thrust plate
18 Washer
19 Washer
20 Washer
21 Washer
22 Washer
23 Washer
24 Bumper
25 Upper strut mounting plate
26 Bush
27 Balljoint
28 Bolt
29 Bolt
30 Bolt

14 Refit the roadwheels, remove the jacks and stands, lower the car to the ground and tighten the roadwheel nuts or bolts.

6 Front strut assembly – dismantling and reassembly

1 With the strut assembly removed from the car as described in Section 2 give it a good clean to remove all the road dirt and grime.
2 Using a proprietary strut spring compressor compress the spring evenly.
3 Unlock the tab on the upper nut shock absorber, use a peg spanner (2 teeth) or similar to hold the retaining ring and remove the nut.
4 Lift off the washer and mounting plate and thrust plate from the strut stem (Fig. 11.5).
5 Release the spring compressor completely.
6 Withdraw the coil spring with the gaskets and upper seat and bumper.
7 The strut, coil spring and components are not stripped as far as would normally be necessary and the whole assembly can be examined for wear. The strut, if worn should be replaced by an exchange unit. The coil spring can be renewed separately as can the bumper, mounting plate etc.
8 Reassembling the strut and it components is done in the reverse order to dismantling. Fit the new parts in the appropriate places as necessary. Lubricate the thrust plate with grease. Tighten the upper damper nut to the specified torque before releasing the spring compressor.

7 Front wishbones – renewing the bushes

1 Remove the wishbone as described in Section 3.
2 To remove the bushes you will require the special Lancia tool No 88042121 (Fig. 11.6). This tool works in two ways as it is used both for withdrawing the old bush and fitting the new one. If you cannot borrow or hire a suitable tool then the wishbone will have to be taken to your local Lancia dealer to have the bushes renewed.
3 Once the bushes have been removed from the seatings, give the seatings a good clean.
4 Fit the new bushes to the wishbone inner pivots using the same tool (Fig. 11.7).
5 To ensure that the bush is correctly located in the wishbone the tool must be screwed in fully.

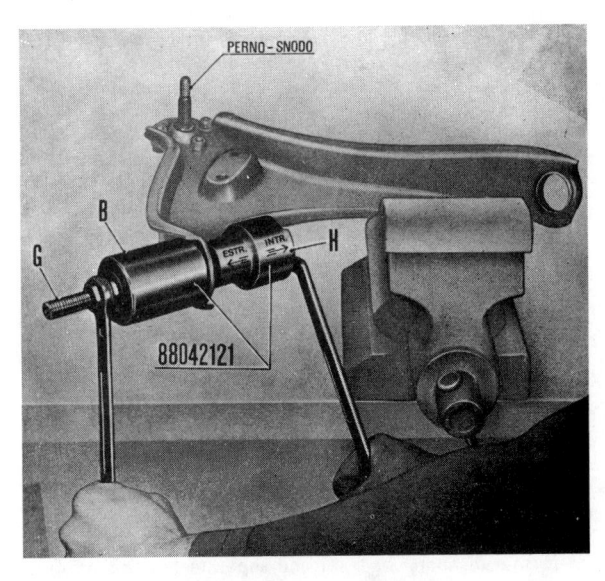

Fig. 11.6 Withdrawing the bush from the arm by using removing – installing tool 88042121

Perno snodo = Ball joint

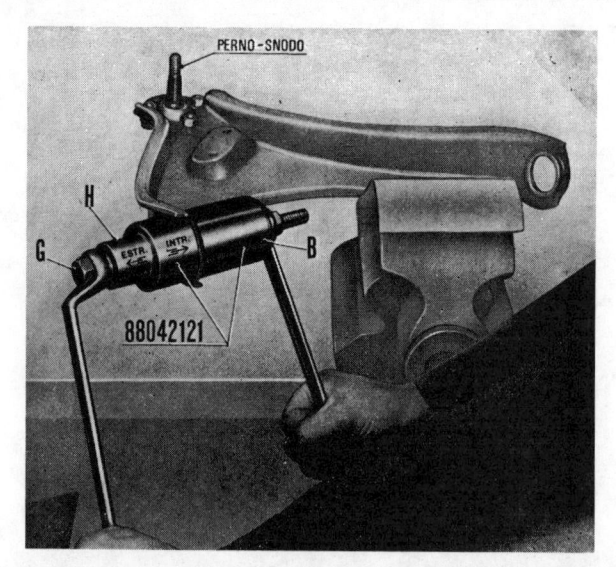

Fig. 11.7 Fitting the bush to the arm by using the removing – installing tool 88042121

G, H, B – The three main parts of the tool

6 Another check on the fitting must now be carried out using more special equipment. The tools required are Lancia tools No 88045126, a dial gauge (88095122) and torque wrench. They are set up as shown in Fig. 11.8, with the torque wrench set at the specified torque for the mounting bolts.

7 With the wishbone clamped in a vice, insert the Lancia tool and hand tighten the nut. Set the dial gauge to zero. Screw in the nut using the torque wrench until it reaches the specified torque setting. Then undo the nut and check that the dial gauge returns to zero. If it doesn't then it means that the bush has been pulled out of the wishbone mounting by the amount shown on the dial gauge, thus indicating that the wishbone is worn and will have to be renewed as well.

8 If the test is alright and the dial gauge returns to zero, carry out the same procedure with the other bush. When both have been checked, the wishbone can be refitted to the car, as described in Section 3.

8 Front wheel hub and carrier – removal and refitting

1 Apply the handbrake, loosen the roadwheel nuts or bolts. Jack up the front of the car and support it on stands placed under the side members of the subframe.
2 Remove the relevant roadwheel.
3 Undo the outer constant velocity joint to the hub assembly, as described in Chapter 8.
4 Remove the front brake caliper and secure it out of the way, as described in Chapter 9.
5 Undo the bolts and remove the brake disc, as described in Chapter 9.
6 Undo the nut which secures the track rod end balljoint to the steering arm on the hub carrier. Pull back the dustguard and separate the two components using a balljoint separator or split wedge (see Section 28).
7 Undo the nut which secures the bottom end swivel of the hub carrier to the wishbone balljoint and separate the two using a split wedge or balljoint separator.
8 Undo the two nuts and withdraw the bolts from the hub carrier upper mounting to the strut. The hub, carrier and bearing can now be lifted away.
9 Refitting is the reverse procedure to removal, but tighten all the nuts and bolts to their specified torques.

9 Front hub and bearings – renewal

1 Remove the front hub carrier assembly as described in the previous Section.
2 In order to separate the hub and the bearing races from the carrier

Fig. 11.8 Checking the fitting of the suspension bushes (Sec 7)

1 Socket
2 Sleeve
3 Dial gauge – 88095122
4 Special tool – 88045126
5 Torque wrench

it will be necessary to have access to a fly-press. Therefore it will be necessary to take the complete assembly to your local Lancia dealer or to a local engineering firm who can do the job properly. Before doing so obtain the necessary bearings as the new ones will have to be pressed in to the carrier and on to the hub, and then the whole job can be done in one go.

3 Before taking the hub assembly to the local engineering firm, undo the crimped ring nut which holds the hub bearing to the carrier. It is exceptionally tight.

4 When the nut has been removed the procedure to be adopted in separating the parts is as follows. First press the hub out of the carrier. The inner bearing race may come out attached to the hub, in which case it will have to be driven off or pulled off the hub. Then press the outer bearing race out of the hub carrier, using a suitably sized steel rod. Lancia use a special tool, Fig. 11.9.

Fig. 11.9 Pressing the bearing out of the hub carrier (Sec 9)

Chapter 11 Suspension and steering

5 For whatever reason, if a bearing is removed, then it must be renewed.
6 Clean all the parts thoroughly in paraffin and then dry them off and inspect them. If the wheel hub is worn it will have to be renewed.
7 To reassemble the hub start by pressing the new outer bearing into the hub carrier. Then press in the new inner bearing race.
8 Before the hub is fitted to the carrier, the locking ring nut must be refitted. Invert the carrier and hold it in a vice with protective jaws and screw in the ring nut. It must be tightened to the specified torque. Then crimp the ring.
9 Take the hub and carrier back to the press and with the carrier supported so that the hub can be pressed fully home, place the hub in position and press it into the bearings and carrier assembly (Fig. 11.10).
10 The hub and hub carrier assembly can now be refitted to the car as described in Section 8.

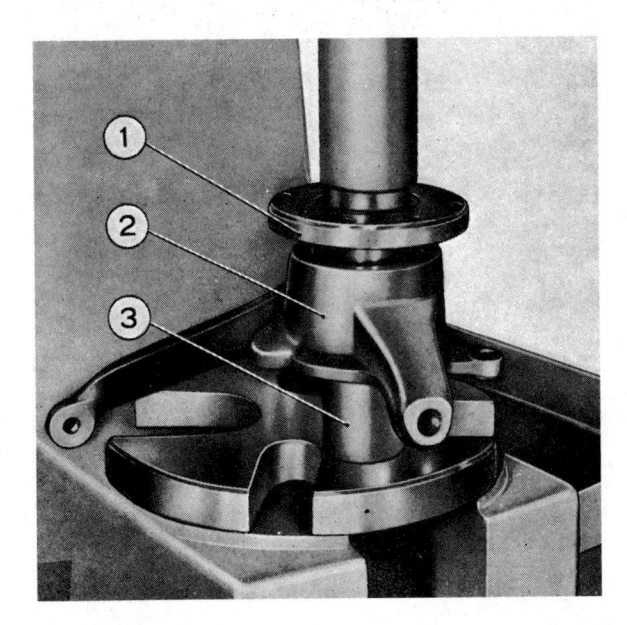

Fig. 11.10 Pressing the hub into the carrier assembly (Sec 9)

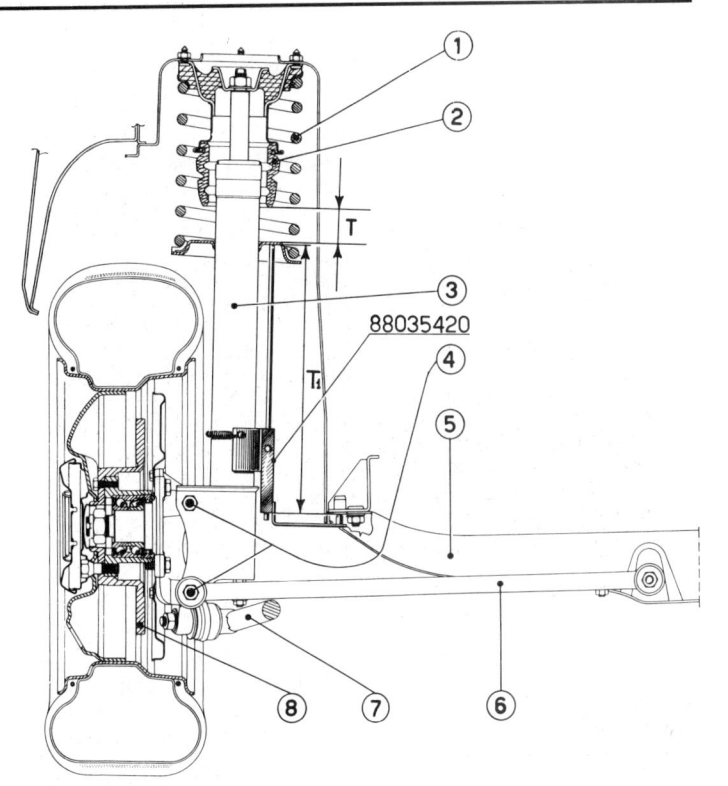

1 Hub
2 Hub carrier
3 Spacer

Fig. 11.11 Rear suspension cross-section (Sec 10)

1 Coil spring
2 Bumper
3 Strut
4 Strut-to-stub axle fixing bolts
5 Transverse links mounting crossmember
6 Transverse link
7 Anti-roll bar
8 Brake disc

Part No – 88035420 – Device for loading the suspension before tightening the suspension mounting bolts and nuts

10 Rear suspension spring and strut – removal and refitting

1 Chock the front roadwheels, remove the hub cap (where it is fitted) and loosen the roadwheel nuts or bolts on the relevant rear wheel.
2 Jack up the rear end of the car and support it on stands under the body sills.
3 Remove the roadwheel.
4 Place a bottle jack or blocks under the end of the anti-roll bar in order to support it and the stub axle assembly. This will ensure that when the strut is detached the brake pipes are not strained by the stub axle dropping.
5 Loosen the upper nut and bolt which secure the stub axle to the strut (Fig. 11.11).
6 In early models with adjustable transverse link rods loosen the nut and bolt which secures the transverse links to the stub axle and strut (Fig. 11.12). Remove the bolt.
7 On later models with non-adjustable transverse links, scribe the relationship of the mounting bolt to the stub axle. Loosen the nuts at each end of the offset bolt or shaft (photo). You will notice when the links are removed from the shaft that the rear end section is offset to the main shaft. This is to allow for adjustment of the toe-in alignment as described in Section 19.
8 With the transverse links removed from either end of the mounting shaft (later models), remove the shaft itself. If it is tight loosen it by applying a spanner to the hexagonal hub on the rear section.
9 Withdraw the upper bolt which has already been loosened. The bottom end of the strut can now be separated from the stub axle. Make sure when this is carried out that the stub axle and anti-roll bar remain well supported. If in doubt tie the hub and caliper assembly to the bodywork.
10 Having freed off the bottom end of the strut it is now time to release the upper mounting bolts. With the Saloon model it is necessary to remove the rear seats and then the backrest in order to raise the rear shelf trim panel to get at the rear strut upper mounting points. With the Coupe the same operation is required, but also the spare wheel must be removed from the boot and the boot trim removed on whichever side is concerned. If the right-hand strut is

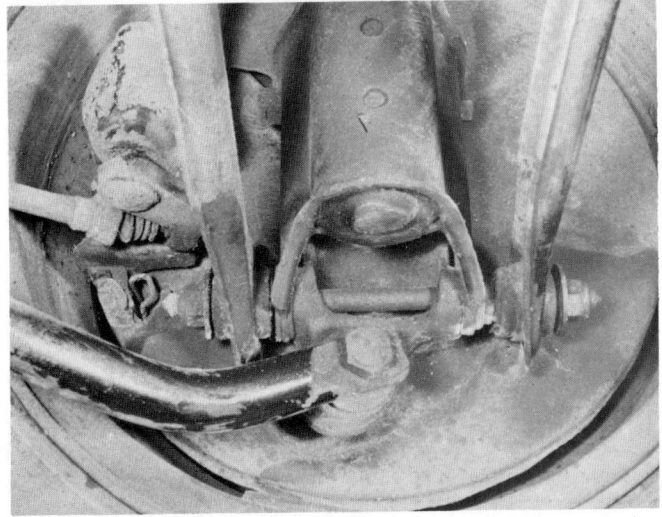

10.7 Transverse links outer end mounting shaft on later models is like this (HPE model shown here)

being removed then it will also be necessary to remove the fuel recovery tank from its mounting bracket in order to remove the nuts. Because of the location of the mounting turret it is necessary to remove two of the nuts and washers from inside the car and the other two from inside the boot.

11 With the HPE model, the task is quite straight forward. Fold down the rear seats. Unscrew the seat back retaining catch knob and remove the trim panel on the required side of the luggage compartment. The mounting plate is located by two of the strut retaining bolts and is easily removed (photo).

12 Once the four mounting nuts have been removed the strut assembly can easily be withdrawn down and out through the wheel arch.

13 To refit the strut to the car start by offering it up from underneath. Make sure the lower mounting bracket is facing the right way round. This task is more easily done by two people, as one can offer the strut up and the other working inside the car can guide the mounting bolts through the holes in the turret and refit the washers and nuts. Tighten the nuts to the specified torque. In the case of the Coupe this is a tricky task because of the location of the turret.

14 Remember in the case of the HPE that there is a mounting plate which is retained by two of the nuts.

15 Refit the components and trim that was removed to reach the upper mounting bolts and nuts.

16 Refit the upper mounting bolt and nut which retains the stub axle to the strut and tighten it to the specified torque.

17 Refit the transverse links to the stub axle. On the early models the bolt is fitted from front to rear. Refit the nut but do not tighten it completely.

18 On the later models, refit the mounting shaft for the transverse links to the stub axle, and align the marks previously scribed. Then refit the outer ends of the transverse links to the mounting shaft. Do not forget the washers and spacers that were removed. Refit the nuts and washers to either end but do not tighten them.

19 The next step is to load the suspension. Lancia specify special equipment for this but it can be done more simply. The object is to achieve the effect of having a laden vehicle, that is with 4 occupants in it in the case of the Saloon and HPE models, or 3 occupants in the case of the Coupe and Spider. For the purposes of the exercise 1 occupant is equal to 70 kg (154 lb).

20 Make up an L-shaped bracket, to take the place of the Lancia tool No 88035420 shown in Fig. 11.11. The vertical length of the bracket must be 284.5 mm (11.20 in) for the Saloon and HPE and 290.5 mm (11.44 in) for the Coupe and Spider. A simple piece of heavy gauge wire will be sufficient. This needs to be fastened to the strut with the top end resting agsint the lower spring mounting plate, as shown in the illustration.

21 Use a jack placed under the hub or the end of the anti-roll bar and raise the suspension until the horizontal part of the L-shaped bracket is in line with the rear suspension crossmember mounting bolt as in Fig. 11.11.

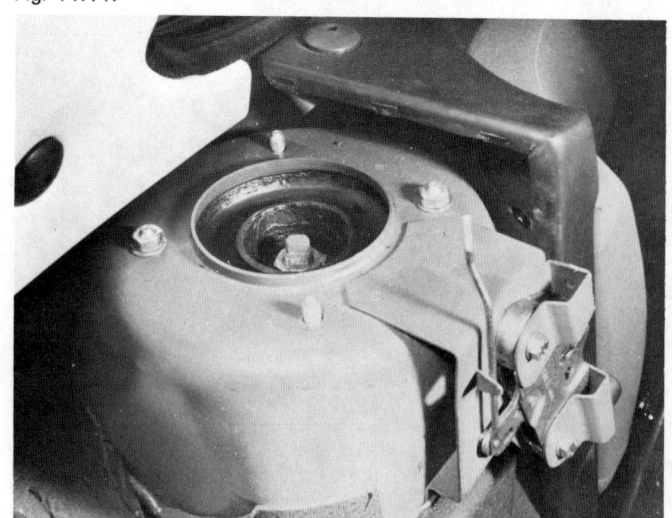

10.11 Remove the last two nuts and the strut can be removed from underneath

22 With the suspension thus loaded, tighten up the transverse link mounting bolt and nut to the specified torque in the case of the early models. With the later models the nuts at either end of the mounting shaft have to be tightened to the specified torque, but while doing so the hexagonal nut on the rear section off the shaft must also be held still. Make sure when doing this that the scribe marks stay aligned or the toe-in will be changed.

23 Lower the suspension and remove the L-shaped bracket and jack.

24 Lower the roadwheel, remove the stands, tighten the roadwheel and refit the hub cap (where fitted).

25 Complete the operation by checking and setting the rear wheel toe-in geometry. This is covered in Section 19.

Fig. 11.12 Rear suspension layout – early type (Secs. 10, 11, 12, 13 and 14)

1 Anti-roll bar mounting nut
2 Anti-roll bar
3 Anti-roll bar support
4 Brake rear limiting valve control shackle bar bolt
5 Brake rear limiting valve control spring rod
6 Brake rear limiting valve support bracket
7 Brake rear limiting valve
8 Brake rear limiting valve mounting bolts
9 Dust guard
10 Brake rear limiting valve control spring rod mounting clamp
11 Fuel tank guard
12 Transverse link bolts
13 Transverse links
14 Crossmember
15 Transverse link mounting bolts

11 Rear suspension transverse links – removal and refitting

1 Block the front wheels, remove the hub caps and loosen the rear roadwheel bolts, then jack up the car and support it securely on stands.
2 Remove the roadwheels.
3 Remove the guard plate from the centre of the crossmember so as to be able to reach the inner mounting bolts and nuts for the transverse links (photo).
4 On cars fitted with the self-levelling headlamp aiming device disconnect the link between the rear levelling unit and the left-hand front transverse link (photo).
5 Remove the nuts from the rear ends of the inner mounting bolts for the transverse links.
6 In early models with adjustable transverse links remove the nuts on the rear ends of the outer mounting bolts for the transverse links (Fig. 11.12).
7 Later models have a nut on the front end and another on the rear end of the outer mounting shaft. Before removing these nuts scribe the position of the shaft in relation to the stub axle as described in the previous Section. Then remove the nuts.
8 Undo the crossmember to body mounting nuts; on early models there is one nut at each end of the member, later models have 2 nuts at each end.

Chapter 11 Suspension and steering

11.3 Transverse links inner mounting bolts and nuts (later models)

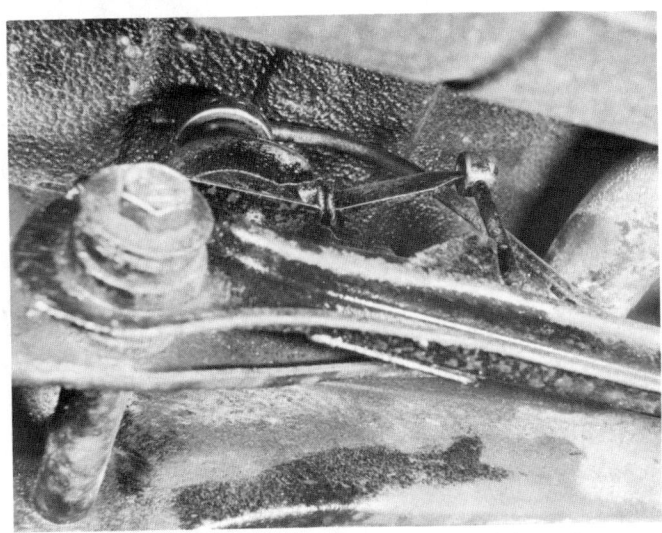

11.4 The rear headlamp self-levelling unit is mounted above the crossmember and is connected to the forward left-hand transverse link as shown here

9 Lower the crossmember away from the body and withdraw the two inboard mounting bolts for the transverse links. Take care not to lose any spacers or washers.
10 With the bolts removed either temporarily refit the crossmember to its mounting bolts or lower it and place it to one side.
11 Remove the transverse links outer mounting bolts in the case of early models or with the later type withdraw the transverse links from the outer mounting shafts. The links are now free and can be placed to one side. Take care not to lose any of the washers and spacers.
12 Refitting is the reverse procedure to removal. Start by attaching the links to the outer mounting points. Note that in early models the adjustable link is at the rear on each side and that on later models the open side of the channel sections all face forward. All the mounting bolts are fitted from front to rear, except the adjustable offset outer mounting shaft which fits in from the rear. Do not tighten up the outer or inner mounting nuts or bolts until the crossmember has been refitted and its mounting nuts tightened to the specified torque.
13 Refer to the previous Section as it is necessary to load the suspension before the inner and outer transverse link mounting nuts are tightened to the correct torque. This is covered from paragraph 18 onwards.
14 Refit the centre guard plate to the cross member and reconnect the headlamp self-levelling unit to the front left-hand transverse link.
15 To complete the operation check the rear wheel toe-in as covered in Section 19 and adjust it if necessary.

12 Rear suspension crossmember – removal and refitting

1 Chock the front wheels and jack up the rear of the car under the centre of the crossmember using a block of wood between the jack and car.
2 Working underneath the car slacken off the transverse link inner mounting bolts and nuts.
3 Undo the retaining nuts for the fuel tank guard plate.
4 Undo and remove the crossmember retaining nuts. On early models there is one nut at each end and on later models there are two (photo and Fig. 11.12).
5 Where the car is fitted with a self-levelling headlamp system the rear unit, which is mounted above the left-hand side of the crossmember, must be disconnected from the transverse link and removed, complete with its mounting bracket.
6 Lower the jack slightly so that the crossmember drops off its mounting bolts. Remove the nuts from the transverse link inner mounting bolts and withdraw the bolts. Do not lose the washers and spacers.
7 The crossmember is now free and can be removed.
8 Refitting is the reverse sequence to removal. Remember that the transverse link inner mounting bolts are fitted from front to rear. The mounting bolts and nuts cannot be fully tightened until the suspension

12.4 On later models the crossmember is retained by two nuts and bolts at each end

has been loaded, as described in Section 10.
9 Start by refitting the transverse links to the crossmember, making sure that the crossmember has been fitted the right way round. Fit the crossmember to the body and tighten those mounting nuts to the specified torque. Load the suspensions and tighten up the inner mounting bolts for the transverse links.
10 Refit the headlamp levelling unit and connect it to the left-hand front transverse link.
11 Refit the tank guard plate.
12 Remove the jacks and check the rear wheel toe-in as described in Section 19 and adjust it if necessary.

13 Anti-roll bar (rear) – removal and refitting

1 Chock the front wheels, jack up the rear of the car and support it on stands.
2 Undo the anti-roll bar and retaining nuts. On early models the end of the anti-roll bar is retained by a locknut after it had passed through a balljoint attached to the rear stub axle (Fig. 11.12). On later models the rear end of the anti-roll bar is attached directly to the stub axle by a bolt and nut through a bush (photo). With the nut removed from the bolt can be withdrawn.
3 The brake pressure limiting valve is actuated by movement of the anti-roll bar and is connected to it by torsion rod and linkage. Scribe

Chapter 11 Suspension and steering

the position of the arm to the anti-roll bar. Undo the U-bolt nuts and disconnect the linkage from the anti-roll bar (photo).

4 On early models the anti-roll bar is secured to the front mountings by a cap and two bolts. Knock back the lock tabs, remove the bolts and the cap (3) in Fig. 11.12.

5 The anti-roll bar is now basically free but the ends of the bar have to be withdrawn from the balljoints (1). Free the front of the bar from the front mounting blocks and drive the bar forward to release the ends. Be careful not to damage the threads.

6 In later models the removal process is slightly easier. The bar is connected to the front mounting arms by two U-bolts on each. Undo the nuts, remove the U-bolts and the anti-roll bar is free (photo).

7 Once the anti-roll bar is free all that remains is to extricate it from above the exhaust pipe and remove it.

8 Refit the anti-roll bar in the reverse order to that in which it was removed. Before locking up the front mounting cap bolts on the early type assembly to the specified torque, check that the anti-roll bar is correctly aligned to the front mounting block as shown in Fig. 11.13.

9 With the later type assembly the two U-bolts on each side automatically ensure that the anti-roll bar is correctly aligned, and the nuts can be tightened to the specified torque.

13.2 The later type anti-roll bar fitting is like this

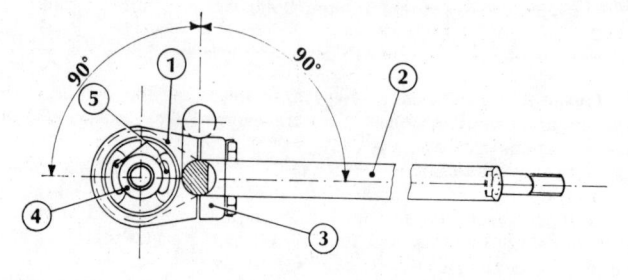

Fig. 11.13 Alignment of rear anti-roll bar and front mounting arm (Secs 13 and 14)

1 Mountings
2 Anti-roll bar
3 Front mounting cap
4 Bolt
5 Bush

14 Anti-roll bar bushes (rear) – removal and refitting

1 Follow the instructions in the previous Section for removing the anti-roll bar but don't actually withdraw it from the car.

2 Undo the nuts and withdraw the bolts from both the pivots of the anti-roll bar support arms then the support arms and bushes can be withdrawn.

3 The rear ends of the anti-roll bar are also mounted in brushes, on the later models, and with the bar removed, these can be withdrawn and renewed.

4 The front supports however have to have their bushes pressed out and new ones fitted using special tools. Take the support arms to your local Lancia dealer and get them to do it for you, unless of course you can borrow the tool No 88032411. For early models the base type A is used and for later models the base type B is used, with the tool for removing and fitting the bushes. Refer to Fig. 11.13 and note that the bushes are fitted to the support arms with the recesses (5) in a horizontal plane.

13.3 The brake pressure limiting valve linkage is connected to the anti-roll bar by a U-bolt (arrowed)

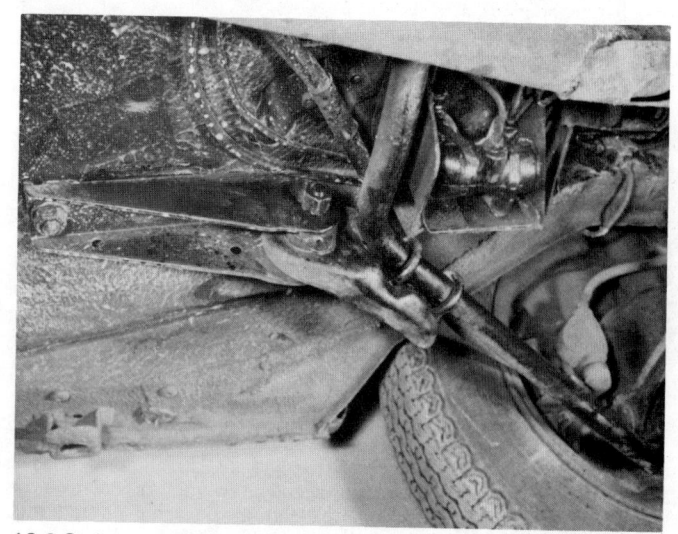

5 Refit the support arms to the mounting brackets and refit the mounting bolts and nuts, but do not tighten them up.

6 Refit the anti-roll bar to the front support arms and locate the rear ends, if removed. Before tightening the front cap bolts, on early models, check the alignment of the support arms and anti-roll bar as in Fig. 11.13.

7 When the alignment is correct tighten the cap mounting bolts to the specified torque and turn up the lock tabs.

8 Tighten the anti-roll bar end mounting bolts to the specified torque.

9 Now the suspension has to be loaded in order to tighten the front support pivot bolts to the specified torque. Refer to Section 10 and follow the instructions from paragraph 19 to paragraph 22. Then tighten the pivot bolts.

10 Unload the suspension and remove the L bracket and jack.

11 Refit the brake pressure limiting valve linkage to the anti-roll bar.

13.6 On later models the anti-roll bar is connected to the front support arms by U-bolts

Chapter 11 Suspension and steering

12 Remove the stands supporting the rear end of the car and lower the rear end. Remove the blocks in front of the front wheels and the job is completed.

15 Transverse link bushes – renewal

1 As with all the suspension bushes fitted to the Lancia Beta Series these bushes have to be pressed out using a fly-press and special punches. The seats have to be reamed out and the new bushes have to be pressed into the ends of the transverse links.
2 Therefore it is considered wiser to remove the transverse links and then take them to your local Lancia dealer and have the bushes renewed by them.
3 When renewing the bushes it must be remembered that both links on the same side must be done together. Also it is worth noting that there are two different types of bush which are available for the later non-adjustable transverse links. Both types (part Nos 82316215 and 82310364) do the same job but all four bushes (when renewed) on one side must be of the same type and must not be mixed (Fig. 11.14).

16 Rear suspension strut – dismantling and reassembly

The dismantling and reassembly procedures for the front suspension strut are given in Section 6. The procedure for the rear suspension is basically the same. By following the procedure in Section 6, with reference to Fig. 11.15, the job should present no difficulties.

17 Rear wheel hub and stub axle – removal and refitting

1 Chock the front wheels, jack up the rear end of the car and support it on stands or blocks, having first of all loosened the roadwheel studs.
2 Remove the roadwheel on the relevant side.
3 Remove the hub centre cap. It may need the use of a slide hammer with gripping attachment to achieve its removal.
4 The hub is secured to the stub axle by a crimped ring nut. Prise out the crimped sections and remove the nut.
5 Remove the brake caliper as described in Chapter 9 and tie it up out of the way, taking care not to strain the brake hoses.
6 Loosen the brake disc mounting bolts and tap the disc outward to free it. Use the bolts as guides and repeat the process until the disc is free.
7 Undo the anti-roll bar and mounting bolt and nut and withdraw the bolt. In the case of early models undo and remove the nut on the inside end of the balljoint which is attached to the lower end of the stub axle assembly. Pull back the dust cover and use a split wedge balljoint separator to free the balljoint from the stub axle. It is helpful in some instances to slacken off the nut on the end of the anti-roll bar.
8 Refer to Section 10 and follow the instructions from paragraph 5 to paragraph 9 and stop at the end of the second sentence.
9 The stub axle and hub can now be removed.
10 Refitting the stub axle assembly is basically the reverse procedure. Begin by locating the bolt which secures the stub axle to the strut. Then refit the transverse links to the stub axle and strut. Refer to the

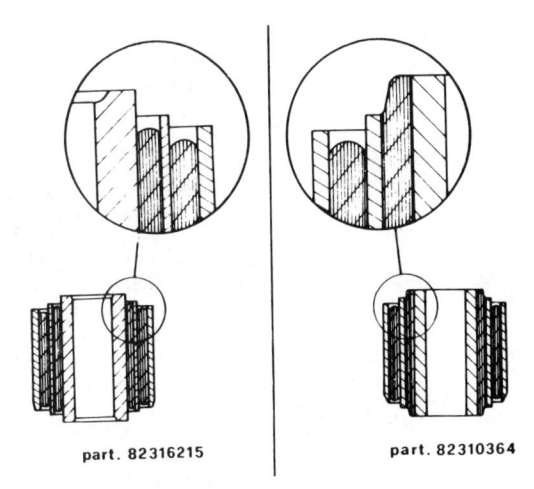

Fig. 11.14 These are the two different types of bush available for later models (Sec 15)

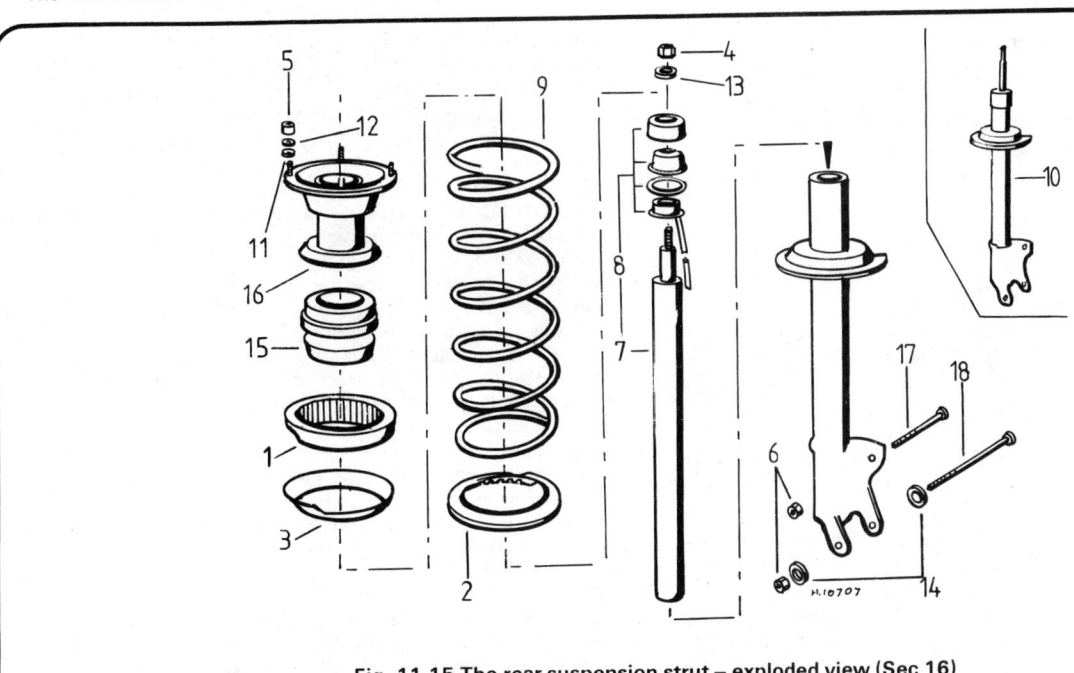

Fig. 11.15 The rear suspension strut – exploded view (Sec 16)

1 Damper ring
2 Gasket
3 Gasket
4 Nut
5 Nut
6 Nut
7 Stem
8 Damper seals and cap
9 Coil spring
10 Inset view of complete strut
11 Washer
12 Washer
13 Washer
14 Washer
15 Bumper
16 Upper seat
17 Mounting bolt
18 Mounting bolt

instructions in Section 6 paragraphs 16, 17 and 18 for relevant procedure.

11 Now the suspension needs to be loaded to tighten the nuts, so refer back to Section 10 and follow the instructions from paragraph 19 to 23.
12 Refit the anti-roll bar, to the lower end of the stub axle, or, to the balljoint if it is an early model. Tighten the nuts to the specified torque.
13 Refit the brake disc and caliper and the wheel hub retaining nut and tighten the nuts to the torques specified for them. After tightening the wheel hub nut crimp the ring section.
14 Refit the hub centre cap.
15 Refit the roadwheel, remove the stands, unblock the front wheels, tighten the roadwheel nuts and refit the hub cap (if fitted).

18 Rear hub and bearings – renewal

1 Remove the stub axle assembly as described in Section 17.
2 To separate the hub from the stub axle required the use of a fly-press, as does the removal and refitting of the bearing. This will mean taking the assembly to a local engineering firm who can do the job for you. Before doing so, obtain the new bearing and take it with you. However, there are complications so read the instructions first.
3 To press the hub off the stub axle place the assembly face downwards on blocks so that the hub can be pressed down and out (Fig. 11.16). This is achieved by either using a special Lancia tool No 88052017 as shown in the illustration or by using two rods of the appropriate size bolted to a base plate for rigidity and placed in the holes in the inner end of the stub axle to drive the hub out.
4 The hub, when pressed out, will come complete with the bearing, which is held in place on the hub by a locking ring nut.
5 To remove the ring nut, which is extremely tightly secured, insert two bolts into the threaded holes in the outer face of the hub and hold the assembly in a vice by the bolts. Uncrimp the locking ring and remove the locking ring nut using either a special socket (Lancia tool No 88051168) or a suitable alternative. It will be necessary to use a long extension bar to release it (Fig. 11.17).
6 Take the assembly back to the press and press the hub bearing out.
7 With the hub, bearing and stub axle dismantled clean all the parts in paraffin. The bearing, once it has been pressed out, cannot be re-used and must be renewed.

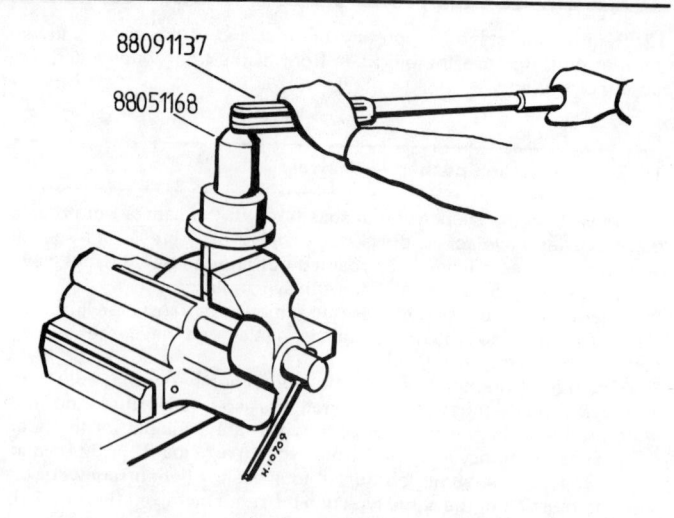

Fig. 11.17 Undoing the hub bearing locking ring nut (Sec 18)

88051168 – special socket to fit the ring nut
CL 88091137 – torque wrench

8 Inspect the hub for wear.
9 Fit a new bearing to the hub in the reverse way to which it was removed. Lock the ring nut tight to the specified torque and crimp the ring sections.
10 Refit the hub to the stub axle and use a press to drive it home.
11 The whole assembly can then be refitted to the car as described in Section 17.

19 Rear wheel toe-in – adjustment

1 The rear wheel toe-in is adjusted by different methods depending on the age of the car.
2 Early models have a system of adjustable transverse links. The ends can be unclamped and the rod lengthened or shortened by rotating the rod or link thereby altering the toe-in adjustment (Fig. 11.18).
3 On later cars the transverse links are made out of simple steel channel sections, and are not adjustable. The outer ends of the transverse links are mounted to the stub axles by offset shafts and the adjustment of the toe-in is achieved by rotating and altering the position of the offset section at the rear of the mounting shaft. This will effectively push the rear end of the wheel out or pull it in. (Fig. 11.19).
4 When adjusting the toe-in the car must be on level ground and laden. That is to say with the equivalent of 4 occupants in the Saloon or HPE models and 3 occupants in the Coupe or Spider. For these purposes 1 occupant is equal to 165 lbs or 70 kg.
5 To actually check or adjust the toe-in it is necessary to have the specialised equipment to carry out the operation properly. Do not attempt this unless the equipment is to hand.
6 If there is any doubt about the toe-in adjustment of the rear wheels of your car it is much safer and simpler to take it to your local Lancia dealer and get him to check it with his light beam equipment.

20 Roadwheels and tyres – general

1 Whenever the roadwheels are removed it is a good idea to clean the insides of the rims to remove accumulations of mud, oil, grease and disc pad dust.
2 Check the condition of the wheel for damage or rust and paint it if necessary.
3 Examine the wheel stub holes. If there is a tendency for the holes to become elongated or for the dished recesses, in which the studs seat, to become over compressed then the wheel needs to be renewed.
4 Pick out any stones or flints from the tread pattern, and check the sidewalls for any signs of bulging.
5 Check the depth of tread on the tyres using a tread depth gauge. If the tread depth is 1 mm or less then the tyre is not legally usable on the road and must be renewed.

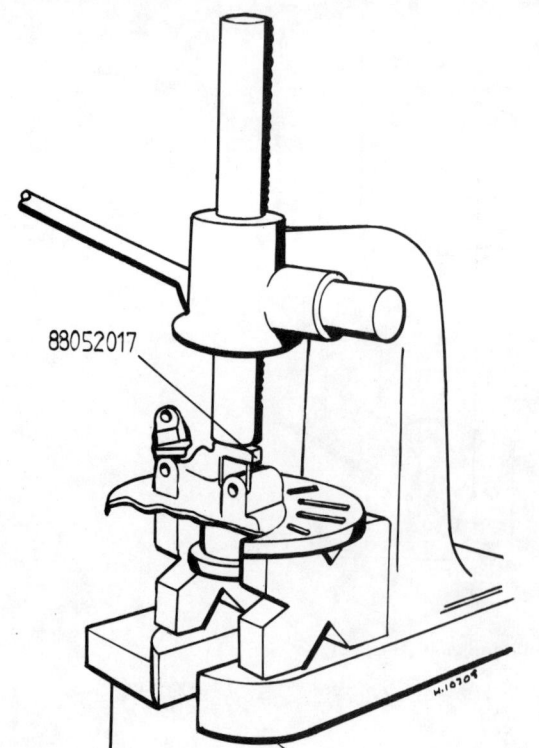

Fig. 11.16 Pressing the rear wheel hub out of the stub axle (Sec 18)

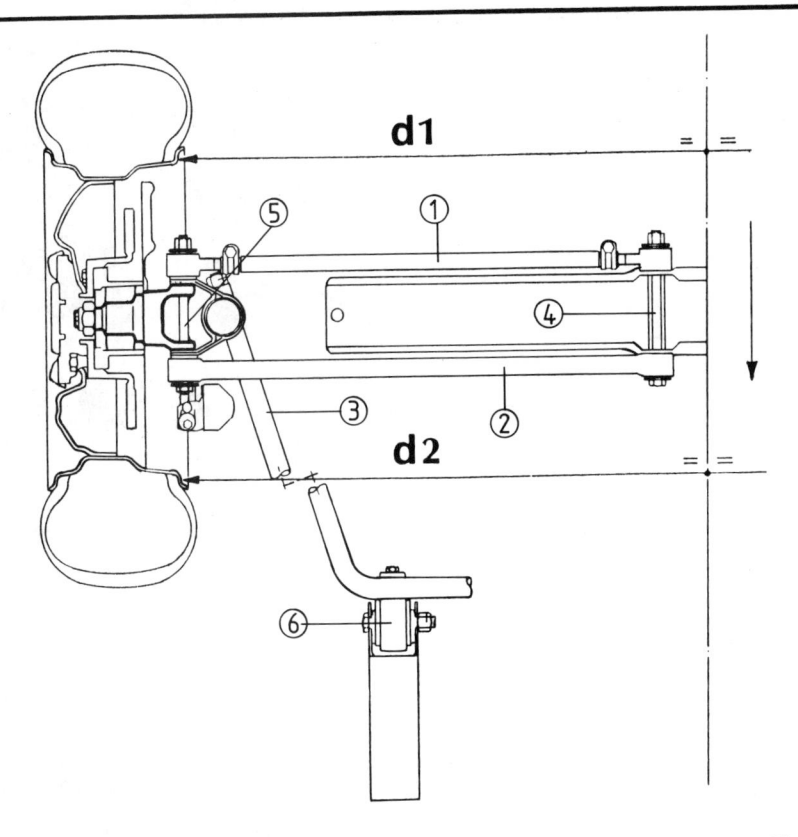

Fig. 11.18 Toe-in adjustment on right-hand side rear of an early model (Sec 19)

1. Rear adjustable link
2. Front link
3. Anti-roll bar
4. Inner mounting bolt
5. Outer mounting bolt
6. Anti-roll bar front mounting arm

Toe in = $d_1 - d_2$

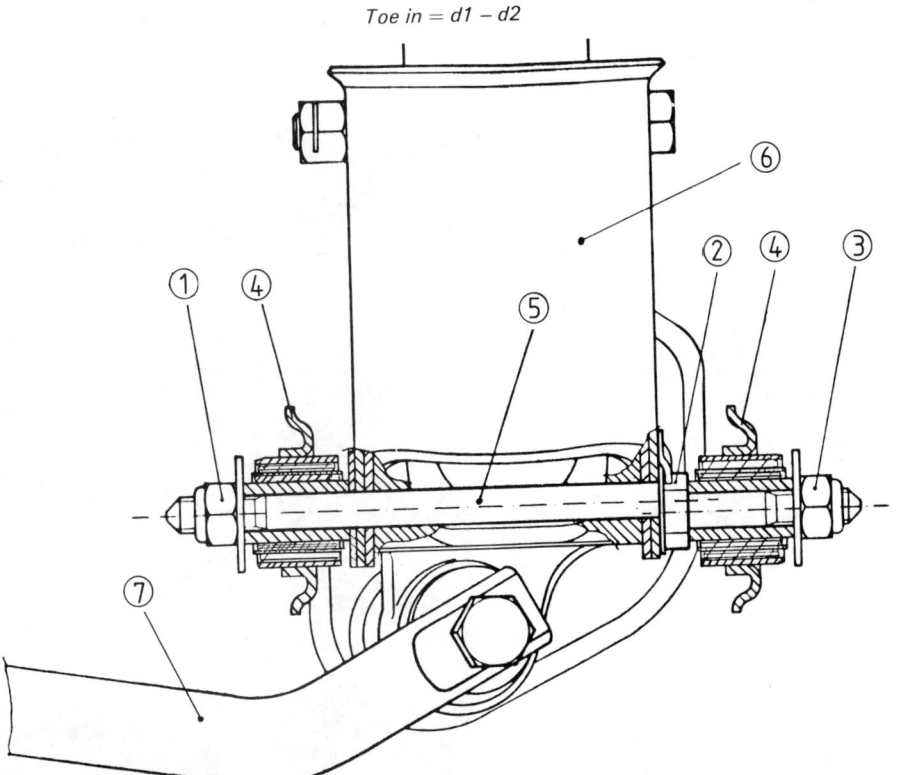

Fig. 11.19 Toe-in adjustment on later models is achieved by rotating the transverse link outer mounting shaft (Sec 19)

1. Front mounting nut
2. Hexagonal nut for altering the toe-in
3. Rear mounting nut
4. Transverse links
5. Mounting shaft
6. Strut
7. Anti-roll bar

Chapter 11 Suspension and steering

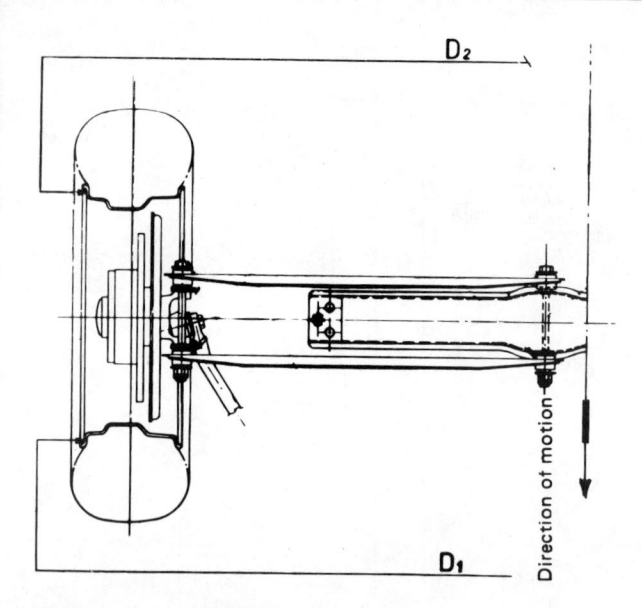

Fig. 11.20 Toe-in adjustment/measurement on later models
(Sec 19)

Toe in = D1 – D2

6 If the wheels have been balanced off the car then the wheels can be moved from front to rear in order to even out the wear.
7 Where the wheels have been balanced on the car they cannot be changed round, unless they are rebalanced each time. When a wheel has been balanced in this way the stud fixing positions should be marked, if a wheel has to be removed. This will ensure that the wheel is refitted in the same relationship to the hub and is therefore in its 'balanced' position.
8 Wheels should be rebalanced halfway through their lives in order to compensate for wear.
9 Always keep the tyres inflated to the recommended pressures and refit the dust caps after checking the pressures. Do not forget to include the spare type in this regular servicing procedure. Pressures are best checked when the tyres are cold.

21 Steering wheel – removal and refitting

1 Disconnect the battery negative cable.
2 Set the steering wheel so that the roadwheels are in the straight ahead position.
3 Remove the horn push. There are different methods of doing this depending on the model and year of manufacture. Early Saloon models have screws underneath the steering wheel spokes which hold the horn control in position. On the Coupe, Spider and HPE models which have a small centre pad and thin metal bars below the spokes, it is simply a matter of releasing the lugs at the rear of the steering wheel and pushing out the centre pad (photo).
4 Remove the horn push spring and the steering wheel retaining nut.
5 Scribe the relationship of the steering wheel to the steering column.
6 Remove the steering wheel. This is a simple operation on cars which have holes already drilled in the steering wheel into which a puller (Lancia tool No 88062038) can be fitted. All later models have holes in the steering wheel when they come from the factory. On early models without the holes refer to Figs. 11.21 and 11.22 and drill and tap the holes to the dimensions shown. Drill the holes carefully. Use a 4 mm drill to start with, then a 6.6 mm drill and finally tap a thread for the M8 x 1.25 mm bolts. Fix the puller to the steering wheel and remove it.
7 To refit the steering wheel, start by lining up the scribe marks on the wheel and column and push the steering wheel onto the shaft as far as it will go.
8 Refit the washer and nut and screw it down thus pressing the steering wheel home. Tighten the nut to the specified torque.
9 Refit the remainder of the components in the reverse order to that in which they were removed.

21.3 When the lugs are released the steering wheel boss and horn push can be removed (HPE, Coupe and Spider)

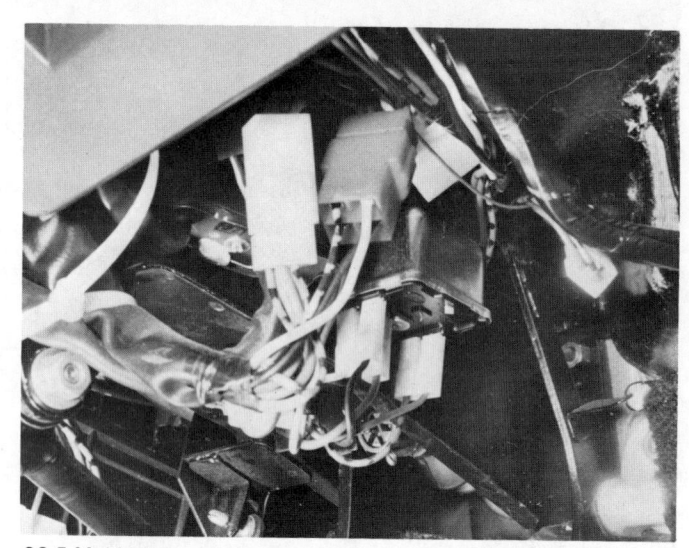

22.5 Multi-plug connectors under dashboard

22.6 The steering column jacket is secured to the plate by four nuts

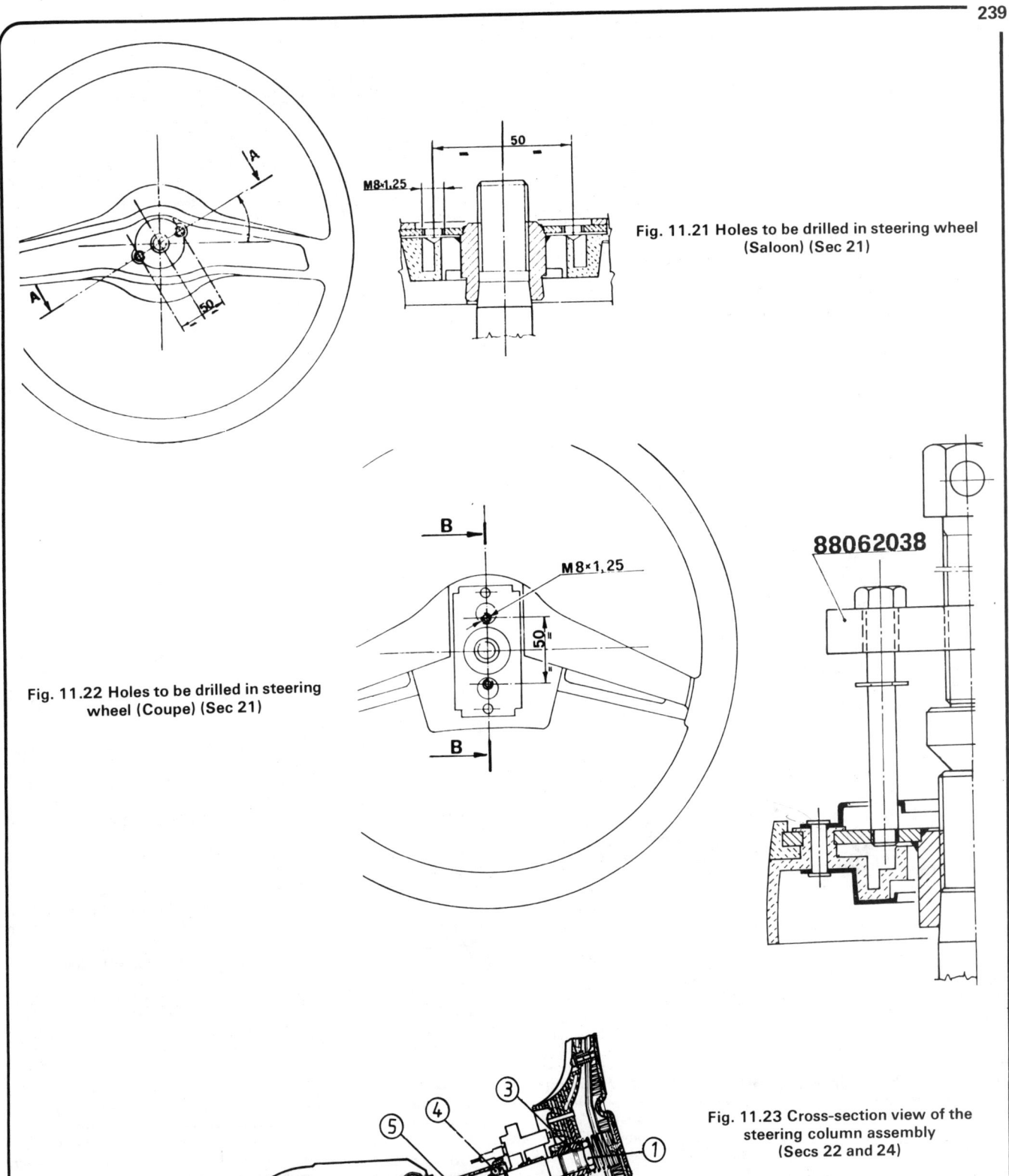

Fig. 11.21 Holes to be drilled in steering wheel (Saloon) (Sec 21)

Fig. 11.22 Holes to be drilled in steering wheel (Coupe) (Sec 21)

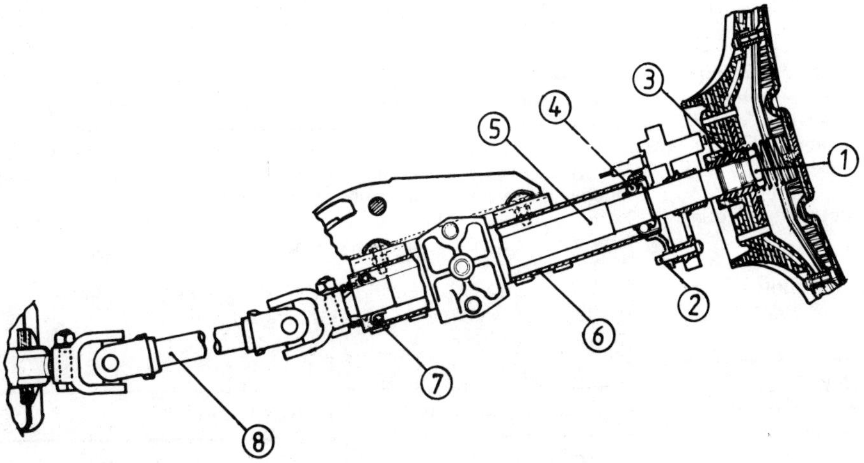

Fig. 11.23 Cross-section view of the steering column assembly (Secs 22 and 24)

1. Steering wheel mounting nut
2. Lights and windscreen wiper controls mounting screw
3. Steering wheel
4. Upper bearing
5. Rigid shaft
6. Steering column jacket
7. Lower bearing
8. Jointed shaft

Chapter 11 Suspension and steering

22 Steering column – removal and refitting

1 Disconnect the battery negative lead.
2 Set the roadwheels in the straight ahead position.
3 Undo the lock bolt on the lower joint of the steering column where it goes through the floor (Fig. 11.24).
4 Remove the screws retaining the steering column moulded casings and then remove them both.
5 Disconnect the multi-lug connectors for the ignition, horn, lighting switches and indicators (photo). Note which fits where.
6 Remove the nuts which secure the steering column jacket to the mounting plate. There are four of them (photo).
7 The steering column, complete with steering wheel, controls and ignition lock, can now be lifted away. As you do so make sure all the wiring is disconnected from the assembly.
8 Refit the steering column in the reverse order, but make sure that the roadwheels are in the straight ahead position and that the steering wheel spokes are either horizontal or level before reconnecting the bottom joint to the pinion shaft.
9 Tighten all the mounting nuts and bolts to their correct specified torques.

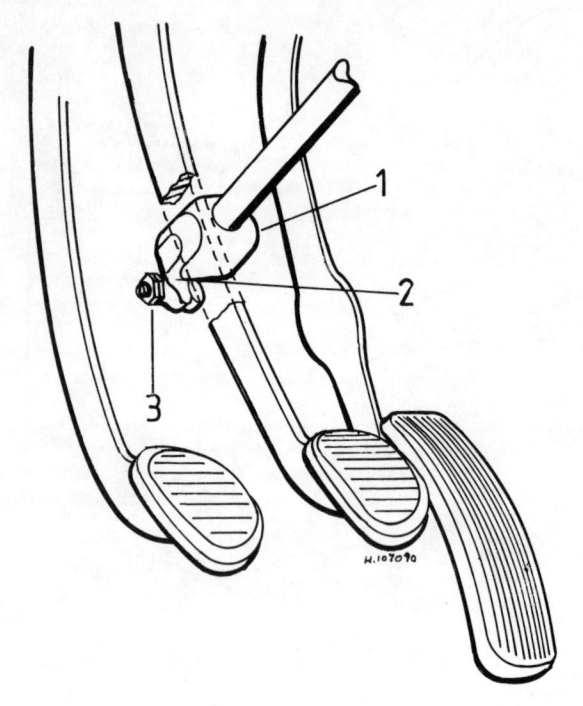

Fig. 11.24 The steering column lower coupling (Sec 22)

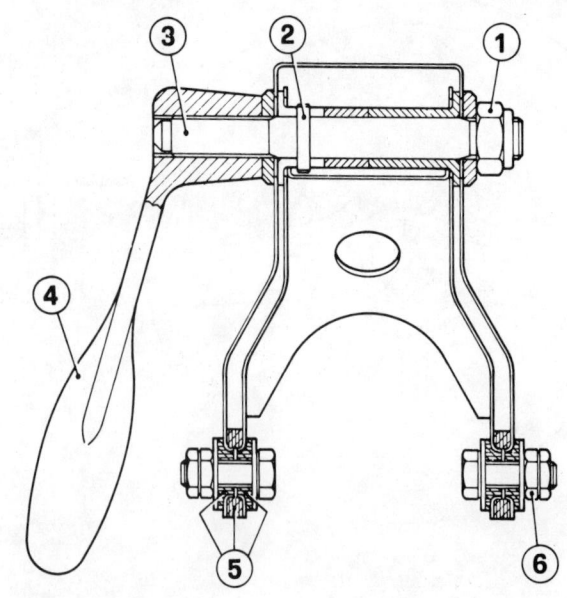

1 Universal joint
2 Pinion shaft
3 Lock bolt

Fig. 11.25 Steering column adjustable bracket (Sec 23)

1 Nut
2 Pin
3 Bolt
4 Locking handle
5 Pedal bracket bush
6 Locknut

3 Remove the steering wheel from the other end of the column as described in Section 21.
4 Undo the screws and remove the light switch and wiper control switch assembly, which is in one piece.
5 Remove the upper bearing retaining circlip and drive off the lower bearing by tapping the upper end of the steering column with a hide hammer.
6 Remove the steering shaft from the jacket and re-insert it upside down to drive off the upper bearing.
7 With the steering column components dismantled inspect them for wear and renew any which are worn.
8 To reassemble the components, start by refitting the upper bearing to the steering column jacket.
9 Then insert the steering column and refit the lower bearing. Tap it into place. Refit the retaining circlips.
10 Next refit the lower jointed section of the steering column and don't forget the spring. This section must be fitted as shown in Fig. 11.23 with the lock bolts in a vertical plane. The upper column itself must be fitted so that the indicator cancellation control notch is located facing the right-hand side. Tighten the lock bolt to the specified torque.
11 Next refit the lights and wiper control switch assembly, followed by the steering wheel, and then refit the whole assembly to the car. Tighten all nuts to their specified torques.

23 Steering column adjustable bracket – removal and refitting

1 Remove the steering column as described in the previous Section.
2 Remove the instrument panel as described in Chapter 10.
3 Remove the nut on the end of the handle locking bolt. Then withdraw the bolt and handle (Fig. 11.25).
4 Undo and remove the nuts on the lower ends of the pivoting bracket where it joins the pedal mounting bracket.
5 The assembly is now free to be removed.
6 Refit in the reverse order to removal, but be sure to tighten all the mounting bolts and nuts to their specified torques.

24 Steering column – dismantling and reassembly

1 Remove the column assembly complete as described in Section 22 and place it on the workbench.
2 Remove the upper locking bolt on the top joint where it meets the main column and remove the jointed section and spring.

25 Steering column anti-theft lock – removal and refitting

1 Remove the steering column assembly as described in Section 22.
2 Drill out the anti-theft lock securing bolts using a 6 mm drill. Then punch out the remnants and lift away the two halves of the lock.
3 Fit the lock to the steering column and tighten the shear head bolts. Ensure that the lock works smoothly before shearing the heads off the bolts.
4 The lock is now securely fitted to the steering column.
5 Refit the whole assembly to the car.

26 Steering rack (manual) – removal and refitting

1 Apply the handbrake, remove the hub caps (where fitted) and loosen the front roadwheel bolts.
2 Set the wheels in the straight ahead position and remove the lock

Chapter 11 Suspension and steering

26.6 The steering rack is secured to the bulkhead with four bolts (top left-hand bolt arrowed). Engine removed for clarity

27.1 The damper is mounted above the rack like this in right-hand drive cars (engine removed for clarity)

bolt in the lower steering joint inside the car. Scribe the relationship of the lower joint to the pinion shaft.
3 Jack up the front of the car and support it on stands placed under the side members of the subframe.
4 Remove the roadwheels.
5 Undo the track rod end balljoint nuts and use a balljoint separator to remove the balljoints, as described in Section 28.
6 Undo the four bolts which secure the steering rack housing and assembly to the rear bulkhead (photos). In most models this is an extremely difficult operation due to the lack of room and the fact that the steering rack mounting clamps are so well hidden that it is almost impossible to get at them. By removing the rubber protective boots on the inside of the wheel arches this may make it slightly easier, and the bolts can just about be reached through the steering rod openings in the engine bay side panels.
7 With the mounting clamps removed the steering rack now has to be pulled forward slightly in order to free the pinion shaft from the lower coupling inside the car and from the bulkhead.
8 To remove the rack assembly from the car it must be manoeuvred sideways through one of the steering rod openings complete with damper unit and with both steering rods still attached. (Note – only the ZF type has a damper).
9 Refitting the rack is the reverse procedure to removal. First set the rack so that it is centred. This is done quite simply by counting the number of revolutions the pinion makes moving the rack from lock-to-lock, and then dividing that figure by two.
10 Manoeuvre the rack through the wheel arch and into position. To actually do this it is best to employ two people, one at each side. Then for the next stage one person can guide the pinion shaft through the bulkhead and refit it to the lower steering column coupling whilst the other locates the rack in position on the bulkhead.
11 Refit the steering rack damper on each side and the steering

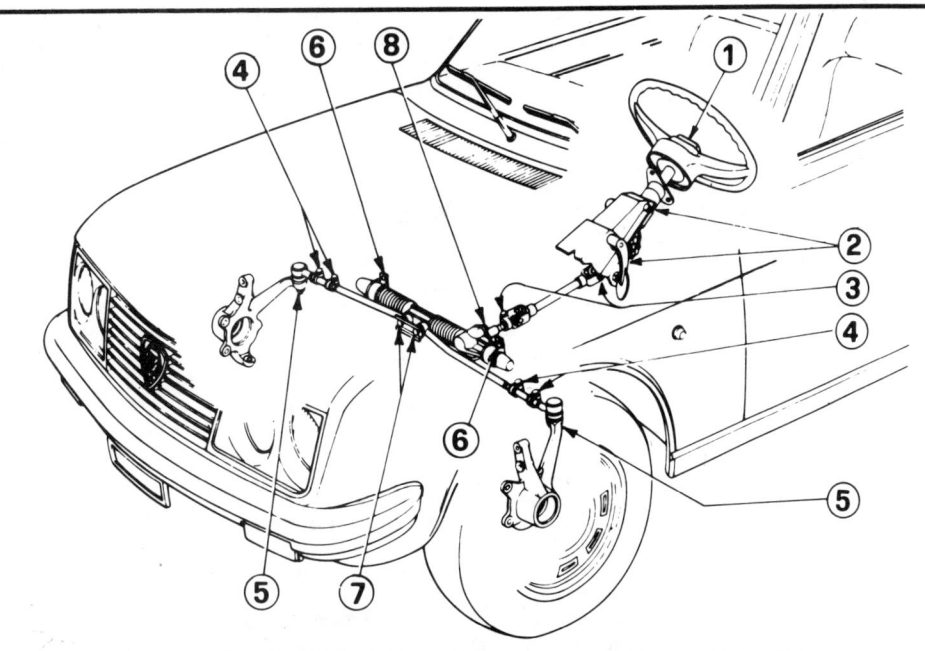

Fig. 11.26 Manual steering gear layout (TRW rack) (Sec 26)

1 Steering wheel
2 Steering jacket mounting bolts
3 Steering column universal joints
4 Track rod end clamp bolts
5 Track rod end balljoint
6 Steering rack mounting bolt
7 Steering rod mounting bolts
8 Cover plate for pinion shaft

column lower joint locking bolt and nut inside. Refit the track rod end and balljoints to the steering arms. Tighten all the nuts and bolts to their specified torques.
12 With the steering rack refitted it is now time to check the steering geometry; in this case the toe-out as described in Section 29.

27 Steering rack damper – removal and refitting

This type of damper is only fitted to the ZF type steering rack.
1 The steering rack damper is mounted between the rack itself and a bracket attached to the steering tie-rod mounting plate on the rack (photo). In effect it cushions the operation of the steering. This has led to comments on the heavy tendency of the steering at low or parking speeds.
2 To remove this damper is almost as difficult as removing the steering rack itself due to the limited amount of space in the engine bay. It was only possible to photograph it when the engine assembly had been removed.
3 Operating from the right-hand side of the car, with the car jacked up, and supported on stands, and the roadwheels removed, remove the steering rod gaiter to the inner wheel arch.
4 Loosen the bolt which secures the end of the steering damper to the rack. In the case of a left-hand drive model this assembly will be reversed. On USA models access is improved if the coolant expansion tank is first removed (see Chapter 2).
5 Remove the steering gaiter on the left-hand inner wheel arch panel and undo and remove the mounting nut on the other end of the damper. An earth cable is also retained by the nut.
6 Go back to the right-hand side withdraw the mounting bolt and the damper.
7 Refitting the stering rack damper is the reverse procedure to removal.

28 Steering tie-rod balljoint – removal and refitting

1 At the outer end of the steering tie-rod there is a balljoint which secures the tie-rod to the steering arm on the hub carrier. The balljoint is also attached to the tie-rod by a double threaded sleeve which is clamped in position by two clamp bolts (photo).
2 By turning the threaded sleeve the alignment of the roadwheel and therefore the toe-out can be altered. This is covered in Section 29.
3 If a balljoint is worn or damaged and needs to be renewed, firstly jack up and support the relevant side of the car and then remove the roadwheel.
4 Undo the balljoint nut which secures it to the steering arm.
5 Pull back the rubber dust guard and use a balljoint separator or split wedge to separate the balljoint from the steering arm.
6 Undo the outer tie-rod sleeve clamps so that the balljoint can be unscrewed. Carefully count the number of turns required to remove the old balljoint, as this will give correct fitting, and setting for the toe-out, for the new balljoint.
7 Fit the new balljoint in exactly the reverse order to that in which the old one was removed. When the balljoint has been screwed in the correct number of turns and is located with the threaded end pointing downwards, lock the clamp on the sleeve.
8 Refit the balljoint to the steering arm and tighten the nut to the correct torque.
9 Refit the roadwheel and lower the car to the ground.
10 Check the toe-out alignment as described in Section 29.

29 Steering geometry – checking the front wheel alignment

1 The steering geometry needs to be checked after any operation involving repair or adjustment of the steering components.
2 This is a job that is best carried out by your local Lancia dealer who has the correct equipment to do it. It requires either the use of a base bar or optical light-beam equipment. To carry it out inaccurately will only result in poor steering, which in many cases may be dangerous, and also bad tyre wear. It is very easy to scrub a tyre through incorrect steering geometry alignment and also in this age of high inflation a very expensive exercise.
3 Fig. 11.27 shows the way in which the toe-out should be measured. To achieve the correct toe-out value, as given in the

28.1 The steering-rod balljoint is attached to the steering arm by a nut and clamped to the rod by a threaded sleeve

specifications, the d1 measurement is subtracted from the d2 measurement.
4 Adjustment of the toe-out adjustment is made by turning the double threaded sleeve on each side by equal amounts as described in Section 28.
5 When carrying out the adjustment in its laden state the car must comply with the following conditions. For the Saloon or HPE models it must have the equivalent of 4 occupants and for the Coupe and Spider moedls 3. For the purposes of this exercise the occupant is equal to 154 lb or 70 kg.
6 At the same time as your car is in the garage get them to check the complete front suspension and steering geometry; not only the toe-in, but also the camber and king-pin inclination.

30 Steering rack (manual) – overhaul

1 No adjustment can be made to the steering gear unless it is removed from the car. With it removed, it is as well to thoroughly strip and check the components before adjustments are carried out. This could save a lot of time later on if there is wear which has not been detected. As can be seen from Section 26, removal of the steering rack assembly is not the easiest of tasks. If there is evidence of bad wear, then it is advisable to get an exchange reconditioned unit.

TRW steering rack
2 With the assembly on the bench loosen the clip which secures one of the rubber end caps; remove the cap and drain the oil into a container.
3 Undo the bolts which secure the steering rods to the rack. There is a lock tab plate which has to be knocked down first. Place the rods to one side.
4 Remove the mounting clamp buffers, slacken the clips and remove the rubber cap and guard from the rack.
5 Undo the bolts which retain the locating support cover in place and remove the struts, seal, spring and rack locating support (Fig. 11.28).
6 Unlock the tab and remove the pinion cover plate. Then remove the seal, shims and drive pinion shaft. The upper bearing will come out with the pinion shaft.
7 Now it is possible to slide the rack and plastic bush out of the housing.
8 Finally remove the lower pinion bearing.
9 Thoroughly clean all the parts with paraffin and then dry them off. Examine all the components carefully. Check the rack and pinion teeth for wear, unevenness, chipping, scoring or cracks. Check the ball bearing races for wear or excessive play. Check the locating support for wear, especially if there was little or no oil in the housing, and for its sliding fit in the housing. It must not bind. Ensure that the rubber gaiters and end caps are all in good condition. It is wise to renew the

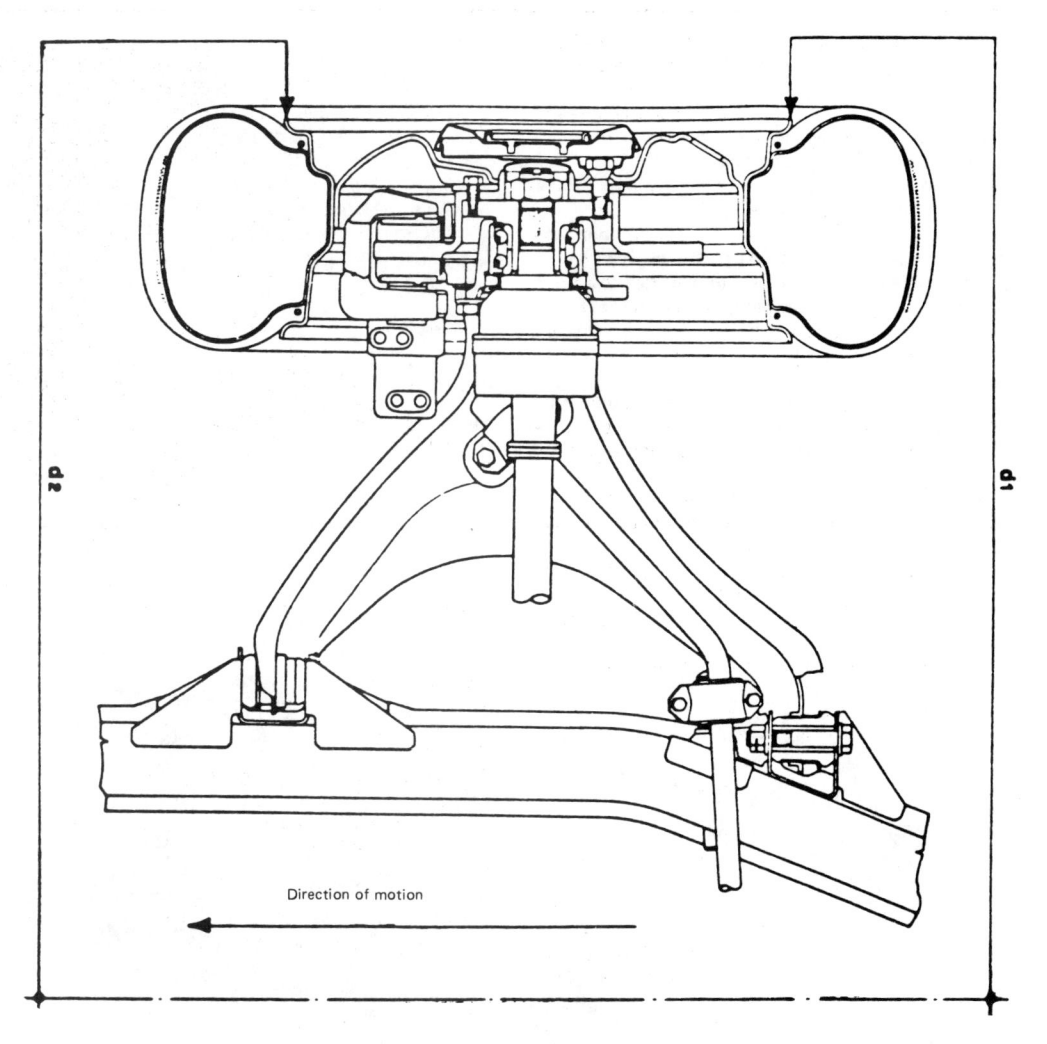

Fig. 11.27 Toe-out measurement diagram (Sec 29)

Toe-out = d2 − d1

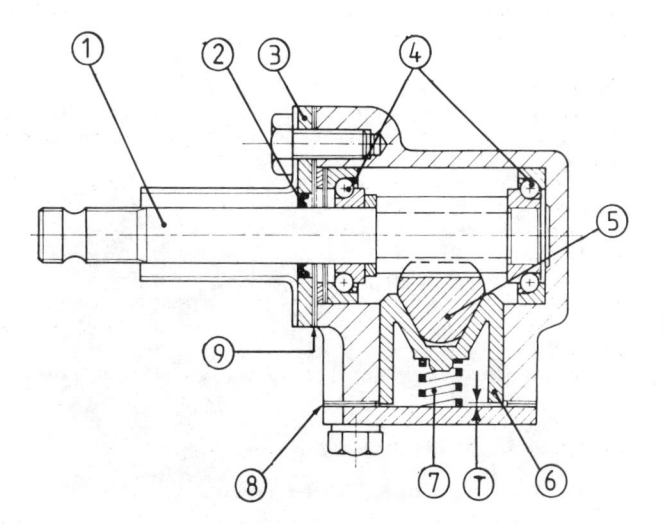

Fig. 11.28 Cutaway sectional view of TRW steering rack and pinion assembly (Sec 30)

1 Drive pinion shaft
2 Seal
3 Cover
4 Pinion bearings
5 Rack
6 Rack locating support
7 Spring
8 Support adjusting shims
9 Seal

Chapter 11 Suspension and steering

rubber caps anyway. Check all the seals and renew them anyway.

10 Check the steering tie-rod inner bushes and have them renewed if worn or damaged. They have to be pressed in and out using a special Lancia tool.

11 Once all the component parts have been checked and renewed where necessary the reassembly process can begin.

12 Start by fitting the plastic bush for the rack to the housing. Make sure that the three locating dogs seat correctly.

13 Refit the pinion shaft lower bearing, the drive pinion together with the upper bearing and the upper bearing thrust plate and washer.

14 The next step is to calculate the shims required to get the right preload setting on the pinion. To do this add shims of measured thickness to the pinion shaft so that they protrude just above the surface of the housing as shown in Fig. 11.30.

15 Then place the pinion cover in place, but without its seal, and measure the gap G. Subtract the measurement from the total thickness of the shims which have been fitted. Add the thickness of the seal and the value of the prescribed pre-load. The resulting figure will give you the thickness of the shims that actually need to be fitted to give the correct pre-load setting.

Example

Total thickness of shims fitted = 5 mm
Measurement G = 2 mm
Thickness of seal = 1 mm
Preload value = 0.130 mm
Total thickness of shims fitted −G = 5 mm − 2 mm = 3 mm
3mm + thickness of seal = 3mm + 1 mm = 4 mm
4 mm + preload value = 4 mm + 0.130 mm = 4.130 mm
Thickness of shims required = 4.130 mm

Shims are available in the following sizes:

Part No	mm	in
82293074	0.13	0.00511
82293075	0.19	0.00748
82293076	0.25	0.00905
82293077	2.30	0.0905

16 Remove the cover and shims and washer and fit the correct thickness of shims as determined by the calculation.

17 Refit the washer, seal and cover and tighten the cover retaining bolts to the specified torque.

Fig. 11.29 Inserting the steering rack bush (Sec 30)

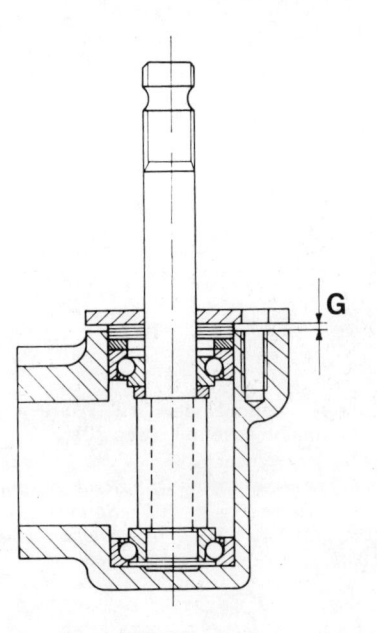

Fig. 11.30 Gauging the pinion bearing pre-load setting (Sec 30)

G = gap for calculation (see text)

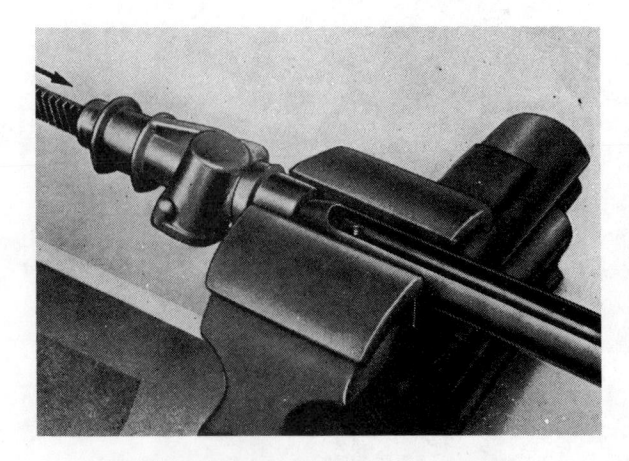

Fig. 11.31 Refitting the rack (Sec 30)

Chapter 11 Suspension and steering

18 Check that the pinion shaft revolves freely and smoothly.
19 Remove the cover and pinion shaft complete with shims and upper bearing.
20 Refit the rack to the housing as shown in Fig. 11.31.
21 Refit the pinion shaft, bearing, shims, washer, seal and cover. Make sure that the pinion meshes with the rack teeth. Finally tighten the pinion cover plate bolts and bend up the lock tab plate.
22 Move the rack in its housing so that it is approximately half way along its travel. Refit the locating support.
23 Refit the cover plate without its seal as in Fig. 11.32. Then measure the gap G in order to calculate the rack to pinion backlash. By adding to that measurement the prescribed amount of backlash of 0.02 to 0.06 mm (0.00078 to 0.0023 in) and then subtracting the thickness of the seal this will give you the thickness of shims that need to be fitted.
24 When this calculation has been carried out and the shims have been selected the spring can be refitted to the locating support followed by the shims, seal and cover. Tighten the bolts to the specified torque, as far as the pinion cover bolts. The bolts should be coated with a sealing compound, and the edges of the shims should also be coated with a sealant. This ensures that the rack stays tight and that there are no oil leaks.
25 Refit the rubber gaiters, clips and end covers and fill the rack with oil.
26 Refit the steering rods to the rack mounting plate and tighten the bolts to the specified torque. Make sure that the rods are fitted to the correct side.
27 Check before refitting that all the boots and caps are securely mounted.

ZF Steering rack

28 Undo the mounting bolt and nut securing the damper to the steering gear and housing and remove it.
29 Loosen the clips and remove the rubber caps from both ends of the rack housing. Drain the oil.
30 Remove the steering tie-rods by knocking down the tabs on the central mounting plate and removing the two bolts. Note that the right-hand rod is different from the left and note also which way they fit. It is possible to fit them upside down.
31 Remove the mounting buffers from either end of the housing and undo the clips and remove the rack gaiters. Then remove the centre sleeve, and the link from the recess in the steering rack housing.

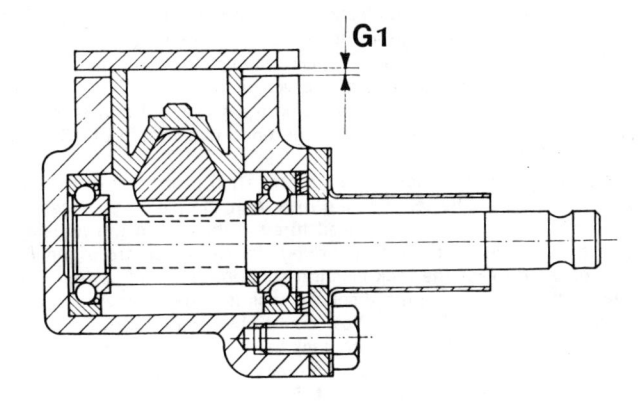

Fig. 11.32 Adjusting the backlash (Sec 30)

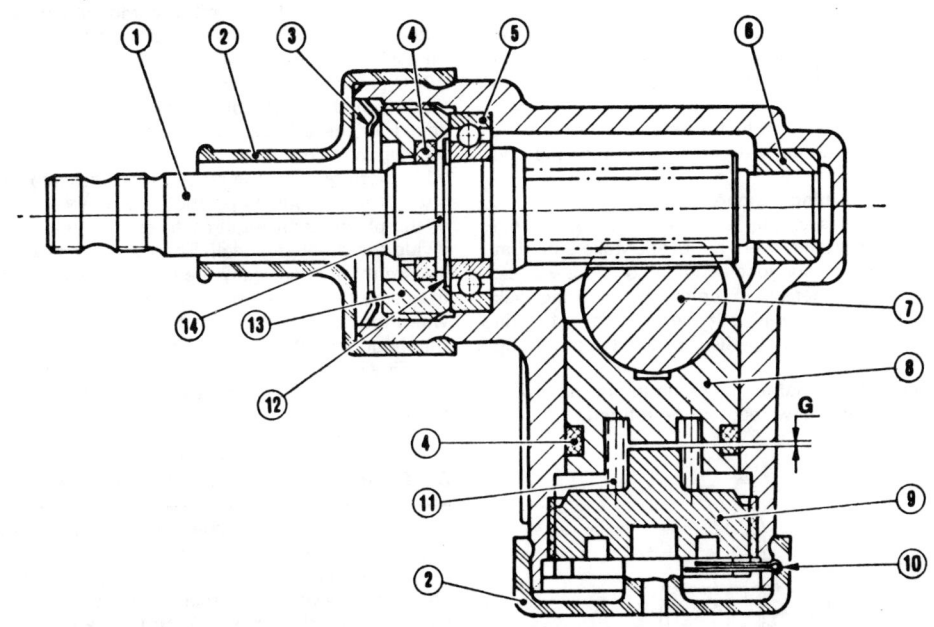

Fig. 11.33 Sectional view of the ZF steering rack and pinion assembly (Sec 30)

1 Pinion shaft
2 Caps
3 Retainer ring
4 Seals
5 Ball bearing
6 Bush
7 Rack
8 Rack locating support
9 Adjuster
10 Cotter key
11 Spring
12 Circlip
13 Pinion adjuster
14 Ring

32 Remove the lower cover plate and the split pin (Fig. 11.33).
33 To remove the adjuster requires a peg spanner, or you can make one out of a suitable sized piece of tubing.
34 Once the adjuster has been withdrawn the spring and rack support can be removed.
35 Now slide the rack out of the housing.
36 Turn the housing so that the pinion shaft is facing upwards and then remove the pinion cover. Lift off the retaining ring and then use the special spanner to withdraw the pinion adjuster.
37 Grip the pinion shaft in a vice with protective jaws and using a hide mallet drive the housing off the shaft so that it comes out with the upper bearing still attached.
38 Remove the plastic bush and also the O-ring for the rack.
39 Unless it is worn and needs to be renewed, the lower bearing for the pinion shaft can stay in position. Do not forget to check its condition however. To remove it requires a puller of the type shown in Fig. 11.34.
40 The upper bearing can be easily pulled off the pinion shaft once the circlip above it has been removed.
41 Check all the components, after cleaning, as covered in paragraphs 9 to 11 of this Section. If there is a considerable amount of wear then it is better to refit a reconditioned rack, which will have the added advantage of being covered by a guarantee.

Chapter 11 Suspension and steering

42 Insert a new O-ring and plastic bush into the housing. Ensure that the 2 locating dogs engage in the seats in the housing.
43 Refit the pinion shaft and upper bearing. They will have to be tapped into the pinion housing and lower bearing. Then fit the spacer and seal and screw in the adjuster using the same tool that was used to remove it. This adjuster must be tightened to a torque of 16 to 19 lbf ft (21 to 25 Nm).
44 Lightly grease the rack and its teeth. Then insert it into the rack housing and mesh the teeth of the rack with the pinion. If necessary a little more grease may be applied through the slot in the housing.
45 Align the steering tie-rods mounting plate at the centre of the slot.
46 Grease and refit the rack locating support, seal and spring. Then screw in the adjuster, with the special tool used for removing it, until the support touches the rack. Now count how many turns the pinion has to make to move the rack from lock to lock. Then halve the number and locate the pinion accordingly. The rack locating support should be tightened to a torque of 4.5 lbf ft (6 Nm).
47 Undo the adjuster slightly if necessary, to insert the split pin.
48 When the adjustments have been made the clearance at G (Fig. 11.33), should be 0.08 mm (0.00315 in) and the revolving torque 0.08 to 0.16 kgf m (0.56 to 1.16 lbf ft). If necessary the setting of the rack locating support adjuster will have to be altered to achieve this figure, which is measured using a torque meter attached to the pinion shaft.
49 Once the steering rack is set up, refit the retaining ring to the pinion adjuster. Fill the slot in the ring with grease and refit the end cap.
50 Refit the cap to the rack locating support housing.
51 Refit the link for the tie-rods to the rack. Refit the rest of the components in the order in which they were removed. Make sure that the rubber gaiters and caps are in very good condition or fit new ones. Just remember how awkward the task is to remove the steering rack assembly in the first place. You don't want to have to remove it again shortly just to renew the gaiters. Tighten all the clips and the mounting bolts to their specified torques.
52 Refit the steering damper, and tighten its mounting nut and bolt.
53 The steering rack assembly is now ready to be refitted.

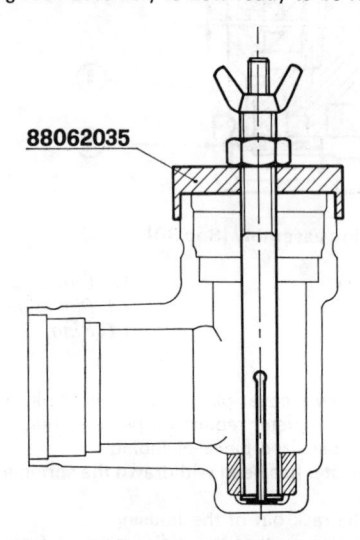

Fig. 11.34 Extracting the lower pinion shaft bearing using special tool 88062035 (Sec 30)

31 Power steering rack assembly – removal and refitting

When dismantling or disconnecting any part of the power steering system it is very important that complete cleanliness is observed. Exposed ends of pipes, hoses or ports should be sealed to prevent the ingress of dirt or foreign matter. Do not at any time start the engine when the fluid reservoir is empty, otherwise the hydraulic system will be seriously damaged.

1 Disconnect the battery negative cable.
2 Remove the lock bolt for the lower jointed section of the steering column where it meets the pinion shaft at the bulkhead.
3 Remove the steering column casings and undo the upper column mounting nuts and lower the steering column so that it rests on the driver's seat. Refer to Section 22.
4 Slacken the front roadwheel mounting bolts, jack up the front end of the car and place stands or supports under the side members of the subframe. Apply the handbrake firmly.
5 Open the bonnet and remove the expansion tank for the cooling system. Refer to Chapter 2 if necessary.
6 Have a container handy to catch the fluid from the power steering rack when the outlet pipe and inlet pipes are removed. This fluid must NOT be re-used.
7 Disconnect the steering tie-rod end balljoints from the steering arms as described in Section 28.
8 Remove the rubber gaiters from the inner wheel arches, through which the tie-rods run.
9 Undo the four bolts which secure the steering rack assembly to the bulkhead. Due to the limited amount of space this can be a very tricky job. Although they are reversed for left-hand drive the photos 26.6 may help you. It is wise to have an assistant to help support the rack at one end whilst the other end is freed. Once the bolts have been removed the whole assembly, complete with steering tie-rods, can be withdrawn from the left-hand side of the car. Once again the help of an assistant is essential.
10 On models produced for the North American market it is necessary to detach the lower piping for the EGR system and the front heat shield for the exhaust system, before the steering rack assembly can be manoeuvred out.
11 Before refitting the steering rack assembly set it so that it is in the straight ahead driving position. To do this, count the number of turns of the pinion shaft required to remove the rack from lock-to-lock and divide the number in half. Then set the rack accordingly.
12 Refitting is basically the reverse operation to removal. Remember to ensure that the steering wheel is kept in the straight ahead position when the lower column joint is refitted to the pinion shaft. Tighten all the mounting nuts and bolts to their specified torque wrench settings.
13 Finally top-up and bleed the power steering system as described in Section 34, and check the steering alignment as covered in Section 29.

32 Power steering rack overhaul – general

1 Whereas it is possible for the competent home mechanic to overhaul the manual type of steering rack, provided the right equipment is available. This is not the case with the power steering rack. The system is very much more complicated and the overhaul should be left to a specialist or your local Lancia dealer.
2 If the steering rack is badly worn then it is advisable to exchange it for a new or reconditioned unit. This will save both time and money and of course a new or reconditioned unit will also have a guarantee.

33 Power steering pump – removal and refitting

1 The power steering pump is mounted underneath the front of the engine and is driven by a belt directly off the crankshaft pulley.
2 To remove the pump first jack up the right front wheel and remove it. Then support the car using an axle stand.
3 Remove the engine splash guard or undertray. On models made for the Swedish or North American market the wheel arch guard will have to be removed as well.
4 Thoroughly clean the pump and banjo connections. Ensure that no dust can get into the pipes.
5 Place a container under the power steering pump and undo the inlet pipe banjo bolt, and let the fluid drain out. This fluid must be disposed of and *not* re-used.
6 The procedure from now on depends upon whether the car has an air conditioning unit fitted or not, as the mounting differs considerably. This could be seen in Figs. 11.36 and 11.37.
7 For a car without an air conditioner carry out the following procedure to paragraph 11. Grip the belt in the middle with one hand and loosen the pump pulley retaining nut with the other.
8 Slacken the belt tensioner retaining bolt (1) and lift the belt off the pump pulley. Then remove the bolt (Fig. 11.36).
9 Undo the pump outlet banjo bolt at the rear of the pump.
10 Slacken the alternator belt tensioner mounting bolt and undo and remove the common mounting pivot bolt (6). The steering pump is now free and can be removed with the pulley still attached.

Chapter 11 Suspension and steering

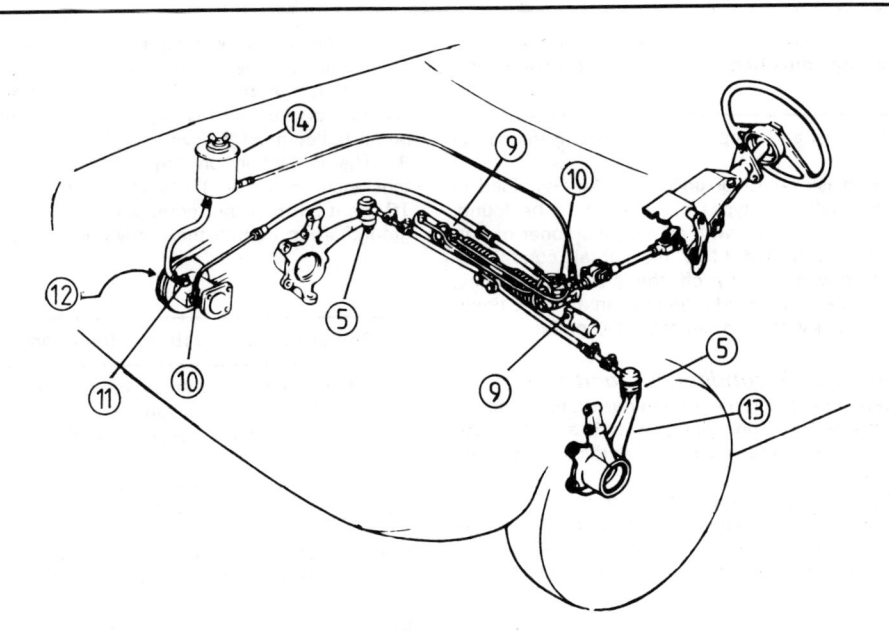

Fig. 11.35 The power steering system layout (Sec 31)

- 5 Track rod end balljoint
- 9 Steering rack clamp
- 10 Banjo union bolts
- 11 Power steering pump
- 12 Power steering pump pulley
- 13 Steering arm
- 14 Reservoir

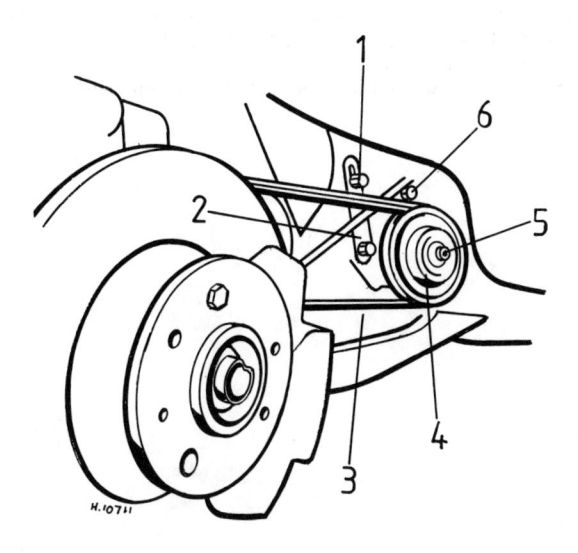

Fig. 11.36 Power steering pump (without air conditioner) (Sec 33)

1. Belt tensioner bolt
2. Belt tensioner
3. Drivebelt
4. Pump pulley
5. Pulley retaining nut
6. Bracket-to-oil filter base fixing bolt

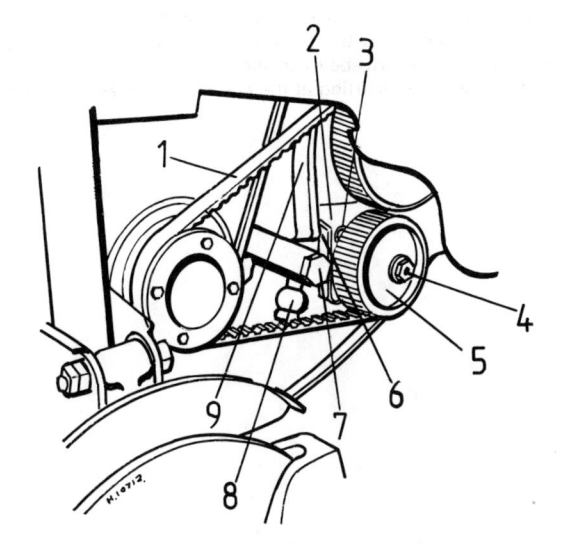

Fig. 11.37 Power steering pump (with air conditioner fitted) (Sec 33)

1. Pump and compressor drive cog belt
2. Pump front support bracket
3. Pump-to-bracket fixing bolt
4. Pulley fixing nut
5. Pump pulley
6. Compressor lower fixing screw
7. Pump
8. Pump fluid inlet banjo bolt
9. Oil filter base

11 Because the pulley retaining bolt was slackened off at the beginning it makes it easier to remove the pulley if necessary.
12 If the car has an air conditioning unit fitted then carry out the procedure from here on. First remove the air cleaner assembly and slacken the upper mounting bolts for the compressor.
13 Then slacken off the steering pump pulley retaining nut (4) and the lower mounting bolt for the compressor (6) (Fig. 11.37).
14 The compressor can be dropped down on its bolts so that the toothed drivebelt can be removed. Once this has been done lift the compressor up again.
15 Remove the steering pump pulley retaining nut and then the pulley itself.
16 Remove the pump outlet banjo bolt from the rear of the pump, and the rear bracket.
17 The pump is held into the front mounting bracket (2) from the rear by bolts (3). Remove these and the pump can be withdrawn. If necessary the bracket can then be removed from the compressor.
18 Refitting is the reverse procedure to removal. Remember to tighten up the banjo bolts and pulley retaining bolt to the specified torque. Always use new washers when refitting the banjo bolts.
19 Refit the drivebelt and tension it correctly as detailed in Section 34.
20 Refill and bleed the power steering system as described in Section 35.

34 Power steering pump drivebelt – removal, refitting and tensioning

1 Jack up the front of the car and support it on stands. Remove the right-hand roadwheel.
2 Remove the engine splash guard or undertray. As described in Section 33, there are two different types of drivebelt to be found, depending on whether the car is fitted with an air conditioner or not. As can be seen from Figs. 11.36 and 11.37 the non-air conditioner type is a simple V-belt driven directly off the crankshaft pulley, whereas there is a toothed belt employed if there is an air conditioner fitted, also driven off the crankshaft from an extended pulley.

Steering pump – without air conditioning unit fitted

3 To remove the drivebelt on a non-air conditioner system refer to Fig. 11.36 and slacken the belt tension adjuster nut (1) and the pivot bolt (6). The pump can then be pivoted slightly upwards and then the belt can be slipped off the pulley.
4 Refit the belt to the crankshaft pulley and then slip it over the steering pump pulley. Pivot the pump downwards so as to tension the belt and then tighten the tension adjuster nut (1). The belt should have a deflection in the centre of its run of approximately 10 to 15 mm (0.39 to 0.59 in).
5 Tighten up the pivot bolt and nut (6).

Steering pump with air conditioning unit fitted

6 Where the car is fitted with an air conditioner first remove the air cleaner assembly. Then slacken the upper mounting bolts for the compressor.
7 Refer to Fig. 11.37, slacken the lower mounting nuts for the compressor (6) and lower the compressor so as to relieve the tension on the drivebelt. The drivebelt can then be removed.
8 Refit the drivebelt starting at the crankshaft pulley, then fit it over the steering pump pulley making sure that the bottom section is tight. Then fit it over the compressor pulley and then lift the compressor up to tension the belt. Tighten the compressor mounting bolts. The sections of the belt leading from the crankshaft pulley and steering pump pulley must be evenly tight.
9 The deflection on the belt on its longest run, when properly tensioned should be 10 to 15 mm (0.39 to 0.59 in).
10 Refit all the other components removed and lower the car to the ground. Then tighten the roadwheel bolts.

35 Power steering system – filling and bleeding

1 Whenever the power steering system has had any work carried out on it always check, fill and bleed it, when it has been reassembled. Never attempt to re-use old fluid that has been drained out during dismantling.
2 Remove the reservoir cap and top up the fluid to the level mark. Use only the specified fluid.
3 Disconnect the HT lead to the coil. Turn the engine over on the starter so that the pump operates and add more fluid as necessary. This is more easily done with two people.
4 When no more fluid needs to be added, reconnect the HT lead to the coil and run the engine at idling speed. Top up the reservoir to the level mark as necessary with the engine running. Turn the steering wheel from lock-to-lock in both directions and top up until no more air bubbles appear.
5 Check all hoses and banjo unions for condition and any sign of leaking.
6 When all the air has been bled out of the system the engine can be stopped. The fluid level will rise 10 to 20 mm (0.5 to 0.75 in) above the level mark in the reservoir.
7 Refit the reservoir cap and the operation is complete.

36 Fault diagnosis – suspension and steering

Symptom	Reason(s)
Steering feels vague, car wanders and floats at speed	Tyre pressures uneven Shock absorbers worn Steering gear balljoints badly worn Suspension geometry incorrect Steering mechanism free play excessive Front suspension and rear suspension pickup points out of alignment Anti-roll bar bushes worn Transverse link bushes worn
Steering pulls to one side or other	Tyre pressures uneven Lower suspension balljoint worn out Suspension struts worn Transverse links damaged
Stiff and heavy steering	Tyre pressures too low No oil in steering gear (TRW type) No grease in steering and suspension balljoints Front wheel toe-out incorrect Suspension geometry incorrect Steering gear incorrectly adjusted too tightly Steering column badly misaligned Bearings in upper section worn or dry
Wheel wobble and vibration	Wheel studs loose Front wheels and tyres out of balance Steering balljoints badly worn Hub bearings badly worn Steering gear free play excessive Front springs weak or broken Anti-roll bar incorrectly located or bushes worn
Noise from wheels	Bearings worn Brakes squealing
Uneven tyre wear	Incorrect wheel geometry Unbalanced wheels Incorrect tyre pressures Transverse links distorted

Chapter 12 Bodywork and subframe

Contents

Air conditioning system – general	46
Air conditioning system – maintenance	47
Bonnet (Coupe, Spider and HPE models) – removal and refitting	21
Bonnet (Saloon models) – removal and refitting	20
Bonnet release cable – removal and refitting	22
Boot lid (Saloon, Coupe and Spider models) – removal and refitting	24
Boot lid lock and catch (Coupe and Spider models) – removal and refitting	26
Boot lid lock and catch (Saloon models) – removal, refitting and adjusting	25
Bumpers – removal and refitting	28
Centre console – removal and refitting	32
Doors – tracing and silencing rattles	6
Door glass (Coupe and HPE models) – removal and refitting	10
Door glass (Saloon models) – removal and refitting	9
Door glass (Spider models) – removal and refitting	11
Door lock, cylinder and handles (Coupe, Spider and HPE models) – removal and refitting	16
Exterior door mirrors – general	42
Facia panel (pre 1979 models) – removal and refitting	30
Front and rear doors – removal, refitting and adjusting	17
Front door handle and lock (Saloon models) – removal, overhaul and refitting	13
Front seats – removal and refitting	33
Front wings – removal and refitting	29
General description	1
Glovebox lid and lock – removal and refitting	23
Heater – removal, dismantling, reassembly and refitting	45
Heating system – general description	44
Interior door handles and control cable (Saloon models) – removal and refitting	15
Interior handles and trim panel – removal and refitting	7
Maintenance – bodywork and underframe	2
Maintenance – upholstery and carpets	3
Major body damage – repair	5
Minor body damage – repair	4
Power window lifts – removal and refitting	12
Radiator grille – removal and refitting	27
Rear door handle and lock (Saloon models) – removal and refitting	14
Rear quarterlight (Coupe and HPE) – removal and refitting	40
Rear quarterlight (Saloon models) – removal and refitting	36
Rear seats – removal and refitting	31
Rear window (Coupe models) – removal and refitting	38
Rear window (Saloon models) – removal and refitting	35
Rear window and folding roof section (Spider models) – removal and refitting	41
Subframe – general	43
Tailgate (HPE models) – removal and refitting	19
Tailgate glass (HPE models) – removal and refitting	39
Tailgate lock and catch (HPE models) – removal, refitting and adjusting	18
Towbar – fitting and wiring	48
Window winder mechanism – removal and refitting	8
Windscreen (Coupe, Spider and HPE models) – removal and refitting	37
Windscreen (Saloon models) – removal and refitting	34

1 General description

All models in the Lancia Beta range are built around a rigid passenger compartment designed to give maximum protection in the event of an accident. The engine and luggage compartments are designed to absorb impacts as well. The whole structure is a pressed steel welded fabrication of many shaped panels to form the whole 'monocoque' bodyshell. At the front end there is a subframe which is a separate unit that carries the engine and transmission assemblies.

All Beta bodies are subjected to extensive anti-corrosion treatment before and during manufacture. Electrophoretic treatment, underbody sealing and cavity injection of underseal in the lower body sections are all part of this process.

It is as well to remember that monocoque structures have no discrete load paths and all metal is stressed to an extent. It is essential therefore to maintain the whole bodyshell both top and underside, inside and outside, clean and corrosion free. Every effort should be made to keep the underside of the car as clear of mud and dirt accumulations as possible. If you were fortunate enough to acquire a new car then it is advisable to have it rust proofed and undersealed at one of the specialist workshops who guarantee their work.

This Chapter describes the 'everyday' measures that can be taken to ensure that your car will look good and be structurally safe to ride in for many years. It does not attempt to describe the methods of structural repair, such tasks have always and will remain outside the scope of the owner 'mechanic'.

Recently the rear mounting points in the body shell for the front subframe have been found to have structural weaknesses. The body sections rust through, causing the rear end of the front subframe to collapse.

This is particularly prevalent in Series A Saloons with chassis designations 828 AB0 (1600 engine), AB1 (1800 engine), 828 AB2 (1400 engine) and Series B Saloons with chassis designation 828 BB3 (1300 engine).

If the rear mounting points in the body front floor pan do rust through the whole subframe will tilt causing the engine to rotate forwards. Should such a condition arise then the car should be taken directly to your local Lancia dealer. Lancia say that although it is not unsafe to drive a car with this problem, the vehicle is technically unroadworthy and may therefore render you liable to prosecution if the vehicle is stopped and examined by the police or Ministry of Transport (DOE) examiners.

There is little that can be done, as the body can not be mended. Later B Series cars have strengthened body to subframe mounting points, as described in Section 43.

2 Maintenance – bodywork and underframe

1 The general condition of a car's bodywork is the thing that significantly affects it value. Maintenance is easy but needs to be regular. Neglect, particularly after minor damage, can lead quickly to

Chapter 12 Bodywork and subframe

further deterioration and costly repair bills. It is important also to keep watch on those parts of the car not immediately visible, for instance the underside, inside all the wheel arches and the lower part of the engine compartment.

2 The basic maintenance routine for the bodywork is washing — preferably with a lot of water, from a hose. This will remove all the loose solids which may have stuck to the car. It is important to flush these off in such a way as to prevent grit from scratching the finish. The wheel arches and underframe need washing in the same way to remove any accumulated mud which will retain moisture and tend to encourage rust. Paradoxically enough, the best time to clean the underframe and wheel arches is in wet weather when the mud is thoroughly wet and soft. In very wet weather the underframe is usually cleaned of large accumulations automatically and this is a good time for inspection.

3 Periodically, it is a good idea to have the whole of the underframe of the car steam cleaned, engine compartment included, so that a thorough inspection can be carried out to see what minor repairs and renovations are necessary. Steam cleaning is available at many garages and is necessary for removal of the accumulation of oily grime which sometimes is allowed to become thick in certain areas. If steam cleaning facilities are not available, there are one or two excellent grease solvents available which can be brush applied. The dirt can then be simply hosed off.

4 After washing paintwork, wipe off with a chamois leather to give an unspotted clear finish. A coat of clear protective wax polish will give added protection against chemical pollutants in the air. If the paintwork sheen has dulled or oxidised, use a cleaner/polisher combination to restore the brilliance of the shine. This requires a little effort, but such dulling is usually caused because regular washing has been neglected. Always check that the door and ventilator opening drain holes and pipes are completely clear so that water can be drained out. Bright work should be treated in the same way as paintwork. Windscreens and windows can be kept clear of the smeary film which often appears, by adding a little ammonia to the water. If they are scratched, a good rub with a proprietary metal polish will often clear them. Never use any form of wax or other body or chromium polish on glass.

2.4 Checking that the body drain holes are clear

3 Maintenance — upholstery and carpets

1 Mats and carpets should be brushed or vacuum cleaned regularly to keep them free of grit. If they are badly stained remove them from the car for scrubbing or sponging and make quite sure they are dry before refitting. Seats and interior trim panels can be kept clean by a wipe over with a damp cloth. If they do become stained (which can be more apparent on light coloured upholstery) use a little liquid detergent and a soft nail brush to scour the grime out of the grain of the material. Do not forget to keep the head lining clean in the same way as the upholstery. When using liquid cleaners inside the car do not over-wet the surfaces being cleaned. Excessive damp could get into the seams and padded interior causing stains, offensive odours or even rot. If the inside of the car gets wet accidentally it is worthwhile taking some trouble to dry it out properly, particularly where carpets are involved. *Do not leave oil or electric heaters inside the car for this purpose.*

4 Minor body damage — repair

The photographic sequences on pages 254 and 255 illustrate the operations detailed in the following sub-sections.

Repair of minor scratches in the car's bodywork

If the scratch is very superficial, and does not penetrate to the metal of the bodywork, repair is very simple. Lightly rub the area of the scratch with a paintwork renovator, or a very fine cutting paste, to remove loose paint from the scratch and to clear the surrounding bodywork of wax polish. Rinse the area with clean water.

Apply touch-up paint to the scratch using a thin paint brush; continue to apply thin layers of paint until the surface of the paint in the scratch is level with the surrounding paintwork. Allow the new paint at least two weeks to harden: then blend it into the surrounding paintwork by rubbing the paintwork, in the scratch area, with a paintwork renovator or a very fine cutting paste. Finally, apply wax polish.

Where the scratch has penetrated right through to the metal of the bodywork, causing the metal to rust, a different repair technique is required. Remove any loose rust from the bottom of the scratch with a penknife, then apply rust inhibiting paint to prevent the formation of rust in the future. Using a rubber or nylon applicator fill the scratch with bodystopper paste. If required, this paste can be mixed with cellulose thinners to provide a very thin paste which is ideal for filling narrow scratches. Before the stopper-paste in the scratch hardens, wrap a piece of smooth cotton rag around the top of a finger. Dip the finger in cellulose thinners and then quickly sweep it across the surface of the stopper-paste in the scratch; this will ensure that the surface of the stopper-paste is slightly hollowed. The scratch can now be painted over as described earlier in this Section.

Repair of dents in the car's bodywork

When deep denting of the vehicle's bodywork has taken place, the first task is to pull the dent out, until the affected bodywork almost attains its original shape. There is little point in trying to restore the original shape completely, as the metal in the damaged area will have stretched on impact and cannot be reshaped fully to its original contour. It is better to bring the level of the dent up to a point which is about $\frac{1}{8}$ in (3 mm) below the level of the surrounding bodywork. In cases where the dent is very shallow anyway, it is not worth trying to pull it out at all. If the underside of the dent is accessible, it can be hammered out gently from behind, using a mallet with a wooden or plastic head. Whilst doing this, hold a suitable block of wood firmly against the outside of the panel to absorb the impact from the hammer blows and thus prevent a large area of the bodywork from being 'belled-out'.

Should the dent be in a section of the bodywork which has double skin or some other factor making it inaccessible from behind, a different technique is called for. Drill several small holes through the metal inside the area — particularly in the deeper section. Then screw long self-tapping screws into the holes just sufficiently for them to gain a good purchase in the metal. Now the dent can be pulled out by pulling on the protruding heads of the screws with a pair of pliers.

The next stage of the repair is the removal of the paint from the damaged area, and from an inch or so of the surrounding 'sound' bodywork. This is accomplished most easily by using a wire brush or abrasive pad on a power drill, although it can be done just as effectively by hand using sheets of abrasive paper. To complete the preparation for filling, score the surface of the bare metal with a screwdriver or the tang of a file, or alternatively, drill small holes in the affected area. This will provide a really good 'key' for the filler paste.

To complete the repair see the Section on filling and re-spraying.

Repair of rust holes or gashes in the car's bodywork

Remove all paint from the affected area and from an inch or so of the surrounding 'sound' bodywork, using an abrasive pad or a wire brush on a power drill. If these are not available a few sheets of abrasive paper will do the job just as effectively. With the paint removed you will be able to gauge the severity of the corrosion and therefore decide whether to renew the whole panel (if this is possible) or to repair the affected area. New body panels are not as expensive as most people think and it is often quicker and more satisfactory to fit a new panel than to attempt to repair large areas of corrosion.

Remove all fittings from the affected area except those which will act as a guide to the original shape of the damaged bodywork (eg headlamp shells etc). Then, using tin snips or a hacksaw blade, remove all loose metal and any other metal badly affected by corrosion. Hammer the edges of the hole inwards in order to create a slight depression for the filler paste.

Wire brush the affected area to remove the powdery rust from the surface of the remaining metal. Paint the affected area with rust inhibiting paint; if the back of the rusted area is accessible treat this also.

Before filling can take place it will be necessary to block the hole in some way. This can be achieved by the use of Zinc gauze or Aluminum tape.

Zinc gauze is probably the best material to use for a large hole. Cut a piece to the approximate size and shape of the hole to be filled, then position it in the hole so that its edges are below the level of the surrounding bodywork. It can be retained in position by several blobs of filler paste around its periphery.

Aluminium tape should be used for small or very narrow holes. Pull a piece off the roll and trim it to the approximate size and shape required, then pull off the backing paper (if used) and stick the tape over the hole; it can be overlapped if the thickness of one piece is insufficient. Burnish down the edges of the tape with the handle of a screwdriver or similar, to ensure that the tape is securely attached to the metal underneath.

Bodywork repairs – filling and re-spraying

Before using this Section, see the Sections on dent, deep scratch, rust holes and gash repairs.

Many types of bodyfiller are available, but generally speaking those proprietary kits which contain a tin of filler paste and a tube of resin hardener are best for this type of repair. A wide, flexible plastic or nylon applicator will be found invaluable for imparting a smooth and well contoured finish to the surface of the filler.

Mix up a little filler on a clean piece of card or board – measure the hardener carefully (follow the maker's instructions on the pack) otherwise the filler will set too rapidly or too slowly.

Using the applicator apply the filler paste to the prepared area; draw the applicator across the surface of the filler to achieve the correct contour and to level the filler surface. As soon as a contour that approximates the correct one is achieved, stop working the paste – if you carry on too long the paste will become sticky and begin to 'pick up' on the applicator. Continue to add thin layers of filler paste at twenty-minute intervals until the level of the filler is just proud of the surrounding bodywork.

Once the filler has hardened, excess can be removed using a metal plane or file. From then on, progressively finer grades of sandpaper should be used, starting with a 40 grade production paper and finishing with 400 grade wet-and-dry paper. Always wrap the abrasive paper around a flat rubber, cork, or wooden block – otherwise the surface of the filler will not be completely flat. During the smoothing of the filler surface the wet-and-dry paper should be periodically rinsed in water. This will ensure that a very smooth finish is imparted to the filler at the final stage.

At this stage the dent should be surrounded by a ring of bare metal, which in turn should be encircled by the finely 'feathered' edge of the good paintwork. Rinse the repair area with clean water, until all of the dust produced by the rubbing-down operation has gone.

Spray the whole repair area with a light coat of primer – this will show up any imperfections in the surface of the filler. Repair these imperfections with fresh filler paste or bodystopper, and once more smooth the surface with abrasive paper. If bodystopper is used, it can be mixed with cellulose thinners to form a really thin paste which is ideal for filling small holes. Repeat this spray and repair procedure until you are satisfied that the surface of the filler, and the feathered edge of the paintwork are perfect. Clean the repair area with clean water and allow to dry fully.

The repair area is now ready for final spraying. Paint spraying must be carried out in warm, dry, windless and dust free atmosphere. This condition can be created artificially if you have access to a large indoor working area, but if you are forced to work in the open, you will have to pick your day very carefully. If you are working indoors, dousing the floor in the work area with water will help to settle the dust which would otherwise be in the atmosphere. If the repair area is confined to one body panel, mask off the surrounding panels; this will help to minimise the effects of a slight mis-match in paint colours. Bodywork fittings (eg chrome strips, door handles etc) will also need to be masked off. Use genuine masking tape and several thicknesses of newspaper for the masking operations.

Before commencing to spray, agitate the aerosol can thoroughly, then spray a test area (an old tin, or similar) until the technique is mastered. Cover the repair area with a thick coat of primer; the thickness should be built up using several thin layers of paint rather than one thick one. Using 400 grade wet-and-dry paper, rub down the surface of the primer until it is really smooth. While doing this, the work area should be thoroughly doused with water, and the wet-and-dry paper periodically rinsed in water. Allow to dry before spraying on more paint.

Spray on the top coat, again building up the thickness by using several thin layers of paint. Start spraying in the centre of the repair area and then, using a circular motion, work outwards until the whole repair area and about 2 inches of the surrounding original paintwork is covered. Remove all masking material 10 to 15 minutes after spraying on the final coat of paint.

Allow the new paint at least two weeks to harden, then, using a paintwork renovator or a very fine cutting paste, blend the edges of the paint into the existing paintwork. Finally, apply wax polish.

5 Major body damage – repair

1 Because the body is built on the monocoque principle, major damage must be repaired by a competent body repairer with the necessary jigs and equipment.
2 In the event of a crash that resulted in buckling of body panels, or damage to the road wheels the car must be taken to a Lancia dealer, or body repairer, where the bodyshell and suspension alignment may be checked.
3 Bodyshell and/or suspension mis-alignment will cause excessive wear of the tyres, steering system and possibly transmission. The handling of the car also will be affected adversely.

6 Doors – tracing and silencing rattles

1 The commonest cause of door rattles is a misaligned, loose or worn striker plate, but other causes may be:

   (a) Loose door handles, window winder handles and door hinges
   (b) Loose, worn or misaligned door lock components
   (c) Loose or worn remote control mechanism

2 It is quite possible for door rattles to be the result of a combination of these faults, so a careful examination must be made to determine the causes of the noise.
3 If the nose of the striker plate is worn and as a result the door rattles, renew it and adjust the plate.
4 If the nose of the door wedge is badly worn and the door rattles as a result, then fit a new door latch assembly.
5 Should the hinge be badly worn then the pivot pin may be renewed; however if with a new pivot pin the hinge is still slack then the door will have to be taken off as described in Section 17 and a new hinge or hinges fitted.

7 Interior handles and trim panel – removal and refitting

1 In order to gain access to any of the mechanisms situated in the door it is first necessary to remove the handles and trim fittings as follows.
2 Remove the plastic trim from the window winder handle, and then remove the trim plugs or trim strip from the arm rest; then remove the interior door handle surround by sliding the trim towards the front end of the door (photos).
3 Remove the small screw at the base of the door locking catch and lift the knob off (photo).

Chapter 12 Bodywork and subframe

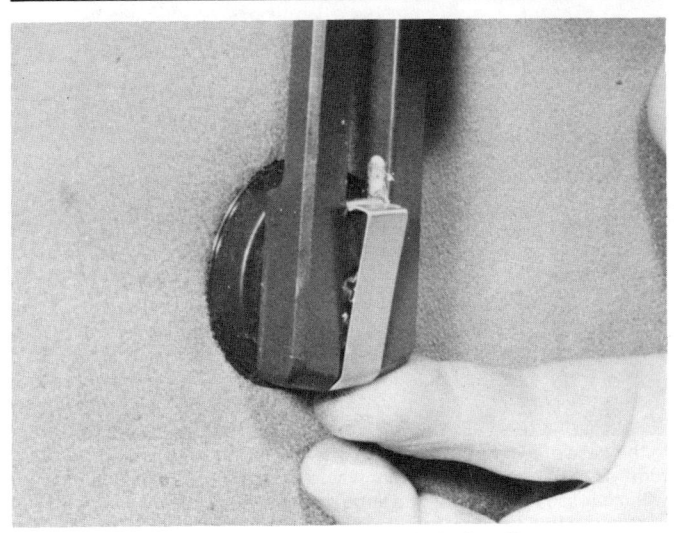

7.2a Removing the trim from the window winder handle

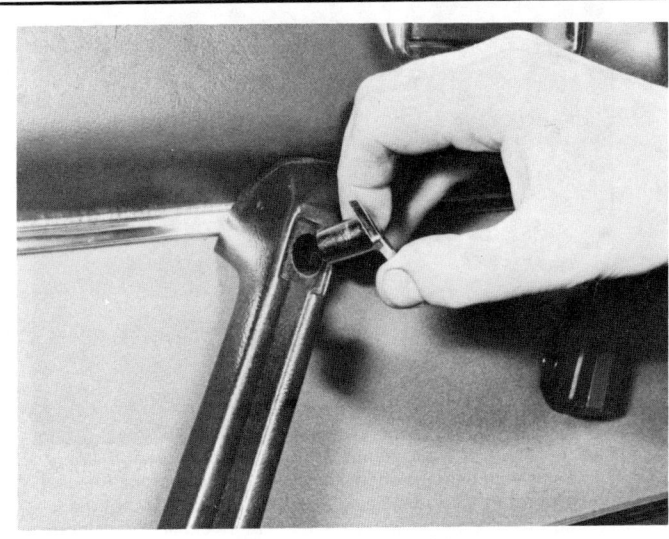

7.2b Removing the trim plug in the top end of the armrest

7.2c The interior door handle trim removed. Lugs show why it needs to be slid horizontally to remove it

7.3 Removing the door lock catch retaining screw

7.4 Removing the window winder retaining screw

4 Undo the crosshead screw securing the window winder in position and remove the handle (photo).
5 Undo the bolts securing the arm rest and remove it.
6 Now, using a wide blade screwdriver, prise off the trim panel (photo). In some models (such as the HPE) it will be necessary to remove the lower chromed retaining strip first (photo).
7 Refitting is the reverse procedure to removal.

8 Window winder mechanism – removal and refitting

1 Refer to Section 7 for instructions as to how to remove the door trim panel, then remove the inner polythene covers.
2 Make sure that the glass is fully raised; this will probably mean refitting the window winder handle temporarily, after bolts (9) or (18) in Fig. 12.1 or Fig. 12.2 have been removed (photo).
3 Then undo and remove bolts (5), (7) or (10) in Fig. 12.1 or all the bolts numbered (5) and bolt number (10) in Fig. 12.2. These bolts secure the window winder mechanism to the door. The mechanism can then be manoeuvred downwards and out. The mechanisms for the Coupe, HPE and Spider models are all basically the same as in Fig. 12.2, but the location of the mounting bolts varies. In essence they all have a 5 bolt mounting which is made up of 2 bolts by the winder handle, 2 bolts at the top of the mechanism and one bolt at the bottom.

This sequence of photographs deals with the repair of the dent and paintwork damage shown in this photo. The procedure will be similar for the repair of a hole. It should be noted that the procedures given here are simplified — more explicit instructions will be found in the text

In the case of a dent the first job — after removing surrounding trim — is to hammer out the dent where access is possible. This will minimise filling. Here, the large dent having been hammered out, the damaged area is being made slightly concave

Now all paint must be removed from the damaged area, by rubbing with coarse abrasive paper. Alternatively, a wire brush or abrasive pad can be used in a power drill. Where the repair area meets good paintwork, the edge of the paintwork should be 'feathered', using a finer grade of abrasive paper

In the case of a hole caused by rusting, all damaged sheet-metal should be cut away before proceeding to this stage. Here, the damaged area is being treated with rust remover and inhibitor before being filled

Mix the body filler according to its manufacturer's instructions. In the case of corrosion damage, it will be necessary to block off any large holes before filling — this can be done with aluminium or plastic mesh, or aluminium tape. Make sure the area is absolutely clean before ...

... applying the filler. Filler should be applied with a flexible applicator, as shown, for best results; the wooden spatula being used for confined areas. Apply thin layers of filler at 20-minute intervals, until the surface of the filler is slightly proud of the surrounding bodywork

Initial shaping can be done with a Surform plane or Dreadnought file. Then, using progressively finer grades of wet-and-dry paper, wrapped around a sanding block, and copious amounts of clean water, rub down the filler until really smooth and flat. Again, feather the edges of adjoining paintwork

The whole repair area can now be sprayed or brush-painted with primer. If spraying, ensure adjoining areas are protected from over-spray. Note that at least one inch of the surrounding sound paintwork should be coated with primer. Primer has a 'thick' consistency, so will find small imperfections

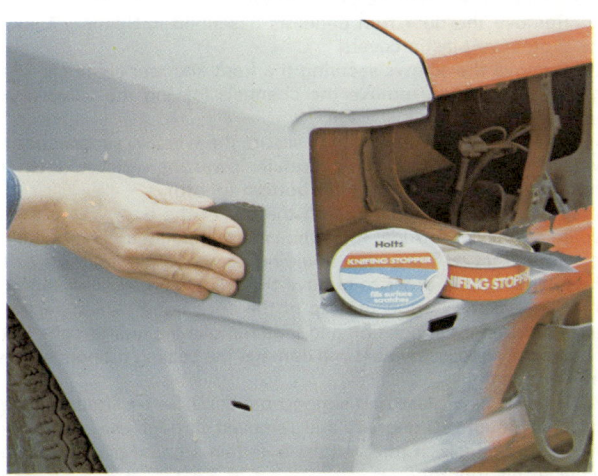

Again, using plenty of water, rub down the primer with a fine grade wet-and-dry paper (400 grade is probably best) until it is really smooth and well blended into the surrounding paintwork. Any remaining imperfections can now be filled by carefully applied knifing stopper paste

When the stopper has hardened, rub down the repair area again before applying the final coat of primer. Before rubbing down this last coat of primer, ensure the repair area is blemish-free – use more stopper if necessary. To ensure that the surface of the primer is really smooth use some finishing compound

The top coat can now be applied. When working out of doors, pick a dry, warm and wind-free day. Ensure surrounding areas are protected from over-spray. Agitate the aerosol thoroughly, then spray the centre of the repair area, working outwards with a circular motion. Apply the paint as several thin coats

After a period of about two weeks, which the paint needs to harden fully, the surface of the repaired area can be 'cut' with a mild cutting compound prior to wax polishing. When carrying out bodywork repairs, remember that the quality of the finished job is proportional to the time and effort expended

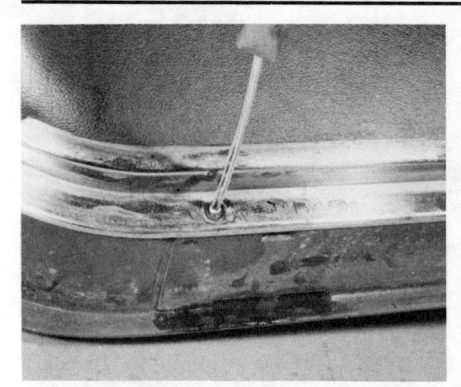

7.6a Removing the lower door panel retaining strip

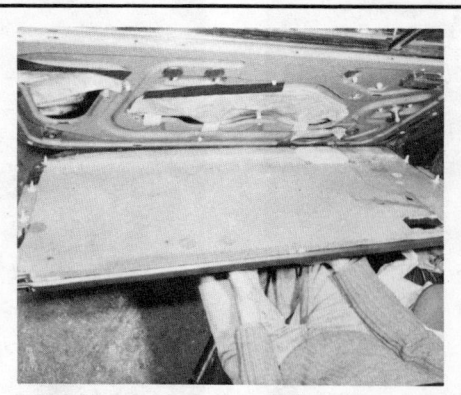

7.6b Lifting the door trim panel away

8.2 Refit the window winder temporarily to fully raise the glass

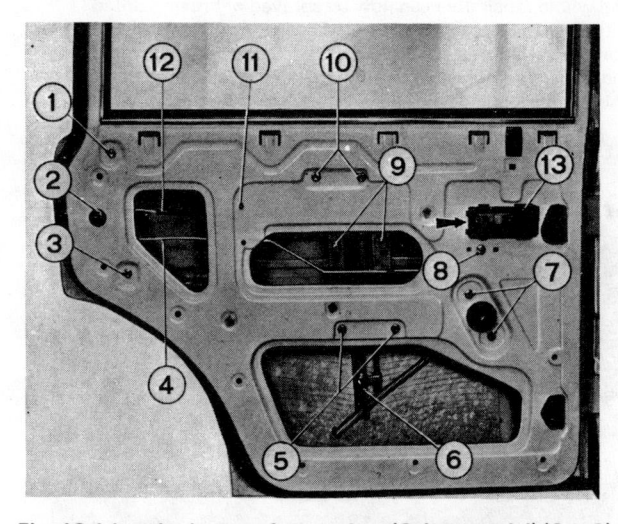

Fig. 12.1 Interior layout of a rear door (Saloon model) (Sec 8)

1. Rear channel fixing screw
2. Opening control cable mounting bolt
3. Rear channel fixing screw
4. Safety control rod
5. Window winder mounting bolts
6. Window winder
7. Window winder mounting bolt
8. Safety control rod idler lever mounting nut
9. Glass lower support mounting bolts
10. Window winder mounting bolt
11. Cable bracket mounting bolt
12. Opening control cable
13. Inner control handle

4 Before refitting the mechanism, in the reverse order to its removal, lightly grease the mechanism. Refit the trim also in reverse order to its removal.

9 Front and rear door glass (Saloon) – removal and refitting

1 Remove the trim panel as described in Section 7.
2 Next remove the plastic protective sheet and then undo and remove the screws, 1 and 3 in Fig. 12.1 or 1 in Fig. 12.2, which secures the rear window channel in place. The channel can then be manoeuvred out through the holes in the panel.
3 Remove the two bolts from the glass support bar, which attach it to the window winder mechanism.
4 Now rotate the glass and lift it upwards and remove it complete with the lower support bar. If the glass is to be renewed, then it can be removed from the support bar and a new glass can be fitted.
5 To refit the window follow the removal procedure in reverse but take care when refitting the channel that the velvety material is evenly spread. With the window attached to the mechanism refit the winder handle temporarily and check that the window runs smoothly in the channels.
6 Then remove the handle and refit the trim etc.

10 Door glass (Coupe and HPE) – removal and refitting

1 This Section covers not only the main window but also the fixed quarterlight, if it needs to be renewed.
2 Remove the door trim panel as described in Section 7 and then the protective plastic covers.
3 Remove the screws securing the front and rear channels (2 and 7 in Fig. 12.3). Then remove the channels leaving the velvet material behind (photo).
4 Remove the bolts (5) which secure the window support bar to the winder mechanism and then carefully lower the window down inside the door so that it rests at the bottom (photo).
5 Then free the mouldings on the top of the door where the glass would normally protrude. Prise the inner moulding up and pull it to the rear to extract it. The outer moulding need only be freed off as far as the quarterlight in order to remove the main window however (photo).
6 Remove the bolt in the top of the door frame which secures the fixed quarterlight in position and then remove the quarterlight, but only if absolutely necessary as it will damage the sealing rubbers unless you are very careful.
7 The window glass and support bar can now be lifted up carefully, rotated through 90 degrees and removed through the gap in the top of the door panels. It may require a certain amount of manoeuvring to achieve this (photo).
8 If the glass and support bar need to be separated this can be done when the assembly has been withdrawn.
9 Refitting is the reverse procedure to removal. However, before refitting the trim panel, refit the window winder temporarily and check that the window runs freely in its channels.

11 Door glass (Spider models) – removal and refitting

1 Remove the trim panel on the appropriate door as described in Section 7, and remove the protective plastic lining.
2 Remove the water guards and lift the rear channel away.
3 Remove the retainer and then the velvet lined mouldings can be removed.
4 Remove the screws (5) in Fig. 12.4 which secure the window support bar to the winder mechanism and the glass can then be lifted straight out.
5 Refitting is the reverse procedure to removal. However, before the trim panel etc are refitted check that the window runs freely in the channels and that the velvet linings are smoothed out.

12 Power window lifts – removal and refitting

1 Remove the door trim panel as described in Section 7.
2 Refer to the appropriate figure for your particular model and remove the mounting screws for the front channel (Figs. 12.1, 2, 3 or 4).
3 Fit the emergency winding handle to the gear after the cover cap has been removed. This comes off when turned in an anti-clockwise direction (Fig. 12.6). Then lower the window (the handle has to be pushed in to engage with the gears) to get at the window lift screws.

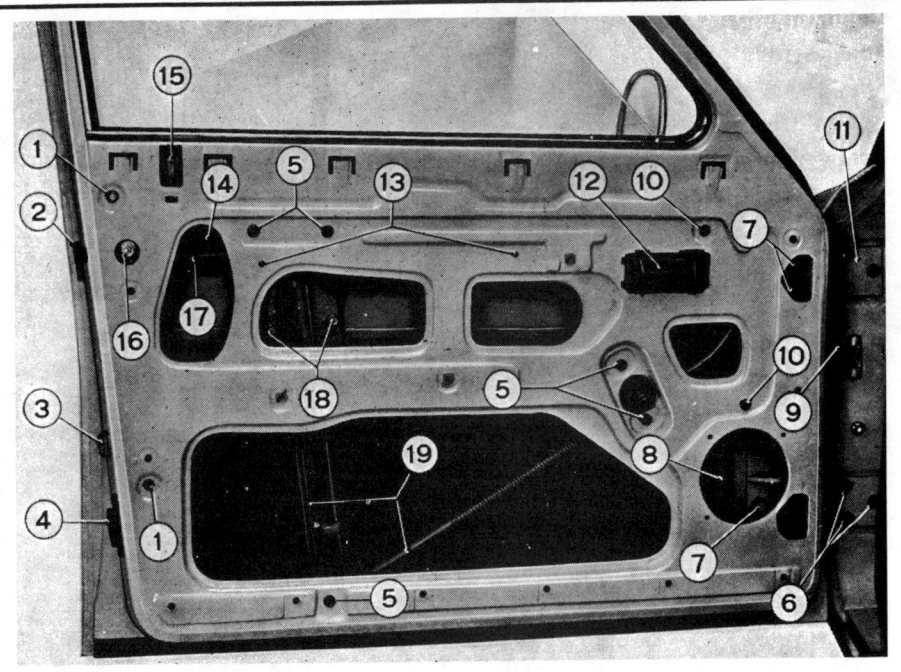

Fig. 12.2 Interior layout of a front door (Saloon model) (Sec 8)

1 Rear channel fixing bolts
2 Lock
3 Safety lamp
4 Bumper seat
5 Window winder mounting bolt
6 Hinge-to-bodywork fixing bolts
7 Hinge-to-door mounting bolts
8 Front channel
9 Check rods
10 Channel fixing bolts
11 Hinge
12 Inner control handle
13 Control cable retainer fixing screws
14 Inside locking knob pawl
15 Inside locking rod
16 Control cable fixing bolt
17 Remote control cable
18 Glass lower support mounting bolts
19 Window winder mechanism

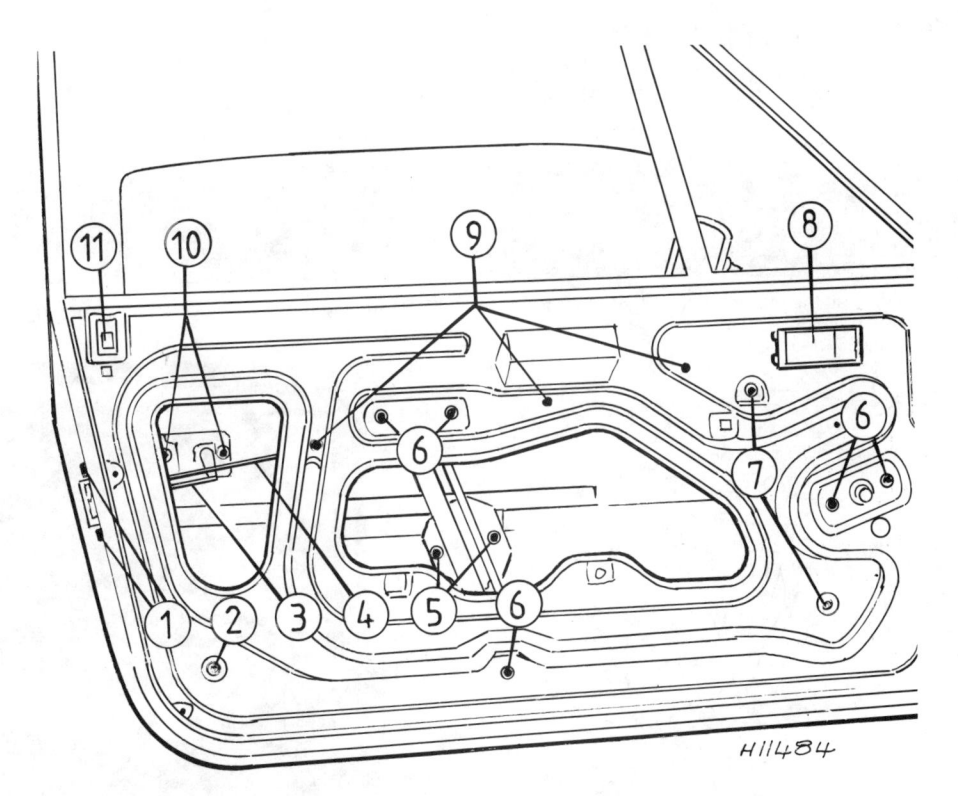

Fig. 12.3 Interior layout of a front door (Coupe and HPE models) (Sec 10)

1 Lock mounting bolts
2 Rear channel fixing bolt
3 Outside control rod
4 Inside control rod
5 Glass-to-window winder fixing bolts
6 Window regulator mounting bolts
7 Front channel fixing screws
8 Inside control handle
9 Control rod retainer
10 Outside handle with lock cylinder fixing bolts
11 Inside locking knob

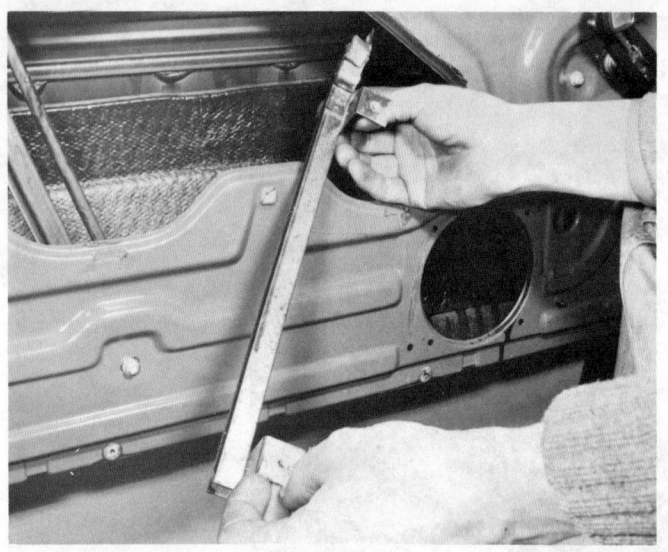

10.3 Removing the front channel

10.4 Undoing the bolts securing the window support bar to the winder mechanism

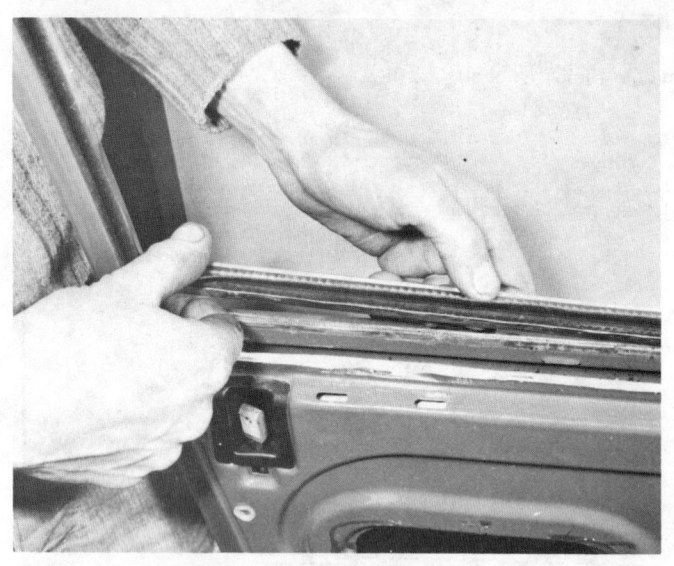

10.5 Freeing the outer moulding on the top of the door (note that the inner moulding has already been removed)

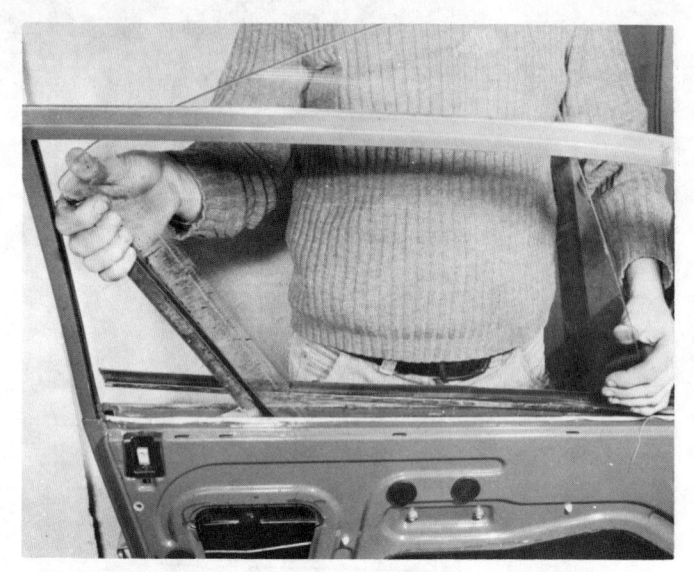

10.7 Removing the door window glass, rotating it to do so

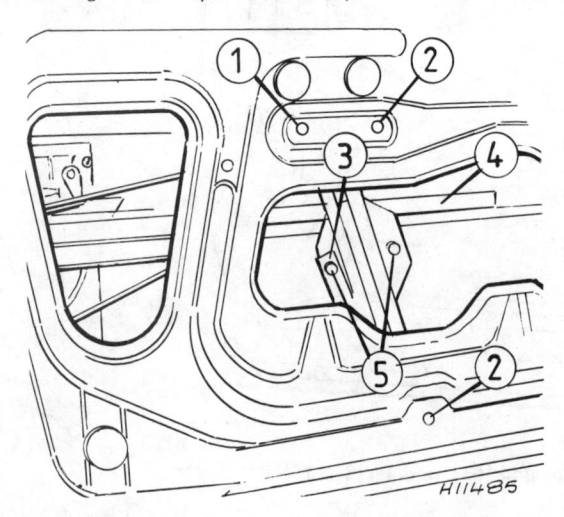

Fig. 12.4 Interior layout of a front door (Spider models) (Sec 11)

1. Window winder mounting bolt
2. Window winder mounting bolt
3. Glass stroke stop
4. Glass support bar
5. Glass to window winder mounting bolts

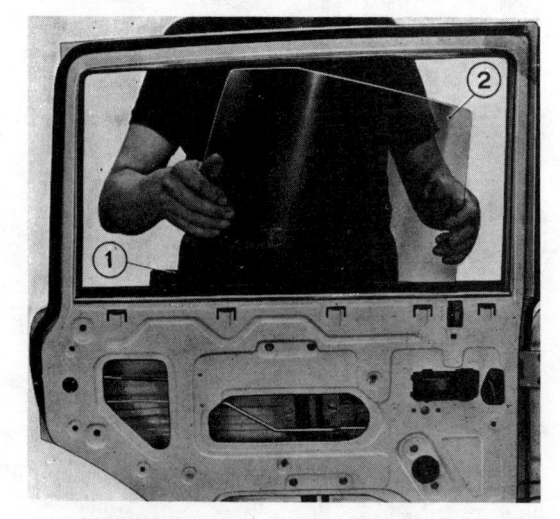

Fig. 12.5 Removing the glass from a rear door (Saloon models) (Sec 9)

1. Glass support bar
2. Glass

Chapter 12 Bodywork and subframe

(These are the equivalent of the window support bar retaining bolts in the manual window models).

4 Remove the two screws, lift the glass to the top of the window and wedge it up.

5 Remove the screws which secure the motor and lift arm to the door. These are the same screws as in the appropriate figure, which hold the window winder mechanism to the door.

6 Remove the bracket from the motor and lift it away.

7 Disconnect the electric supply cable from the motor at the Lucar connectors and manoeuvre the motor and lift arm mechanism out of the door in one go.

8 The motor and lift arm can now be separated and the parts can be renewed as necessary.

9 Refitting is a reversal of the removal procedure (see Figs. 12.8 and 12.9).

13 Front door handle and lock (Saloon models) – removal, overhaul and refitting

1 Remove the front trim panel as described in Section 7, and then remove the protective plastic lining.

2 Refer to Fig. 12.2 and disconnect the inside lock rod from the pawl and then undo and remove the two fixing bolts for the rear channel. The rear channel can now be removed.

3 Undo and remove the lock retaining bolts and push the lock through the door onto the inside and withdraw it.

4 With the lock out of the way insert a spanner through the hole and remove the nuts on the inside of the door which hold the handle in place. Remove the handle.

5 To overhaul or renew the cylinder place the mechanism on the

Fig. 12.6 Removing the emergency winder gear cover (Sec 12)

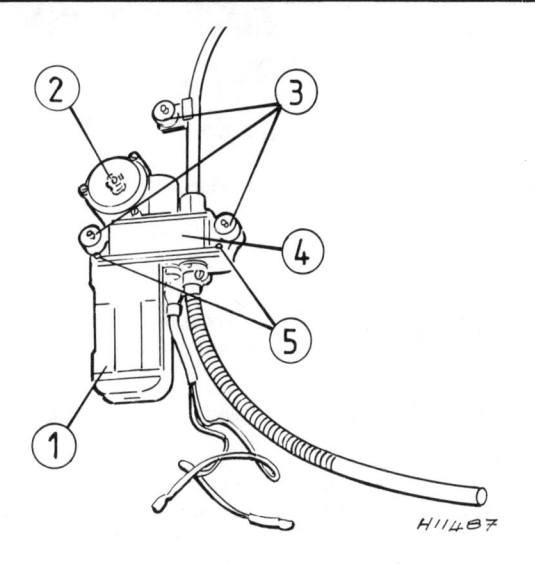

Fig. 12.7 Window power lift motor (Sec 12)

1 Motor
2 Emergency handle fitting gear
3 Motor-to-bracket mounting pads
4 Step-down gear
5 Motor mounting screws

Fig. 12.8 Starting to refit the whole assembly (Sec 12)

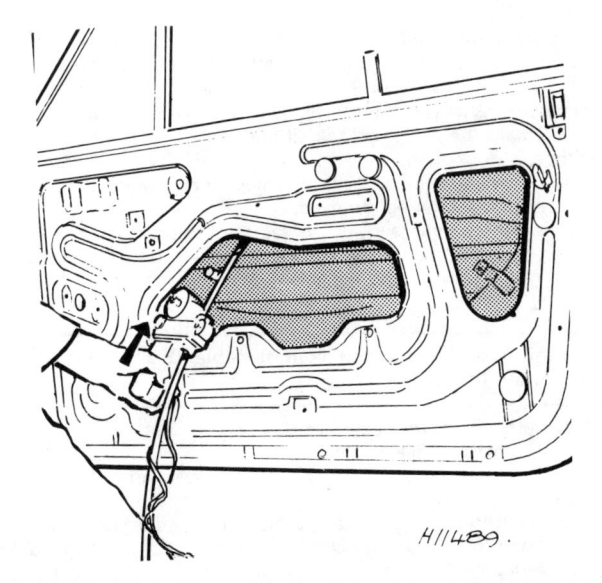

Fig. 12.9 Once the lift mechanism is in place push the motor in and up (Sec 12)

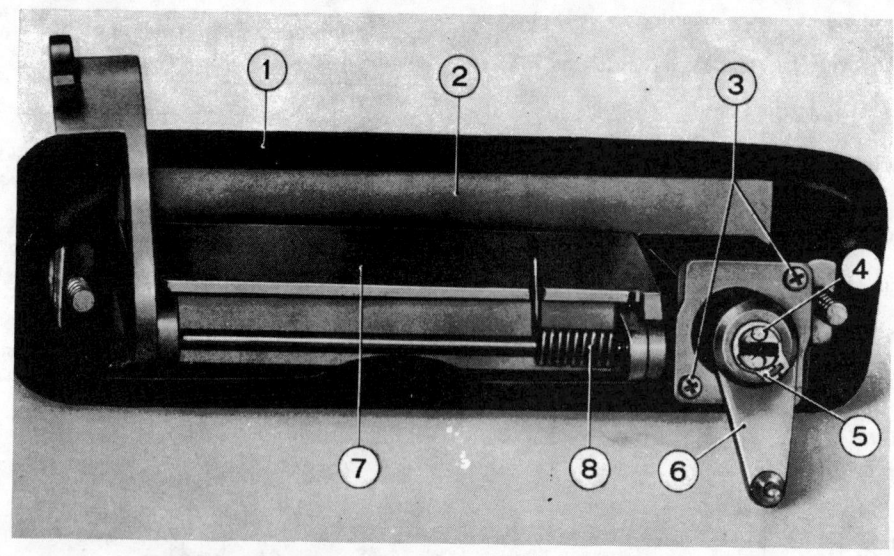

Fig. 12.10 Front door handle and lock (Saloon models) (Sec 13)

1 Gasket
2 Handle stationary portion
3 Lock cylinder fixing screws
4 Lock cylinder
5 Lock ring
6 Lock pawl
7 Handle movable portion
8 Spring

bench and remove the circlip. Then remove the pawl, followed by the cylinder which is held in place by two small screws (Fig. 12.10).
6 Refit the cylinder to its seat and do up the screws. Refit the pawl and secure it with the circlip.
7 Before refitting the handle and lock to the door lightly oil all the moving and hinged parts.
8 Refit the handle to the door and ensure that the sealing ring is correctly fitted to the handle. Then refit the nuts.
9 The rest of the operation is the reverse procedure to removal, but check the operation of the mechanism before refitting the trim panel.

14 Rear door handle and lock (Saloon models) – removal and refitting

1 Remove the door trim panel as described in Section 7, and the protective plastic lining.
2 Refer to Fig. 12.1 and remove the retaining screws for the rear channel.
3 Disconnect the inside locking rod from the lock itself using a screwdriver. Then undo the retaining bolts for the lock, take the lock out and lower it down inside the door without disconnecting the control cable, unless you want to change the lock on its own, in which case disconnect the cable and take it out.
4 With the lock either lowered out of the way, or removed completely, it will now be possible to remove the retaining nuts for the handle, which can then be removed.
5 Refitting the handle and lock is the reverse process to removal. Check that the cables and rods operate correctly and freely and adjust them if necessary before refitting the door trim panel.

15 Interior door handle and control cable (Saloon models) – removal and refitting

1 Remove the door trim panel as described in Section 7, and then remove the protective plastic lining.
2 Refer to Figs. 12.1 or 12.2 depending on whether it is a front or rear door that is being dealt with.
3 Disconnect the interior control cable from the door lock mechanism by undoing the retaining bolt (2 in Fig. 12.1 or 16 in Fig. 12.2), release the retaining clip.
4 Release the interior door handle from the inner panel by prising out the hinge and pull the handle towards the forward edge of the door (see arrow in Fig. 12.1).
5 Remove the handle with the control cable still attached.
6 Release the cable from the handle.

7 Refit the cable to the handle and insert the cable through the mounting aperture for the handle. Locate the end with your other hand and feed the cable rearwards.
8 Secure the interior door handle in position with its clip and reconnect the cable to the lock and refit the bolt.
9 Check that the mechanism works correctly before refitting the trim panel.

16 Door lock, cylinder and handles (Coupe, HPE and Spider) – removal and refitting

1 Remove the trim panel from the door as described in Section 7, and remove the protective plastic lining.
2 Refer to Fig. 12.3 and 12.4 for the respective models and remove the rear window channel.
3 Refer to Fig. 12.11 and disconnect the outside control rod (2) from the exterior door handle mechanism (5), undo and remove the handle retaining nuts ((10) in Fig. 12.3) and withdraw the handle mechanism complete with the locking cylinder (photo).
4 The locking cylinder can then easily be removed from the handle mechanism.
5 With the outside locking rod disconnected the lock can be removed by undoing the retaining bolts and releasing the control rod retainers from the inside panel (photo).
6 Then pull the inside door handle forward and push the lock into the door cavity, as shown by the arrows in Fig. 12.3 (photo).
7 With the lock in an accessible position disconnect the interior control rod from it and it can be withdrawn.
8 With the lock disconnected from the rear end of the interior control rod, the inside door handle can be levered out by releasing the lugs in the panel and the rod can be withdrawn and renewed if necessary.
9 Refitting of the whole or any part of these assemblies is the reverse procedure to removal. However, before refitting the interior trim panel to the door make sure that all the mechanisms work correctly and that the rods etc are adjusted properly.

17 Front and rear doors (all models) – removal, refitting and adjusting

1 Where the car has warning lamps in the trailing edges of the front doors the negative battery cable must first be disconnected.
2 Remove the interior door trim panel as described in Section 7.
3 Disconnect the door warning lamp wiring (photo).
4 Separate the door check rod from the front pillar.

Chapter 12 Bodywork and subframe

16.3 The outside control rod and one of the door handle retaining nuts are clearly visible in this picture

16.5 The four door lock retaining bolts are easily removed

16.6 It is easy to see from this photo how the inside door handle should be fitted. It slides forward and out quite easily

17.3 Removing a door rear edge warning lamp

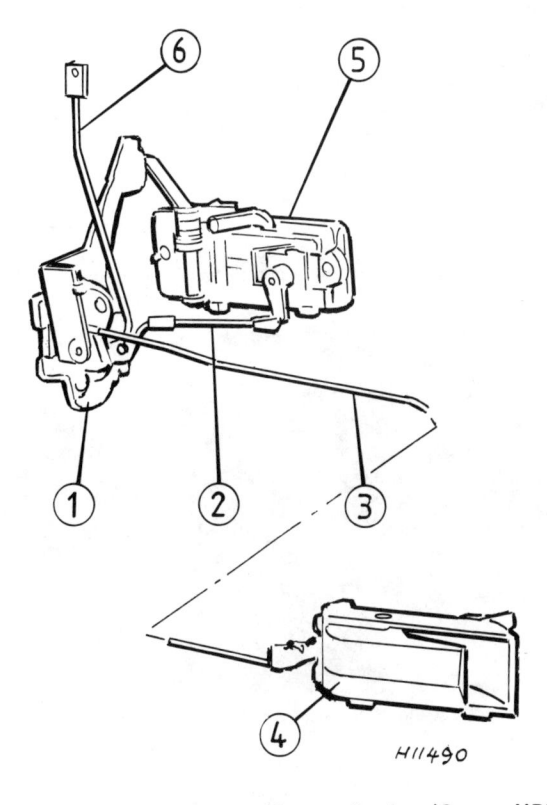

Fig. 12.11 Door handles and locking mechanism (Coupe, HPE and Spider models) (Sec 16)

1 Lock
2 Outside control rod
3 Inside control rod
4 Inside remote control handle
5 Outside control handle with lock cylinder
6 Inside locking control

5 Scribe the positions of the hinges on the front pillar, so that the door will hang properly when it is refitted.
6 Undo and remove the three screws securing the top hinge to the front pillar and the three screws securing the bottom hinge to the front pillar. One person needs to support the weight of the door whilst the other person undoes and remove the screws. The door can then be lifted away.
7 Refitting the door is again a two man job. One person needs to

hold the weight of the door whilst the other person refits the three screws to each of the two hinges. Don't tighten them up completely first of all.

8 With the hinges held by the screws check that the door opens and closes easily, and that it does not foul any part of the surrounding bodywork. The clearance around the door should be the same all the way round. Also the door should not protrude or be recessed. If the fitting is not correct open the door, slacken the hinge mounting screws and alter the position of the door slightly.

9 If the door is difficult to close, even when the clearance and flush fitting of the door seem to be right then check the striker plate fitting. Loosen the fixing bolts and move it slightly to ensure that the door shuts correctly.

10 When all the faults have been rectified tighten the striker plate retaining bolts and the six hinge mounting screws.

11 Refit the rest of the assembly in the reverse order to removal.

18 Tailgate lock and catch (HPE models) – removal, refitting and adjusting

1 Open the tailgate and remove the interior trim.
2 Disconnect the balljoint at the end of the control rod from the locking arm, then unscrew the three lock mounting screws and withdraw the lock and rod (photos). The catch can be removed from the tailgate by unscrewing the three retaining bolts.
3 If necessary remove the two Allen bolts to detach the striker from the rear body panel (photo).
4 Refitting is the reverse procedure to removal, but check that the mechanism works correctly before fitting the trim.
5 Check the lock with the key for ease of operation and effectiveness. If necessary the rod can be lengthened or shortened as required. The adjuster can be seen in photo 18.2a.
6 Close and open the tailgate to check that it catches and releases properly. If necessary adjust the striker by slackening the retaining screws and move it slightly then tighten the screws and check the ease of operation.

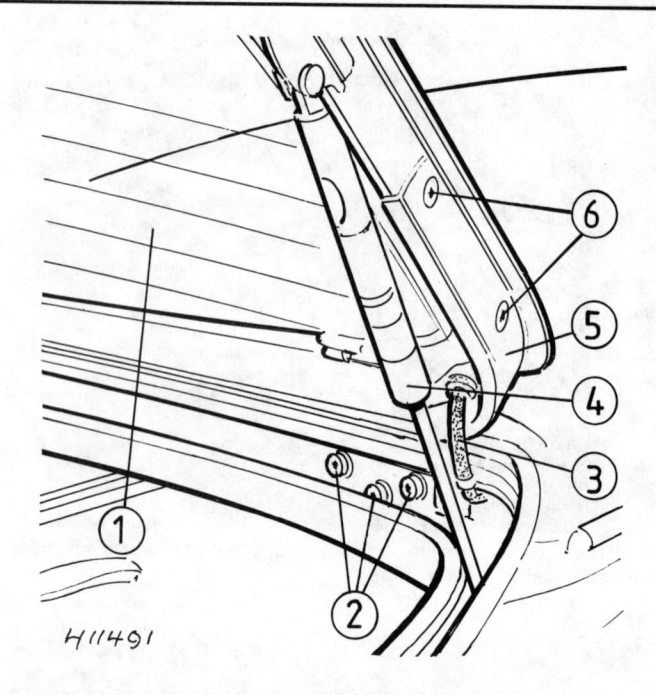

Fig. 12.12 Tailgate (HPE) – hinges and strut layout (Sec 19)

1 Louvre sun shade
2 Hinge-to-roof panel fixing screws
3 Electric cables tube and clip
4 Strut
5 Hinge
6 Hinge-to-tailgate fixing screw

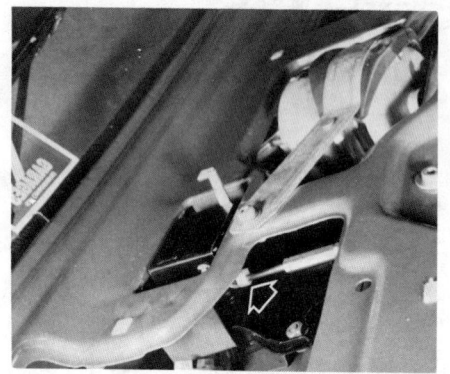

18.2a Disconnect the control rod balljoint (arrowed)

18.2b Unscrew the three lock retaining screws

18.3 The tailgate striker

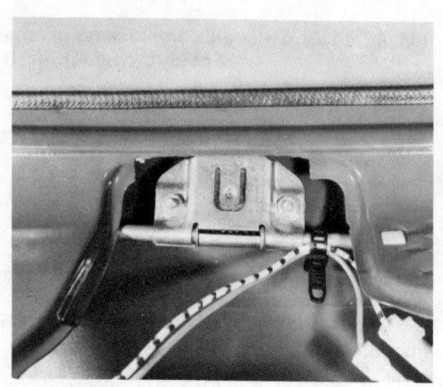

19.3 The wiring connectors for the rear windscreen wiper motor

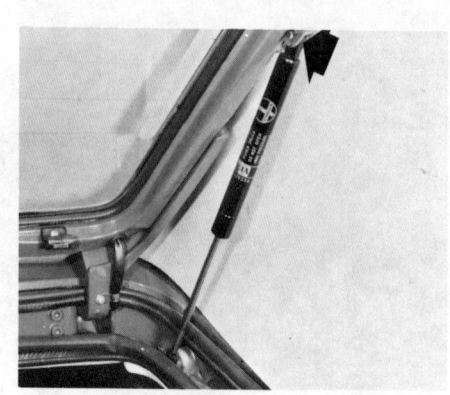

19.5 Remove the clip (arrowed) and push out the pin to disconnect the strut from the tailgate

Chapter 12 Bodywork and subframe

19 Tailgate (HPE models) – removal and refitting

1 Open the tailgate and remove the louvred panel. Then disconnect the battery negative lead.
2 Lift away the inside trim panel on the lower section of the tailgate in order to reach the wiring for the wiper motor and heated rear window.
3 Disconnect the wiring connectors for both systems and make a note of which wire fits where (photo). Then bind a long piece of string, to the end of the leads for each system using tape and pull the leads out of the top of the tailgate after you have removed the cable sheath and clips. Disconnect the pieces of string from the leads and leave them in position in the tailgate for refitting purposes.
4 Scribe the relationship of the hinges to the tailgate itself and loosen the screws which are fitted into the top sides of the tailgate. At this stage one person should support and take the weight of the tailgate.
5 Remove the clips from the inner end of the strut mounting pin at the tailgate, push out the pin and lower the strut so that it rests in the window surround (photo).
6 Remove the loosened screws and lift the tailgate away.
7 Refit the tailgate in the reverse order, making sure that the hinges are correctly lined up. Do not tighten the screws until the closing and opening operation of the tailgate have been checked.

20 Bonnet (Saloon models) – removal and refitting

1 Remove the radiator grille as described in Section 27.
2 Lift the bonnet up and get one person to hold it while the other person removes the circlips from the hinge pins.
3 Expand the bonnet stay so that it slips over the retainer and fold it down out of the way.
4 Withdraw the pins from the hinges and lift the bonnet away.
5 To store it, place it upright against a wall where it won't be damaged. Place rags under the bottom corners so that the paintwork is not chipped..
6 Refitting the bonnet is the reverse procedure.

21 Bonnet (Coupe, HPE and Spider models) – removal and refitting

1 Open the bonnet. While one person supports the bonnet the other can lever out the stay rod retaining spring (photo). By releasing the stay from the slot it can be folded down.
2 Scribe the positions of the hinges on the front panel, then remove the three bolts retaining each hinge (photo).
3 The two people can now remove the bonnet to a safe place. Store it as instructed in the last section, paragraph 5.
4 Refitting is the reverse procedure.

22 Bonnet release cable – removal and refitting

1 Disconnect the cable from the bonnet lever inside the car (photo).
2 Undo the outer cable retaining clips in the engine compartments and remove the cable from the locking mechanism, by undoing the clamp nut (photos). The cable runs from the lever to the left-hand lock and a secondary cable joins the two locks.
3 Some models have two bonnet release levers inside the car. In this case the main one is mounted on the left-hand side under the dashboard, and the emergency one is on the right-hand side under the dashboard.
4 On models which have two bonnet locks, two levers are fitted. Each lever has two cables.
5 Refitting is the reverse procedure to removal, but always start at the lock end. Check that the system works efficiently when the new cable is fitted.

23 Glovebox lid and lock – removal and refitting

1 Open the lid.
2 To remove the lock, withdraw the circlip which retains the plastic spring latch to the lock. Remove the retainer and lift the lock away from the lid.
3 To remove the lid first undo the lid support arm retaining screw and remove it. Pull the hinge pins in towards the centre and remove the lid. Remove the pins together with the springs, but take care that the springs don't force the pins to fly off.
4 Refitting of both the lock and lid are the reverse procedure to removal.

24 Boot lid (Saloon, Coupe and Spider models) – removal and refitting

1 Scribe the position of the hinges to the boot lid.
2 On Saloon models release the number plate lighting cable from the clips on the right-hand hinge and from the cable connector plug (Fig. 12.13).
3 Undo the four retaining nuts; two on each hinge, and lift the boot lid away.
4 When refitting the lid line up the scribe marks after all the units have been refitted but not tightened. Check that the boot lid closes and opens properly, and adjust the lid on the hinges if the fit is not correct.
5 Finally tighten all the nuts. Then refit the wiring as necessary.

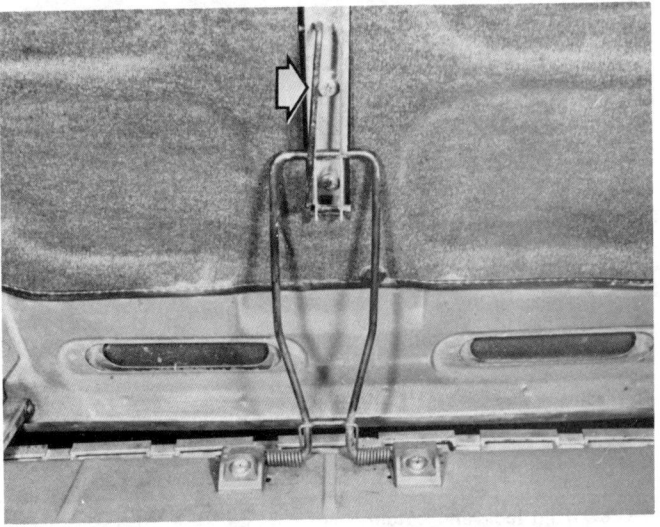

21.1 The bonnet stay rod and spring is kept in place by the retainer spring (arrowed)

21.2 Scribe the hinge positions then remove the three screws

Chapter 12 Bodywork and subframe

22.1 On the HPE models the bonnet release lever is on the right-hand side under the dash

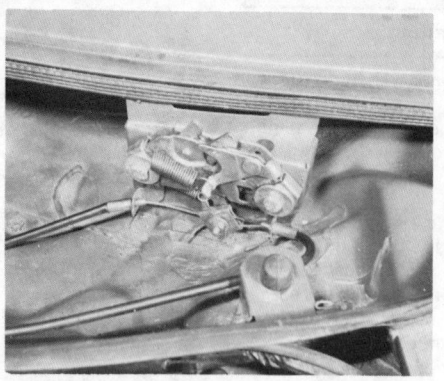

22.2a The bonnet locking mechanism on the left-hand side showing the cable ends clamped by the nut and plate

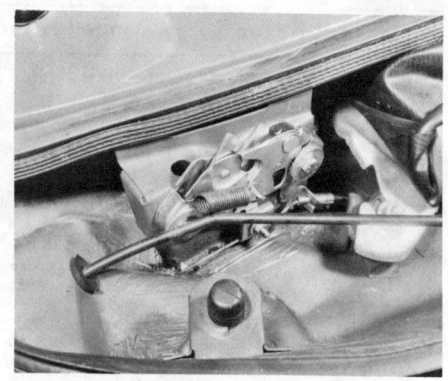

22.2b The right-hand bonnet lock with the main operating cable in the foreground coming through the bulkhead from the lever

25 Boot lid lock and catch (Saloon models) – removal, refitting and adjusting

1 Open the boot lid and work from inside.
2 Undo and remove the retaining bolts for the control arm (Fig. 12.14).
3 By moving the control arm bracket to one side it is now possible to get at the boot lock retaining nut and remove it. The lock can then be pushed through the hole, which has been created by removing the control arm.
4 Release the vertical control rod from the control arm and place the arm to one side.
5 Remove the catch retaining nuts at the bottom edge of the lid and it is then possible to remove the catch and the control rod.
6 Refitting is basically the reverse procedure to removal, but refit the control arm and bracket to the lid before attaching the control rod to it. If the lock has been removed then this must be refitted first of all. Do not tighten up the control arm retaining bolts until the adjustments have been made.
7 Check that the catch operates freely and that the lock plunger is in contact with the control arm. This can be adjusted by moving the control arm bracket slightly. When the adjustment is correct tighten the retaining bolts.
8 If the catch does not work easily after adjusting the mechanism then the striker needs to be moved. Loosen the bolts and move it forwards and backwards until the lid opens and shuts correctly. Tighten the bolts. With the boot lid closed the lock button should have a maximum of 2 mm (0.08 in) of free travel.

26 Boot lid lock and catch (Coupe and Spider) – removal and refitting

1 Open the boot lid and undo the bolt and nuts which secure the lock to the lid. The catch and lock cylinder can then be removed.
2 Refitting is the reverse procedure to removal.

27 Radiator grille – removal and refitting

1 Remove the screws which attach the top of the grille to the front body panel.
2 In the case of the Saloon models the grille can then be lifted out. The bottom edge has lugs on it which locate in the lower body panel.
3 The grille fitted to the Coupe, Spider and HPE models is mounted in the same way but it cannot be removed, without first removing the

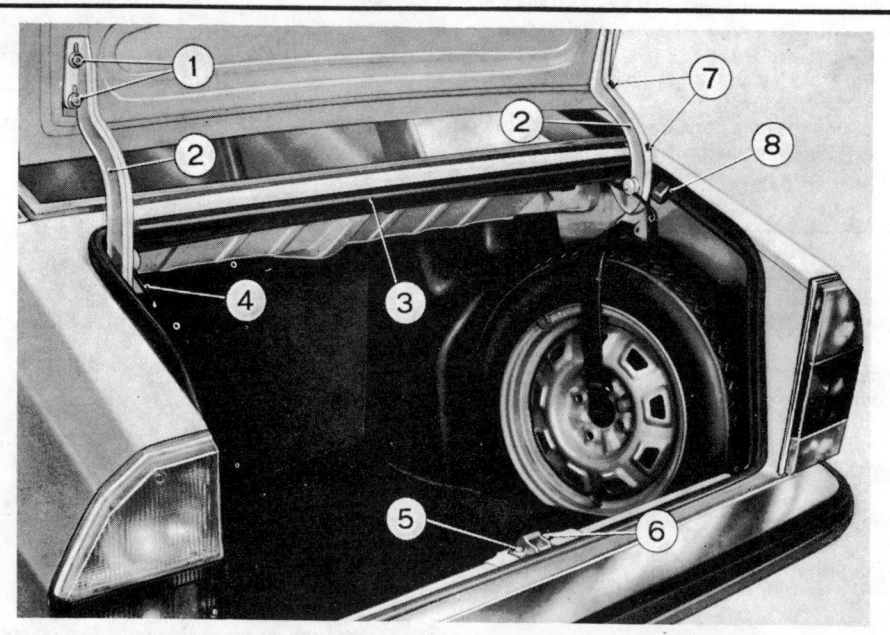

Fig. 12.13 Boot lid and hinge layout (Saloon models) (Sec 24)

| 1 Hinge mounting nuts | 3 Weatherstrip | 5 Lock mounting bolts | 7 Electric cable retainers |
| 2 Hinges | 4 Torsion rod prop | 6 Lock | 8 Locating pads |

Chapter 12 Bodywork and subframe

headlights as described in Chapter 10, as the bottom corners of the grille are trapped under the headlight units and cannot be lifted up through to allow the lugs to come free from the bottom panel.
4 Refitting is the reverse procedure to removal.

28 Bumpers – removal and refitting

1 Before attempting to remove the bumper begin by disconnecting the electrical wiring, to the side lights and flashers in the case of the front bumper, and to the number plate lights in the case of the rear bumper fitted to the Coupe, Spider and HPE models.
2 With the wiring to the front bumper it is generally easier to undo the nut and bolt at the rear of each light unit, bend the bracket slightly outward and withdraw the complete light unit and wiring. Then let it hang down without straining the cables.
3 In the case of the rear bumper (Coupe, Spider and HPE) the bulb holders simply pull out.
4 To remove the front bumper first remove the bolts from inside the front wings. These secure the ends of the front bumper to the wings, in the case of the Saloon, or on other models undo the bolt and nut which retain the end of the bumpers to the stay (photo).
5 Remove the two bolts at each side which secure the bumper to the brackets or shock absorbers and lift the bumper way.
6 To remove the rear bumper the process is similar, but the bolts which secure the stays at the ends of the bumpers in the Saloon model are located inside the boot and the spare wheel and interior trim will need to be removed first. Take care when doing this that the spacers are not lost. If necessary the bumper mounting brackets can easily be removed from inside the luggage area. Each is secured by two bolts (photo).
7 Refitting is the reverse process to removal in all cases.

29 Front wing – removal and refitting

1 Remove the front bumper as described in Section 28.
2 From inside the engine compartment unscrew the thumb bolts and detach the headlamp door from the wing.
3 From inside the wheelarch disconnect and remove the side repeater lamp.
4 Detach the sill moulding from the wing.
5 Working inside the car remove the kick panel then lift up the matting and remove the bolt from the front door pillar.
6 Unscrew all the mounting nuts and bolts and withdraw the front wing panel from the body.
7 Refitting is a reversal of removal.

30 Facia panel (pre 1977 models) – removal and refitting

1 Disconnect the battery negative lead.
2 Disconnect the choke control cable from the carburettor (see Chapter 3).
3 Disconnect the speedometer cable from the gearbox.

Saloon models
4 Remove the instrument panel as described in Chapter 10 and, working through the aperture, identify and disconnect the leads to the heated rear window switch, windscreen wiper switch, choke cable head, cigarette lighter, and glovebox light.
5 Prise out the passenger side air vent with a screwdriver and unscrew the exposed facia panel nut.
6 Unscrew the upper facia panel retaining nuts.
7 Disconnect the hand throttle cable from the accelerator pedal and bracket.
8 Unscrew the lower retaining nuts and withdraw the facia panel from the car.

Coupe, HPE, and Spider models
9 Lower the steering wheel and remove the upper and lower steering column casings, then remove the instrument panel and auxiliary switch panel as described in Chapter 10.
10 Remove the upper grille and clock (see Chapter 10).

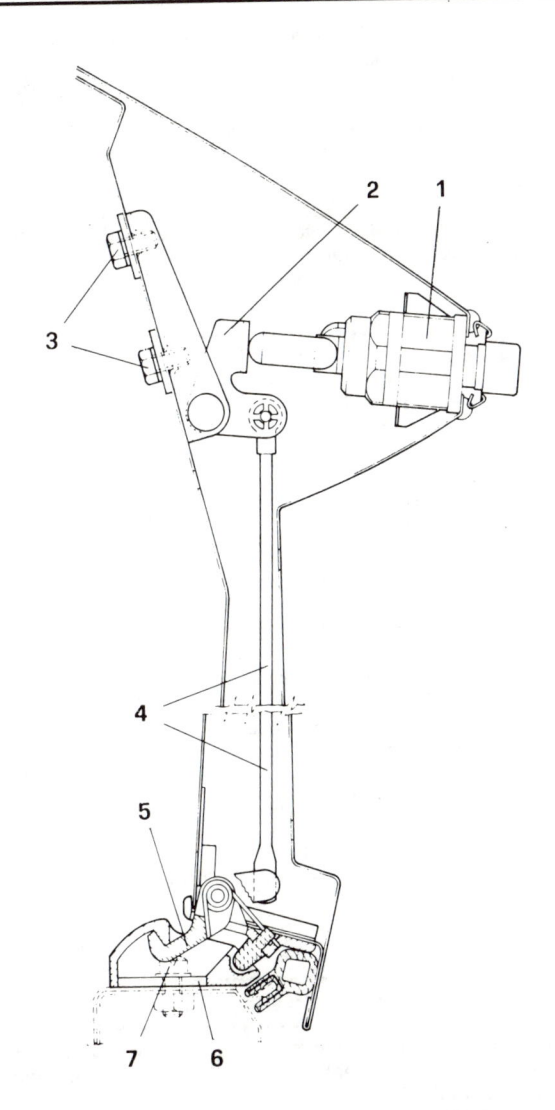

Fig. 12.14 Boot lid lock and catch mechanism (Saloon models) (Sec 25)

1 Lock
2 Control arm
3 Mounting bolts
4 Control rod
5 Catch
6 Striker plate
7 Striker plate fixing bolts

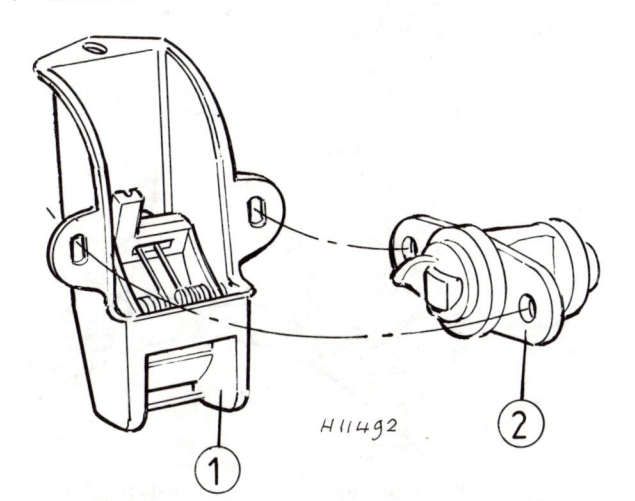

Fig. 12.15 Boot lid lock and catch (Coupe and Spider) (Sec 26)

1 Catch
2 Locking cylinder

266 Chapter 12 Bodywork and subframe

11 Prise out the side vents with a screwdriver and unscrew the exposed facia panel bolts.
12 Disconnect the hand throttle cable from the accelerator pedal and bracket.
13 Unscrew the lower retaining nuts and withdraw the facia panel from the car.

All models
14 If necessary the various switches, vents, and auxiliary components can now be removed and transferred to the new panel.
15 Refitting is a reversal of removal.

31 Rear seats – removal and refitting

1 On Coupe, HPE, and Spider models unscrew the backrest securing nuts from the rear of the seat.
2 On all models lift out the rear seat cushion and unbolt the backrest from the floor.
3 Raise the backrest and withdraw it from the car.
4 Refitting is a reversal of removal.

32 Centre console – removal and refitting

Saloon models
1 Remove the gear lever knob and lift off the boot, which is attached to the base plate (Fig. 12.16).
2 Remove the mounting screws and lift the console away.
3 Refitting is a reversal of the removal procedure.

Coupe, Spider and HPE models
4 Remove the gear lever knob and lift off the boot.
5 Remove the handbrake boot.
6 Unscrew the centre console mounting screws; there are three on each side. The console can then be lifted out.
7 Refitting is the reverse procedure to removal.

33 Front seats – removal and refitting

1 Slide the seat forwards and remove the two rear seat runner mounting bolts. Slide the seat to the rear and remove the two front seat runner mounting bolts.
2 The complete seat assembly can now be removed from the car.
3 To refit the seat, place it in position and line up the mounting holes. Then refit the mounting bolts in the same way that they were removed. Only tighten up the first pair after the second pair have been located and refitted.

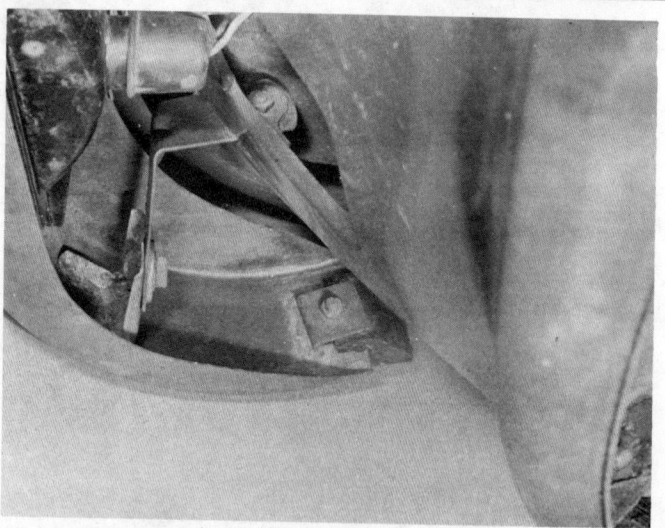

28.4 The bumper stay with its retaining bolt can be easily separated from the outer end of the front bumper

28.5 The front bumper retaining bolts

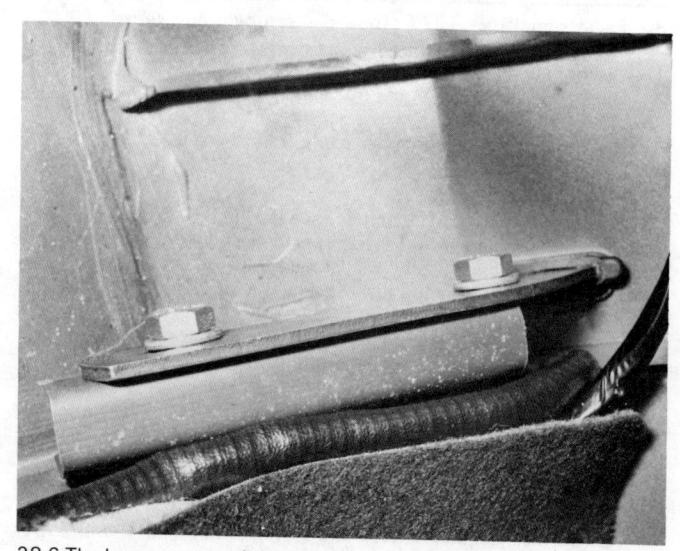

28.6 The bumper mounting brackets are retained by two bolts (HPE model shown)

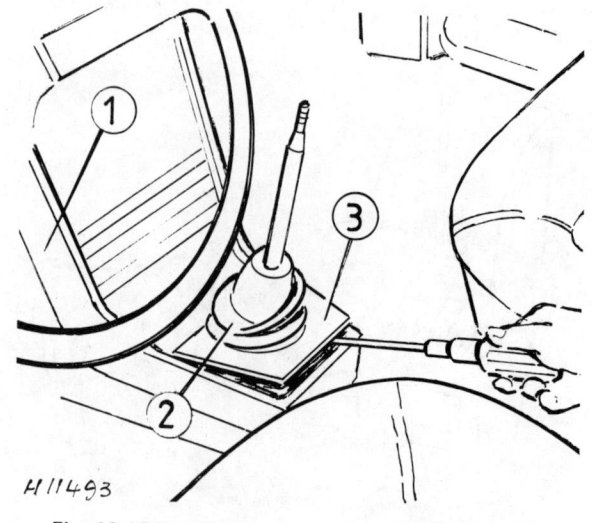

Fig. 12.16 The Saloon type centre console (Sec 32)

1 Console
2 Boot
3 Base plate

Chapter 12 Bodywork and subframe

34 Windscreen (Saloon models) – removal and refitting

1 Refer to Chapter 10 and remove the instrument panel.
2 Disconnect the choke cable from the carburettor, and from the knob in the dashboard.
3 From behind the dashboard disconnect all the leads to the various switches and the glovebox light. Label all the leads as they are disconnected.
4 Undo and remove the upper nuts on the drivers side of the car, which secure the facia panel to the body; these are accessible through the hole where the instrument panel fits.
5 Lever off the passengers side air vent in the dashboard using a screwdriver so as to reach the dashboard retaining nut behind it.
6 Disconnect the hand throttle cable from the accelerator pedal and bracket, and remove the lower retaining screws so that the complete dashboard can be lifted carefully away.
7 The next stage is to detach the headlining at the front above the windscreen. In order to achieve this it is necessary to first remove the grab handle, sunvisors, rear view mirror complete with arm, the cap on the left-hand side of the roof in order to get at and remove the screw underneath,. and the front courtesy light. In all models, except the 1400, it is necessary to remove the screw underneath the front courtesy light as well.
8 Remove the windscreen pillar mouldings.
9 For the next part of the operation it is necessary to have two people, one inside the car and one outside. The person inside pushes the windscreen outwards using his hands and the person outside retrieves the screen, and stops it jumping out uncontrollably. Then both people can work from outside and lift out the screen complete with its weatherstrip and mouldings.
10 Now go back to the car and check that the water drain holes in the lower corners of the windscreen surround are clear and then clean off the whole of the windscreen surround using petrol.
11 The windscreen will have been removed for two main reasons, either the glass is damaged and needs to be renewed or the weatherstrip is leaking and needs to be renewed. If the weatherstrip is in good condition it can be reused but it is advisable to renew it anyway.
12 First of all remove the top and bottom moulding caps on the windscreen trim and then remove the two moulding halves. The weatherstrip can then be prised off. Do this carefully if it is to be used again.
13 If the glass is to be refitted and the weatherstrip is being renewed then thoroughly clean the edges of the glass using petrol.
14 If the weatherstrip itself is being reused then thoroughly clean it with methylated spirit before fitting it to a new windscreen.
15 With the weatherstrip fitted to the glass, the mouldings should next be fitted from the bottom of the screen and secured in position with the upper and lower caps.
16 In order to make sure that the weatherstrip seats correctly in the windscreen frame insert a length of string in the outer groove of the weatherstrip. The string must be long enough to ensure that it goes twice round the weatherstrip and the ends must protrude about 12 inches (300 mm) from the top corners. In other words start at the top left-hand corner of the screen with a tail and wrap the string completely around the screen to where you started and then go on until the other end reaches the top right-hand corner where another tail is required.
17 Now it is time to fit the windscreen, in its prepared state, to the car. Using two people as before lift the screen up and offer the lower edge of the weatherstrip to the lower edge of the windscreen frame.
18 Lower the screen into position and using the hand, tap all round the edge of the screen.
19 With a person seated inside the car, pull the ends of the string, in turn, in order to ensure that the weatherstrip seats correctly. It will be helpful if the person outside goes on tapping the edge of the glass during the operation as the string is pulled round (Fig. 12.18).
20 Once the string has been pulled out completely the windscreen should be properly seated and watertight.
21 Now the interior trim parts and dashboard that were removed initially can be refitted in the reverse order to removal.

Fig. 12.18 Fitting the windscreen – although this figure shows two people outside it can be done by one person (Sec 34)

35 Rear window (Saloon models) – removal and refitting

1 Disconnect the battery negative lead.
2 Remove the rear seat squab and backrest, which is secured to the floor pan by bolts. It can then be lifted up and away.
3 Remove the rear shelf and disconnect the heated rear window cables.
4 Prise the rear courtesy light off using a screwdriver and disconnect the wiring to it.
5 Release the retainers to remove the rear window upper padding, except on the 1400 model.
6 Refer to the previous Section and use exactly the same procedure as for the removal and refitting of the windscreen (paragraphs 9 to 20).
7 Take care when refitting the glass to the body that the slots in the weatherstrip and the body drain holes are lined up correctly.
8 Refit the trim panel and mouldings and seats in the reverse order to removal.

36 Rear quarterlight (Saloon models) – removal and refitting

The procedure for removing and refitting the rear quarterlight glass is basically the same as given in Section 34 for the windscreen, but

Fig. 12.17 Check that the windscreen surround drain holes are clear (Sec 34)

1 Left-hand drain hole

there is no trim to be removed first. Start at paragraph 9 and follow through to paragraph 20.

37 Windscreen (Coupe, Spider and HPE) – removal and refitting

1 Because of the way in which the different types of windscreen are fitted it is necessary to have a special locking jig to refit a windscreen in either a Coupe, Spider or HPE model.
2 It is therefore not recommended that this operation be carried out by the home mechanic, but the vehicle should be taken to a competent Lancia dealer.

38 Rear window (Coupe) – removal and refitting

1 The rear window is held in position by sealant, like the windscreen, and requires a special jig to lock it in place.
2 The rear window can be removed in the same way as the Saloon model rear window, as described in Section 35. However, refitting would have to be carried out by a Lancia dealer using the special jig.

39 Tailgate glass (HPE) – removing and refitting

The tailgate glass, like the windscreen, requires a special fitting jig to secure it in place when it is refitted. Therefore it is advisable to have the complete operation carried by your local Lancia dealer as it is not a straightforward job.

40 Rear quarterlight (Coupe and HPE) – removal and refitting

1 Undo the quarterlight latch so that the window is free (photo).
2 Remove the weatherstrip from the door rear pillar so that you can get at the hinge bolts. Remove the bolts and lift off the quarterlight.
3 On the HPE model the window is removed by slipping it off the slots in the pillar.
4 The window can then be placed on the workbench and the glass can be changed. First remove the mouldings and then remove the glass and weatherstrip from the frame. Next remove the weatherstrip and clean it with methylated spirit.
5 The glass and window are then refitted in the reverse order to removal.

41 Rear window and folding roof section (Spider) – removal and refitting

1 Undo the units in the boot, behind the lining, which holds the rear seat backrests in place; the spare wheel must first be removed.
2 Remove the rear seat backrests from inside the car and the seat squabs.
3 Unlock the rear hood clamps but do not fold the hood down.
4 Undo and remove the pivot bolts for the hood frame on each side and also the roof section mounting brackets (two bolts on each side).
5 The folding roof section can now be removed complete with the lower weatherstrip.
6 Before refitting the folding roof check that the weatherstrip at the bottom edge of the hood is in good condition. If it is not, renew it.
7 When refitting the folding rear section first re-align the weatherstrip and hood lower edge and refit the bolts to the mounting brackets.
8 Refit the pivot bolts on each side and then check that the hood closes properly. If the fasteners do not meet properly then the bracket(s) on which the hood frame pivot is mounted needs to be adjusted.
9 To do this remove the side panel in the rear of the car, then loosen the mounting bolts and adjust the bracket. Then check that the hood fasteners meet and operate correctly.
10 Replace the trim panel and refit the seat squabs and backrests in the reverse order to the removal sequence.

42 Exterior door mirrors – general

1 Depending on the model and age will largely depend whether the vehicle has one or two exterior mirrors. Up to early 1979 the type fitted was basically the same on all models. It was only adjustable from the outside and had a ball and nylon socket friction mounting giving infinitely variable adjustment.
2 This type of mirror can easily be removed and replaced. It is mounted on a base plate which is screwed to the outer door panel. To remove the mirror assembly unscrew the Allen screw in the rear of the base and it can then be lifted away (photo).
3 Later (Coupe, Spider and HPE) models from February 1979 onwards have their exterior door mirrors adjustable from inside the car. The mirror is mounted in the bottom rear corner of the quarterlight with a small operating lever on the inside (Fig. 12.19).

Fig. 12.19 Later type exterior mirror (Sec 42)

1 Adjuster knob

43 Subframe – general

1 The front subframe carries the whole of the engine and transmission assemblies, together with the majority of the front suspension.
2 It is mounted to the front body shell by four mounting bolts, one at each corner (photos).
3 The subframe is made up of four steel channel sections which are welded together to form a roughly square assembly (Fig. 12.20).
4 The front crossmember has the front gearbox mounting fitted into it and the rear crossmember has a mounting bracket as an integral part of it which carries the gear linkage idler arm. The rear gearbox/differential unit mounting is also attached to the rear subframe crossmember (see Chapter 11, photo 27.1).
5 It can be removed from the vehicle only by removing the complete engine/transmission assembly. There are two ways of achieving this.

40.1 The quarterlight catch on the HPE models can be freed at either end

Fig. 12.20 The front subframe dimensions (Sec 43)

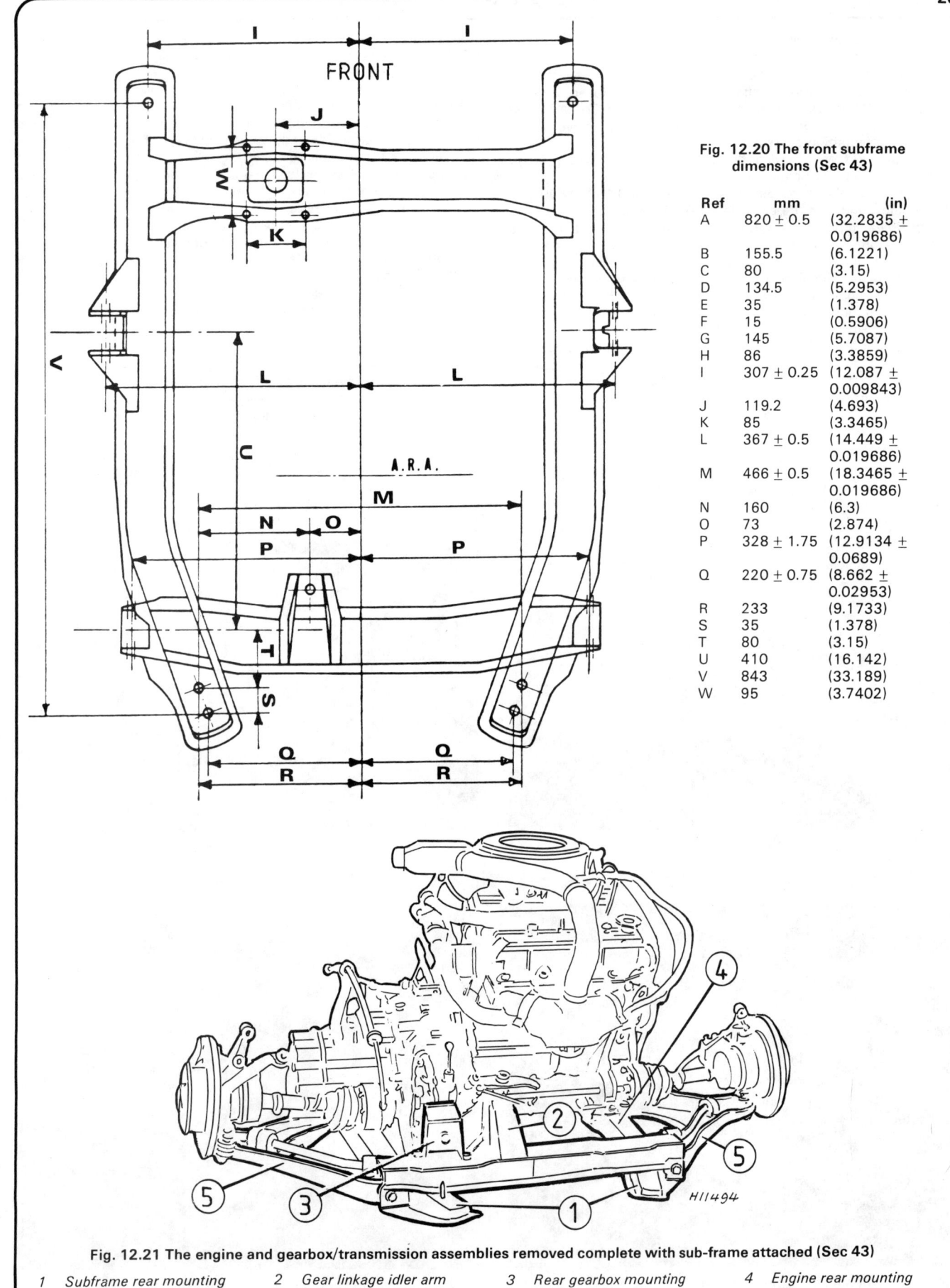

Ref	mm	(in)
A	820 ± 0.5	(32.2835 ± 0.019686)
B	155.5	(6.1221)
C	80	(3.15)
D	134.5	(5.2953)
E	35	(1.378)
F	15	(0.5906)
G	145	(5.7087)
H	86	(3.3859)
I	307 ± 0.25	(12.087 ± 0.009843)
J	119.2	(4.693)
K	85	(3.3465)
L	367 ± 0.5	(14.449 ± 0.019686)
M	466 ± 0.5	(18.3465 ± 0.019686)
N	160	(6.3)
O	73	(2.874)
P	328 ± 1.75	(12.9134 ± 0.0689)
Q	220 ± 0.75	(8.662 ± 0.02953)
R	233	(9.1733)
S	35	(1.378)
T	80	(3.15)
U	410	(16.142)
V	843	(33.189)
W	95	(3.7402)

Fig. 12.21 The engine and gearbox/transmission assemblies removed complete with sub-frame attached (Sec 43)

1 Subframe rear mounting points
2 Gear linkage idler arm bracket
3 Rear gearbox mounting bracket
4 Engine rear mounting
5 Wishbones

Either the engine and transmission unit are removed first and the subframe is then detached from the body shell (after the steering, suspension and self-levelling headlamp system have been disconnected as well) or the whole subframe assembly is removed from the car complete with the engine and gearbox etc attached. The subframe can then be detached from the different assemblies. Whichever way it is done it is a lengthy process (Fig. 12.21).

6 The operation of removing the complete engine and gearbox assembly is covered in Chapter 1.

7 On early models of the Series A Saloon with 1400, 1600 and 1800 engines, and also the early Series B 1300 Saloon, there have been considerable problems with rusting of the rear mounting points for the subframe. It is the body section that rusts and not the subframe, and there is no way in which this problem can be overcome by welding in new body sections. The only way to overcome this problem is to acquire a new bodyshell and fit the components and body parts to it, although this is somewhat expensive to do.

8 Later body shells were all built with strengthened mounting points for the rear end of the front subframe and no deterioration has been found in later models.

9 Regular checks of the subframe rear mounting points are advisable and if any weakness is noted then the vehicle should be taken directly to your local Lancia dealer for consultation.

44 Heating system – general description

1 The heating system is mounted directly beneath the facia panel in the centre of the car with controls in the centre section of the top of the console. Vents fed by ducts are located in the centre console, the footwells and the outside ends of the dashboard. There are also ducts and vents for the rear passengers.

2 The heater itself consists of a small radiator core which is supplied with hot water from the engine cooling system. This can be shut off when the heater is not required, by a valve on the inlet pipe.

3 The air supply to the interior of the car is either directed by a ram jet effect of air being pushed through the heater matrix and being fed through the ducting, or, it is boosted by a fan, controlled by a switch on the dashboard.

4 On some models the interior heater blows hot air all the time. This is due to a poor linkage system and the valve on the inlet pipe opening fractionally.

45 Heater – removal, dismantling, reassembly and refitting

1 The heater radiator and motor unit is situated underneath the

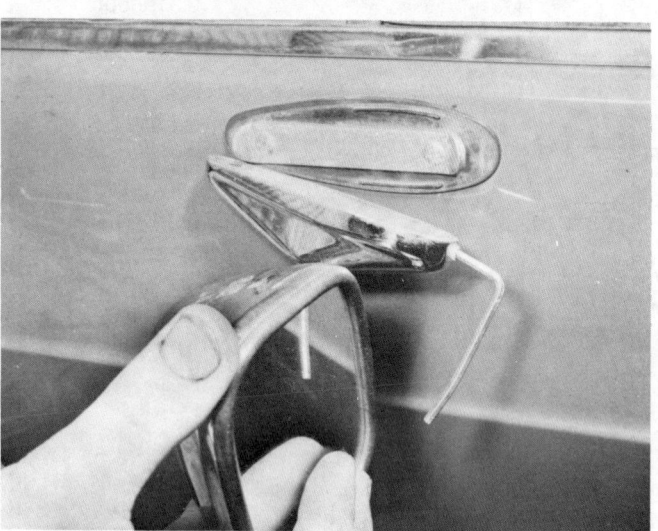

42.2 Removing an exterior door mirror

43.2a Front right subframe mounting point

43.2b Rear right subframe mounting point

45.21 The left-hand fresh air duct is adjustable for vertical and horizontal air flow

Chapter 12 Bodywork and subframe

dashboard and is covered by the centre console. Flexible ducts take the air from the heater unit to the various outlets for demisting and air 'conditioning'.

2 It will be necessary to remove the instrument panel and centre console as directed in Chapter 10 and Section 32 of this Chapter in order to gain access to the heater and controls. The knobs on the ends of the control levers will also have to be removed.

3 The heater unit comprises an upper casing bolted to the bodywork behind the facia panel, a radiator and valve assembly which slots into the base of the upper casing. The fan motor and lower housing is clipped to the underside of the upper casing.

4 If you intend to work on the air deflectors, electric motor or fan, there is no need to disturb the radiator and therefore no need to drain the cooling system.

Removal of heater motor and fan only

5 Having removed the centre console, disconnect the single control cable attached to the lower air deflector. Then prise off the four spring clips which retain the upper and lower halves of the heater casing together. Remove the lower casing which will contain the motor and fan. There will be a third moulding between the two halves; this surrounds the fan and ducts the warm air down to the lower casing.

6 Once the lower casing has been removed the motor electrical connections can be detached, and the motor/fan unit unclipped from the inside of the lower casing.

7 If the motor is faulty there is no point trying to dismantle it, only complete motors are available as spares. The fan may be detached from the motor shaft after the single retaining nut has been undone, and the special hollow screw loosened in the centre of the fan boss.

8 Refitting the motor and fan follows the reversal of the removal procedure. Remember to fix the rubber spacers each side of the motor when it is reassembled into the lower casing.

Removal of heater radiator

9 It will be necessary to drain the whole cooling system as detailed in Chapter 2, remove the centre console, and then remove the lower heating casing together with the motor and fan as described in paragraphs 5 to 8 of this Section.

10 Continue by detaching the control cable from the water flow valve and the two water hoses from the ends of the metal radiator pipes.

11 The radiator may now be eased down out of the upper heater casing. There is a soft rubber strip fitted around the radiator between it and the inside surface of the upper heating casing.

12 With the radiator assembly free of the car, the pipes and valves may be detached as necessary. Do not try and dismantle the valve. Lancia supply whole valves only. It is possible to carry out minor repair work on the radiator with proprietary sealing compounds, but it is advisable to entrust repair work to the local radiator specialist. They will have the expertise to tackle a soldering job on the radiator.

13 Refitting the radiator follows the reversal of the removal procedure; remember as always to use new gaskets at flange joints of the radiator supply pipes. Do not use sealing compounds instead of new gaskets, it will make it extremely difficult to break the joint next time it is to be dismantled.

The heater controls

14 The three heater controls are connected to the lower air vent, hot water supply and fresh air inlet control and are situated in the forward central console, and therefore the control links are short.

15 The cables may be detached from the operating levers by simply

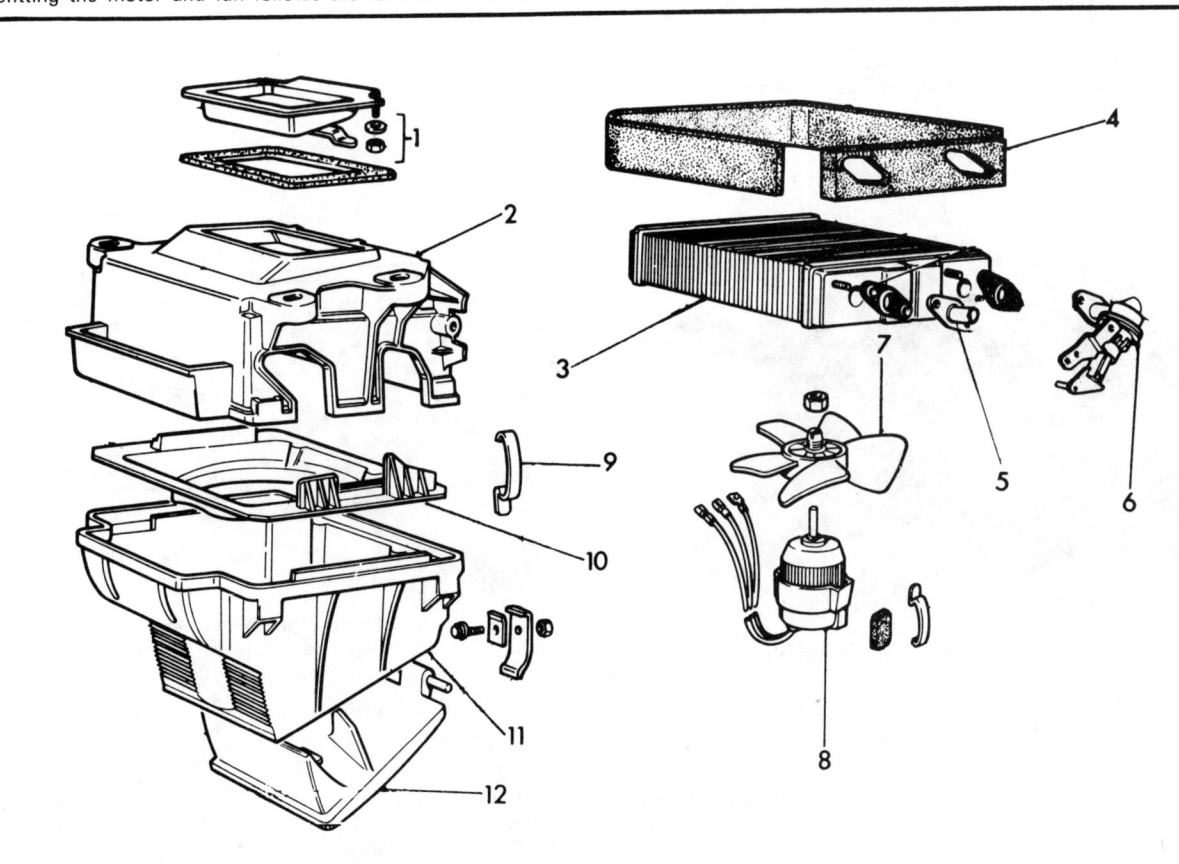

Fig. 12.22 The heater components exploded view (Sec 45)

1 Air inlet control flap, and gasket
2 Heater upper casing
3 Heater radiator
4 Radiator periphery seal
5 Water outlet pipe
6 Water inlet control valve
7 Fan
8 Heater motor
9 Spring clip holding upper and lower casings together
10 Centre moulding which surrounds the fan to duct the warmed air down to the distribution passages
11 Heater lower casing
12 Warm air control flap

unhooking from the holes in the base of the levers. The forward ends of the cables, which attach to the valve and flap actuation levers, may be removed from those levers equally easily. The control levers pivot on steel pins retained in the bracket assembly with split pins. Once those pins have been removed the pivot pins can be extracted to free the operating levers. Retrieve the spacers and washers.

The heater upper casing

16 The whole heater has been designed so that all maintenance and repair operations on the heater components may be carried out without needing to remove the upper casing. Study the exploded views of the heater components and the text in this Chapter and then if the upper casing must be removed proceed as follows.
17 Remove the centre console, the lower heater casing together with the motor and fan, and finally remove the radiator itself.
18 The heater upper casing is secured to the bodywork behind the facia panel with four nuts and bolts, all of which are quite accessible. Retrieve the soft gasket between the casing and the mounting.

Air ducting

19 Air from the outside of the car is taken into a chamber between the engine compartment and facia panel through vents in the engine hood. The heater unit mounted beneath this chamber draws in air as required through the inlet flap and deflectors. The purpose of the deflectors is to prevent rain water from being taken into the heater. At each end of this chamber, near the sides of the car, there are ducts which take cool fresh air directly to controllable vents at each end of the dashboard in the car.
20 Little is required in the way of maintenance to these ducts; the terminal vent flap assemblies are screwed to the reverse side of the facia panel and flexible ducting connects the terminal vents to the feed ducts.
21 The fresh air ducts can easily be removed from the facia panel by squeezing them at the edges to release the lugs which hold them into the dashboard itself (photo).

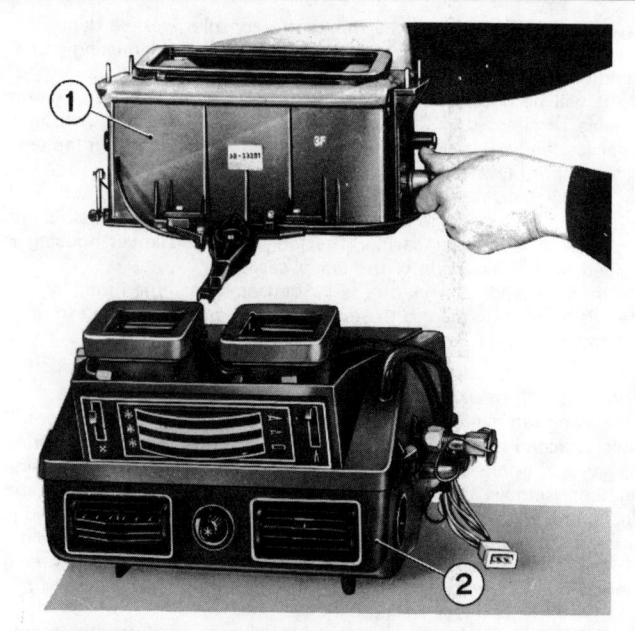

Fig. 12.24 The heater unit is mounted above the air conditioner evaporator (Sec 46)

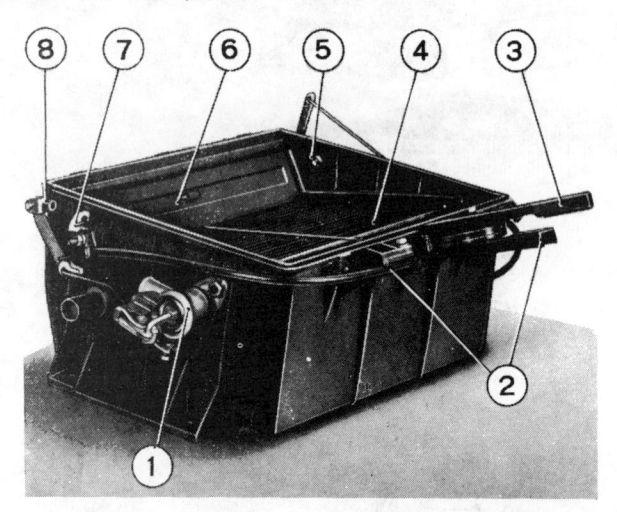

1 Heater unit
2 Evaporator unit

Fig. 12.23 The heater removed complete (Sec 45)

1 Water valve
2 Valve and air blending flap control lever and cable
3 Air intake control lever
4 Core
5 Spring retainers
6 Air blending flap
7 Spring retainer inner pin
8 Valve operating lever mounting screws

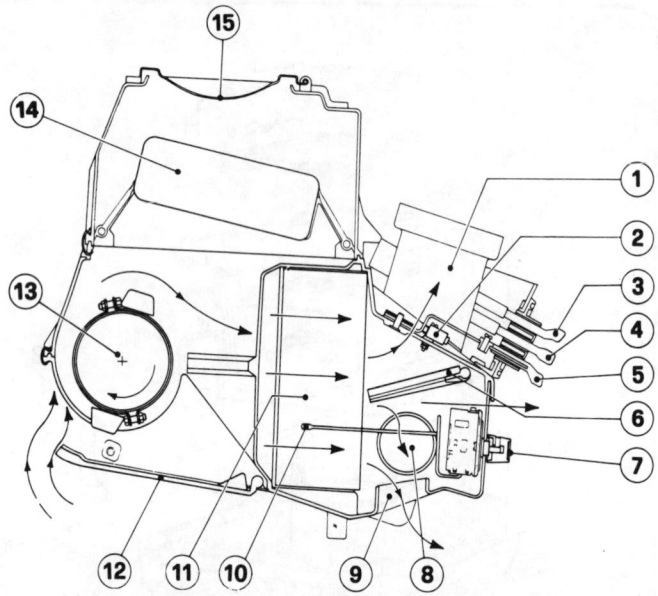

Fig. 12.25 Sectional view of the heater and air conditioner unit (under the dashboard) (Sec 46)

1 Air pipe to windscreen
2 Switch on recirculation lever
3 Heater valve control lever
4 Outside air inlet control lever
5 Air recirculation inlet control lever
6 Air distribution inlet
7 Thermostat
8 Side air outlets
9 Rear seat air pipe
10 Thermostatic sensor
11 Evaporator radiator
12 Air recirculation inlet
13 Heater air conditioner fan
14 Heater radiator
15 Outside air inlet

46 Air conditioning system – general

Some Lancia Beta models are fitted with an air conditioning unit. This unit is in addition to the existing heating system and is attached to it (Fig. 12.24).

The purpose of an air conditioning unit is to cool the interior of the car. An air conditioning unit is basically the same in idea as a domestic refrigerator, which uses a pressurised system filled with a non-toxic gas; in this case Freon 12.

Beneath the heater unit in the car is another unit with another radiator core. The difference between the two is that one is designed to be hot and the other cold. Like the heater the air conditioning unit also uses the fan to assist the cooling effect as the weather or interior temperature of the car rises. There is a thermostat control knob on the facia in addition to the normal control levers (Fig. 12.26).

As can be seen from the diagram in Fig. 12.27 the main

Chapter 12 Bodywork and subframe

components of the air conditioning plant are a compressor, a condenser, a filter, and expansion valve and an evaporator and fan.

The compressor is driven by a toothed drivebelt from the crankshaft pulley and is mounted on the right-hand front side of the engine above the alternator.

The condenser and fan are mounted alongside the radiator for the cooling system.

Because the air conditioning unit is a pressurised system there is very little that can be done to it by the home mechanic. If there is a leakage in the system then it will have to be attended to by a specialist and re-charged with special refrigeration equipment. Maintenance is covered in the next section.

47 Air conditioning system – maintenance

1 Although no stripping and inspection can be carried out on the air conditioning plant there are several regular maintenance procedures that should be attended to.
2 Clean all the flies, leaves etc from the condenser radiator on a regular basis, especially during the summer months. This should then be done monthly.
3 Check the compressor drivebelt every six months for both tension and wear. If it is worn then slacken the compressor mounting bolts and lower the compressor on its bracket so that the belt can be renewed. Fit a new belt and then lift the compressor up until the belt is tensioned. Then tighten the mounting bolts.
4 During the winter months run the unit periodically to prevent the coolant seals in the compressor from going hard and thereby losing refrigerant.

Fig. 12.26 Heater/air conditioner controls (Sec 46)

1 Booster control switch
2 Air recirculation flap control lever
3 Fresh air inlet control lever
4 Cock and blender control lever
5 Air flap control
6 Temperature control knob (thermostat)

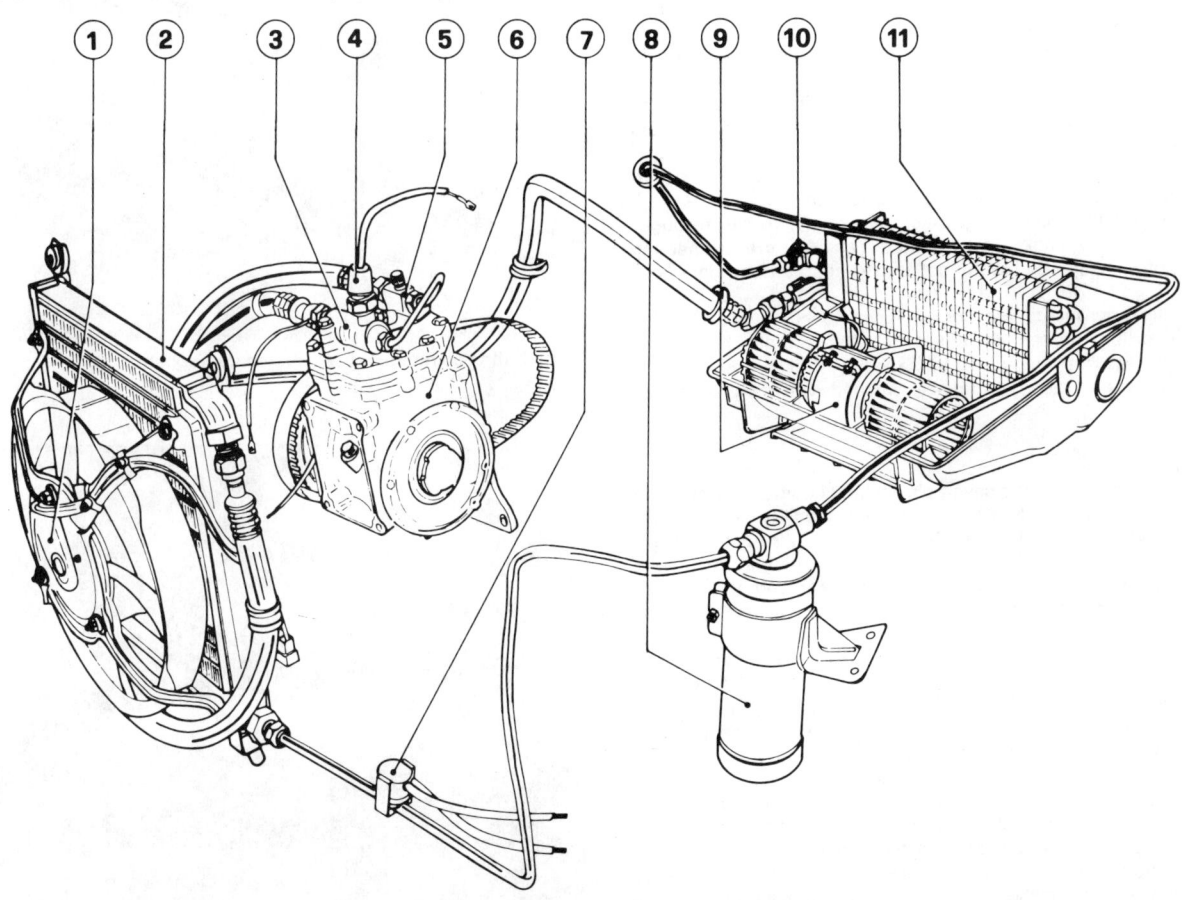

Fig. 12.27 Air conditioning system (Sec 46)

1 Condenser fan
2 Condenser radiator
3 Isobaric valve
4 Minimum pressure switch
5 Delivery valve
6 Compressor
7 Fan connection to thermostat
8 Filter
9 Evaporator fan
10 Expansion valve
11 Evaporator radiator

Chapter 12 Bodywork and subframe

48 Towbar – fitting and wiring

The procedure below relates to the fitting of a Witter towbar kit specifically designed for the Lancia Beta. Other makes of towbar may be supplied with their own fitting instructions, which should be followed where they differ from those given below.

Fitting

1 Pull the bulb holders out of the number plate lamp. Remove the spare wheel and the load space carpets, undo the two bumper bracket securing bolts on each side and remove the bumper.
2 Raise and securely support the rear of the car with ramps or axle stands.
3 Remove the lash-down bracket and fit the forward end of the towbar central member in its place (photo). Do not tighten the nuts fully yet, just nip them up.
4 Provisionally fit the spacer plate, connecting member, socket plate and ball hitch to the rear end of the central member (photo). (Depending on manufacturing tolerances, the spacer plate may not be needed). Again, do not tighten the nuts fully yet.
5 Offer up the external transverse member to the top of the connecting member and temporarily secure it with two nuts and bolts. Take care not to trap the number plate light wires where they leave the body (photo).
6 If all is well, the external transverse member should be in contact with the rear of the car. Adjust the position of components having slotted holes, if necessary, and add or remove spacers at the ball hitch end to achieve this. When satisfied, mark the position of the four bolt holes which will be used to secure the transverse members. Remember that the external holes will have corresponding holes inside the car.
7 Remove the external transverse member, then centre-punch and drill the holes. Take care to drill the internal holes in line with the external ones! Enlarge the internal holes, drilling from the inside, to accept the tubular spacers. Remove any burrs and paint the bare metal with rust-inhibiting primer.
8 Press the bolts through the holes in the internal transverse numbers, fit the spacers to the bolts, and pass the bolts and spacers through the inside holes so that the ends of the bolts protrude through the outside holes. This is slightly tricky and care must be taken not to lose the spacers inside the double-skinned section.
9 Fit the external transverse member, securing it to the through-bolts with nuts and shakeproof washers, and to the connecting member (photo). Again, be careful not to trap the number plate light wires.
10 When all components are correctly positioned, without undue strain or distortion, tighten all nuts and bolts securely. If it is not wished to proceed with the wiring immediately, refit the bumper and lower the car to the ground; otherwise, proceed with the instructions below.

Wiring

11 Wiring connections to the seven-pin socket should conform to the internationally agreed standard to ensure compatibility between different towing vehicles and trailers. Refer to Fig. 12.29 for the correct connections. As far as the rear fog lights are concerned, at the time of writing these are obligatory only if *both* the car and trailer were manufactured after 1st October 1979 or first used after 1st April 1980.
12 The Type 12N socket shown is intended for trailer or caravan external lighting. If a caravan is towed and it is wished to provide power for caravan internal lighting, auxiliary battery etc, a Type 12S socket should be fitted in addition. The standard pin connections for this type of socket are shown in Fig. 12.30. The two types of socket appear similar but they are not interchangeable.
13 In order to avoid overloading the direction indicator flasher unit, and to conform with the legal requirement that a dashboard warning light shall indicate to the driver when the trailer indicator lights are operating, either a slave relay unit or a heavy duty flasher unit must be fitted. A heavy duty flasher unit should be supplied with wiring instructions and will be used instead of the existing flasher unit; a typical wiring diagram is shown in Fig. 12.31. The instruction below relate to the fitting of a Hella 'Transflash' slave unit.
14 The relay must be mounted with its terminals downwards in a place where it will not be damaged by luggage etc. On the project car

48.3 Fit the forward end of the towbar central member in place of the lash-down bracket

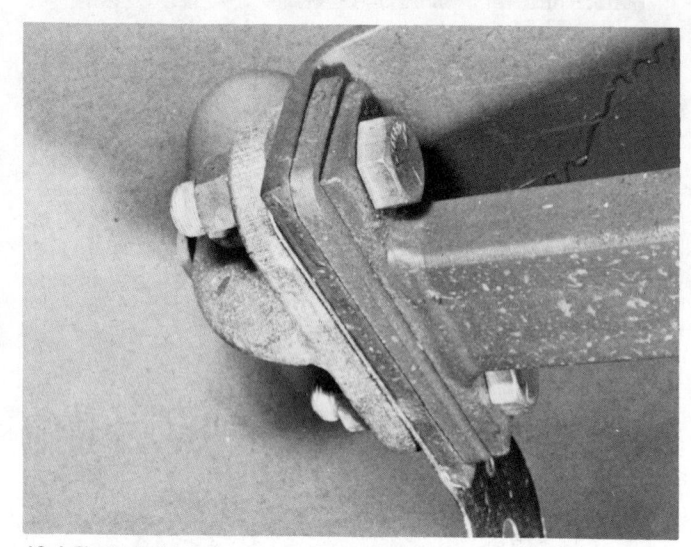

48.4 Fit the ball hitch and associated components

48.5 Take care not to trap the number plate light wires

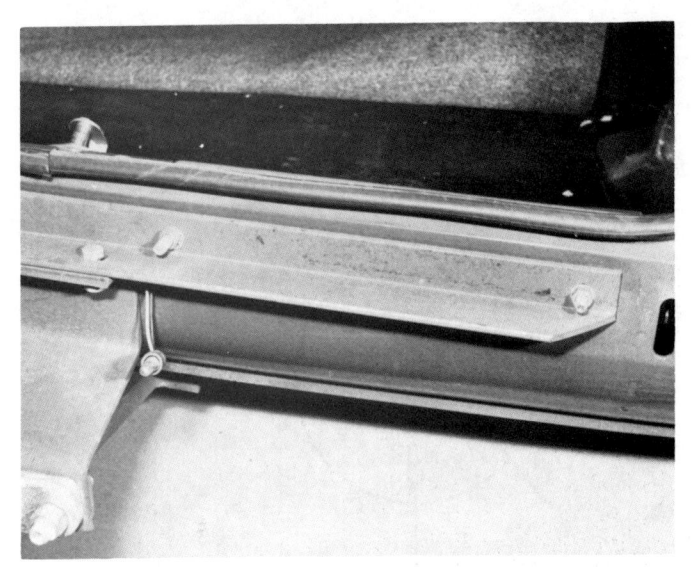

48.9 Secure the external transverse member

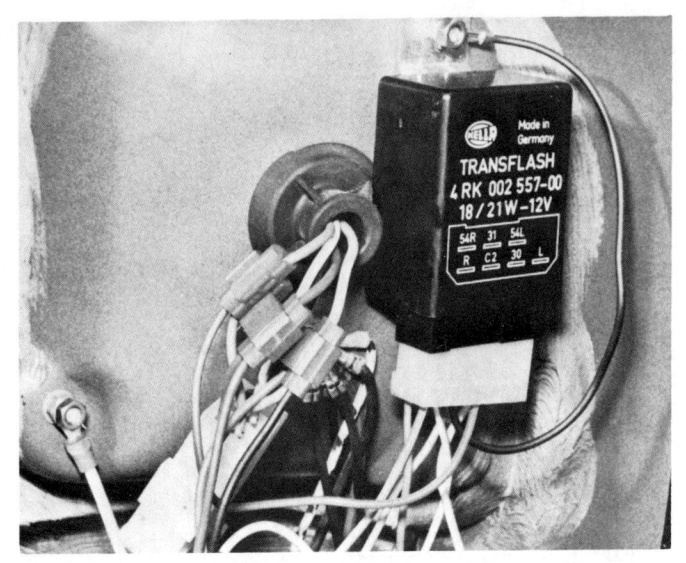

48.14 The relay mounted on the rear lamp cover. Note earth tags beneath cover securing nuts

48.17 Feed the cable through the grommet

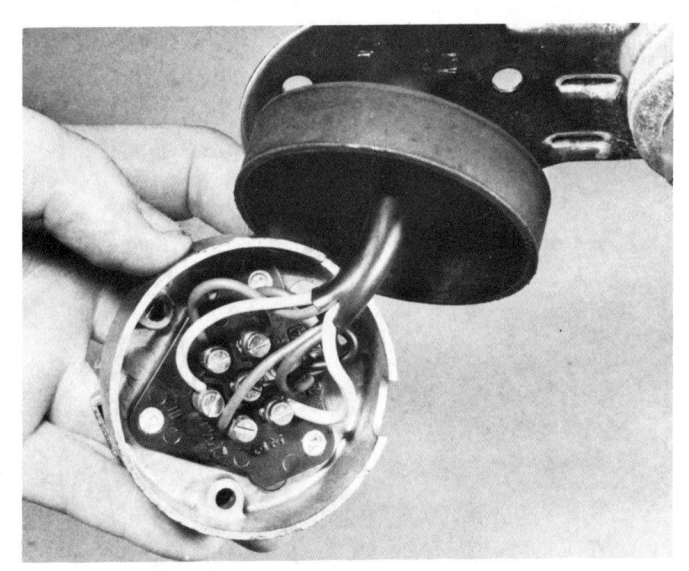

48.19a Connect the wires to the socket ...

48.19b ... then bolt the socket to the bracket

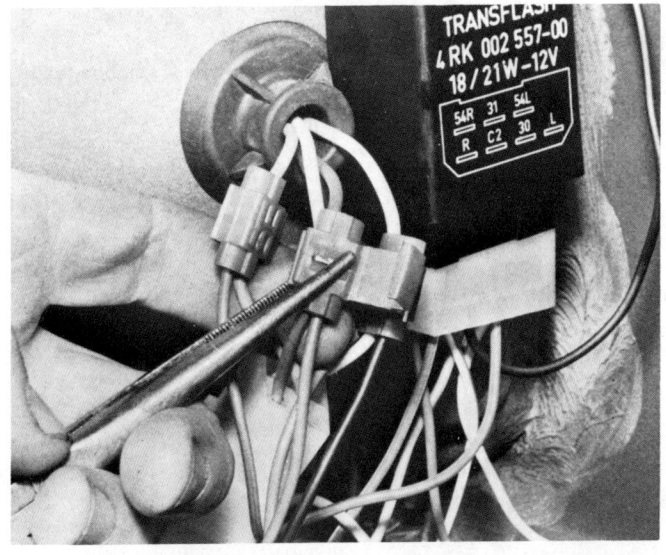

48.23 'Scotchlok' connector blade must be snapped home with pliers

Fig. 12.28 Witter towbar components (Sec 48)

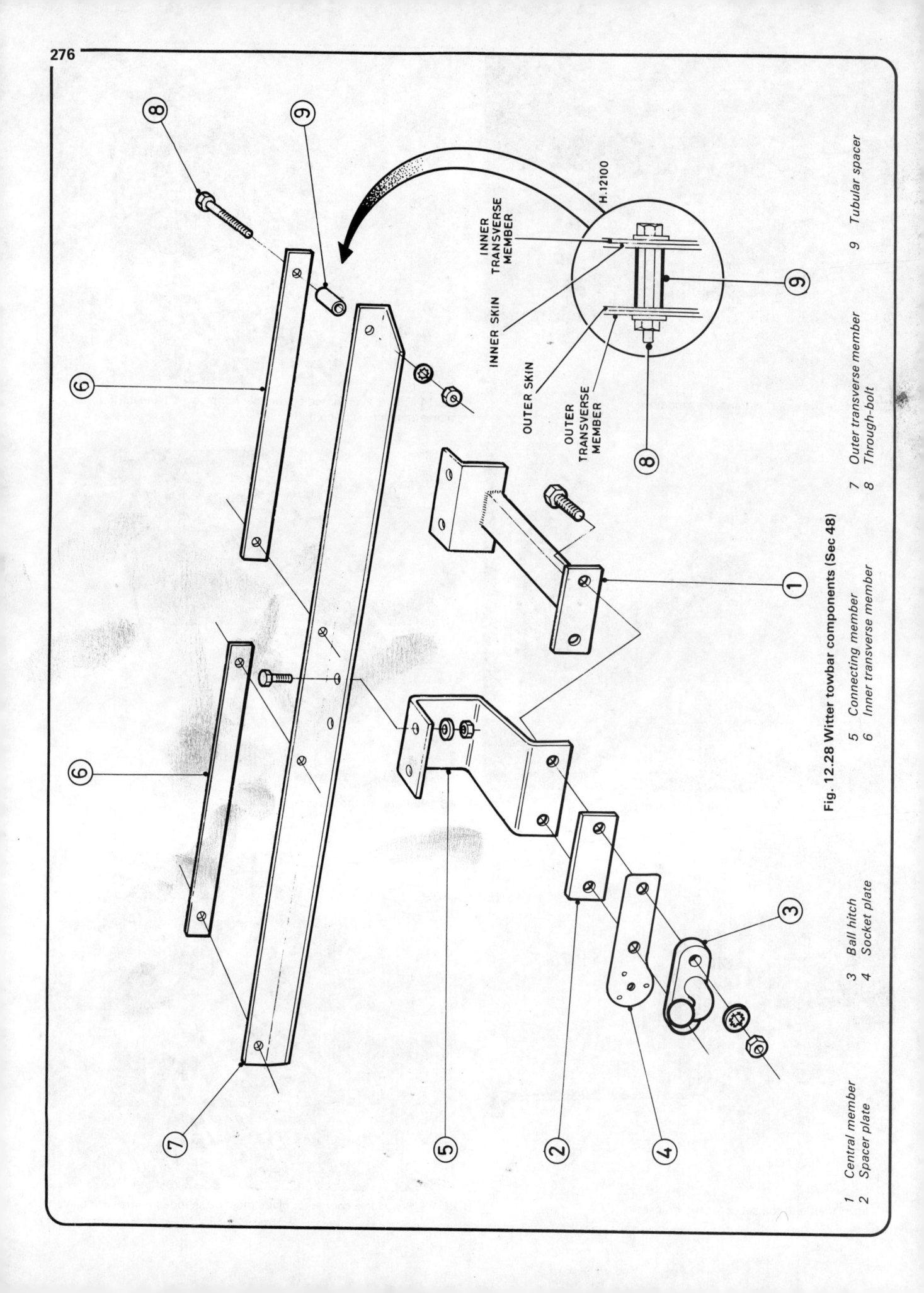

1 Central member
2 Spacer plate
3 Ball hitch
4 Socket plate
5 Connecting member
6 Inner transverse member
7 Outer transverse member
8 Through-bolt
9 Tubular spacer

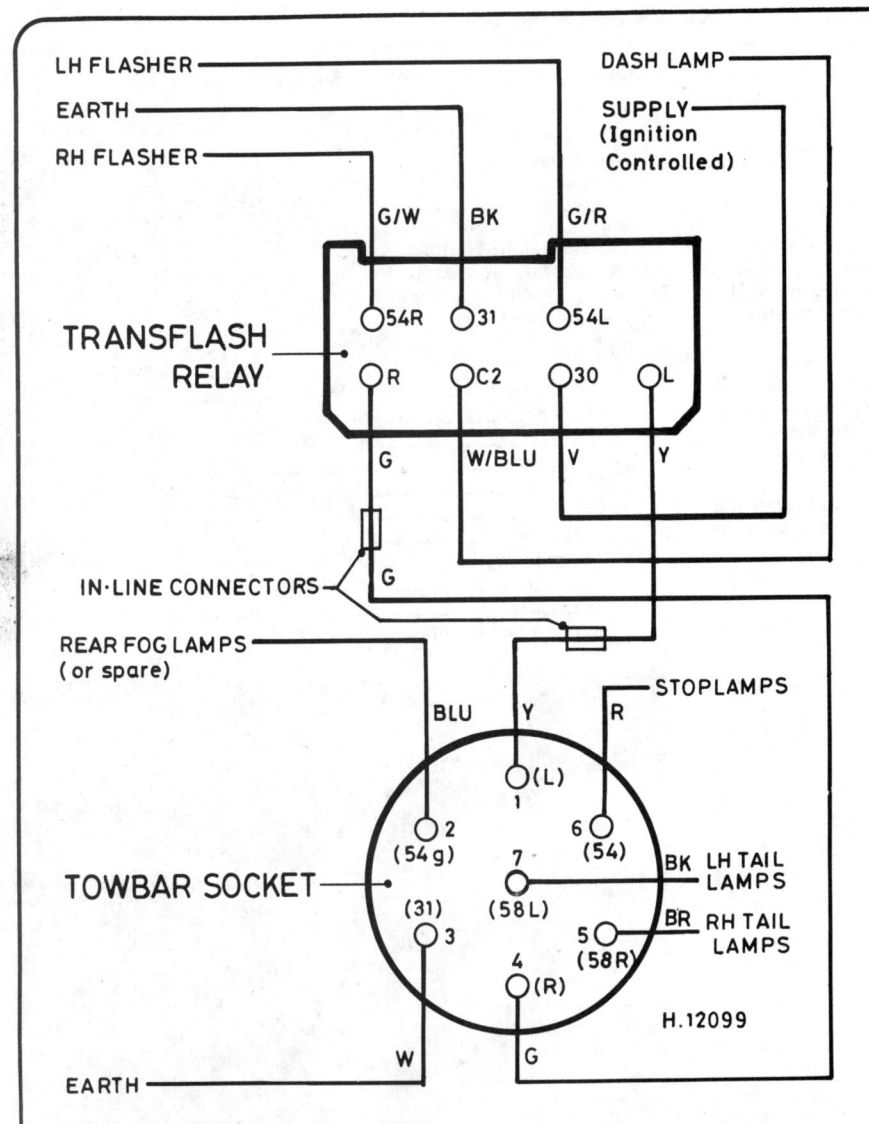

Fig. 12.29 Wiring diagram for connecting standard 7-pin towbar socket (Type 12N) to 'Transflash' relay and car wiring. Socket is viewed from wire entry side (Sec 48)

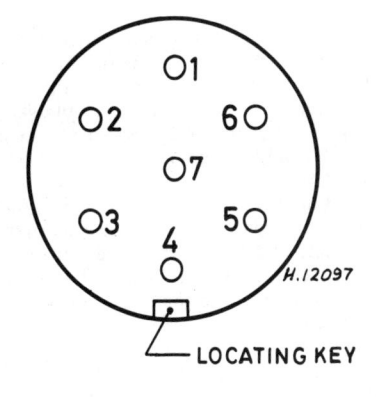

Fig. 12.30 Standard pin junctions and cable colours for auxiliary (Type 12S) 7-pin socket. Socket is viewed from wire entry side (Sec 48)

1 Yellow — Reversing lamp and/or reversing catch
2 Blue — Spare
3 White — Earth
4 Green — Power supply
5 Brown — Warning lamp for reversing catch
6 Red — Power supply
7 Black — Spare

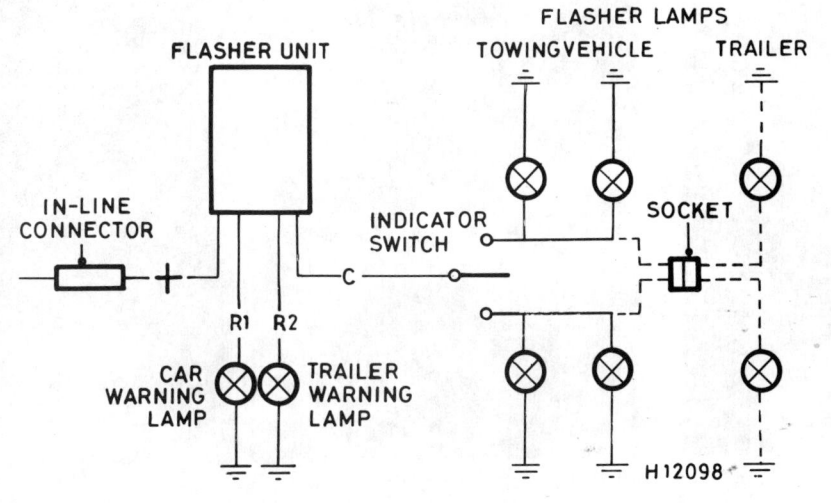

BK	Black
BLU	Blue
BR	Brown
G	Green
R	Red
V	Violet
W	White
Y	Yellow

Fig. 12.31 Typical wiring diagram for heavy duty flasher unit. If the old flasher unit was a 2-pin type, terminal R1 is not used (Sec 48)

Terminal	Function	Original designation
+	Live feed	+, + 49, 15, 15A, 15X, H, X, B
R1	Car warning lamp	C, C1, K, K1, KP, KBL, P, T, Rep
R2	Trailer warning lamp	None (new connection)
C	To indicator switch	49A, 49S, 54, 54L, S, Com, O, L, CL

Chapter 12 Bodywork and subframe

it was decided to mount the relay on the left-hand rear lamp cluster cover (photo), since the spare wheel is stowed on the right. One of the cover nuts was used to secure the relay; if this is unacceptable, drill a small hole in the desired location and secure with a self-tapping screw.

15 Remove the load space floor to gain access to the existing cable runs.

16 Drill a hole in the car floor large enough to accept the 7-core cable and a grommet. Note that the floor immediately around the towbar forward attachment is double-skinned for strength, and that various other areas of the floor are also double-skinned. Choose a single-skinned section for the sake of simplicity, or be prepared to use two grommets. De-burr and paint the edges of the hole.

17 Fit the grommet. Feed the 7-core cable through into the car, using a litle liquid detergent as a lubricant if necessary (photo).

18 Pass the outside end of the cable through the hole in the socket mounting plate and fit the protective rubber boot. Slit the cable outer sleeve at the end to expose the individual wires, then strip about ½ in (13 mm) of insulation off the end of each wire.

19 Connect the wires to the 7-pin socket in accordance with the wiring diagram (photo). Make sure that the connections are tight, then bolt the socket to the bracket (photo).

20 Determine the cable route inside the car to the slave relay. Trim off any excess cable, then expose the individual wires.

21 Consult the main wiring diagram if necessary and identify the following leads at the rear of the car:

LH flasher
LH tail lamp
RH flasher
RH tail lamp
Stoplamp (either side)

22 Disconnect the battery. Connect the relay wires and the 7-pin socket wires to the appropriate car wires or to each other as detailed in the table below.

Car	Relay	Socket
	R (Green)	4 (Green)
	L (Yellow)	1 (Yellow)
RH flasher	54R (Green/white)	
LH flasher	54L (Green/red)	
LH tail lamp		7 (Black)
RH tail lamp		5 (Brown)
Stoplamp		6 (Red)
Live supply	30 (Violet)	
Warning lamp	C2 (Blue/white)	
Earth	31 (Black)	
Earth		3 (White)
Rear fog lights (if fitted)		2 (Blue)

48.24 Follow existing cable runs where possible

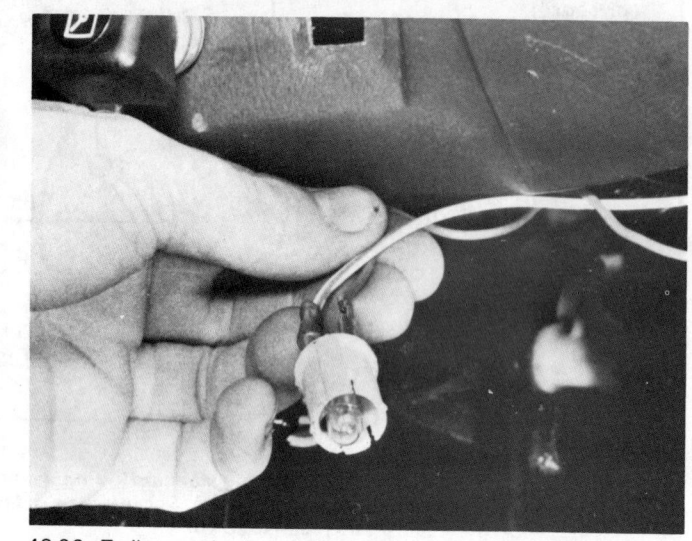

48.26a Trailer warning lamp bulb holder is connected by means of spade terminals ...

48.26b ... and fits into spare hole in dashboard

48.28 Secure cable with ties

Chapter 12 Bodywork and subframe

23 All connections must be sound and well-insulated. Wires twisted together and bound with insulating tape will not do! The following types of connector can be used:

(a) *In-line connectors. These are used to join two wires end-to-end. The wire ends are bared and the connector crimped on using pliers or a special crimping tool*

(b) *'Scotchlok' connectors. These are used to tap into the existing car wiring. No baring of the cables is necessary as the connector blade cuts through the insulation (photo). Trouble can occur if too many strands of the conductor are severed by the blade*

(c) *Terminal block. This can substitute for either of the above. The wire ends are bared and secured in the block with screws. Movement of the conductors can result in fatigue fractures; screws can work loose*

(d) *Spade connectors. These are crimped onto bare wire in the same way as in-line connectors and used to connect onto spade terminals – eg on the warning lamp bulb holder.. 'Piggy-back' terminals are available to enable additional connections to be made to the same terminal*

(e) *Earth tags. These too are crimped onto bare wire and then bolted to a suitable part of the vehicle. Good metal-to-metal contact is essential*

24 Follow the existing cable runs when positioning wires to run from side to side or from front to rear of the car (photo). A flexible probe is useful – curtain wire or an old speedometer inner cable will be suitable.

25 The ignition-controlled power supply for the slave relay is most easily taken from under the dashboard. The direction indicator flasher unit is easily accessible and connection can be made to its power supply terminal using a 'piggy-back' spade connector.

26 The trailer flasher warning lamp can be mounted in a spare hole in the dash if one is available (photos), or on its own bracket. One side of the bulb holder must be connected to earth.

27 When the wiring is complete, reconnect the battery, plug a trailer board into the 7-pin socket and check the various light functions. If trouble is experienced, check that all connections are clean and tight – especially the trailer earth connection – and examine 'Scotchlok' connectors closely if these have been used.

28 When satisfied, secure the new cable runs with tape or ties (photo), and refit floor panels, carpets etc, where these were removed.

Conversion factors

Length (distance)
Inches (in)	X	25.4	= Millimetres (mm)	X 0.039	= Inches (in)
Feet (ft)	X	0.305	= Metres (m)	X 3.281	= Feet (ft)
Miles	X	1.609	= Kilometres (km)	X 0.621	= Miles

Volume (capacity)
Cubic inches (cu in; in³)	X	16.387	= Cubic centimetres (cc; cm³)	X 0.061	= Cubic inches (cu in; in³)
Imperial pints (Imp pt)	X	0.568	= Litres (l)	X 1.76	= Imperial pints (Imp pt)
Imperial quarts (Imp qt)	X	1.137	= Litres (l)	X 0.88	= Imperial quarts (Imp qt)
Imperial quarts (Imp qt)	X	1.201	= US quarts (US qt)	X 0.833	= Imperial quarts (Imp qt)
US quarts (US qt)	X	0.946	= Litres (l)	X 1.057	= US quarts (US qt)
Imperial gallons (Imp gal)	X	4.546	= Litres (l)	X 0.22	= Imperial gallons (Imp gal)
Imperial gallons (Imp gal)	X	1.201	= US gallons (US gal)	X 0.833	= Imperial gallons (Imp gal)
US gallons (US gal)	X	3.785	= Litres (l)	X 0.264	= US gallons (US gal)

Mass (weight)
Ounces (oz)	X	28.35	= Grams (g)	X 0.035	= Ounces (oz)
Pounds (lb)	X	0.454	= Kilograms (kg)	X 2.205	= Pounds (lb)

Force
Ounces-force (ozf; oz)	X	0.278	= Newtons (N)	X 3.6	= Ounces-force (ozf; oz)
Pounds-force (lbf; lb)	X	4.448	= Newtons (N)	X 0.225	= Pounds-force (lbf; lb)
Newtons (N)	X	0.1	= Kilograms-force (kgf; kg)	X 9.81	= Newtons (N)

Pressure
Pounds-force per square inch (psi; lbf/in²; lb/in²)	X	0.070	= Kilograms-force per square centimetre (kgf/cm²; kg/cm²)	X 14.223	= Pounds-force per square inch (psi; lbf/in²; lb/in²)
Pounds-force per square inch (psi; lbf/in²; lb/in²)	X	0.068	= Atmospheres (atm)	X 14.696	= Pounds-force per square inch (psi; lbf/in²; lb/in²)
Pounds-force per square inch (psi; lbf/in²; lb/in²)	X	0.069	= Bars	X 14.5	= Pounds-force per square inch (psi; lbf/in²; lb/in²)
Pounds-force per square inch (psi; lbf/in²; lb/in²)	X	6.895	= Kilopascals (kPa)	X 0.145	= Pounds-force per square inch (psi; lbf/in²; lb/in²)
Kilopascals (kPa)	X	0.01	= Kilograms-force per square centimetre (kgf/cm²; kg/cm²)	X 98.1	= Kilopascals (kPa)

Torque (moment of force)
Pounds-force inches (lbf in; lb in)	X	1.152	= Kilograms-force centimetre (kgf cm; kg cm)	X 0.868	= Pounds-force inches (lbf in; lb in)
Pounds-force inches (lbf in; lb in)	X	0.113	= Newton metres (Nm)	X 8.85	= Pounds-force inches (lbf in; lb in)
Pounds-force inches (lbf in; lb in)	X	0.083	= Pounds-force feet (lbf ft; lb ft)	X 12	= Pounds-force inches (lbf in; lb in)
Pounds-force feet (lbf ft; lb ft)	X	0.138	= Kilograms-force metres (kgf m; kg m)	X 7.233	= Pounds-force feet (lbf ft; lb ft)
Pounds-force feet (lbf ft; lb ft)	X	1.356	= Newton metres (Nm)	X 0.738	= Pounds-force feet (lbf ft; lb ft)
Newton metres (Nm)	X	0.102	= Kilograms-force metres (kgf m; kg m)	X 9.804	= Newton metres (Nm)

Power
Horsepower (hp)	X	745.7	= Watts (W)	X 0.0013	= Horsepower (hp)

Velocity (speed)
Miles per hour (miles/hr; mph)	X	1.609	= Kilometres per hour (km/hr; kph)	X 0.621	= Miles per hour (miles/hr; mph)

*Fuel consumption**
Miles per gallon, Imperial (mpg)	X	0.354	= Kilometres per litre (km/l)	X 2.825	= Miles per gallon, Imperial (mpg)
Miles per gallon, US (mpg)	X	0.425	= Kilometres per litre (km/l)	X 2.352	= Miles per gallon, US (mpg)

Temperature

Degrees Fahrenheit (°F) = (°C x $\frac{9}{5}$) + 32

Degrees Celsius (Degrees Centigrade; °C) = (°F − 32) x $\frac{5}{9}$

*It is common practice to convert from miles per gallon (mpg) to litres/100 kilometres (l/100km), where mpg (Imperial) x l/100 km = 282 and mpg (US) x l/100 km = 235

Index

A

Air cleaner – 76
Air conditioning
    system – 60
    general – 272
    maintenance – 273
Alternator
    brush renewal – 162
    drivebelt – 162
    general description – 161
    regulator – 161, 162
    testing and maintenance – 161
Anti-freeze mixture – 67
Anti-roll bar
    removal and refitting (rear) – 233
    removal and refitting (front) – 228
    bushes, removal and refitting – 234
Automatic transmission
    description – 127
    fault diagnosis – 129
    gear linkage – 123, 124
    mainshaft – 116
    removal and refitting – 111
    synchromesh – 119

B

Battery
    charging – 161
    maintenance and care – 160
    removal and refitting – 160
Bodywork and fittings
    air conditioning – 272, 273
    bonnet, boot and tailgate – 261, 264, 262, 274
    bumpers – 265
    description – 250
    doors – 187
    door glass – 256
    door locks – 259, 260
    front seats – 266
    front wings – 265
    glovebox – 263
    heater – 270, 271
    heater system – 270
    maintenance – 250, 251
    major body repairs – 187
    minor body repairs – 184
    radiator grille – 264
    rear seats – 266
    rear windows – 267, 268
    subframe – 268
    tailgate 262, 263, 268, 280
Brake discs – 150
Brake servo – 153, 154
Braking system
    bleeding the hydraulic system – 145
    brake and clutch pedal assembly – 153
    brake calipers – 147, 148, 149
    brake discs – 150
    description – 145
    fault diagnosis – 156
    handbrake adjustment – 152
    handbrake cable – 152
    handbrake lever – 153
    master cylinder – 149, 150
    pads – 147, 149
Bumpers – 265

C

Carburettor
    air cleaner – 76
    automatic choke – 81, 84
    choke cable – 77
    description – 77
    emission control – 87, 88
    fault diagnosis – 93
    fuel pump – 86
    fuel tank – 86
    in-line fuel filter – 85
    inlet manifold – 85
Camshaft – 37
Clutch
    assembly – 103, 105
    cable – 103
    description – 102
    fault diagnosis – 108
    pedal – 102, 103
    release mechanism – 107
Conversion factors – 280
Cooling system
    description – 65
    draining, flushing, filling – 66, 67
    drivebelt – 70
    electric fan – 67, 68
    expansion tank – 69
    fault diagnosis – 71
    radiator – 67, 68
    temperature transducers – 70
    thermostat – 69
    water pump – 70
Crankshaft – 47
Cylinder bores – 42
Cylinder head
    reassembly – 50
    refitting – 53
    removal – 59

D

Differential unit
    fault diagnosis – 137
    general description – 130
    removal and refitting – 130
Disc brakes
    calipers – 147, 149
    inspection, removal and refitting – 150
    pads – 147, 149
Distributor
    condenser – 96
    contact breaker points – 96, 97
    overhaul – 96
    removal and refitting – 96, 97
Door locks – 259, 260, 271
Driveshafts and CV joints
    centre sections – 140
    dismantling – 140
    fault diagnosis – 143
    general description – 138
    removal and refitting – 139

E

Electrical system
    alternator – 161, 163

Index

auxiliary switches, and panel – 172, 173, 174
battery – 160, 163
courtesy lights, switches – 171, 172, 176
fault diagnosis – 184
flasher units – 177
front parking and flasher lamps – 170
fuses – 166
general description – 160
glovebox light – 172
headlamps – 167, 168
horns – 178
ignition switch – 176
instrument panel – 174
instruments, removal and refitting – 175
radio and tape players – 180, 181, 182, 183, 184
rear lamp assembly – 170
reversing light – 176
starter motor – 162, 164, 165, 166
windscreen wipers and washers – 178, 179, 180

Engine
ancillary components – 50
ancillary driveshafts – 46, 49
big-end bearings – 47, 59
cylinder bores – 42
cylinder head – 37, 50, 53, 59
crankshaft – 47
dismantling – 32, 33
fault diagnosis – 64
flywheel – 47, 50
general description – 19
gudgeon pin – 45
initial start-up after overhaul – 58
major operations possible with engine in-situ – 26
major operations possible with the engine removed – 26
notes for cars fitted with air conditioning – 60
notes for USA models – 60
oil filter – 42
oil pump – 42, 49
oil sump – 49, 59
pistons, connecting rods and rings – 45, 49
reassembly – 47
small-end bush renewal – 46
subframe, engine and gearbox removal and refitting – 31
timing belt – 53, 56
valve/tappet clearance setting – 53

Exhaust
general description – 75, 88
manifold gasket – 93
renewal – 91, 93

F

Fault diagnosis
automatic transmission – 129
braking system – 156
clutch – 106, 107
cooling system – 71
differential unit – 137
driveshafts and CV joints – 143
electrical system – 184, 185
engine – 64
fuel system – 93
ignition system – 101
manual gearbox – 129
suspension and steering – 249

Flywheel – 47, 50
Front suspension – 226
Fuel system
air cleaner – 76
automatic choke – 81, 84
carburettor – 77, 78, 81, 84, 85
choke cable – 77
emission systems, UK and USA – 87, 88
fuel tank – 86
fuel tank sender unit – 86
general description – 75
in-line fuel filter – 85
throttle cable – 76
Fuses – 166

G

Gearbox (manual)
dismantling – 111
fault diagnosis – 129
gear linkage – 123, 125
general description – 111
main shaft – 116
primary shaft – 118
removal and refitting – 111
synchromesh assemblies – 119

Gearbox (automatic)
fault diagnosis – 129
fluid level – 128
general description – 127
oil seals – 128, 129
removal and refitting – 111
selector linkage – 128
separation from engine – 128

General dimensions and weights – 6

H

Handbrake
handbrake – 168
bulbs – 168
self-levelling system – 168
unit – 167, 168

Heater
general description – 270
removal, dismantling, reassembly and refitting – 270, 271, 272

Horn – 178

I

Ignition system
coil – 98
condenser – 96
distributor – 96, 97, 98
electronic ignition – 98
fault diagnosis – 101
general description – 95
routine maintenance – 95
spark plugs and hT leads – 100, 101
timing, static – 100

Instruments
panel – 174
removal and refitting – 176

J

Jacking and towing – 12

L

Lights
brake stop light switch – 176
courtesy lamp – 171, 176
front parking and flasher lamp – 170
glove box light – 172

Index

headlights – 167, 168
luggage compartment light – 171, 176
number plate lamps – 172
rear lamp assembly – 170, 171
reversing lamp switch – 176
side repeater lamps – 170
under bonnet light – 171
Lubricants, recommended – 13
Lubrication chart – 13

M

Maintenance, routine – 15
Manual gearbox see **Gearbox (manual)** – 110
Master cylinder – 149, 150

O

Oil filter and housing – 42
Oil pump
  overhaul – 42
  refitting – 49
Oil sump
  refitting – 49
  removal and refitting in-situ – 59

P

Pistons and piston rings
  examination and renewal – 45
Pistons, connecting rods and big-end bearings
  refitting – 49
Power steering pump
  removal and refitting – 246
Power steering pump drivebelt
  removal, refitting and tensioning – 248
Power steering rack assembly
  removal and refitting – 246
Power steering rack overhaul
  general – 246
Power steering system
  filling and bleeding – 248
Power window lifts
  removal and refitting – 246

R

Radiator
  inspection and cleaning – 68
  removal and refitting – 67
Radios and tape players – 180, 181, 182, 183, 184
Rear brake caliper block
  removal and refitting – 151, 149
Rear brake compensator device
  description – 154
  overhaul – 154
  removal, refitting and adjusting – 154
Rear brake pads
  removal, inspection and refitting – 149
Rear lamp assembly
  removal and refitting – 170, 171
Rear suspension see **Suspension and steering**
Routine maintenance – 15

S

Spare parts, buying – 8
Spark plugs and HT leads – 100
Starter motor
  circuit testing – 164
  dismantling, repair and assembly – 165
  general description – 162
  removal and refitting – 164
Starter motor drive pinion
  inspection and repair – 166
Subframe, engine and gearbox
  removal and refitting – 31
Subframe
  general – 268, 270
Sump, oil
  refitting – 49
  removal and refitting, engine in-situ – 59
Suspension and steering
  anti-roll bar (front) – 228
  anti-roll bar (rear) – 233
  anti-roll bar bushes (rear) – 234
  fault diagnosis – 249
  front hub and bearings – 230
  front strut assembly – 229
  front suspension spring and strut – 226
  front suspension wishbones – 226
  front wheel hub and carrier – 230
  front wishbones – 229
  general description – 226
  power steering pump – 246
  power steering drivebelt – 248
  power steering rack – 246
  power steering system – 248
  rear hub and bearings – 236
  rear suspension crossmembers – 233
  rear suspension spring and strut – 231
  rear suspension strut – 235
  rear wheel toe-in – 236
  road wheels and tyres – 236
  steering column – 240
  steering geometry – 242
  steering wheel – 238
  transverse link bushes – 235
  wishbone outer balljoints – 237

T

Tailgate – 262, 263, 268
Temperature transducers – 70
Thermostat – 69
Tools – 10
Tow bar – 274, 278, 279
Towing – 12

U

Under bonnet light
  removal and refitting – 171, 172

V

Valve
  clearance settings – 53
  gear and camshafts – 37

W

Water pump
  removal and refitting – 70
Window winder mechanism – 253
Windscreen
  all models – 267, 268
  washer pump – 178
  wiper arms and blades – 178
  wiper motor and mechanism – 179
Wiring diagrams – 186 to 223

Printed by
Haynes Publishing Group
Sparkford Yeovil Somerset
England